Treatise on Concrete

louis i. kahn
towards the zero degree of concrete
1960-1974

Other books by Anna Rosellini

Editor of *Arte e Architettura. Le cornici della Storia*, with F. Bardati (2005), Bruno Mondadori Editore

Le Corbusier. Béton Brut and Ineffable Space, 1940-1965, Surface Materials and Psychophysiology of Vision (2011), with R. Gargiani, EPFL Press (German translation, 2014, Edition Detail)

Le Corbusier e la superficie, dal rivestimento d'intonaco la béton brut (2013), Aracne

Treatise on Concrete
Directed by Roberto Gargiani

anna rosellini

louis i. kahn towards the zero degree of concrete 1960-1974

Translated from the Italian by Stephen Piccolo

EPFL Press

A Swiss academic publisher
distributed by Routledge

EPFL Press

The EPFL Press is the English-language imprint of the Foundation of the Presses polytechniques et universitaires romandes (PPUR). The PPUR publishes mainly works of teaching and research of the Ecole polytechnique fédérale de Lausanne (EPFL), of universities and other institutions of higher education.

Presses polytechniques
et universitaires romandes
EPFL – Rolex Learning Center
Post office box 119
CH-1015 Lausanne, Switzerland
E-mail: ppur@epfl.ch
Phone: +41 (0)21 693 2130
Fax: +41 (0)21 693 4027

www.epflpress.org

ISBN 978-2-940222-77-3 (EPFL Press)
Printed in Italy

Routledge
Taylor & Francis Group

Taylor & Francis Group Ltd
2 Park Square, Milton Park
Abingdon, Oxford, OX14 4RN, UK

Routledge is an imprint of Taylor & Francis Group, An informa business.

Simultaneously published in the USA and Canada by Routledge, 711 Third Avenue, New York 10017

www.routledge.com

Library of Congress Cataloging-in-Publication Data
A catalog record for this book is available from the Library of Congress.

The graphical layout of the book and the cover design are by Paola Ranzini Pallavicini of Studio Pagina Milan, Italy. The photolithographic services were provided by Walter Bassani, AG-MEDIA, Milan Italy.

This work was translated from the Italian by Stephen Piccolo (www.transiting.eu).

Research for this publication has been financed by the European Research Council grant as part of project ERC2008-AdG, Proposal No. 230394, "The surfaces of cement and reinforced concrete: A history of the formwork and processing of the surface."

ÉCOLE POLYTECHNIQUE
FÉDÉRALE DE LAUSANNE

Series Editors for the publication program in Architecture are Professors Jacques Lucan and Luca Ortelli.

The publisher and authors express their thanks to the Ecole polytechnique fédérale de Lausanne (EPFL) for its generous support towards the publication of this book.

The author is grateful to the ENAC faculty of the EPFL, who provided additional support.

Treatise on Concrete

Directed by Roberto Gargiani

The volumes of the *Treatise on Concrete* intend to offer a new and documented vision of the evolution of construction techniques and of the creative potential and formal expressions of concrete, from antiquity to the present time. Each book in the series will be based on specific archival research; the analysis of scientific, technical and iconographic original sources; and the investigation into documents made available by companies, engineers, architects and artists.
The full spectrum of the use of concrete will be covered, including the building of foundations, ports, moles, bridges, vaults, domes, fortresses, bunkers, houses and monuments, as well as the manufacture of artificial stones and sculptures. In addition, the volumes will address the theoretical issues of *béton brut*, Brutalism and the true nature of matter.
All volumes focus on the techniques and materials for the manufacture of formwork, from wood to fabric; on the meaning of the imprint; on the surface processing of concrete by hand or with machine; and on the composition of the mixture.
The series of books aims to provide the scientific community with an unique and updated reference on the historical and contemporary uses of concrete, which will open a new chapter in the knowledge of this crucial material for architectural construction, as well as for masterpieces of engineering and art.

Volumes published:
Concrete, From Archeology to Invention, 1700–1769
Louis I. Kahn, Exposed Concrete and Hollow Stones, 1949-1959

Volumes in preparation:
Tadao Ando, Concrete Spaces
Mnemosyne in Concrete
The Concrete of Pier Luigi Nervi
Research on the Roman Cement, 1770-1779
The Origins of Modern Concrete Building in Germany, 1800-1877
Concrete Sculpture, from 1950 to Today: Visions of a Material
Contemporary Ornament in Concrete
Bared Concrete of Natural and Artificial Pozzolanas, 1780-1817
Formless: Beyond Formwork
The Concrete of Perret and Pupils

Table of Contents

chapter one
the nature of concrete: the "liquid stone" of the salk institute

Slabs of stone and concrete in the ocean coast landscape

At the start of the 1960s the works of architecture in *béton brut* inspired by the masterpieces of Le Corbusier and the question of Brutalism are at the center of the international debate, and the work of Louis Isadore Kahn becomes the focus of critics investigating the nature of his concrete. Henry Russell Hitchcock questions the definition of "Neo-Brutalist" for the Richards Laboratories and the Yale Art Gallery;[1] Jones Causton reveals how Kahn uses columns "ornamented only by the marks left by the wood molds," in the manner of Auguste Perret;[2] Reyner Banham dubs Kahn, again speaking of the Richards Laboratories, "one of the Brutalists' favourite architects,"[3] while Vincent J. Scully, Jr., associates Kahn with the "Brutalists" and in particular the "English" ones, discussing the influence of the Maisons

[1] "Kahn's work appeals less to the eye than the mind, but unlike the Neo-Brutalists, whether Le Corbusier in his late work or his English juniors, Kahn does not set out to exacerbate the eye. His concrete elements in the laboratories, being pre-cast (the relatively small amount of in situ concrete is mostly brick-faced), are highly finished by steam-curing" (Henry Russell Hitchcock, *Notes of a Traveller: Wright and Kahn*, in, "Zodiac," 1960, No. 6, (pp. 15-21), p. 20). See also Hiroki Onobayashi, *Louis Kahn, Order for Concrete*, in, "Kokusai Kentiku," vol. XXVII, June 1960, pp. 49-53. For the general overview of Kahn's work, see the fundamental essays by David B. Brownlee, David G. De Long, *Louis I. Kahn. In the Realm of Architecture*, New York, Rizzoli, 1991; and Robert McCarter, *Louis I. Kahn*, London, Phaidon, 2005. See also the essay, destined to open a new chapter in the American historical study of Kahn, by Sarah Williams Goldhagen, *Louis Kahn's Situated Modernism*, New Haven, London, Yale University, 2001. Again on the work of Kahn, see also Mateo Kries, Jochen Eisenbrand, Stanislaus von Moos, ed., *Louis Kahn. The Power of Architecture*, Weil am Rhein, Vitra Design Museum, 2012. On the construction processes and structures of Kahn, see the fundamental essay by Thomas Leslie, *Louis I. Kahn. Building Art, Building Science*, New York, G. Braziller, Inc., 2005.

[2] Jones Causton, *Architecture Today & Tomorrow*, New York, McGraw-Hill, 1961, p. 182. In the caption of the photograph of the staircase of the Yale Art Gallery, it is emphasized that there are "undisguised marks of the wood framework into which concrete was poured" (*ibid*, p. 181).

[3] Reyner Banham, *Guide to Modern Architecture*, London, Architectural Press, 1962, p. 74.

1-2. Salk Institute for Biological Studies, La Jolla, California, 1959-65.

1. Model showing the three units: Laboratory, Meeting House and Housing (LIK, 030.IV.A.540.14.80).

2. "Plot Plan," 5 January 1961 (LIK, 030.I.C.540.4).

Jaoul of Le Corbusier[4] and certain details of the Esherick House,[5] while indicating similarities between the concrete of the Yale Art Gallery and that of Le Corbusier: "The concrete was left rough with the marks of its forms upon it, as Le Corbusier had already treated that material [...]."[6]

When the book *The New Brutalism: Ethic or Aesthetic?* was published in 1966, where eleven years after his article in "The Architectural Review" Banham takes stock of the main international developments, Kahn's work – still represented by the Yale Art Gallery alone – is again described against the backdrop of the question regarding the possibility of inserting it, or not, in the "concept of the New Brutalism," due to its being so "devoted to the ground-rules of academic classicism."[7] Only the staircase of the Yale Art Gallery seems to possess, in Banham's view, an expressive force capable of taking *béton brut* to extreme development:

"The concealment of the stairs within an almost unpierced drum of concrete, heightens one's awareness of the experience of ascending the stairs, three short flights arranged in a triangle between one floor and the next, rising between sheer walls of concrete, unmodulated by anything beyond the vertically planked shutter-pattern of the concrete, the impress of the fixing studs that held the shuttering together, and the horizontal joint at each floor-level, marking the height of one lift of the shuttering. It is a classic demonstration of absolute Brutalist truth to a particular method of construction, and has the added historical importance of being the most extended demonstration to date, by anyone other than Le Corbusier, of the aesthetics of 'béton brut.'"[8]

But well prior to the publication of Banham's essay, Kahn is already making a decisive move beyond the forms of expression of *béton brut* as it was deployed in the extraordinary sequence of his works extending from the Yale Art Gallery to the First Unitarian Church and School. Starting in the early 1960s, he conducts experimentation on the nature of concrete and the processes of its fabrication, extending to the control of the compound and the system of the formwork itself. Until the end of the 20th century, this was to be the only path capable of progressing beyond the *béton brut* of Le Corbusier and beyond any New Brutalism, laying the groundwork for an idea of exposed concrete architecture seen as control of the entire construction process. After all, Kahn had proven that he was able to free himself of the aesthetic constraints of *béton brut* above all with the works in prefabricated concrete and those made with concrete blocks.

One of Kahn's staff, William S. Huff, who is supervising the worksite of the Tribune Review Building in Greensburg in 1960, recalls a significant anecdote regarding the formwork made with wooden boards used in the United States, and also by Kahn, following the fashion launched by Le Corbusier. "When Corbu made one of his few visits this country, he asked, 'Why do you people have all those fussy little boards in your concrete work?' The reply was, 'Because you do it, master.' Corbu then said, 'But you have all that wonderful plywood that comes in great sheets; in Europe we are poor and must use old floor boards, taken up from buildings being demolished, for our formwork.' Kahn knew the story. He did use floor boards,

[4] Vincent J. Scully, Jr., *Louis I. Kahn*, New York, G. Braziller, 1962, pp. 19-20.
[5] *Ibid*, p. 35.
[6] *Ibid*, p. 21. The photograph of the Yale Art Gallery interior closes the series of images published in, Peter Collins, *Concrete: the vision of a new architecture. A study of Auguste Perret and his precursors*, London Feber and Feber, 1959, fig. 104.
[7] Reyner Banham, *The New Brutalism: Ethic or Aesthetic?*, Stuttgart, Bern, Krämerm Verlag, 1966, p. 44.
[8] *Ibid*. For some notes on Kahn and Brutalism see also, Alison and Peter

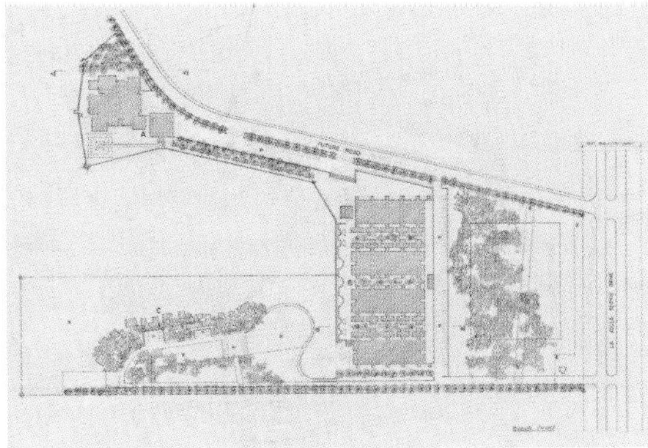

as Corbu used them, on several of his early buildings – such as the Yale Art Gallery addition. Then, when we did the Salk project, Lou really went all out to employ those large plywood planes in his forms."[9]

The process through which Kahn decides to stop using wooden boards and to switch to plywood panels is a bit more complex than Huff would have it. But the value of his story is borne out by the fact that the fundamental passage in Kahn's way of making formwork takes place precisely in the years of the construction, in 1961-63, of the Carpenter Center for the Visual Arts for Harvard University, in Cambridge, Massachusetts, where Le Corbusier, also thanks to the knowledge of José Luis Sert regarding American techniques and materials for the construction of concrete, uses large plywood panels assembled with V joints. The example of that work, for Kahn, marks the start of a new phase of research on the nature of concrete, which will take him far from the image of *béton brut*. Though nothing had come of the enlightening criticisms of C. Clark Macomber against Kahn's preference, on the Yale Art Gallery worksite, for wooden boards, which Macomber saw as a "method of the past."

At the start of the 1960s, Kahn is involved in the design and construction of the building that marks the crucial chapter in the evolution of his exposed concrete, the Institute of Biology at Torrey Pines, later the Salk Institute for Biological Studies, commissioned in 1959 by Jonas Edward Salk, known for his discovery of a polio vaccine, and built in the state of California, at La Jolla, near San Diego, in the town of Torrey Pines, on a promontory facing the Pacific Ocean – the building is completed in 1965.

The design of the imprints on the concrete is not just a direct result of the technical logic of assembly of the formwork, as stated on occasion by Kahn regarding this work. Precisely the most important exposed concrete of the architecture of Kahn bears the secret traces of all the phases of the project, when the surfaces of the Salk Institute are characterized by slabs of stone, as if the face of the concrete onto which the colors and changing light of the ocean weather were crossed by a sequence of layers of textures of different kinds, from those generated by the logics of the worksite to those produced

Smithson, *The Space Between*, in, "Oppositions," October 1974, No. 4, (pp. 75-78), p. 78.

[9] William S. Huff, *Louis Kahn: 'Sorted Recollections and Lapses in Familiarities.' Interview with Jason Aronoff*, in, "Little Journal," vol. V, September 1981, No. 1, now in, Alessandra Latour, ed., *Louis I. Kahn, l'uomo, il maestro*, Roma, Edizioni Kappa, 1986, (pp. 405-431), p. 423.

3. Cordova Shell stone (photograph by the author).

through mnemonic artistic processes. The very nature of the concrete compound and even its color and grain are not immune to Kahn's initial passion for the stone slabs and the view of the work in the landscape.

After Salk's visit to the Richards Laboratories in December 1959, and after Kahn's trip to La Jolla in January 1960, the first version of the project for the Salk Institute for Biological Studies took form, resembling an archipelago, as documented by the model dated March 1960. The laboratories are like four circular islands, each containing different functions of the program. Two of these islands feature towers similar to those of the Richards Laboratories. The lodgings for the researchers, staff and guests are scattered in the landscape, while the center for scientific conferences, with the auditorium, conference rooms and library inserted in a rectangular plan, is set on the extremity of the point towards the ocean.

The issue of materials is not addressed in the drawings or other documents until 1961, when the different parts of the project, also after a series of studies for a symbolic clustering of cells, take on the configuration destined to become the second variant of the Salk Institute (starting in January, August E. Komendant is on board for the structures; during the course of the year, Fred S. Dubin also begins to work on the physical plant systems).

In the project studied in 1960-61 and developed after Jonas Edward Salk has prepared, on 17 May 1961, his "Statement of Aims" specifying that the Institute should be a "center for research" for "work in an atmosphere of intellectual imagination,"[10] the complex is composed of three entities: the Laboratory Building or Laboratory Center, the Meeting House and the Residence of Fellows or Housing Facilities. Concrete and stone are indicated as the fundamental materials of the complex, each deployed in keeping with its physical and mechanical characteristics, but related in such a way as to avoid making the stone into mere cladding for the concrete. It is this contamination of materials in pursuit of a new constructive unity that concerns Kahn until the middle of 1962. Only concrete, or concrete with stone, are solutions Kahn considers in the light of the two methods of fabrication of the reinforced concrete structures: poured in place and precast.[11]

While the nature of the concrete compound is only addressed for the first time in 1962, until June 1961 Kahn specifies the type of stone for the slabs, namely Travertine or limestone. The type generically indicated as limestone, when specified, turns out to be the stone known as Cordova Shell, a limestone containing many fossil seashells, which when cut in slabs has a surface tattooed by many hollow imprints. It is worth noting that the *béton brut* of the Unité d'Habitation in Marseille was also decorated with the imprints of seashells placed in the formwork. Both Travertine and Cordova Shell have surfaces marked by irregular cavities of different kinds.

[10] Jonas Edward Salk, "Institute for Biology. Statement of Aims," 17 May 1961, LIK, 030.II.A.89.14.
[11] On Kahn's way of conceiving the design in terms of concrete, it is worth re-reading the statement of Komendant precisely when the Salk Institute is about to be completed: "When Kahn designs, I don't think he actually considers the material. He may say that he does, but I don't think he actually does so. It seems to me that the material most closely responding to his architectural philosophy is concrete and thus his selection is instinctive. I base this belief on the fact that his design preliminaries are almost always brilliant for concrete. He does not split the surface, and his architecture is 'heavy-duty.' And so is concrete. I doubt that Kahn's architecture could be executed in marble, because I don't think of marble as a heavy-duty material. If Kahn tries to use a light material, he betrays his own basic philosophy and gets confused. When he uses another material to protect concrete, his decision is based on erroneous evidence, mishaps through previous misuse of concrete, or some other contradictory factor" (August E. Komendant, *Komendant on concrete*, in, "Progressive Architecture," XLVII, October 1966, No. 10, (pp. 208-214), p. 208). On Kahn's concrete, from the Salk Institute to his later works, see Réjean Legault, *Louis Kahn and the Life of Materials*, in, Kries,

Kahn prefers the Cordova Shell, which he even considers using with rugged workmanship, allowing the saw marks to remain on the surface from the cutting of the slabs from blocks - "finish: sawed face."[12]

The stone slabs serve Kahn to generate the character of a monument for the most representative parts of the complex, the Meeting House and the Residence of Fellows, while the lack of the stones indicates the functional character of the Laboratory Building, where exposed concrete predominates. Certain touches of stone inserted in the Laboratory Building, in any case, create links among the parts of the complex immersed in the coastal landscape, playing the same role as the brick introduced in the Richards Laboratories to attune the building to the edified context of the campus and the city of Philadelphia. The sheets of stone also contribute to create that "atmosphere of intellectual imagination" desired by Salk, and they are Kahn's preferred pictorial tool to insert the complex in the profile of the ocean coast, which in June 1961 he indicates as the main source of inspiration of the project:

"The choice of the site of Torrey Pines, La Jolla, overlooking the sea and protected by surrounding park and University property is the first inspiring act towards creation of the environment for the Institute of Biology. From the presence of the uninterrupted sky, the sea and the horizon, the clear and dramatic configuration of weather-beaten land spare of foliage, the buildings and their gardens must find their position in deference to Nature."[13]

The colors of the rocky landscape where the Salk Institute is to stand are described in the geological report in October. The vegetation is sparse and low, the earth sandy, and the sandstone has an ochre hue ("porous silty sand" and "hard sandy clay"). The canyon that opens at the foot of the Laboratory Building, and around which Kahn positions the Meeting House and the Residence of Fellows, is composed of "Rose Canyon Formation," with a pinkish color. The geological layers of the rocky shore are "Linda Vista Formation," with a brownish color, and "Beach Ridge Unit," composed for the most part of "basal conglomerate."[14] A few months after reading the geological report, Kahn says that he prefers, among the chromatic varieties of Cordova Shell, the one with a "cream" color.[15]

In the Meeting House, right from the versions of the project studied in 1960-61, there are true open-air masonry shells, protecting glazed rooms for reading and seminars, next to the dining hall. These perforated masonry envelopes, or "visible shells of space,"[16] are detached from the body of the building to protect the library and seminar rooms from sunlight and heat, thanks to the shady interspace open to the sky. Those shells are derived from the screens envisioned for the United States Consulate Buildings; Kahn calls them "walls which modify glare,"[17] "wall for glare"[18] or "screen walls,"[19] and they have different configurations: "round walls" for the library and

Eisenbrand, Von Moos, *op. cit.*, pp. 219-233.
[12] Arthur Hannafin, letter to the Texas Limestone Company, 22 June 1961, LIK, 030.II.A.108.25. For some hypotheses of reconstructions of the first designs of Kahn for the Salk Institute see, Kent Larson, *Louis Kahn, Unbuilt Masterworks*, New York, The Monacelli Press, 2000, pp. 49-77.
[13] "Abstract of program for the Institute of Biology at Torrey Pines, La Jolla, San Diego," n.d., LIK, 030.II.A.27.16.
[14] Woodward, Clyde, Sherard and Associates, "Soil Investigation for the Proposed Institute for Biology," 16 October 1961, LIK, 030.II.A.107.4.
[15] See specifications, 17 January 1962, LIK, 030.II.A.108.11.
[16] Brown's expression refers to Roman architecture (Frank E. Brown, *Roman Architecture*, New York, G. Braziller, 1961, p. 15).
[17] "Abstract of program for the Institute of Biology at Torrey Pines, La Jolla, San Diego," n.d., LIK, 030.II.A.27.16.
[18] *Cf.* Jan C. Rowan, *Wanting to be ... the Philadelphia School*, in, "Progressive Architecture," XLII, April 1961, No. 42, (pp. 131-163), p. 141.
[19] The term is found in study drawings (see, among others, the drawing "Wall Sections," 24 July 1961, LIK, 030.I.C.540.008).

"square walls" for the seminars. In the points of contact between the rooms and the "screen walls" Kahn inserts the pillars of the floor systems, called "column points."[20] "[...] the wall has a bold opening to the sky and an opening for the view," Kahn explains. "Thus the glare will be modified by the lighted wall and the view not shut off. The glass will be away from the wall."[21] "Screen of columns" in front of an "open space" is the definition for the peristyles of Roman houses, used by Frank E. Brown in the essay *Roman Architecture*, published in 1961.[22] In fact, more than one aspect of Kahn's project for La Jolla evokes images of Roman monuments published by Brown, like those of the Sanctuary of Fortuna Primigenia at Prenestina.

Stone face and concrete nucleus: towards a new opus caementicium

Describing the role to be played by concrete in the Salk Institute, to be poured in place, prefabricated and used in blocks to make certain walls, according to the project of 1961, Kahn puts the accent not on the formal qualities of the material, but simply on the potential of the structural capacity of the large beams of the Laboratory Building, precast and equipped with cavities for the physical plant conduits – he writes of "use of concrete in the performance of the hard task of spanning uninterrupted areas."[23] The other two fundamental materials of the Salk Institute, namely glass and stone, are assigned a specific role in viewing, depending on the vantage point of the observer: the glass creates large surfaces for the viewing of landscapes; the stone slabs are placed in the concrete walls to create the color of the envelopes in the landscape. "[...] the use of cut stone, preferably Cordova Shell or Travertine for all enclosing walls and the walls which modify glare," Kahn writes.[24] The range of essential materials for the architecture of the complex is completed by the various gardens, with their "fresh" air and "waterways."

It is worth paying attention to the symbolic framework which in Kahn's outlook should regulate every part of the complex, if we want to fully grasp the meaning attributed to the materials, beyond the technique of their installation. Precisely one of the parts of the Salk Institute most characterized by the envelope of stone slabs set into the concrete core, the Meeting House, a complex of the size of a grand monastery or a palace with a courtyard, is presented by Kahn as a "building which symbolizes the purposes of the Institute." The plan of the large square hall, at times drawn without the roof, and around which the various other buildings are grouped, is significantly a variation on that of the church in Rochester, with the auditorium surrounded by classrooms. In the differing geometries of those spaces, Kahn stages the metaphor of human beings gathered to spread scientific knowledge. Water and its sounds mark the route that takes the researcher to the Meeting House, from the "square pool," devised so as to create "a noisy

[20] Rowan, *op. cit.*, p. 141.
[21] Notes for the preparation of "Abstract of program for the Institute of Biology at Torrey Pines, La Jolla, San Diego," n.d., LIK, 030.II.A.27.16.
[22] Brown, *op. cit.*, p. 21. The drawings of plans by Francesco di Giorgio published by Wittkower may also have influenced the invention of the screen walls.
[23] "Abstract of program for the Institute of Biology at Torrey Pines, La Jolla, San Diego" cit. Certain generic indications on concrete, "poured-in-place" or "pre-cast," and on the various "exposed concrete ceilings" are contained in, "Footnotes to accompany cost analysis," n.d., [1961], LIK, 030.II.A.Salk La Jolla.
[24] "Abstract of program for the Institute of Biology at Torrey Pines, La Jolla, San Diego," cit.
[25] Notes for the preparation of "Abstract of program for the Institute of Biology at Torrey Pines, La Jolla, San Diego," n.d., LIK, 030.II.A.27.16.
[26] Arthur Hannafin, letter to the Texas Limestone Company, 22 June 1961, LIK, 030.II.A.108.25.

fountain," along a "watercourse" as far as the panoramic terrace bordered by a portico, known as the "Colonnaded Ambulatory," where water springs forth in a "quiet fountain."[25] Pools of water, terraces for contemplation, slabs of stone with a concrete core are more than clues to Kahn's passion for the great Roman monumental complexes, like Hadrian's Villa at Tivoli, which is one of the fundamental points of reference of the Salk Institute, together with the basilica and monastery of San Francesco in Assisi. Precisely as in the Roman wall, where the face and nucleus of *opus caementicium* constituted an inseparable unit, so Kahn wants to achieve a constructive system with similar characteristics to build his "ruin" on the Pacific coast.

If the stone slabs are introduced for reasons of pictorial image of the Salk Institute in the spectacular coastal landscape, during the course of 1961 Kahn sets out to also give those slabs a fundamental technical role in the fabrication process of the concrete, imagining their use as permanent formwork. The question of stone slabs for the formwork is addressed in June, when members of Kahn's staff request information from the Texas Limestone Company, based in Austin, to learn if the rectangular stone slabs, measuring 2 x 3 feet, with a thickness of 3 or 4 inches, could form the panels to be left in place on one or both sides of a formwork. They also order samples of the Cordova Shell, to evaluate its "normal range of texture, color and composition."[26]

In the version of the design of the Meeting House corresponding to this phase of the selection of the materials (many drawings are made in July 1961), Kahn envisions two different systems of construction in reinforced concrete for the walls and the "screen walls."

The walls are in concrete poured in place, 8 inches thick. While the concrete of the walls inside the Meeting House is poured in wooden formwork, that of the articulated rind of the Meeting House is poured into a formwork system with two different faces: on the external face, the panel is composed of slabs of stone to be left in place; the other face is made with wooden boards. The thickness of these walls with the cladding-formwork incorporated remains at 8 inches (5 of reinforced concrete and 3 of stone). The circular wall of the tower to contain the books – the "Library Stacks" – also has incorporated stone slabs. In these walls, the concrete remains exposed only along the ends of the floor slabs that form horizontal bands in the facades. The other lines would be those of the pattern of joints of the stone slabs.

Kahn envisions another construction system for the "screen walls" that protect the reading and seminar spaces. These "screen walls" are thin shells in prefabricated reinforced concrete, with a thickness of 4 inches. The pillars inserted in the "screen walls" to support the floor slabs of the rooms are also

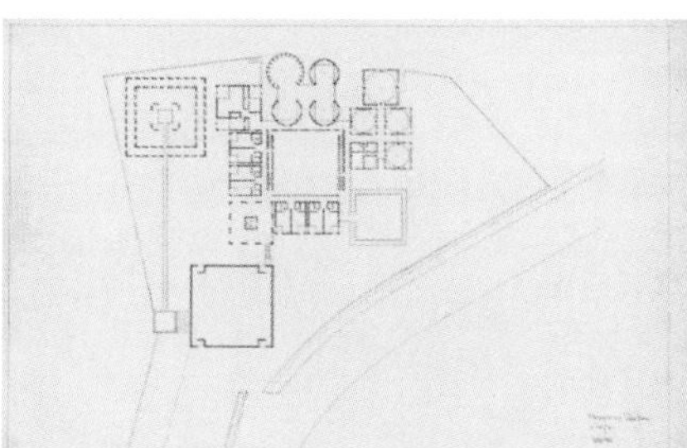

4. Salk Institute for Biological Studies, La Jolla, California, 1959-65. Meeting House, "Presentation Plan," 10 July 1961 (LIK, 030.I.C.540.8).

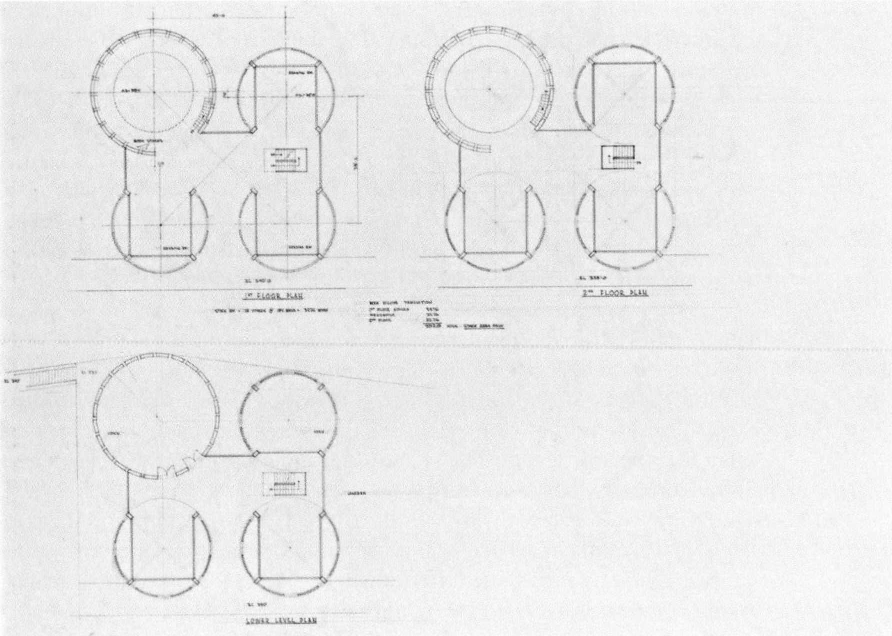
1st FLOOR PLAN
2nd FLOOR PLAN
LOWER LEVEL PLAN

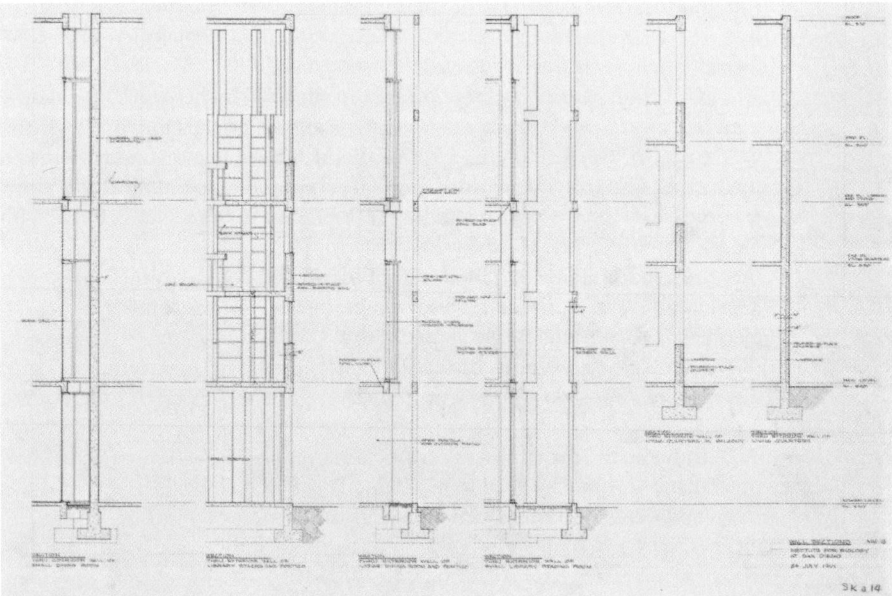
Sk a 14

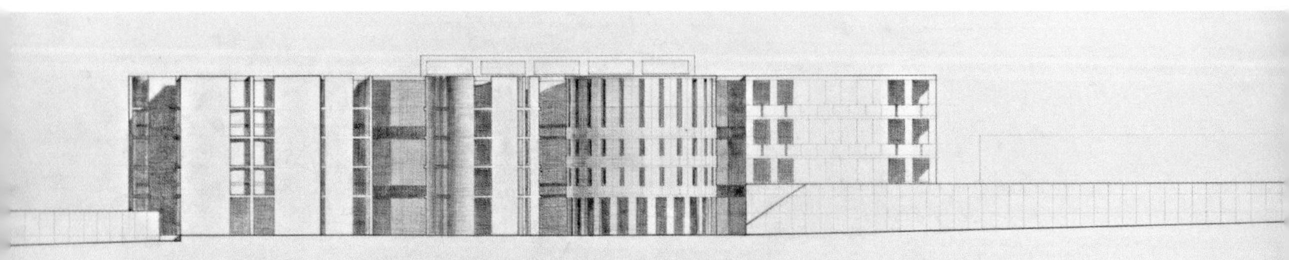

5-7. Salk Institute for Biological Studies, La Jolla, California, 1959-65.

5. Meeting House, "Library Studies," 12 July 1961 (LIK, 030.I.C.540.8).

6. Meeting House, "Wall Sections," 24 July 1961 (LIK, 030.I.C.540.8).

7. Meeting House, "West Elevation," 14 July 1961 (LIK, 030.I.C.540.8).

prefabricated. The drawings indicate that in all the prefabricated parts, the concrete is left exposed, free of any imprints of formwork or of joints resulting from the assembly of parts. The concrete of the floor slabs poured in place is also to be left exposed, corresponding to indications written in documents in 1961 and referring to certain parts of the Salk Institute, calling for an "exposed concrete ceiling." The stone slabs incorporated in the reinforced concrete walls are also found in the construction system for the apartments of the Housing Facilities and the researchers' quarters connected to the Laboratory Building.

Thick stone rinds, slender screens of prefabricated concrete and concrete walls poured in place would have formed the different visages of the variegated landscape of buildings of the Salk Institute for Biological Studies, in keeping with the project of 1961.

Kahn's staff investigate various solutions for the use of the stone in the construction system of the "screen walls," ranging from isodomic laying of blocks to assembly of large slabs. One fundamental change is made in October-November, when the thin prefabricated elements are replaced by thick masonry masses, to build using a technique that is the evolution of the one envisioned for the earlier 8 inches walls with incorporated stone cladding. Now Kahn foresees installation of the stone slabs in such a way as to form – thanks to horizontal metal bracing between the two faces and vertical metal brackets inserted between the stone slabs – a permanent claddinf formwork, with an interspace for the pouring of the concrete. In the invention of this mode of implementation of concrete, the masonry faces bonded to the concrete core in ancient Roman construction, which Kahn had studied during his stay in Italy, seem to live on. It is no coincidence that when the "screen walls" are no longer prefabricated shells and become concrete structures poured into stone rinds, Kahn introduces arches in the large openings at the top, crowning a sequence of passages with Roman overtones.

Precisely in 1961, in the book *Roman Architecture*, Brown describes the Roman construction methods in pithy but significant terms: "Roman architects were moved to seize on the arch as the formal substitute for post and lintel, and on the vault as the means of closing the shell of space in a continuous curve. In the heat of building, under the spur of their new incentives, they perfected old materials and invented new methods. From rubblework came concrete, laid between masonry forms, which yielded its permanent faces. The plastic wall with the help of centering becomes the complete plastic shell."[27]

When they become thicker and take on faces of stone slab, the "screen walls" of Kahn can gain greater independence from the structure of the rooms, becoming true open-air ruins. Not by chance, the columns that in the prefabricated version connected the "screen walls" to the floor slabs now disappear. In the studies for the "screen walls" towards the end of 1961, the pattern of joints of the stones no longer reflects masonry systems, and in the design of the elevations the stone slabs return to having measurements closer to those of the plywood panels of the formwork.

Wrapped in slabs of Cordova Shell stone, the Salk Institute for Biological Studies would become a great fossil concretion facing the ocean, a body with skin marked by scars, those that time leaves on the skin of old men, in keeping with the poetic description provided by Le Corbusier for the *béton brut* of Marseille, or the scars caused by burns on the face of Kahn himself.

[27] Brown, *op. cit.*, pp. 19-20.

8-11. Salk Institute for Biological Studies, La Jolla, California, 1959-65.

8. Meeting House, study design of the installation of the stone formwork (LIK, 030.I.C.540.12).

9. Meeting House, "Library. Large Screen Wall" (LIK, 030.I.C.540.13)

10. Meeting House, "Library. Partial Plan First Level," 25 October 1961 (LIK, 030.I.C.540.9).

11. Meeting House, "Library. Wall Elevation," 25 October 1961 (LIK, 030.I.C.540.9).

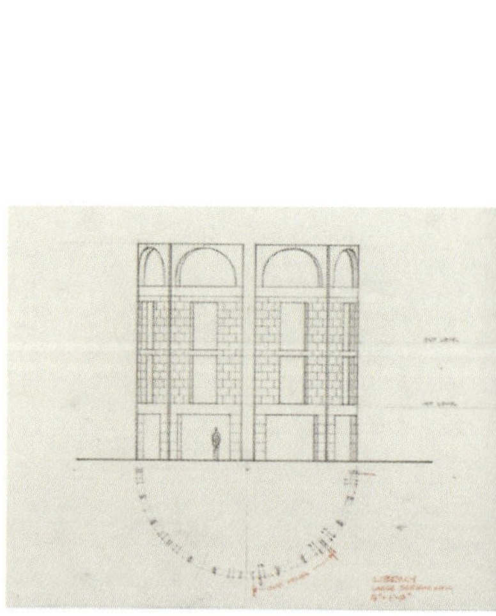

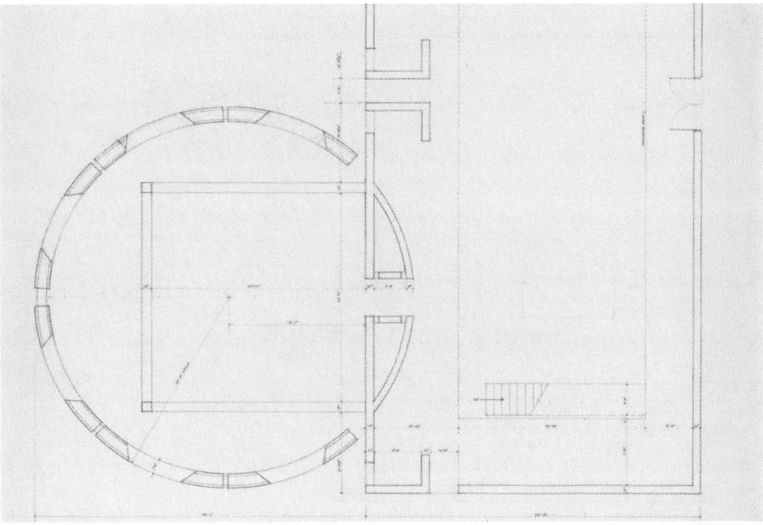

The fact that Kahn does not choose Cordova Shell in the color gray reveals his intent to underline that the skin of the Salk Institute is not made of exposed concrete, but is another geological layer of the promontory. These are indispensable premises in order to understand the complex nature of the concrete that was to be the final material of the work.

"Limestone-concrete wall": the prefabricated panels

The question of the various methods of fabrication of the concrete for the Salk Institute, addressed by Kahn and his staff during the various project phases, takes on apodeictic tones in the expression "poured-in-place versus precast" used by Kahn on 10 January 1962 to indicate the "concrete construction methods" on the agenda, examined with the aim of reducing construction costs.[28] Kahn discusses these problems with the technicians of the Kiewit Sons' construction company, based in Arcadia, California, initially selected for the job.

[28] Louis I. Kahn, letter to C.J. Pankow, 10 January 1962, LIK, 030.II.A.108.25.

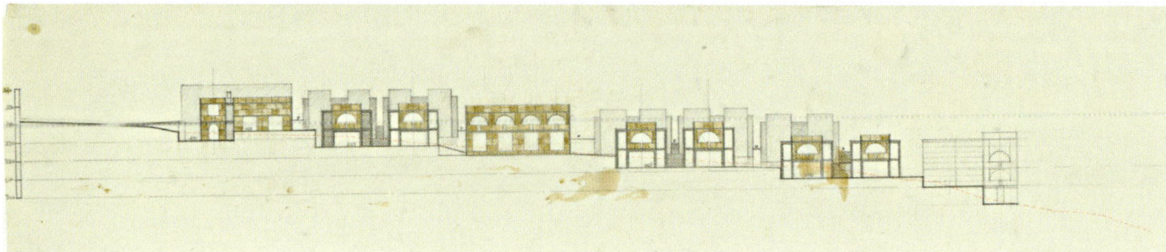

12-15. Salk Institute for Biological Studies, La Jolla, California, 1959-65.

12. Meeting House, elevation (LIK, 030.I.C.540.12).

13. Housing, elevation (LIK, 030.I.C.540.16).

14-15. Meeting House, plans and elevation, 17 January 1962 (LIK, 030.I.C.540.9).

In the project in January the concrete is still to be made in two ways, poured in place and precast. The "Outline Specifications" prepared on 17 January contain sketchy but significant indications on the implementation of concrete and the treatment of its surfaces. In the case of the concrete poured in place, Kahn wants to make an exposed concrete that is still of the type of the Yale Art Gallery, the Tribune Review Building and the First Unitarian Church, namely poured in formwork made with panels composed of wooden boards, joined like the planks of wood flooring – "narrow t & g [tongue and groove] flooring boards."[29] Thanks to these panels, Kahn creates the pattern of the concrete surface, especially in the rooms of the Salk Institute – "use generally on interior."

The same specifications indicate that the plywood, destined to become the fundamental panel of the formwork of the completed work, is still viewed in this project phase as an economical material, on a par with the "rough boards" or "other forms as approved," also to be used in service zones like "tunnels, tranches, mechanical rooms, fan rooms, unexposed locations." Precisely the change of the role of plywood in the design of the formwork becomes the most eloquent clue to grasp the meaning of the exposed concrete used by Kahn for the Salk Institute. Wooden boards would not only have left imprints on the concrete, in the project of January 1962; they would have also been used to make sliding "wood shutters" for the windows and "wood paneling" to place in the Studies and the Meeting House, and some of the floorings of the Meeting House itself.

Also in the January project, the visible concrete surfaces of the floorings are to have a finish made with a "steel trowel" or "wood float"; the same surfaces would have to be polished where tile floorings are to be used. The range of colors of the floorings also includes slabs of that gray granite that increasingly recurs in Kahn's works in exposed concrete, to be installed with rudimentary workmanship – "sawn finish."

The specifications also cover the "Mortar Materials" to use in the walls of concrete blocks and bricks, depending on the rooms, to create a hierarchy

[29] "Outline Specification for Construction of the Salk Institute for Biological Studies San Diego California," 17 January 1962, LIK, 030.II.A.108.11. The concrete to pour in place is of the "lightweight" type.

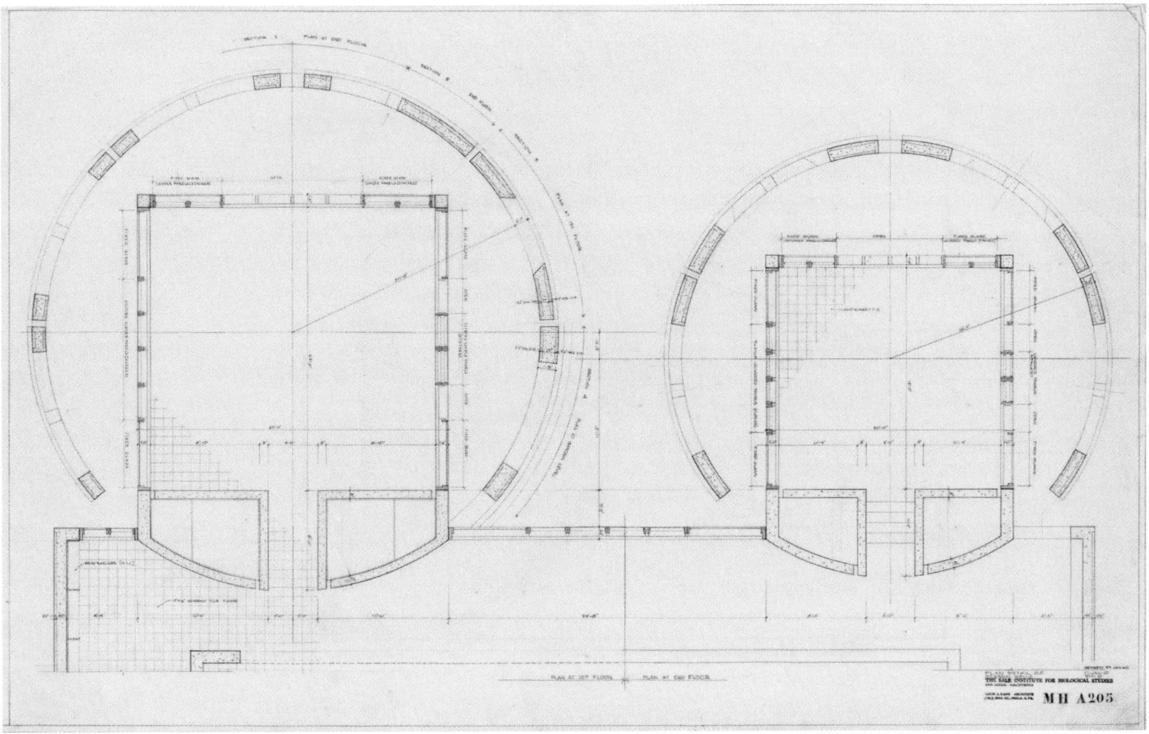
PLAN AT 1ST FLOOR
PLAN AT 2ND FLOOR
THE SALK INSTITUTE FOR BIOLOGICAL STUDIES
MH A205

MH SK 32

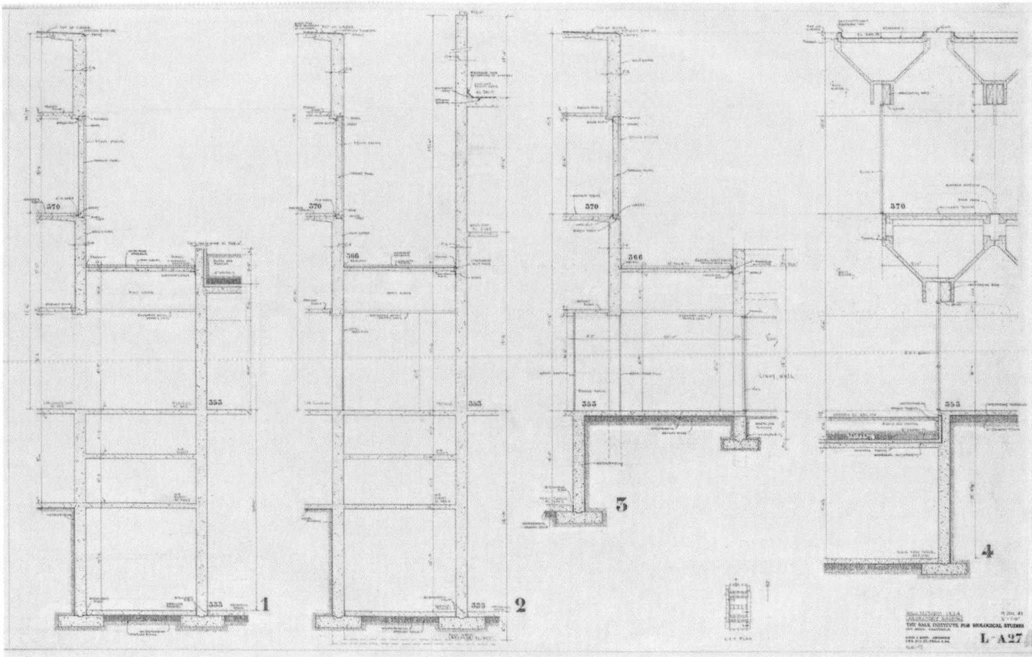

16-17. Salk Institute for Biological Studies, La Jolla, California, 1959-65.

16. Laboratory, "Wall Sections 1.2.3.4," 17 January 1962 (LIK, 030.I.C.540.2).

17. Meeting House, "Meeting Hall Wall Sections," 17 January 1962 (LIK, 030.I.C.540.10).

of chromatic values corresponding to that of the spaces. Kahn envisions three types of cement – "Portland cement," "white cement" and "masonry cement" – and sands of different colors, "for various uses"; he also enriches the palette of mortars with the addition in the compound of "pure mineral pigments."[30]

The use of stone slabs is involved in both the poured in place and the precast concrete. The presence of large surfaces clad with these slabs justifies the indication, in the specifications, regarding the use of panels of wooden boards for the formwork of the interior walls of the Salk Institute. In the specifications of the concrete to pour in place, the stone slabs are included in the list of the formwork, alongside the panels of wooden boards or plywood, confirming Kahn's intention to incorporate the cladding in the constructive process of the concrete. The appearance of the definition "concrete wall facing" should not mislead us when it comes to the effective role of the stone slabs.

In the project of January 1962, Kahn has definitively given up on the hypothesis of the use of Travertine. The slabs are to be in the Cordova Shell marketed by Texas Quarries, Inc., with a "cream" color, to be installed with a sawn surface, or "sand sawed."

In the specifications, the section on "Precast Concrete" is terse: "conform to UBC requirements; stone facing where shown." The prefabrication is no longer limited to the beams of the Laboratory Building, but also involves the masonry sectors with the stone slabs, as can be seen in the drawings. The use of prefabrication for these masonry sectors means Kahn has to revise the

[30] *Ibid.* For the walls exposed to weather silicon waterproofing is envisioned.

[31] Louis I. Kahn, letter to Peter Kiewit Sons' Co., 10 January 1962, LIK, 030.II.A.108.25.

[32] "Outline Specification for Construction of the Salk Institute for Biological Studies San Diego California," 17 January 1962, cit.

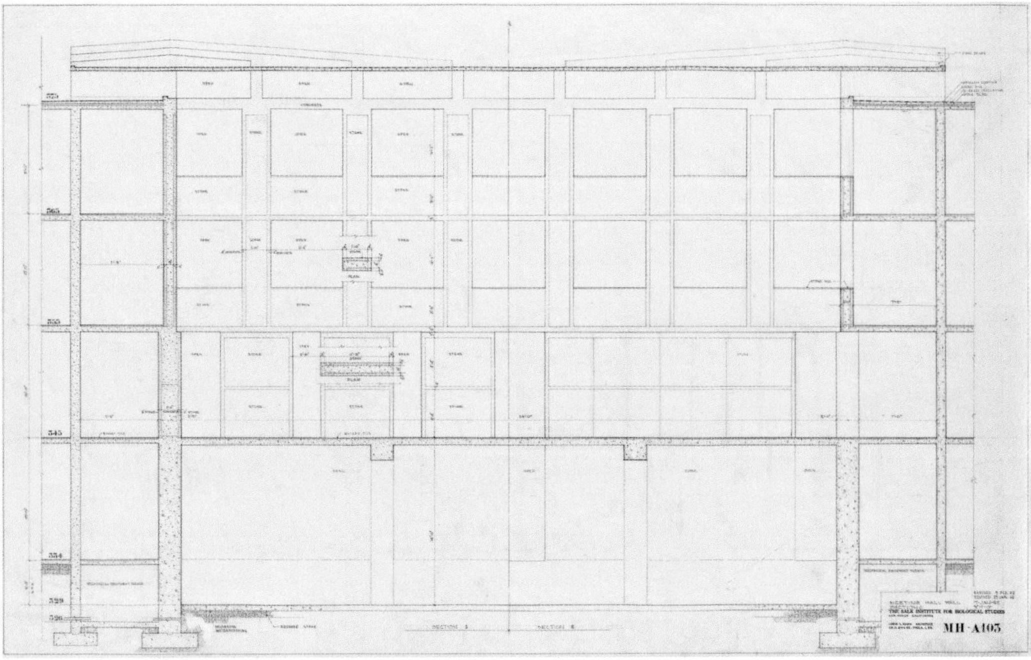

technical details of the installation of the stone slabs, in order to conserve a type of cladding that is still incorporated in the body of the concrete, as in the project of 1961.

The walls with the stone slabs called for in the previous project are broken down into panels to be precast at the factory and assembled at the worksite. This operation has an impact on the design of the laying of the slabs, which no longer form the formwork for the pouring of the concrete. Now Kahn works on an expressive design of the new construction procedure, which happens through molds in which to install the slabs and pour the conglomerate, obtaining structural panels. "Limestone-concrete wall"[31] or "concrete wall facing"[32] are the new definitions to indicate the walls of the Salk Institute with incorporated stone slabs. The fact that the arrangement of the stone slabs has changed to adapt to the prevalent prefabrication system means that those slabs have in any case become decorative masonry applications.

The more complex system of prefabricated panels, in the project of January 1962, is the one with which Kahn plans to make the "screen walls."

In the "screen walls" the two faces of the panels that form the "limestone-concrete wall" are equipped with stone slabs, also because they are both facades (the slabs of the ground floor have a thickness of 3 inches; those on the other floorings are 2 inches thick). Precisely in the wake of the transformation from permanent formwork to decorative slab, in the masonry system of precast panels Kahn introduces the distinction between stone slabs and panels, between the thin grooves of the adjoining stone slabs and the thick bands of concrete left at the joints between the panels. This distinction is necessary, in his view, to explicitly display the fact that one panel can be clad by multiple slabs. Another indication of the process of gradual transformation of the role of the stone slabs from permanent formwork to decorative

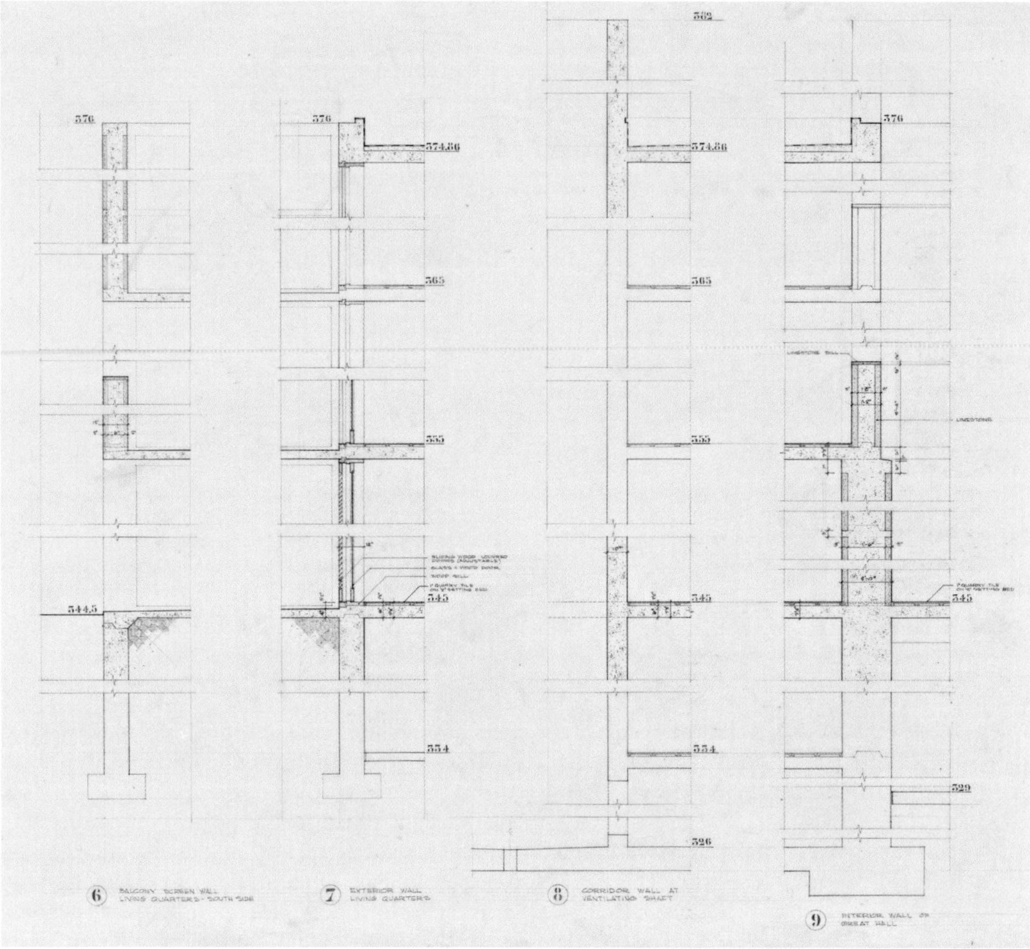

panel can be seen in a drawing, where Kahn eliminates, in the intrados of the openings, the narrow vertical stone slab that closed the formwork, replacing it with an exposed concrete surface, given the fact that the pouring would be done in a mold at the workshop. The evolution of this detail of the intrados confirms Kahn's desire to make the core of the panel increasingly visible, since in other drawings from January he calls for a slab that is no longer flush with the intrados but set back, so that the whole thickness of the panel is in concrete, creating a band or border in elevation.

In the "screen walls" the panels are to be precast with two faces made of stone sheets, mounted both vertically and horizontally through a male-female interlock. In the "screen walls" with a circular plan the curvature of the panels is provided by the visible faces of the stone slabs. The band left by Kahn to make the panels visible once installed on the worksite actually has a degree of ambiguity, because it does not exactly border the panel – in the horizontal bands the real joint coincides with the upper edge (fig. 15).

In the Hall of the Meeting House, Kahn calls for panels precast according to the same criteria as the "screen walls." Instead, in the other volumes of the Meeting House, the stone slabs are installed in reinforced concrete walls poured in place, incorporated in the concrete wall as segments

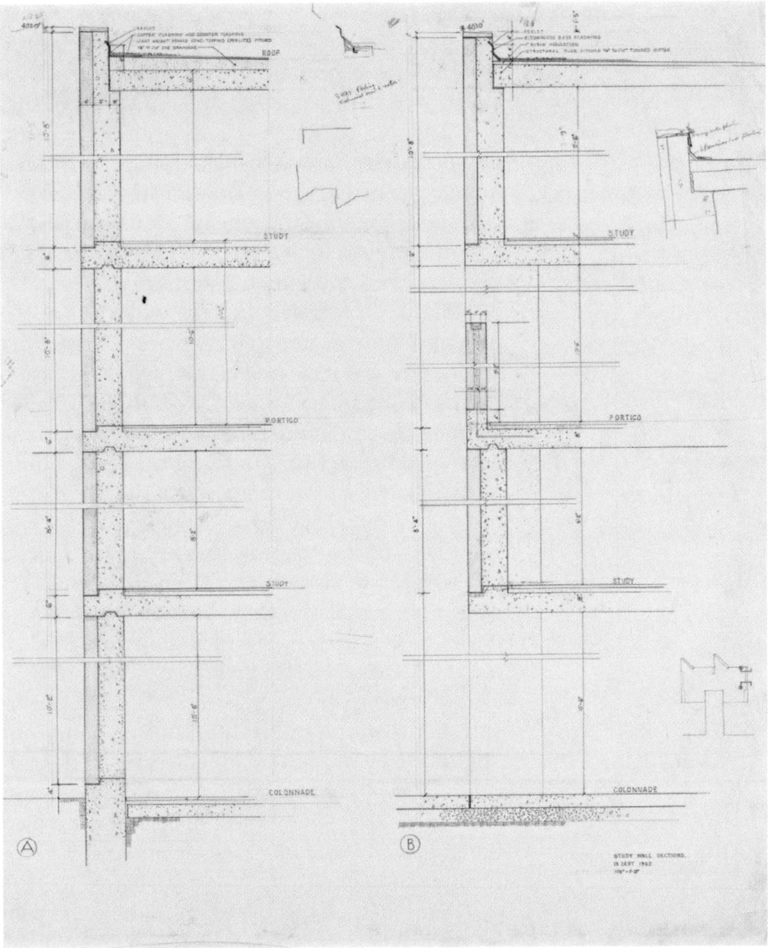

18-19. Salk Institute for Biological Studies, La Jolla, California, 1959-65.

18. Sections of Laboratory walls, 1 May 1962, (LIK, 030.I.C.540.10).

19. Laboratory, "Study Wall Sections," 13 September 1962 (LIK, 030.I.C.540.23).

of formwork to remain in place. The lodgings of the fellows in the Housing Facilities and the Studies placed up against the Laboratory Building are to be made with similar precast panels in concrete, again with incorporated stone slabs. Here again, Kahn distinguishes the joints between the slabs from those between the panels, as in the "screen walls." Other precast panels, but with the stone slabs placed only on the outer sides (the borders are still there), are specified to close off certain sectors of the Laboratory Building.

The border bands in exposed concrete, in spite of their degree of ambiguity, not only allude to the construction system of precast panels, but also serve to indicate that the stone slabs are not merely the cladding of a concrete wall, but an integral part of the process of fabrication of the panel. The border bands, for Kahn, represent the most evident sign of the change of function in the pouring process, with respect to the previous design. Reduced to framed compartments, the slabs wind up being decorative features of the concrete surface, as a great stone metopes. In spite of Kahn's efforts to find a constructive logic for the stone slabs, the abandonment of the idea of using them to create permanent formwork leads to the gradual elimination of the system.

Poured in place versus precast

The project at the start of 1962 is discussed on 17 and 18 January with the Kiewit Sons' Company, Komendant, Dubin and two members of Salk's staff,[33] and then refined in the months to follow, leading to a new meeting with Salk on 27 March. The processes for working the concrete and stone are indicated in the documents by Kahn's office. On 15 March, the "preliminary concrete specifications" are ready.[34]

Following the meeting on 27 March and the criticisms voiced by Salk regarding the plan to make the Laboratory Building in four buildings with two separated courtyards, Kahn works on a new scheme with a single courtyard flanked by two buildings, with the Laboratories and the Studies. In May and June the new project enters its definitive phase. The section of the laboratories has increased in height, and the beams are of the Vierendeel type. Now each building contains three floors of laboratories, still without intermediate pillars, alternating with service spaces for physical plant system, contained inside the Vierendeel trusses. To bring natural light to the laboratory located in the basement of each building, Kahn excavates large patios to be used as gardens.

The principle of providing a view of the ocean, which guided the studies on the angles of the facades of the various Studies, is confirmed in the final project (the gable wall of each study bends to create an opening towards the ocean).

The whole complex, including the Laboratory Building with the Studies, the Meeting House and the Residences of Fellows, continues to get its character from the use of Cordova Shell. The perspective views of the courtyard between the two volumes of the laboratories show the walls of the Studies clad with stone slabs assembled in keeping with the geometry of the isodomic wall, but larger than those of the blocks – each level contains two rows of slabs.[35] The project development concentrates on the Laboratory Building with the Studies, which would also become the only part of the Salk Institute to be built.

The question that has emerged throughout the various phases of the project regarding the nature of the concrete, poured in place or precast, reaches a turning point only in the spring of 1962. It is at this point in the project process that Kahn imagines the Laboratory Building, including its great Vierendeel beams, made with concrete entirely poured in place, solving the riddle formulated at the start of January: "poured-in-place versus precast." Now all the stone slabs have to have the function of permanent formwork – the prefabricated panels with the inserted stone slabs have been eliminated.

Formwork and concrete according to the programs of 1962

The development of the project, the contracts with the suppliers and the George Fuller Company, called in for the construction in place of Kiewit Sons' Company, with Keast & Hood to collaborate with Komendant on the structural calculations, and the start of excavation for the construction of the Laboratory Building, in July 1962,[36] usher in a decisive phase in the definition of the technical details for the fabrication of the reinforced concrete. In 1962 and 1963 the specifications and drawings indicate an evolution in Kahn's idea for the construction of the claddings. But prior to the definitive version that appears in 1963, the project of the Salk Institute, even when the Laboratory Building has been modified and the Vierendeel beams appear, continues to stick to the guidelines of the rough specifications prepared in January 1962 for the details of the fabrication of the concrete.

The manuscript "Specifications" with the sections on "Concrete" and "Cut Stone" dating back to 1962 forms the basis for the other main specifications, until 1963. The variations introduced in the successive versions correspond to important passages in the study of the technical details of the implementation of the concrete.

The page of the "Concrete" section of the manuscript opens with the "Special Requirements," which contain an apologia of exposed concrete construction that might even seem out of place here if we consider the fact that many surfaces of the Salk Institute are to be clad with stone slabs, as in the project of January 1962. But the apodeictic statement that the Salk Institute will be a "concrete building" can be assessed in the light of the question posed immediately after it, that of the formwork. In the section on the "forms" Kahn introduces the stone slabs, which also in the specifications are therefore simply another aspect of the quality of the concrete. The writing of the specifications confirms that Kahn sees concrete as a material to shape with formwork, whose truth and form cannot be separated from the materials of the formwork itself.

"Since this is a concrete building," we read, "the quality of the concrete work will largely determine the quality of the architecture. Accordingly, concrete must have not only the required strength but also the desired appearance. In order to insure a uniform color for all concrete for the entire project, all concrete materials shall be from approved sources, and the same sources shall be used for the duration of the work. The appearance of form contact surfaces exposed to view in the finished work is fundamental to the architectural design. Accordingly, formwork must be carefully designed and constructed to achieve required finishes. In general, vertical surfaces shall have a definite texture and pattern imparted by board forms, and soffits and bottoms of slabs shall have a plain, smooth finish of the type achieved with well-built plywood forms. The drawings indicate the locations of the various types of forms required. Form contact surfaces of exposed concrete shall not be patched or finished in any way after removal of forms unless particularly otherwise directed by Architect. Any honeycombed concrete shall be patched only if and as directed by Architect or shall be replaced if required."[37]

In the specifications, Kahn and his staff define the nature of the concrete and the methods of control of the compound during the various worksite phases. The compound should be made with Portland cement and lightweight aggregate; indications are also made for additives to delay setting, to prevent evaporation from leaving cavities (Pozzolith or Plastiment), and to waterproof the surface (Sika). It should be noted that these specifications do not include indications of particular criteria for the selection of aggregate and the type of cement, as if the color of the concrete were not a central concern at this time. The composition sets out to obtain a "dense, uniform concrete, the surface of which shall require no finishing when forms are removed and which shall be free of voids or honeycombs." The pouring of the concrete will be accompanied by vibration, using special machinery (later specifications indicate other operations to carry out during pouring).

[33] *Cf.* Louis I. Kahn, letter to Peter Kiewit Sons', Co., 10 January 1962, cit., and Louis I. Kahn, letter to August E. Komendant, 25 July 1962, LIK, 030.II.A.89.3.

[34] "Preliminary concrete specifications," 15 March 1962, LIK, 030.II.A.107.13

[35] See "Laboratories: Studies, schematic, freehand, perspective," 1962, LIK, 030.I.A.540.3, and "Laboratories and Plaza, preliminary freehand, perspective from west," 1962, LIK, 030.I.A.540.3.

[36] August E. Komendant, letter to Louis I. Kahn, 24 August 1962, LIK, 030.II.A.89.3.

[37] Specifications, n.d., [1962], LIK, 030.II.A.108.27.

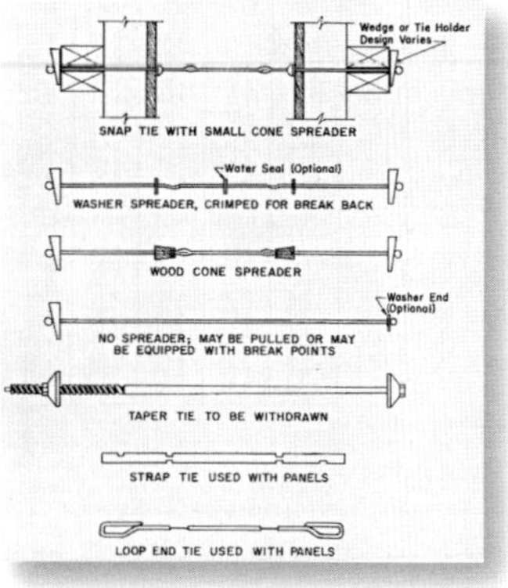

20. Some typical single member ties (M. K. Hurd, *Formwork for Concrete*, Detroit, American Concrete Institute, 1963, p. 52).

Kahn wants to ascertain, prior to the fabrication, the formal characteristics of the concrete and its surfaces after stripping, by making samples, or true wall sectors. Precisely these samples are used, throughout the construction, as the parameters of quality of the concrete to be obtained in the execution of the various phases of the work.

The section on the formwork in the specifications is divided into areas: the first, on "Forms" (divided into "general requirements," "forms for exposed vertical surfaces" and "forms for concrete not exposed to view in finished work"), the second on "Form Coating" and the third on "Limestone Facing."

In the "general requirements" Kahn simply indicates certain instructions for the construction of the wooden formwork, of importance for the quality of the concrete to be obtained. Apart from the stipulations on the tolerance of the measurements of pillars, beams or walls after pouring, the most important specification has to do with the ban on the use of chamfers to apply in the corners of the formwork. The same ban was imposed on the worksite in Rochester. Once the decision has been made to do away with the expedient of the strip that nevertheless made it possible, in an economical way, to avoid flawed corners, Kahn adds recommendations regarding the corners of the formwork: "Forms' four outside corners shall be reasonably tight to prevent forming of rough fins."

In "forms for exposed vertical surfaces" we find confirmation of a type of exposed concrete marked by boards joined with the "tongue and groove" system. The type of wood is generically indicated as local conifers. For the spreaders system, Kahn specifies the current type, known as "cone fast," with steel rods with coating and conical plastic caps, produced by Superior Concrete Accessories, Inc. The wooden boards and form ties have to be assembled in "a neat, regular pattern in accordance with approved shop drawings." The holes left by the spreaders, once the formwork is stripped, remain unfilled in the rooms, and filled in the "exposed exterior wall faces." We should remember that in the cylindrical wall of the staircase in the Yale Art Gallery the holes with the ends of the metal tubes were left without sealing because not being exposed to weather they would not run the risk of making rust stains on the concrete.

Concealing the holes left by the ties on an exposed concrete surface calls for extremely accurate technical procedures. Kahn and his staff explain how to operate in the filling of the holes, in order to avoid visible marks on the surface ("white cement should be used with the gray cement to match the color of the wall"). The procedure calls for smoothing the surface with different tools made with different materials, and for the application of moist packs for days on each filled hole.

The fact that in this phase of the project Kahn decides to fill the holes left by the spreaders with invisible patching needs careful evaluation. Erasing one of the fundamental features of the concrete construction process from view means betraying the idea of the truth of the process, and the tenet stated at the start of the specifications on the crucial importance of the "appearance of form contact surfaces exposed to view in the finished work." Kahn closes the holes to prevent the metal of the ties from marking the concrete surfaces that come into contact with rain. But the definitive technical solution is found precisely in keeping with the principle of truth of the concrete fabrication system.

The fact that the solution of the filling of the holes is commented on with a question ("why," reads a note on the specifications), bears witness to discussions on the degree of expressivity of the various parts of the formwork.

For the "forms for concrete not exposed to view in finished work" other materials are chosen, confirming the fact that for Kahn wooden boards are still the preferred tools for shaping only the concrete of the main walls. For the formwork in secondary spaces, plywood is indicated; as an alternative steel is mentioned, but with certain doubts, as indicated by a question on the margin near this proposal. The spreaders of the panels are of the type with a simple coated iron tie. Kahn indicates that these ties have to be cut to a depth of at least 1 inches below the concrete surface. All the surfaces of any kind of formwork must produce a "smooth, even surface for concrete."

In the section on "Form Coating" various treatments are described for the surfaces of the formwork. Confirming Kahn's desire to obtain the most perfect surface possible for the concrete, the formwork is coated with a plastic film or varnish, depending on whether the exposed concrete is for main or secondary structures. Apart from the filling of the holes, once the formwork is stripped Kahn calls for intervention on the surface involving only the removal of impurities generated during the pouring.

The fact that the formwork has become, also in the wake of the international success of the *béton brut* of Le Corbusier, the essential factor determining the aesthetic quality of exposed concrete, is confirmed in the section of the specifications on the decisive technical drawings for the concrete, precisely those of the formwork: "Formwork for exposed concrete, showing size and arrangement of panels, direction of boards, location of form ties and other features that will in any way leave a mark in the finished concrete."

Technical details for the stone slabs as permanent formwork

One section of the specifications of the formwork in 1962 for the Salk Institute is on the stone slabs. The fact that the inner surfaces of the formwork have cladding takes on a particular significance when this operation happens, as planned by Kahn, through the application of stone slabs. The term used in the specifications for this type of cladding of the formwork is "form liner," a term that enters current usage to indicate any type of decorated lining to apply to the surfaces of the formwork that come into contact with the poured concrete. It is worth mentioning that the *bonhommes* in wood applied by Le Corbusier to the boards of the formwork to obtain the *sculptures moulées* preview the technical criteria of the implementation on Kahn's part of stone slabs as "form liners," with the substantial difference that while the *bonhommes* and the usual "form liners" imprint decorations on the concrete, the "form liner" of Kahn is destined to remain in place, determining the character of the concrete to which it remains bonded.

"Limestone facing where indicated," we read in the "Limestone facing" section of the specifications, "shall be set in place before concrete is poured to serve as form liner for poured-in-place walls and precast wall panels. Cooperate with stone setters to facilitate accurate placement of stone. Provide all necessary bracing, supports, spreaders and incidentals required to maintain stone securely in position until concrete work is completed. Cover backs of butt joints in stone with non-staining, leak-proof sealing tape before pouring concrete. Make all provision necessary to prevent concrete from leaking and staining face of stone. Drench stone thoroughly with clean water immediately before pouring concrete."[38]

Having interpreted the form liner as a technique for a cladding incorporated in the pouring means that Kahn intends to undermine the very essence of the *béton brut* of Le Corbusier and its various international

[38] *Ibid.*

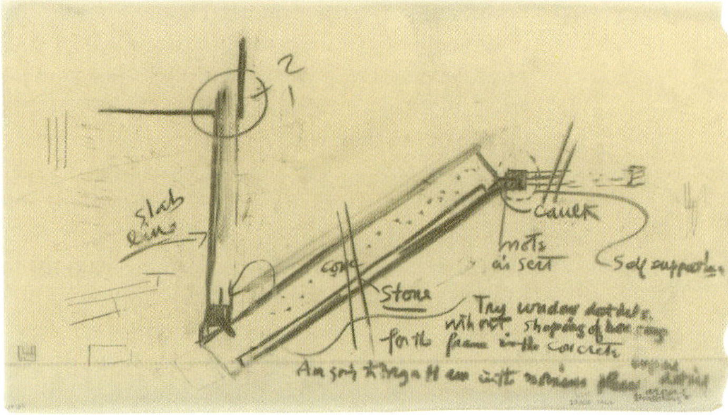

21-23. Salk Institute for Biological Studies, La Jolla, California, 1959-65.

21. Study sketch for panels in concrete and stone (LIK, 030.I.A.540.99).

22. "Middle Level Plan," 17 January 1963 (Salk Institute Archive, La Jolla).

23. Sectional perspective of the North Laboratory Wing (LIK, 030.IV.C.540.1.1).

derivatives, given the fact that his form liner eliminates the concept of the decorative imprint and replaces it with the panel itself, left in place. It is clear that Kahn is attempting to erase the decorativist aspects of *béton brut*, but without failing to explore all the possible degrees of formal characterization of the concrete surface through the technical process.

In the part of the specifications on "Cut Stone" explanations are provided of the geological characteristics of Cordova Shell Limestone, the quality of the surface of its slabs, the type of "smooth machine finish," with the presence of irregular cavities – "open fossil shell marking of varying sizes, shapes and depths," – the transport of the slabs and their storage at the worksite, and the installation, with recommendations to avoid marking the surface of the stone during pouring. Each slab will be fitted, in fact, with metal anchors, applied in special holes, numbered on the back and then installed: "Set stone accurately in accordance with approved setting drawings true to line and level with all joints and anchor holes full. All stone shall be clean when laid. Drench stone with clean water just before laying. Build in all anchors, dowels, other accessories. Point joints to an approved finish as the work progresses. Keep face of stone free of mortar at all times. Clean up splashes and stains as the work progresses."[39]

Once laid, the stone slabs have to be washed with "fiber bristle brushes and water," and also with "soap powder" if necessary.

Further notes in the specifications have to do with other components, always depending on concrete, such as the fabrication of the concrete blocks for the walls, indicating that they should be of "lightweight aggregate, but not cinders," and the type of working: "units shall have true faces, straight, sharp edges and no objectionable defects."[40]

Kahn studies the detailing of the inclined masonry sector of a study, decorated with a stone panel. The band of concrete that frames the stone turns out to be decisive for the positioning of the window casement. The note on one page is precisely about the stone, its relationship with the concrete wall and the position of the casement.[41] But the insertion of the stone slabs in the formwork, the eventual means of attaching them and the system of the formwork itself when the slabs are used are all details that are not covered, not even in the specifications.

Allen M. Katz, an employee of the George A. Fuller Company and the Project Manager sent to La Jolla to supervise the worksite, develops an alternative solution to that of Kahn, for a different "method of installing stone faces on concrete wall in the Studies."[42] Rather than applying the stone slabs to the formwork, Katz proposes attaching them to a framework of prefabricated reinforced concrete parts, or "precast concrete members," in order to obtain formwork, or "precast fillers," composed of slabs of stone with joints sealed by a waterproofing product to prevent material from emerging during the pouring. Katz's proposal implies elimination of the concrete bands that frame certain groups of slabs in Kahn's system, since the formwork is no longer the support on which to mount the slabs themselves. But as long as he continues to envision the use of stone slabs, Kahn continues to favor his own system. The method suggested by Katz is the subject of discussions between Kahn and Dale L. Harvey, a member of the Salk Institute.

"You will note," Harvey writes on 20 September, "that the main provision is for installing precaste [sic] concrete members in a horizontal plane in lieu of poured-in-place concrete as indicated on your details. This method makes job construction much easier due to the fact that there is solid support provided for all stonework prior to the pouring of concrete. In your consideration, do not be concerned with the anchorage to the poured concrete nor the method used for form ties. The end result will be a fairly tight joint between natural stone and precaste [sic] concrete fillers with one of several types of waterstops used in the joints to prevent leakage of grout upon the surface of the stonework. The anchorage of stonework and precaste [sic] concrete liners will be worked out to the satisfaction of the San Diego Building Department. The method and type of form ties proposed will be submitted to you later. This same method of precaste [sic] fillers can be used for those in a vertical position and might conceivably be used in a similar fashion for a vertical end condition."[43]

[39] *Ibid.*
[40] *Ibid.*
[41] *The Louis I. Kahn archive: personal drawings. The completely illustrated catalogue of the drawings in the Louis I. Kahn Collection, University of Pennsylvania and Pennsylvania Historical and Museum Commission*, New York, Garland, 1987, vol. II, p. 78, No. 540.99; see also *ibid.*, pp. 78-81, Nos. 540.100-06.
[42] Dale L. Harvey, letter to Louis I. Kahn, 20 September 1962, LIK, 030.II.A.107.36.
[43] *Ibid.*

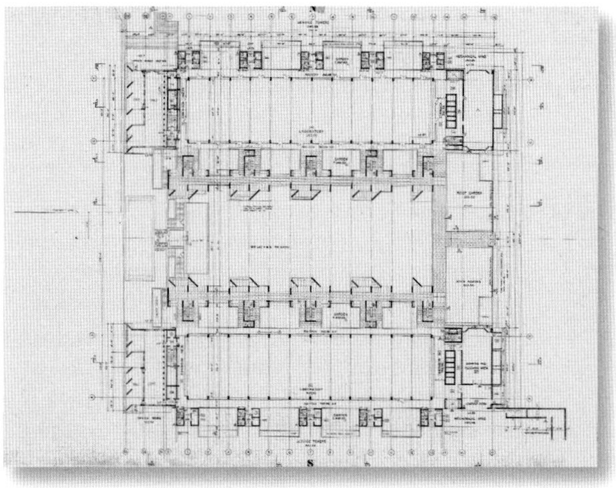

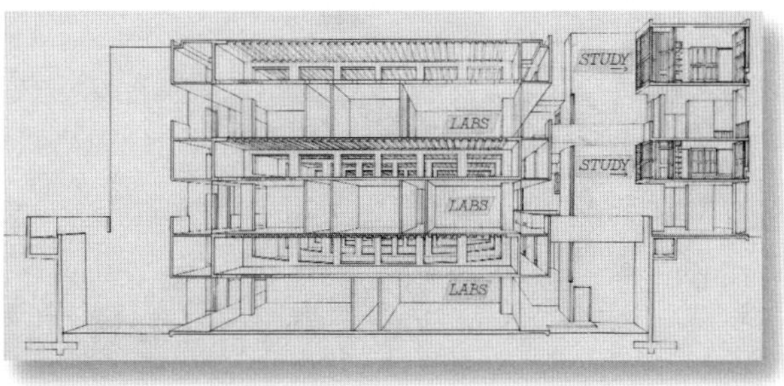

Formwork according to the specifications of October 1962

Other pithy specifications, also regarding the concrete and its formwork, are prepared in October 1962 and document the way the definition of certain technical details of the Salk Institute is moving in the direction of resolution of the theoretical defects contained in the previous constructive choices.

Besides confirming that Kahn's office will supply the drawings for "all formwork for exposed concrete," and that "sample panels" will have to be made in concrete, of large size, it is also specified that "all concrete materials" should come "from same sources of supply for entire work to insure uniform appearance." Then, with respect to the earlier specifications, additional indications are provided about the formwork: the forms for the "exposed walls" are "story height panels" of about 1 × 4 feet, with wooden boards "run vertically."

The spreader is confirmed as the one in plastic, but the treatment of the holes changes, given the fact that instead of attempting to conceal their presence with filler, as previously envisioned, now Kahn describes a seal recessed from the surface by about 1 feet and formed by "epoxy grout" or "lead plugs." The decision to leave the holes visible is destined to be confirmed, to the point of becoming one of the main characteristics of the concrete of the Salk Institute. For the ceilings the choice goes to plywood formwork. Plywood is also indicated for "forms for concealed concrete." Boards, spreaders or plywood panels are all elements to arrange according to a pattern supplied by Kahn.

24. Ieoh Ming Pei and Associates, Kips Bay Plaza, New York, 1957-63 (photograph by the author).

25-26. Salk Institute for Biological Studies, La Jolla, California, 1959-65.

25. Sample walls panels built near the construction site, 6 December 1962 (LIK, 030.IV.A.540.8.51).

26. Sample walls panels built near the construction site (Salk Institute Archive, La Jolla).

Another important clarification found in this version of the specifications, confirming Kahn's constant reflection, and that of his staff, on the technical details for an ever more accurate implementation of the concrete, has to do with the inclusion of the methods of pouring in the overall design of the concrete, with the aim of also controlling the eventual seams between the different phases. Thus, to the patterns of the parts of the formwork, that of the phases of the pouring is added, thanks to signs engraved on the formwork themselves to indicate where to stop the pouring, leaving a groove imprinted on the surface. "Pour Joints. Only at approved locations. Form notches at pour joints where indicated," is written in the specifications.[44] On the concrete surfaces exposed to weather a protective layer of silicon is recommended.

The color of Pei's concrete

In July 1962 the first traces of bulldozers appear on the worksite for the construction of the Salk Institute, while in October the excavation for the foundation is already in progress. In this phase of the refinement of the technical details, another issue at stake, along with that of the formwork, was the question of the composition of the concrete.

Since August, it has been advised to include in the specifications the procedures for the making and laboratory testing of samples for the "concrete

[44] "Specifications," 8 October 1962, LIK, 030.II.A.108.16.
[45] Galen H. Schlosser, letter to Allen Katz, 16 August 1962, LIK, 030.II.A.108.31.
[46] At the start of December, along the unpaved road leading to the worksite, several "sample walls" have been built to study the nature of the concrete, its composition, strength and color (see the worksite photograph, dated 6 December 1962, LIK, 030.IV.A.540.8). Other "sample walls" will be built during 1963, again with the same objectives.
[47] Dale L. Harvey, letter to Louis I. Kahn, 25 October 1962, LIK, 030.II.A.107.36. For the mix design used in the Kips Bay Plaza see Edward L. Friedman, *Search*, in, "Progressive Architecture", XLI, October 1960, (pp. 160-75), p. 165.
[48] *Ibid.* "Salk's idea, which is now taking shape in Kahn's stark, volcano-colored concrete, called for a new kind of environment" (Wolf Von Eckardt, *A 'Scientific Monastery' That Picasso Would Love*, in, "The Washington Post," 26 February 1967).
[49] Dale L. Harvey, letter to Louis I. Kahn, 25 October 1962, LIK, 030.II.A.107.36. On the Friedman expertise in the field of concrete see, Edward L. Friedman, *Cast-in-Place Technique Restudied*, in "Progressive Architecture," XLI, October 1960, pp. 160-175.

mixes." Kahn's office asks Katz to prepare a compound that will produce a "light color for concrete": "you are to use your judgment in the selection of sands, aggregates and Portland cement to be used by the testing laboratory."[45] On the worksite, concrete samples are made to test the color of the compound.[46] The assessment of these samples by Salk is influenced by the visit, on 24 October, to the complex designed by Ieoh Ming Pei, the Kips Bay Plaza, on Second Avenue, near 33th Street, in New York, characterized (according to Salk's own impression) by "an exposed concrete job in which the concrete color has a very warm, beige appearance."[47] After this visit, Kahn's concrete trials seem to have an unsatisfactory color, in Salk's view. It is significant that Salk judges the color of the concrete against the backdrop of the landscape where the building is to stand, as Harvey reports to Kahn: "Dr. Salk has informed me that he does not like the color of the sample concrete wall in La Jolla. He would prefer a warmer color to blend with the natural soil and surroundings."[48] In the project documents, this is the first time we find such a direct, clear desire to obtain a concrete with some relationship to the landscape at La Jolla. The information gathered by Harvey from a member of Pei's staff, Edward L. Friedman, regarding the composition of the concrete of the Kips Bay Plaza is made available to Kahn with the aim of convincing him to conduct other trials for the concrete of the Salk Institute. Friedman indicates that the compound used by Pei was supplied by the Haller Testing Laboratories of New York, and obtained without additives, with natural aggregates and Saylors cement produced in Lehigh Valley, Pennsylvania. From the same source, Kahn learns of the possibility of coloring the cement through intervention during the fabrication process itself – a fact that may have influenced the direction of future developments in his research on the color of the concrete for the Salk Institute. "[…] Friedman informed me," Harvey reports, "that Penn-Dixie Cement Corporation claims that they can produce the color of cement desired by a change in the burning process. This color would be constant […]."[49]

The specifications of December 1962

The excavation, in December 1962, has reached the stage of drainage at the location of the Retaining Walls of the Laboratory Building, and new structural calculations and drawings have to be prepared. That same month other specifications are outlined, delivered to Kahn's office on 6 December, which will be further revised in the early days of 1963.

The "specifications for concrete" of 6 December are similar in many points to those of the manuscript version. The opening page, more synthetic than in the handwritten version in its defense of exposed concrete, contains the statement of the importance, for the "architectural design," of the "various finishes required for exposed concrete surfaces." Therefore any type of photograph of the concrete is forbidden after the stripping, in the masonry sectors marked by a "definite texture and pattern imparted to the surface by the formwork," and technical precautions are also recommended for the proper attachment, where required, of the "limestone facing."[50]

The document confirms that Kahn's office will supply the drawings for all the details of the formwork, from the "direction of boards" to the "location of form ties and any other features that will in any way leave a mark in the finished concrete." The list now also includes the drawings for the formwork of "floors and Vierendeel truss." As in the handwritten specifications, those of 6 December indicate the need to make "sample panels of concrete," adding that this is an operation preliminary to "obtaining approval for formwork of exposed concrete." The complexity of the system of concrete walls with stone slabs prompts Kahn and his staff to include among these sample panels the making of a "portion of stone-faced concrete wall with both a horizontal and a vertical joint." This item was not included in the previous specifications.

With respect to the Portland Cement chosen in the prior specifications, now a "durable and light colored" cement is indicated, while the aggregates now include "quartz sand" and "light-weight" aggregate with maximum size of 3/4 inch (an additive is chosen to improve the workability of the compound, the Plastiment produced by Sika, discarding the Pozzolith). It is clear that Kahn is after a concrete surface that is as pale as possible, also in the wake of Salk's observations on the concrete of Pei. Kahn will intervene with other recommendations to achieve this same goal.

The fabrication of the concrete is further described in the section "Concrete," where to ensure quality and uniformity, or the "color and texture of all exposed, unpainted, concrete surfaces," the preparation prior to pouring of samples subject to Kahn's approval is imposed. The question of the color is again emphasized, specifying that it must be "adjusted to his satisfaction by use of the proper cements and aggregates." The inclusion of the word "unpainted" in the list of adjectives to indicate the surfaces underlines the fact that the color of the concrete must be that of the compound.

Among the drawings provided by Kahn, we find the significant addition of a "pouring plan showing sequence of pouring and locations of construction joints." With the goal of obtaining, after stripping, surfaces that are "even and dense, free from aggregate pockets or honeycomb," the "Concrete" section requires pouring that avoids "segregation of the concrete" and the formation of hardened concrete deposits on the formwork, with pouring accompanied by mechanical vibration operations and remixing by hand, done while the workers strike the formwork with rubber mallets, wooden clubs or electrical hammers applied to the spreaders – "chipping guns." It is also specified that the concrete should be poured in layers, without time gaps, to avoid "seams and planes of weakness."[51] The pursuit of a continuous surface not marked by flaws is so careful that the importance is again emphasized of introducing joints to correspond to the phases of pouring: "If a section cannot be placed continuously, construction joints may be located at points provided for in the drawings or approved by the Architect." Again, the eventual "curing compounds" have to be made with Kahn's approval.

In the section on the formwork, at the paragraph entitled "Forms for textured exposed concrete," the recommendations of the handwritten version of the specifications are confirmed, from the boards to the spreaders with plastic cones. The question of the sealing of the holes left by the cones of the spreaders continues to be a concern for Kahn's office. While the text of the specifications confirms the prescriptions of the manuscript version, for sealing with cement grout and moist packs, beside the term "holes" the section of a hole sealed with a disk of lead is drawn, with the question "lead plugs?," thus pointing however tersely to what had been indicated in the specifications of October 1962. For the formwork of the "concrete not exposed to view in finished work," steel continues to appear, alongside the plywood. But with respect to the handwritten specifications, in which this had been posed as a question, now the word steel is crossed out, thus excluding the possibility of using metal formwork.

The text of the "Limestone Facing" section is the same as in the handwritten specifications (it should be noted that the specifications on "Cut Stone" are missing). On the other hand, the section on "Inspecting & Testing" has been substantially modified, expanded point by point to indicate the modes and types of trials regarding "concrete mixes," the preparation of samples for each compound to be poured, and the responsibilities of the engineers and the construction company to control the quality of the pouring during the various construction phases.

Retaining Wall, "tentative form pattern" and elimination of the stone

When the specifications of December 1962 are being prepared, Kahn and his staff are working on the Retaining Wall of the Salk Institute, to be placed against the terrain of the excavation to create large patios providing ventilation and lighting for the first level of the Laboratory Building. Starting precisely with this wall, Kahn studies the imprint patterns left by the formwork and spreaders. The fact that it is a structural work makes the Retaining Wall part of the category to be built with plywood formwork.

The drawing of 18-21 December 1962, entitled "Retaining Wall," documents the study of the formwork or, as is tellingly indicated, the "tentative form pattern."[52] It shows the subdivision of the wall into three rows of rectangular panels, placed vertically, each with six spreaders, and an upper row corresponding to the overhanging masonry sector and functioning as a parapet – here the panels are quadrangular, and each contains four spreaders. The four rows of panels are separated by three horizontal bands, while the panels touch, side by side, with just a thin vertical imprint line.

[50] "Specifications," n.d., received from Kahn's office on 6 December 1962, LIK, 030.II.A.107.36.
[51] *Ibid.*
[52] See the drawing "Retaining Wall details" made on 18 December and revised on 21 December 1962, LIK, 030.I.C.540.047.

27. Salk Institute for Biological Studies, La Jolla, California, 1959-65. "Retaining Wall Details," 18-21 December 1962 (LIK, 030.I.C.540.47).

Starting with this study, we see a clear focus on obtaining, with the essential formwork elements, a design of the wall in which the holes left by the spreaders play a decisive role. Thus the sketch added to the specifications of 6 December regarding a type of seal that leaves the hole visible is confirmed in the drawing of the Retaining Wall pattern. Kahn's staff also examines the question of marks left on the wall in the case of pouring done in two phases, as we have also seen in the specifications of 6 December.[53]

It is probable that the specifications being prepared on 27 December 1962 ("we are to have completed drawings and specifications next week"[54]) are the typewritten version, not dated but traced to 1963,[55] based on the handwritten document and the version of 6 December, of the previous specifications. The opening page of the new specifications is a replica of the defense of the "concrete building" contained in the handwritten specifications (in other points the text is also taken from this version). The descriptions of formwork, composition and working of the concrete come substantially from the version of 6 December, with certain significant simplifications. The most important for the character of the concrete surfaces has to do with the spreaders, whose model is still left unspecified because Kahn wants to examine it in relation to the sealing of the holes. The question of the sealing of the holes left behind by the cones, unresolved in the previous specifications, is clarified in this new version. It is now indicated that the finish of the holes on the "exposed exterior wall faces" should be "plugged with lead driven into a depth slightly back from face of concrete." It is clear that the device of the spreaders will be developed precisely in keeping with the choice of closing the holes with a metal that does not rust, and which is also in tune with the color of the concrete surface. The study of the "Retaining Wall details" turns out to be decisive for the technical choice of the filling of the holes. Otto Wagner had also covered in lead the heads of iron nails used to attach the granite cladding of the Postsparkasse.

For the materials of the formwork to use in the surfaces "not exposed to view," the option of metal is definitively eliminated. The formwork should be only in plywood, with spreaders of the same type foreseen in the earlier specifications, to be left recessed from the exposed concrete surface. In the new specifications, points are extracted from the "Concrete" section of the version of 6 December regarding the modes of implementation, and are gathered under the heading "Placing Concrete," with the premise that "placement of concrete shall be in conformity with approved pouring plans."[56]

But the most significant change in the specifications for the character of the surfaces of the Laboratory Building has to do with the stone slabs. With respect to the earlier texts, the section on "Limestone Facing" has been removed.[57] The idea of transforming the stone slabs into parts of permanent formwork, which has proven for some time to be complicated and costly, is definitively discarded. Since he does not want to reduce the stone to mere cladding, Kahn opts for its complete elimination from the walls of the building. At this point the Laboratory Building is to be a work with walls only in exposed reinforced concrete.

[53] "We will proceed with the subdrain at the footing of Retaining Walls [...] as shown on drawing SK-12-18-62, revised 12-21-62. This drawing has changed the Retaining Walls from those submitted to the City, upon which we received a permit to pour. We can do no detailing nor building of these walls and footings until we receive the engineering drawings for same. These drawings, of course, will have to be approved by the City. If the wall can be designed structurally so that the entire height of the Retaining Wall can be poured at one operation with dowels installed for the cantilever slab to be poured later, then construction will be much easier, and no cold joint will show in the wall at the elevation of the bottom

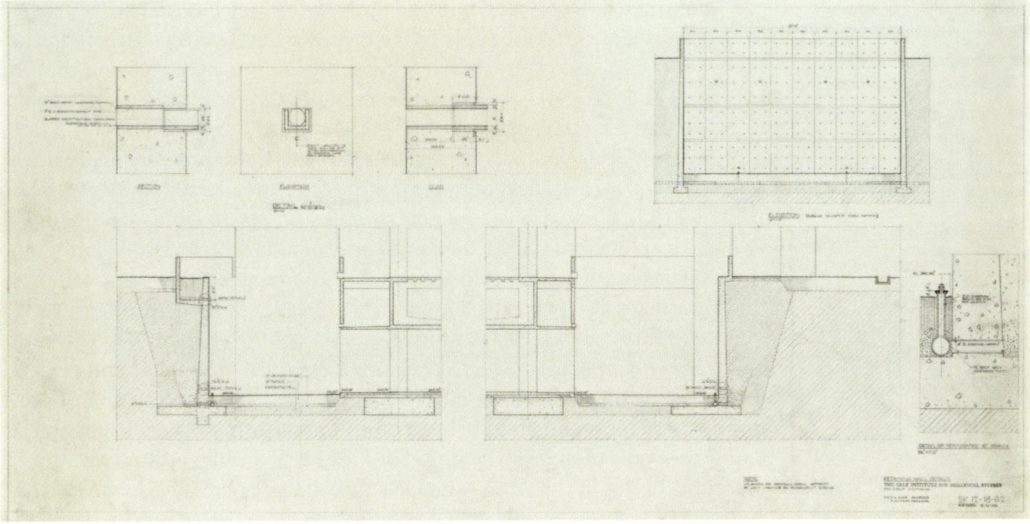

Preliminary design for the formwork of the Studies, January 1963

The elimination of the stone slabs and the choice of making a building entirely in exposed concrete poured in place open up one of the most important chapters in the development of the design for the Salk Institute: the chapter on the nature of the concrete and the design of the formwork details, on which Kahn's office concentrates for the whole year of 1963. The long series of trials for the compound, and the large number of drawings to examine the motifs generated on the various walls of the complex by the formwork, confirm the new, decisive phase of the project.

Kahn and his staff, especially Frederick Langford, John E. MacAllister and Galen Schlosser, in agreement with the client (Salk, Harvey, Earl Walls), the Fuller Company (in particular, Katz and Bill Carleton, the latter making about 200 formwork drawings) and Komendant and his consultants (including the engineering firm Ferver-Dorland & Associates), define the characteristics of the concrete components, to obtain the right color and strength, and the technical and formal characteristics of the formwork.[58] Langford later becomes the protagonist of the development of the various formwork systems applied in many of Kahn's most important works.

From the start of 1963, architects, engineers and technicians work on the constructive characteristics of the east wall of the Mechanical Building, identified as the sector from which to begin the construction of the Salk Institute, and to erect on an experimental basis to make the definitive decisions regarding the formwork design and the concrete compound, then proceeding with the construction of the Retaining Walls. Visible only in the underground spaces, the east wall is destined to be part of the unexposed concrete.

of the cantilever slab" (Dale L. Harvey, letter to John E. MacAllister, 27 December 1962, LIK, 030.II.A.107.36).
[54] *Ibid.*
[55] "Specifications," n.d., [1963], LIK, 030.II.A.107.19.
[56] *Ibid.*
[57] In the excerpt from the specifications of 6 December on the qualities of the exposed concrete surface, the adjective "unpainted" is removed – all the surfaces will be left exposed, and thus unpainted.
[58] In December and January 1963 a "Laboratory mock-up" is made (see worksite photograph, dated 6 December 1962, LIK, 030.IV.A.540.8).

28-29. Salk Institute for Biological Studies, La Jolla, California, 1959-65.

28. Summary table of the materials, 5 January 1963 (LIK, 030.I.C.540.22).

29. Formwork, "Preliminary Details," LA-301, January 1963 (LIK, 030.I.C.540.31).

In no construction work prior to the Salk Institute has Kahn ever devoted such close attention to the study of the formwork system. Precisely in the implementation phase of the Salk Institute, it becomes clear to Kahn that the exposed concrete wall has to have the qualities of graphic texture of masonry walls, those same qualities he had sought for so long with the stone slabs as formwork or as compartments incorporated in the prefabricated panels. The whole project process in its various phases since 1960 has shown Kahn the need to imagine a monumental complex having such formal quality as to make it an integral part of the landscape, quality which now – without the stone slabs with which he had gauged the color and graphic composition of the surface – will have to be sought in the control of the concrete to be left exposed. Many drawings and study sketches confirm the pursuit of an original formwork system in which each part, from the panels to the components for their assembly, is calculated to obtain a controlled modulation of every imprint. Precisely the composition of all the imprints makes it possible to define the walls of the Salk Institute for Biological Studies as fundamental expressions of the 20th-century pursuit of the nature of exposed concrete.

Every part of the formwork and the mutual relationships between the parts, all the drawbacks of the conventional methods and worksite habits become the focus of a study aimed at the invention of a system capable of designing the concrete surfaces with one and the same visible logic.

The formwork system under consideration since the start of 1963 is based on plywood panels, which have been widespread for some time in American reinforced concrete constructions, but not yet made noble as fundamental elements to add character to the surfaces of *béton brut*. Kahn's choice of plywood panels arises after having envisioned, in the projects of 1961-62, formwork with wooden boards and pouring systems with incorporated stone slabs. To understand the reasoning behind this choice we need to remember that the size of the plywood panels is comparable to that of the stone slabs framed in compartments which Kahn would have liked to install.

But it is probably knowledge of recent works of architecture by Le Corbusier in *béton brut* that encourages Kahn to give up the formwork with boards. We cannot rule out the idea that it is precisely the construction of the Carpenter Center, with the concrete in large sectors poured into plywood formwork, that induces Kahn to come to terms with the new formal expressions of *béton brut* of Le Corbusier, as Huff's anecdote about the considerations of Le Corbusier, known to Kahn, regarding the beauty of plywood for formwork might, after all, suggest. Kahn's visit in November 1962 to the Capitol at Chandigarh[59] could also have had important consequences for the developments of his idea of architectural concrete. Observing the monuments of the Capitol, Kahn cannot have failed to grasp the radical breakthrough made by Le Corbusier in the *béton brut*, towards the smooth surface obtained through the use of metal shutters, though the pieces of sheet metal were quite small.

The passage from wooden boards to plywood decided on by Kahn for the Salk Institute leads to careful study of every possible imperfection of texture and color of the surface, no longer optically disguised by the imprint patterns of the boards.

On 5 January 1963 Kahn's office prepares a table summing up the materials to be used in the Salk Institute.[60] Five types of treatment of the concrete for the walls, floor slabs and floorings are listed: "concrete with smooth finish from plywood with approved joint and tie pattern," "concrete

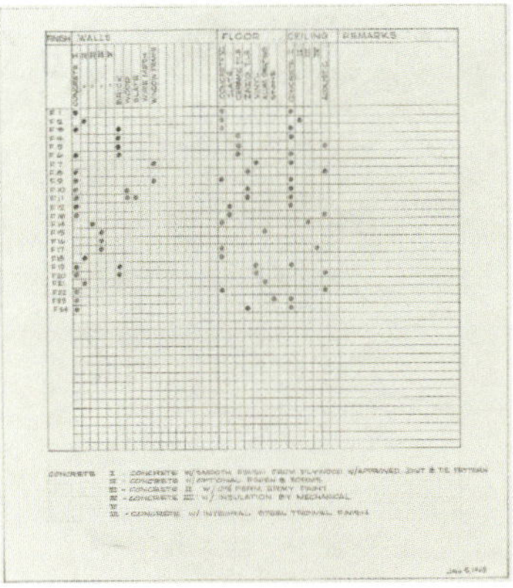

with optional finish and forms," which can also be "with 0% perm.[ent] epoxy paint" and "with insulation by mechanical," and "concrete with integral steel trowel finish."[61] For walls and floor slabs, the choice is mostly of "concrete with smooth finish from plywood with approved joint and tie pattern," with which at this point Kahn foresees determining the character of the design of the exposed concrete.

Halfway through January there is discussion of the design of the imprints on the concrete of the floor slabs of the laboratories, with steel shutters, and therefore also of the layout of the physical plant systems and the ducts to be left in the floor slabs. Kahn envisions "steel pan forms" of 16 inches to obtain a pattern of uniform lines; instead, Komendant and Robert Dorland, of Ferver-Dorland & Associates, propose a pattern with panels of 20 inches and 30 inches, alternating with bands of 5".[62] At the same time, Kahn works on providing final drawings of the profile of the imprints of the panels of the formwork, or "form relief patterns," for the Retaining Walls and the "lower mechanical walls."[63]

During January, Kahn and his staff prepare the first drawings for the patterns to imprint with the plywood formwork on the concrete of the Studies and the Retaining Walls of the esplanade.[64] The graphic conventions and certain technical solutions of the new formwork system correspond to what has been publicized in American specialized literature on formwork, and in particular in the manual by Mary Krumboltz Hurd, *Formwork for Concrete*, published in 1963 and destined to become a text of reference.[65]

In the design of the formwork for the Studies the plywood panels are arranged vertically and pierced by the pattern of the holes of the spreaders. The spreaders are positioned on the surface of the panel not so much to resist the pressure of the pouring, as to produce the desired pattern of holes. Thus the imprints of the sizing of the panel and those of the holes constitute the first visible form on the concrete of Kahn's formwork system.

The drawings of the Studies show that the openings generate panels of different widths.[66] Narrow panels arranged horizontally and a slim strip

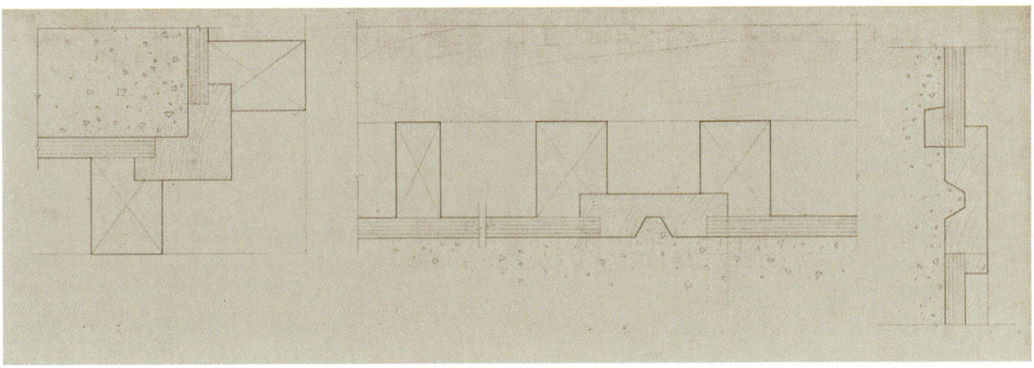

[59] David B. Brownlee, *Assembly… a Place of Transcendence*, in, Brownlee, De Long, *op. cit.*, (pp. 78-91), p. 82.
[60] See the drawing dated 5 January 1963, LIK, 030.I.C.540.022.
[61] The numbering of the concrete types goes from I to VI; but type V is not described, nor is it envisioned for use in any part of the building (see the drawing dated 5 January 1963, cit.).
[62] Dale L. Harvey, letter to Louis I. Kahn, 18 January 1963, LIK, 030.II.A.107.36.
[63] Dale L. Harvey, letter to Frederick Langford, 22 January 1963, LIK, 030.II.A.107.36. "If you just give us information by sketch or letter of what you want, then the Fuller Company will prepare their own form shop drawings," is Harvey's suggestion to speed things up.
[64] See the drawing, "Preliminary details," LA-300, January 1963, LIK, 030.I.C.540.031
[65] Mary Krumboltz Hurd, *Formwork for Concrete*, Detroit, American Concrete Institute, 1963.

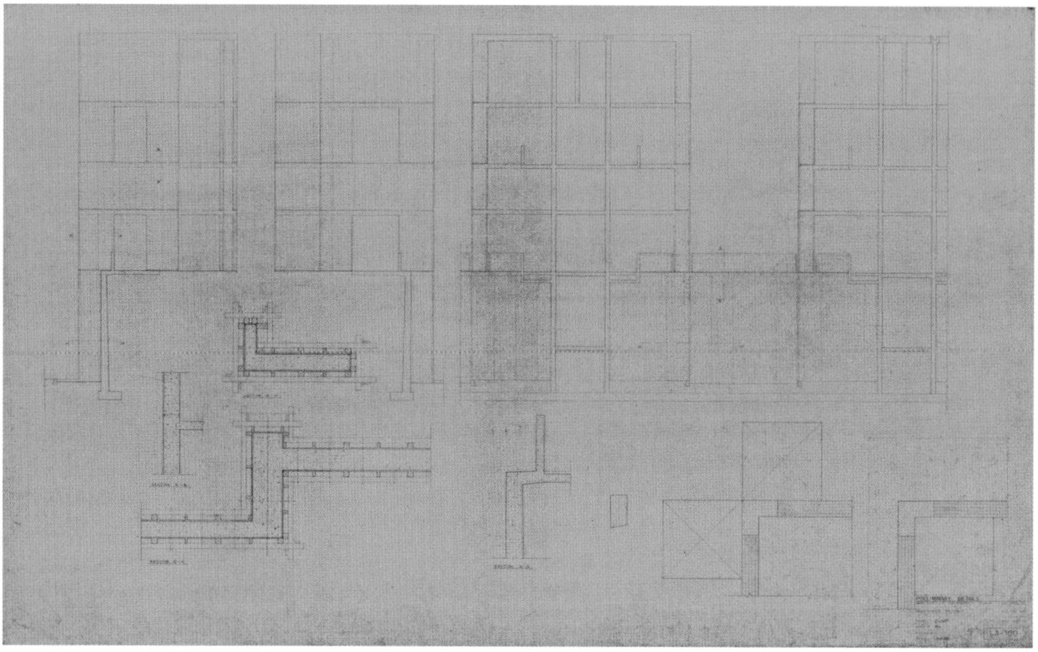

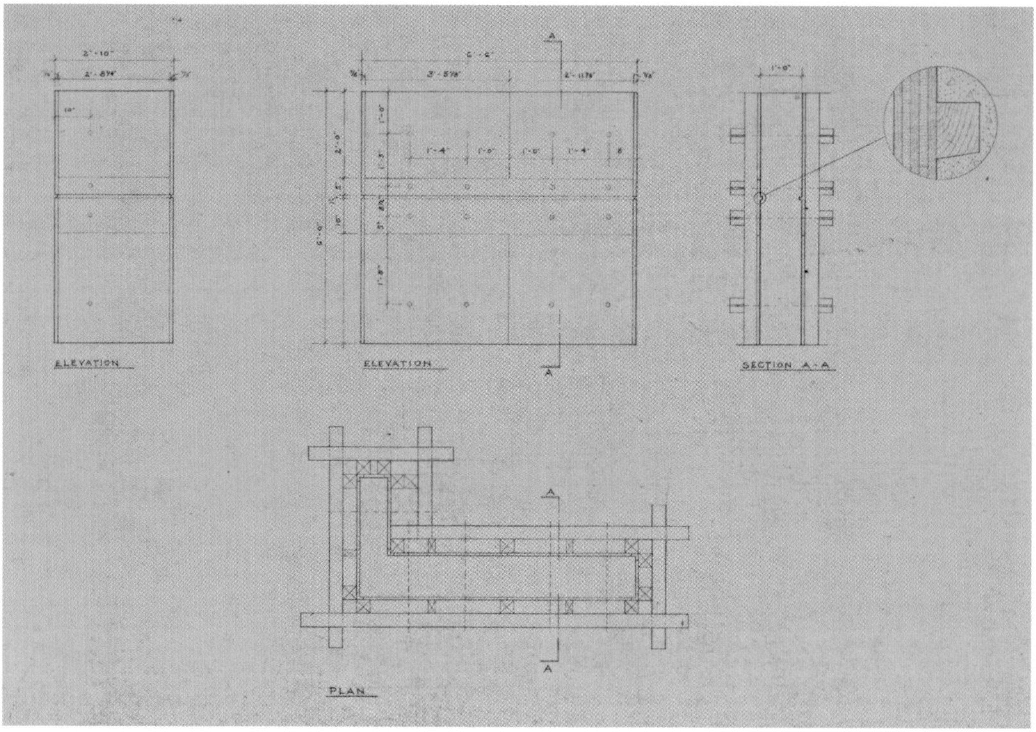

30-31. Salk Institute for Biological Studies, La Jolla, California, 1959-65.

30. Formwork, "Preliminary Details," LA-300, January 1963 (LIK, 030.I.C.540.31).

31. Formwork, "Preliminary Details," LA-302, January 1963 (LIK, 030.I.C.540.31).

placed against them reveal, with an imprint and a groove, the presence of the floor slabs of the various levels. The motif of the imprints of the spreaders is also extended to this horizontal band that tops the walls, corresponding to the position of the floor slabs. This detail is similar to the design imprinted in the *béton brut* of the Carpenter Center. Like Le Corbusier, Kahn too makes use of the groove to indicate the floor slabs, but with one substantial difference that is a harbinger of further variations: while in the Carpenter Center the floor slab is underscored by two grooves that make the thickness of the structure visible, in the drawings for the Salk Institute a single groove is indicated, marking one phase of the worksite operations, that of the pouring of the concrete up to the floor tread level of the slab. The lower surface of the floor slab is identified by the imprint of the joining of the narrow panels arranged horizontally, not with another groove as in the Carpenter Center. Therefore it is clear, right from the first drawings of January 1963, that for Kahn the groove has a dual value: it is simultaneously the pour joint and the line that identifies the floor slab. In any case, this intention to indicate the band of the floor slab by adding the imprint of the groove should probably be seen as Kahn's attempt to interpret certain details of the *béton brut* of the Carpenter Center. In the development of the system, Kahn proceeds coherently, over and above the influence of the work of Le Corbusier.

Also in the drawings for the formwork of the Studies, a vertical imprint appears near each corner; this imprint documents the detail that leads to the invention of the definitive formwork system of the Salk Institute.

One of Kahn's main concerns in the making of the concrete walls is to achieve a sharp corner, no longer beveled thanks to the insertion of a "chamfer strip." "Forming corners in architectural concrete is a critical operation," Hurd warns.[67] Precisely in order to make perfect "corners in architectural concrete," Kahn and his staff work on two special solutions, not documented in the manuals, and used according to a discrimination between corners, in the masonry sectors of the Studies, on the facade or in the interior. For the facade corners the plan is to introduce, in the angles, between two panels of plywood and flush with them, a special strip of solid wood, shaped like an L with symmetrical flanges and held in position with two studs. It is this special strip, placed against the plywood panels, that generates the vertical imprint beside the external corners. For the corners inside, at the ends of the walls, the plywood panel protrudes on both sides, passing beyond the corners, rather than being cut to exactly the width of the wall, as usually occurs. Instead, in the corners of the internal faces that are not those of the end of the walls, the plywood panels are directly juxtaposed, as in current practice.

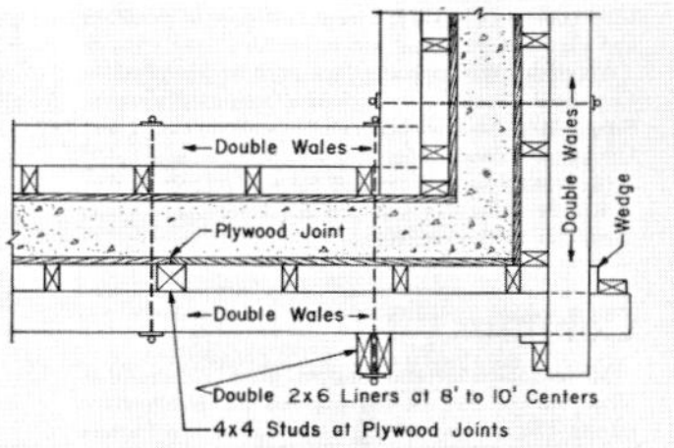

32. Typical details of straight wall formwork showing one method of securing a tight corner. Vertical kick strips are attached to intersecting wales at the outside corner, and the wales are wedged tightly to prevent leakage (M. K. Hurd, *Formwork for Concrete*, Detroit, American Concrete Institute, 1963, p. 231).

The thickness of the studs placed against the plywood panels is calculated, as usual in architectural concrete, according to the presence or absence of a joint. So where the panels touch, a thicker post is inserted, on both sides of the wall. In Hurd's manual this solution is applied only for the outer face. This precaution prevents movement of the panels during pouring. Precisely the constructive detail that prevents movement of the panels during pouring is destined to become another characteristic aspect of Kahn's formwork system, together with that of the corner. The load-bearing structure of the panels itself is reinforced in the corner with wedges and studs, as advised by Hurd. "Vertical kick

[66] See the drawing "Preliminary details," January 1963, cit.
[67] Hurd, *op. cit.*, p. 231.

strips are attached to intersecting wales at the outside corner, and the wales are wedged tightly to prevent leakage," Hurd explains regarding "typical details of straight wall forms showing one method of securing a tight corner."[68]

In the drawings for the Studies we see that the Retaining Wall bordering the large patios of the basement level was to have been marked by other imprints of the same rectangular panels – two rows of panels are called for, overlaid without any feature in between.

The irregular pattern of the holes of the spreaders and the different widths of the panels, both determined by the different measurements of the masonry sectors and the openings, confirm a design process aimed at weaving a netting of lines onto the body of the building, but without arriving at a modular coordination between the parts of the building and the formwork system, unless there is a reconsideration of the measurements of the work, from the heights of the floors to the widths of the openings and walls, against the backdrop of the new system of panel formwork. The final work too is crossed by a regular pattern of imprints and grooves, marked by inevitable exceptions.

The T strip in solid wood as a junction between the plywood panels

The question of the joint between contiguous pieces of the formwork system, previously resolved by Kahn in the case of wooden boards by using the tongue and groove configuration, returns when plywood panels are considered for the formwork of the Salk Institute. Still in January 1963, and after having prepared the preliminary drawings for the Studies, Kahn invents the technical detail that will determine the character of his formwork system and the design of the concrete surfaces of the Salk Institute. While in the "tentative form pattern" of the Retaining Wall and in the formwork of the Studies the plywood panels are simply placed side by side in a vertical arrangement, with an external stud mounted against the joint, thicker than the intermediate studs, now Kahn studies a way to give form to that joint which is also critical in the fabrication process, because it is precisely along that joint that defects can arise during the pouring of the concrete, as in the case of the boards joined with the tongue and groove system. The entire design of the joint between plywood panels is developed by Kahn with the true nature of concrete in mind, as Huff points out: "In seeing concrete as 'liquid stone,' he noted that it wants to burst out of its formwork. So, he designed the joinery of the forms of the Salk so that the pours created purposely bled-out joints, making very beautiful moldings, as it were."[69]

But the invention of what was to become the V joint, like the passage from boards to plywood, happens through a process of technical development that in this case pivots on the meaning of the joint and of the "liquid" or "molten stone," as Kahn defines concrete.

In the joint between two plywood panels, Kahn inserts a strip of solid wood, with a wide T section, so that the flanges can align against the panels and the post can be wedged into the gap and shape the poured concrete with the sculpted design of its face flush with the panels. The origin of this solution can already be seen in the previous studies to obtain perfect corners by using an L strip, again to be inserted between the panels. The idea of making the joint between the two elements visible, which is a constant in the technical poetics of Kahn starting with the "shadow joint," is also reflected in the solution of widening the gap between the plywood panels to insert a special piece.

With the T strip, what was a single imprint of the joint between two adjacent panels is replaced by a double imprint (generated by the juxtaposition of the edges of the face of the strip with the edges of the panels) and a molding with a free design, shaped by the negative profile sculpted on the face of the strip. The difference of material between the plywood panel and the solid wood strip justifies the sculptural value of the molding, which not by chance also takes on a marked relief, as it can be hollowed in a thick block of wood. Precisely thanks to the technical detail of the T strip, the formwork becomes a true work of carpentry rather than a simple worksite too for the pouring of concrete. It is no coincidence that once the joint between the panels has been widened, Kahn's drawing for the piece to insert in that gap reproduces the lines of certain details of his carpentry work, such as the connecting pieces in the staircase and balustrade of the Esherick House, in particular. The evolution of the design of the concrete surface of the Salk Institute has thus progressed into a process of technical refinement that is to transform that surface into an imaginary *boiserie*.

We should not rule out the idea that this wedging of the strip between two panels also represents, for Kahn, an attempt to render visible, in the imprint pattern, one of the fundamental features of the formwork system, otherwise destined to remain hidden behind the panels, namely the stud, though in his solution the strip no longer plays a structural role, but has become the sealing of the joint. The T strip even has to be held in position by two due studs, which makes the formwork system more complex and costly.

In the new imprint pattern, the panel becomes the constituent unit of the mural form of Kahn's concrete. Thanks to its size and its framing, it leaves imprints that resemble the large stone slabs envisioned in the first versions of the project for the Salk Institute, rather than resembling any other expression of *béton brut*, although Kahn has not achieved this result by imitating the stone with the plywood panel.

In Kahn's intentions, the T strip reveals the assembly of the formwork, contributing to create a pattern of lines that express the construction process, while serving to keep the two plywood panels in position, together with the studs. The main optical role is that of shaping a molding that will make any deformations of the concrete surfaces caused by shifting between contiguous panels become imperceptible.

Kahn's studies for the molding to sculpt on the surface of the strip inserted between the panels reflect the concern with finding a profile that expresses the joint and is capable of concealing any possible offsets between the planes of the concrete surfaces. For the profile of the molding, Kahn proposes the typical V shape that is usually generated by the beveling of the edges of the plywood panels. But now that profile is sculpted at the center of the face of the strip, thus becoming independent of the logic of the placement of the panels next to each other that had generated the formwork system. The point of the V that is cut is to display the fact that its origin no longer depends on the process of assembly of the formwork with cut edges.[70] That profile will take on other meanings in the process of definition of the joint between the plywood panels of the formwork for the Salk Institute.

[68] *Ibid.*, captions 11-17.
[69] Huff, *Louis Kahn: 'Sorted Recollections and Lapses in Familiarities'* cit., p. 423. The expression "liquid stone" is Kahn's (*ibid.*, p. 415). "Liquid stone" is a definition that is used as a synonym of concrete in, Francis S. Onderdonk, Jr., *The Ferro-Concrete Style. Reinforced Concrete in Modern Architecture*, Architectural Book Publishing Co., In., 1928, pp. 49, 131, 252.
[70] See the drawing "Preliminary details," LA-301, January 1963, LIK, 030.I.C.540.031.

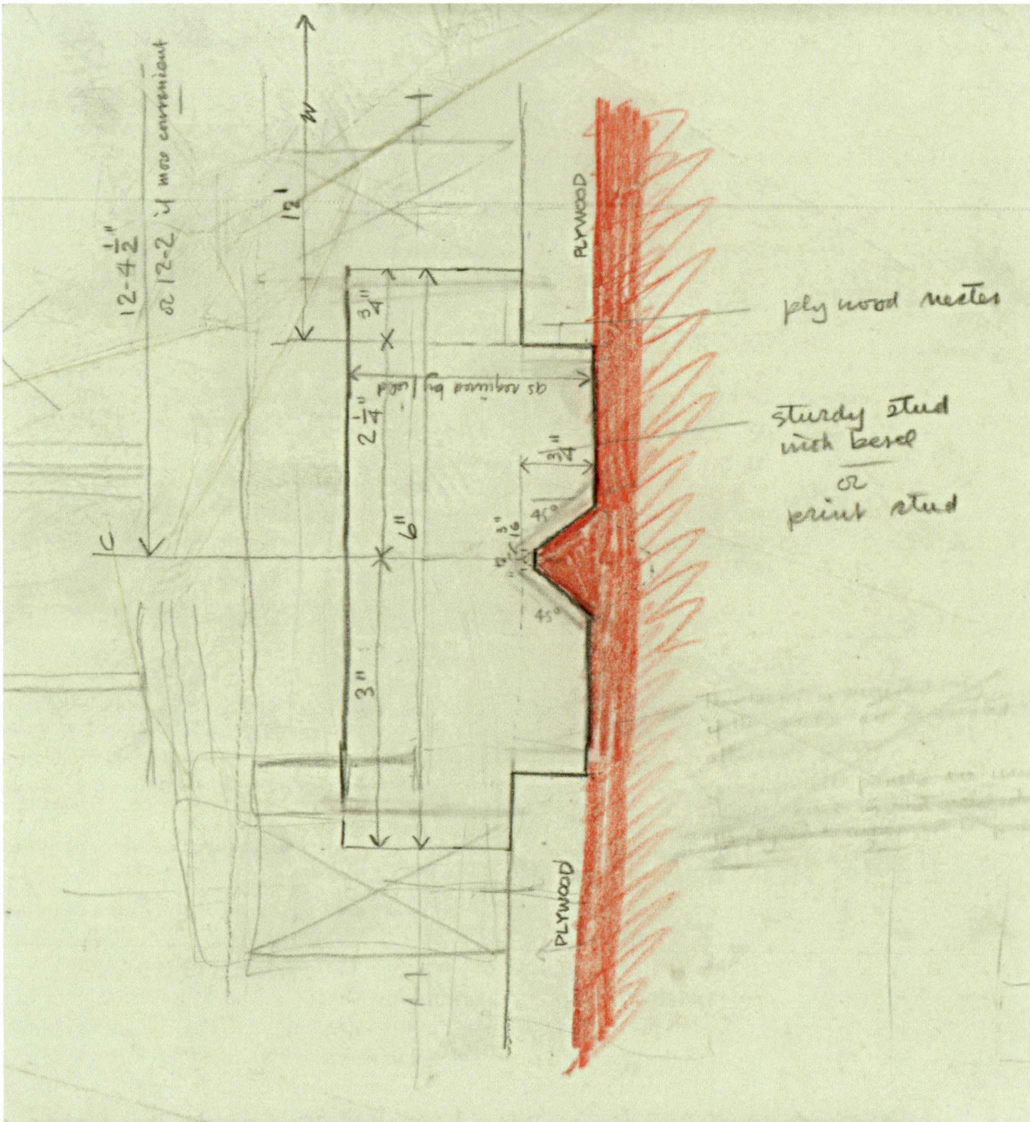

33-37. Salk Institute for Biological Studies, La Jolla, California, 1959-65.

33. Sketch for the juncture strip (LIK, 030.I.C.540.23).

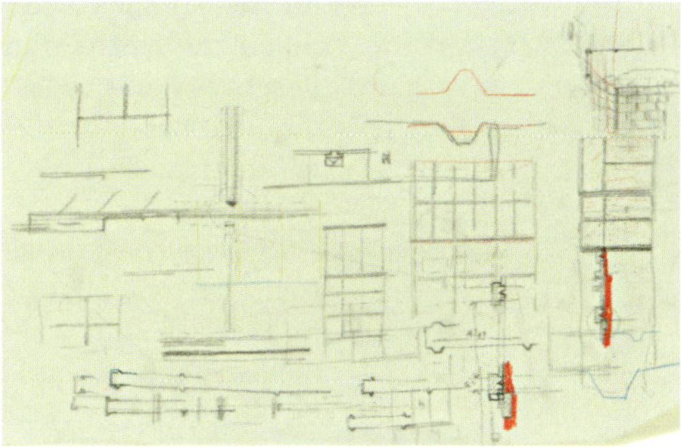

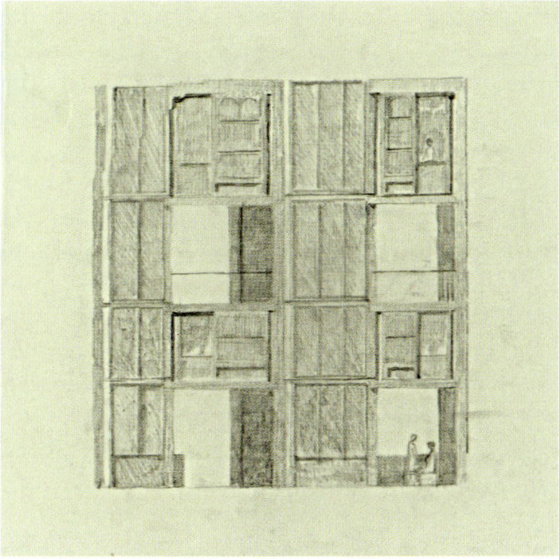

34. Sketch for juncture strip and formwork (LIK, 030.I.C.540.22).

35. Sketch for juncture strip and formwork (LIK, 030.I.C.540.22).

36. Sketch for the Study Towers (LIK, 030.I.C.540.23).

37. Study Towers elevation (LIK, 030.I.C.540.23).

Kahn's formwork system also calls for slender strips of wood to apply on the plywood panels to print grooves in the concrete. These strips have one side slightly inclined to facilitate stripping and to prevent rainwater from accumulating in the groove made on the concrete. The configuration of the L strip with symmetrical flanges to insert in the corners, shown in the previous drawings for the Studies, is also revised in the light of the invention of the T strip, transforming that strip into another piece equipped with flanges to firmly align against the panels. In several drawings of masonry sectors, perhaps for the Studies, the patterns generated by the imprints of spreaders and panels, grooves and moldings are examined.[71]

New studies for the formwork of the Retaining Walls and Service Towers

At the end of January 1963, Kahn and his staff concentrate on the formwork for the Retaining Walls, with the adjacent Service Towers, because these are the parts of the complex to be constructed in the first months of the year. As on other Kahn worksites, the definition of the technical details, which are anything but minor, happens during the course of the construction. Many sketches show the configuration of the solid wood T strip with its molding,[72] the juxtaposition of two strips by means of a special strip inserted between the flanges, so as to create a groove in the concrete between the two reliefs, or solutions of panels placed side by side without the flanged strip and with the edges beveled, to shape the concrete into a V profile.[73]

The design of the concrete for the Retaining Walls and Service Towers has the same panels and framing, but arranged in different patterns. For the Retaining Walls the arrangement of the panels is horizontal, to express a sturdy layering to resist the thrust of the terrain, with aligned vertical joints or staggered vertical joints, where the latter case is intended to replicate the image of a grand fortification. Instead, in the Service Towers a vertical arrangement of the panels accentuates the upward thrust of the volumes. The horizontal or vertical positioning of the panels corresponds to Kahn's design for the project in Rochester, with the wooden boards arranged so as to differentiate the meaning of the various walls by means of vertical or horizontal imprints. Precisely as Wagner had done with the measurements and arrangement of cladding elements in his works, Kahn uses the formwork panels to imprint the meanings of the various parts of the Salk Institute.

When the Salk Institute is under construction, no similar patterns have been documented in the main formwork manuals (in the examples of patterns made with large plywood panels, the arrangement is vertical, as in the Northwest Airlines Engine Test Building in Minneapolis, published by Hurd[74]). Neither are the panels arranged horizontally in the Carpenter Center and the architecture in *béton brut* inspired by the works of Le Corbusier.

38-39. Salk Institute for Biological Studies, La Jolla, California, 1959-65. Sketches for juncture strip and formwork of the lower Mechanical Room, east wall (LIK, 030.I.C.540.22).

Similar reliefs, again obtained with special strips of solid wood inserted between plywood panels, are now developed to also add character to the concrete surfaces of the Studies. Kahn considers different patterns, with vertical or horizontal panels alternating with two squares, grooves of separation between levels and a design of the ends of the walls with the edges framed, along the two parallel faces, by grooves or V-shaped or trapezoidal reliefs obtained with solid wood strips. To achieve the final imaginary *boisierie*, it is significant that in certain sketches the pattern of reliefs and grooves imprinted in the concrete is compared to the patterns of lines of the wooden casements.

The experimental formwork in the walls of the Mechanical Building

From the pattern studies for the panels and the various strip configurations, Kahn selects several solutions that form a range of types of formwork to assemble in the construction of the East Wall of the Mechanical Building of the Salk Institute. On 29 January 1963 the two definitive sets of drawings of the formwork for this wall are ready, prepared precisely to be able to conduct trials through which to assess, on the worksite, the various modes of implementation and the configurations of the solutions. In the drawings, the large rectangular panels framed with different moldings and pierced by the holes of the spreaders are combined in different groups of patterns in various sizes, arranged vertically or horizontally, with staggered or aligned joints, to create large decorated wall sectors. The sets of drawings are like a monumental catalogue of forms for the laying of masonry.

In the drawings entitled "preliminary form details," also dated 29 January, seven different types of joints are proposed, identified by the letters from A to G, all used in the trial project to be conducted along the East Wall.[75] The solutions of joints between panels involving the insertion of a shaped wooden strip include three configurations for the design of the heads, always 4 inches wide: a flat head, without molding, aligned with the two panels (D), and a V-shaped groove, or one with the point cut. In the latter case, two widths are indicated for the face of the cut point (C, 1/4 inch, and E, 1/8 inch). The joint between the panels can also be done with two shaped strips, placed together in such a way that in the cavity generated by the two flanges a special strip can be wedged, also in solid wood, protruding and with sloped sides, to make a groove in the concrete. The pair of shaped strips can have the V-shaped groove of the type E – this is case B; or the strip is of a different type, characterized by a protruding V point, which makes a groove in the concrete – this is case A. In both cases, a large frame is imprinted on the wall, with three parallel moldings. Kahn also considers two conventional solutions of joining between panels, without intermediate strips, with or without cut edges – types G and F. The solution of the V imprint is calculated to obtain a vivid relief, in tune with the other reliefs of the various frames of the panels. If we consider the drawings in January for the patterns of the concrete of the Studies, to obtain with vertical panels simply placed side by side, without beveling, the F solution indicated for the east wall takes on the meaning of a trial for this solution formulated first in this part of the Salk Institute.

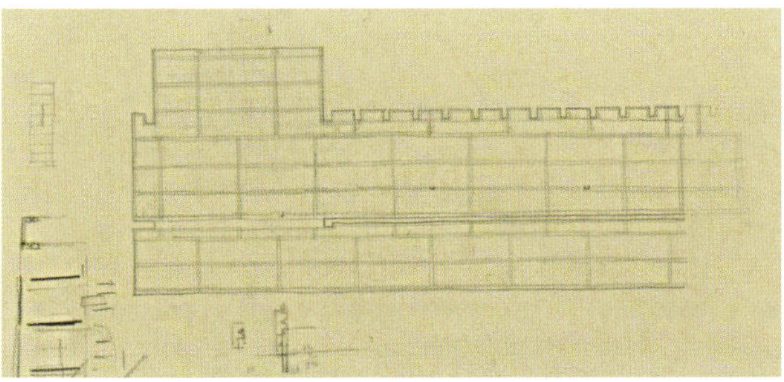

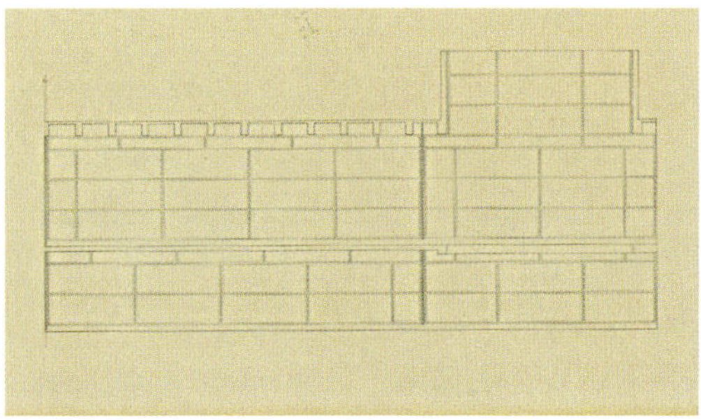

[71] See the drawing "Prelimnary details," January 1963, LIK, 030.I.C.540.031.
[72] The profile of the molding is generally a V-shaped groove with the tip cut (see the sheet of sketches, n.d., LIK, 030.I.540.022); nevertheless variations exist with a protruding V that produces a groove on the concrete.
[73] See the development sketches, n.d., [January 1963], LIK, 030.I.C.540.022.
[74] Hurd, *op. cit.*, p. 225.
[75] See the drawing "Preliminary form details," LA-306, 29 January 1963, LIK, 030.I.C.540.031.

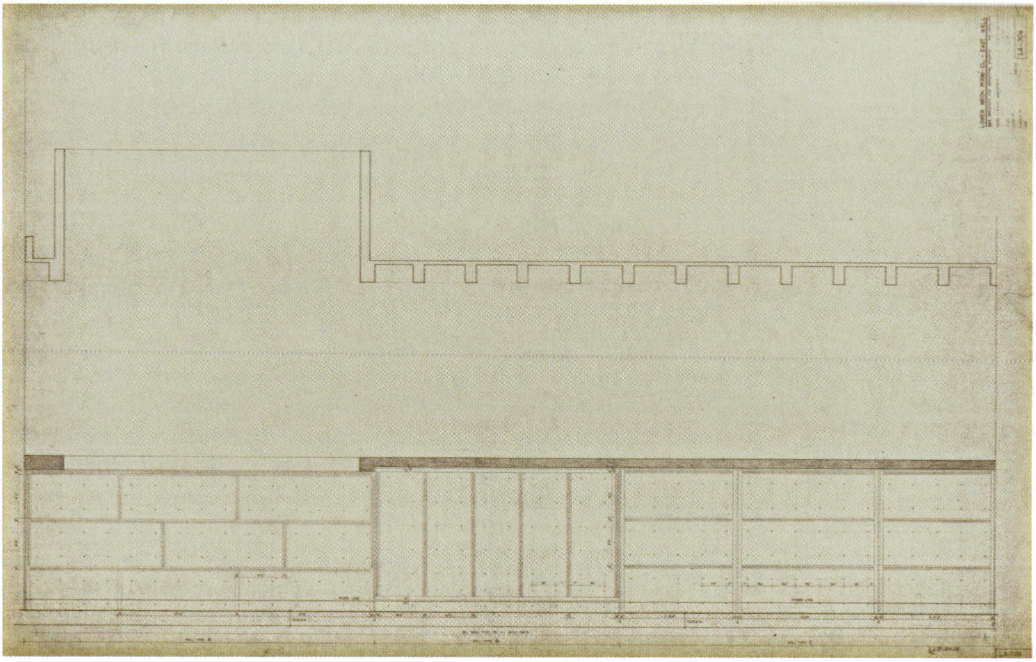

In February, Kahn and his staff develop other patterns to be made, again as trials, on the wall facing the East Wall, namely the West Wall of the Mechanical Building. Again, the set of drawings for the West Wall contains two variants, but for a pattern of only large rectangular panels, still framed and arranged horizontally, with aligned vertical joints to avoid any allusion to laid masonry. In some of these sectors the panels are without any trace of spreaders, as if Kahn had returned to the solution of the filling of the holes (the panels with spreaders are marked by two rows arranged in a square pattern, whose measurements are not indicated).

Precisely in the development of the West Wall, Kahn applies to the project of the face of the "Retaining Wall at Service Tower" the motif of the horizontally arranged panels, configured in three columns with aligned vertical joints and separated by a groove (the holes from the spreaders are visible, arranged in a square design - 12 per panel). Again in this case, as in examples of the East and West Walls, each panel is framed by the imprint of the joint strip. To obtain the vertical grooves, Kahn uses the system of doubling of the strips with a special strip wedged in between, as in cases A and B.[76] In the parapet the groove becomes an opening against the sky between the panels.

In all the definitive drawings for the East and West Walls and for the Retaining Wall, Kahn and his staff trace the lines of the imprints generated by coplanar elements, and which form the framing of the panels; but they never draw the lines of technical joining between the strips, which nevertheless do leave similar imprints on the concrete (for example, in the corners the framing bands of the panels are shown as smooth, even though two strips are joined at those points). This discrepancy between drawing and

[76] See the drawing "Retaining Wall at Service Tower," LA-346, February 1963, LIK, 030.I.C.540.031.

40-42. Salk Institute for Biological Studies, La Jolla, California, 1959-65.

40. "Lower Mech.[anical] Room – El.[evation] – East Wall," LA-304 (LIK, 030.I.C.540.31).

41. "Lower Mech.[anical] Room – El.[evation] – East Wall," LA-305, 29 January 1963 (LIK, 030.I.C.540.31).

42. "Preliminary Form Details," LA-303, 29 January 1963 (LIK, 030.I.C.540.31).

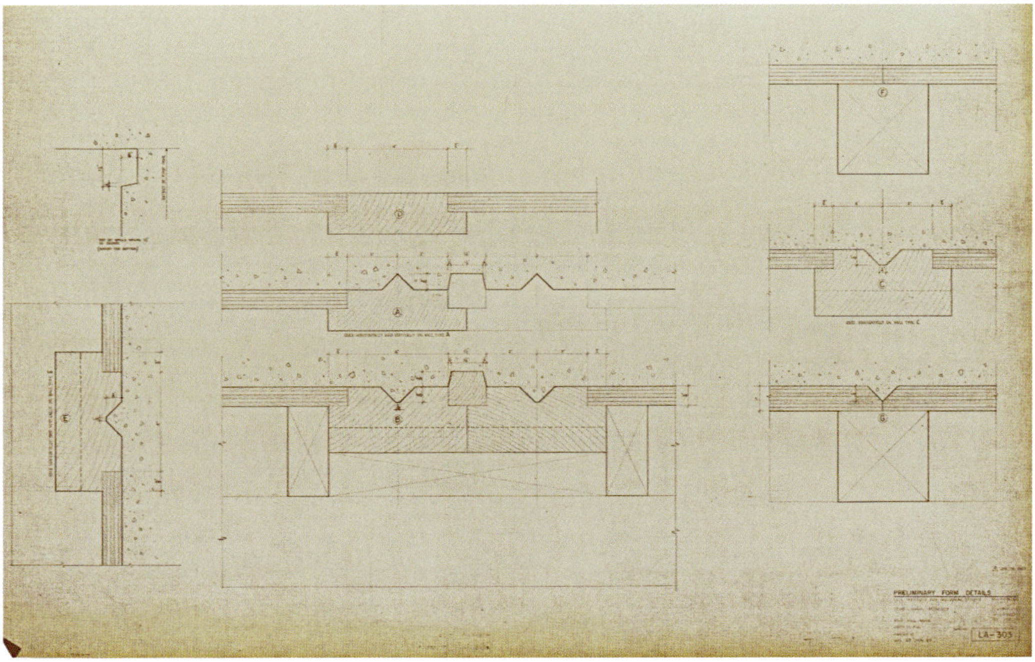

construction, which is surprising if we consider Kahn's focus on the assembly of the parts of the formwork and the imprint on the concrete, will be resolved only after the construction of the East Wall.

The arrangement of the framed panels is defined with different descriptions in the correspondence in January-February 1963, all focusing on the terms "pattern" and "relief" to indicate the degree of artistic elaboration Kahn wishes to grant to the formwork: "form relief patterns,"[77] "form patterns," "relief designs,"[78] "plain forms or relief patterned."[79] All the other sectors of the envelope not in concrete, in the terminology of Kahn and his staff, are indicated as "window wall" or "glass wall," where particular attention is devoted to finding the casements most like to stand up to corrosion,[80] but also the "window wall pattern,"[81] confirming the fact that all the lines – those of the imprints and those of the casements – constitute the compositional filigree of the surface of Kahn's architecture.

Mix design trials for the concrete: the whiteness of pozzolan

The construction of the Salk Institute for Biological Studies represents a crucial phase in the fabrication of concrete, and not only in the work of Kahn. The question of the nature of concrete is not limited to the design of the formwork, but also involves the composition of the compound itself. Already in the specifications we find the description of a composition in order

[77] Dale L. Harvey, letter to Frederick Langford, 22 January 1963, LIK, 030.II.A.107.36.
[78] Dale L. Harvey, letter to Louis I. Kahn, 18 February 1963, LIK, 030.II.A.107.36.
[79] Report on a meeting of 20 February 1963, LIK, 030.II.A.108.46.
[80] Marshall D. Meyers, letter to Frederick Langford, 5 February 1963, LIK, 030.II.A.107.36.
[81] John E. MacAllister, report on a meeting on 20 April 1963, LIK, 030.II.A.107.18.

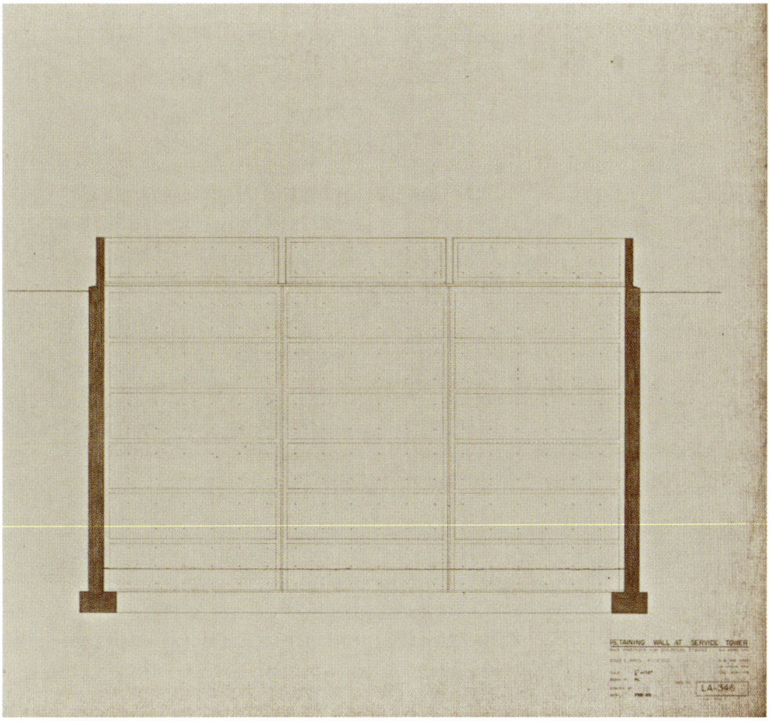

43-44. Salk Institute for Biological Studies, La Jolla, California, 1959-65.

43. "Lower Mech.[anical] Room – El.[evation] – West Wall," LA-306, February 1963 (LIK, 030.I.C.540.31).

44. "Retaining Wall at Service Tower," LA-346, February 1963 (LIK, 030.I.C.540.31).

to obtain the palest color possible. But Kahn's research does not explore only traditional components. The goal of his testing is to obtain a color without the addition of substances extraneous to the conventional mix.

Against the backdrop of the spectacular coastal landscape, Kahn conducts experiments on the nature of concrete. Salk plays a fundamental role in the choice of the ideal color for the concrete to be inserted in that landscape. So the concrete of the Salk Institute becomes a geological material. The many consulting contributions and tests to achieve a particular chromatic result in the concrete can also be explained by the fact that this work is the first in which Kahn leaves poured in place concrete exposed also on the exterior. It is precisely in terms of natural light that Kahn studies the composition of the material and decides to make use of pozzolan.

The studies on the joints of the formwork system are accompanied by trials on the chromatic composition and strength of the concrete, which also enter a final phase starting in the early days of 1963, again in preparation for the construction of the East Wall of the Mechanical Building. Salk's desire to obtain a concrete with given tonal qualities, after having seen the work of Pei, prompts Kahn and his staff to study the individual components of the material. At first the research moves in the direction indicated by Pei's staff, namely the coloring of the concrete by means of its own components. Therefore the choice goes to cement with pozzolan additive, known as Pozzolanic Cement. But the construction of sample walls to test the strength of the concrete demonstrates, in January and February, the problems involved in using Pozzolanic Cement for the Salk Institute, since the trials show that the concrete cannot stand up to the 5000 lb/sq.in. required for the East Wall of the Mechanical Building. This marks the start of a series of attempts to reconcile the pursuit of a concrete with a particular color and the required load-bearing strength of the various sectors of the complex. While the concrete made with Pozzolanic Cement is not strong enough, nevertheless the color of the walls meets with Salk's approval. "[...] the existing sample wall" is "agreeable to Dr. Salk," we read in a document in February.[82]

Following the strength trials of the sample walls, still in February there is discussion, for the color of the concrete of the Salk Institute, of the possibility of obtaining the desired tone by means of the addition of a "colored pigment," while using a concrete strong enough to bear 5000 lb/sq.in.[83] To this end, the Scofield Company of Los Angeles, specializing in products for coloring cement, is ordered to make "packets of colors pigments" to make concrete with a color like that of the sample walls in Pozzolanic Cement approved by Salk.[84] "Sample walls will be poured soon of both lightweight and hardrock concrete with pigmented packets added," Katz reports on 13 February.[85] In this phase of the project, the hardrock concrete is envisioned for most of the complex, while the lightweight concrete is limited to only certain sectors.

At this point of the experimentation, two solutions are emerging to produce a concrete with certain characteristics of strength and color: pigments added to the compound, or the choice of its components themselves. Though there are no documents to directly bear this out, it is probable that Kahn favors the solution of the color generated directly by the compound, given that fact that it is precisely this solution, in spite of the inadequate strength of the resulting concrete, that is emphasized during the course of the various trials.

Katz conducts tests on the color and strength of the concrete, starting with the pigments supplied by the Scofield Company. On 20 February it is decided to build four sample walls in concrete of the hardrock type, with a strength of 5000 lb/sq.in and the addition of pigments. The documents confirm that the color is selected to comply with the wishes of Salk: "Each sample will be a different shade of color desired by Dr. Salk."[86]

For the type of cement with which to build the East Wall of the Mechanical Building, Katz decides not to use Pozzolanic Cement, but the kind produced by the Riverside Cement Company, a division of the American Cement Corporation of Los Angeles, operating in California and based at Oro Grande, near Victorville. The cement of the Riverside Cement Company has also been chosen for its color – "was approved for color."[87] Nevertheless, doubts remain as to whether it can achieve the necessary level of strength, to be verified by means of compression testing.

On 25 February there is discussion of a "change in concrete materials" and the "color of concrete sample walls": the lightweight concrete is extended to the construction of the Service Stair Towers of the laboratories and the complex of the Studies (a decision made by Kahn or Komendant, which leads to a significant cost increase).[88] The making of the four sample walls and the East Wall of the Mechanical Building, the former with pigments added to the compound, the latter only with the concrete of the

[82] Report on a meeting on 13 February 1963, LIK, 030.II.A.108.46.
[83] Report on a meeting on 6 February 1963, LIK, 030.II.A.108.46.
[84] Report on a meeting on 13 February 1963, LIK, 030.II.A.108.46.
[85] *Ibid.*
[86] Report on a meeting on 20 February 1963, LIK, 030.II.A.108.46.
[87] *Ibid.*
[88] Allen M. Katz, letter to Dale L. Harvey, 25 February 1963 ("Change in concrete materials"), LIK, 030.II.A.107.36.
[89] Dale L. Harvey, letter to Jonas E. Salk, 26 February 1963, LIK, 030.II.A.107.36.
[90] Allen M. Katz, letter to Dale L. Harvey, 25 February

Riverside Cement Company, makes it possible in mid-March, when the concrete has set and strength testing can be conducted, to decide which solution to choose for the final work. The total cost of the two solutions is estimated at 45,000 dollars for the Riverside Special Blend Cement and 76,000 dollars for the concrete with added pigments.

Certain negative views are expressed regarding the use of added colored pigments. Harvey fears the possibility of accumulation of pigment during the pouring phase, with consequences for the overall tone: "If the mix using a pigmented integral color is used, I believe it will leave quite a pronounced color shade line at each cold joint. This is caused mainly by a different rate of dehydration, and is typical of any concrete pour but becomes exaggerated in darker colors."[89]

The question of the strength of a concrete chosen above all for its chromatic characteristics continues to be an open one. The tests performed on the concrete of the Riverside Cement Company show that the compound supplied by the company, namely the Riverside Special Blend Cement, according to the documents, does not reach a strength of 5000 lb/sq.in, just as was the case with the Pozzolanic Cement. But given the fact that the East Wall, as the calculations show, has to stand up to 3750 lb/sq.in, Katz proposes proceeding with the pouring, in any case, given the urgent need to begin the operations, and postponing the development of a stronger concrete, with the Riverside Cement Company, to a subsequent phase.[90]

To achieve the color of the concrete desired by Salk, there is also another possible solution, indicated by Harvey, besides that of the colored pigments and that of the Riverside Special Blend Cement: it would be possible to use an ordinary concrete and "then treat the surface, and brush coat with one of the several long lasting cement paints [...]." "This painted surface will last from five to ten years," the hypothesis specifies.[91] There are no traces in the documents of trials conducted to test this solution, which would have contradicted all of Kahn's efforts to make a concrete not requiring intervention after stripping of the formwork.

The trials confirm that pozzolan is more economical than the artificial additives offered by the industry to color cement, and it also remains the additive with qualities that have been proven due to its use over the centuries.

The pozzolan added to the compound of the concrete definitively chosen for the Salk Institute is supplied by the Airox Company, based in Los Angeles, and is quarried at Santa Maria in California. That type of pozzolan mainly contains silica, with variable percentages of aluminium, magnesium, iron and sodium.[92] The advertising of the Airox Company reports that the product mixed in concrete "increases ultimate strengths, limits leaching, decreases permeability, improves workability, minimizes segregation, and lowers heat of hydration."[93]

For the concrete of the Salk Institute, the pozzolan is mixed with Riverside Cement. For the aggregate, those typical of southern California are selected, supplied by Crestlite Products, based in San Clemente. The sand is

1963 ("Color of concrete sample walls"), LIK, 030.II.A.107.36.

[91] Dale L. Harvey, letter to Jonas E. Salk, 26 February 1963, LIK, 030.II.A.107.36.

[92] *Cf.* John N. Faick, *Geology and technology of some natural pozzolans in north central California*, in, "Society of Economic Geologists," vol. LVIII, 1 August 1963, No. 5, pp. 702-719.

[93] See Airox Company, *Airox pozzolan for quality concrete*, Airox Company, 1959.

supplied by Caudell and Johnson of San Diego.[94] Pozzolith 3LAA is also added to the compound. When the mix design of the Salk Institute with the pozzolan has been confirmed as the new recipe for a concrete with a special color, Marshall D. Meyers, a member of Kahn's staff, sends other builders the formula used at La Jolla:

"The Pozzolan was interground with the Riverside gray cement. The mix was an 8.9 sack mix / cu. yd. and the Pozzolan was 12% by weight. The aggregate was 3/4" rock and sand, both the rock and the sand were the same material. They were obtained from the Mission Valley area [...]. 50% of the sand needed was obtained at the site and the remaining 50% was ground from the rock. This aggregate was described as being very dirty (mud) and needed multiple washings before it could be used."[95]

Thanks to the pozzolan, Kahn's concrete emerges from the formwork with smooth surfaces, at times so pale as to create an ideal backdrop for the reflection of the colors of the California sky. In no other construction by Kahn does the concrete have the pale glow emitted by the walls of the Salk Institute. "The color of the concrete is a kind of almost silium color. It is a color of rose," Kahn would explain.[96] In 1973 Kahn again spoke of the color of the concrete of the Salk Institute, which just after construction had appeared to have a rose shade: "Now, after several years, the concrete appears of a lavender tone."[97]

The East Wall of the Mechanical Building: fossils of panels and reliefs

After excavation and the creation of the foundation plinths for the pillars in the two Laboratories, the Salk Institute worksite continues with the creation of the foundations, first of the East Wall and then of the West Wall of the Mechanical Building, both destined to remain underground, and then proceeds with the samples requested by Kahn prior to the construction of the Retaining Walls and Service Towers, with their walls that will be left visible. Halfway through January 1963 only the foundation of the East Wall has been completed.[98] On 12 February all of its reinforcements are in place, while those of the West Wall are still those that emerge from the curb of the foundation.[99]

During a worksite visit at the start of February, Kahn discusses the construction of the patterns to use for the East Wall (probably he has not yet done the drawing of the trial patterns of the West Wall). Only later does Langford also provide Katz with the drawings of "several different panel layouts" to make along the West Wall, to test all the effects of different combinations of

panels and reliefs.[100] While the formwork for the East Wall trials are being assembled by the Fuller Company, Harvey rejects Kahn's new request, on 18 February, to also continue with the tests along the West Wall:

"I have instructed Mr. Katz not to build this second wall as a form mock-up for three reasons. First, the 170 feet of different concrete designs we are now building certainly should be sufficient for you to make up your mind without seeing the same relief patterns in different arrangements. Secondly, on this new design there is a great waste of plywood due to the cutting necessary to produce the patterns. This waste plus the additional cost of detail forms over regular plywood forms does not seem to be justifiable since, you as an architect, should be able to determine from the east wall everything you need to know about the relief pattern effects. Thirdly, now that we have the release on approximately 5000 cubic yards of concrete, we are going to put the job on an accelerated schedule, and cannot stand for any unnecessary delays."[101]

The panels of the formwork of the East Wall are assembled in one week, from about 20 to 27 February,[102] while the pouring of the concrete is slated for 27 February, so the results can be viewed by Kahn on 2 March (but the visit is postponed because the pouring is still in progress on 1 March). Some of the holes left by the spreaders are sealed with lead disks.

The construction of the trials to be left in place in the East Wall of the Mechanical Building generates what will be called the "fossils" visible in the underground spaces of the Salk Institute, as Meyers would later write:

"When Kahn was working out the details of the formwork for the concrete walls at the Salk Institute Laboratories, he experimented with several ideas by having these trials cast in the basement walls in order to review them. In this way, construction could proceed and, most importantly, these attempts would remain in the building as the history of its making. They would be like 'fossils' he would say. And not being in the most public of spaces, they could be discovered by anyone who took the time to seek them out. So the history of the making of the building, the struggle that I first heard him talk about years before at Yale, is there in that building. At Salk this struggle is petrified in the cast concrete much like the continuum of those charcoal marks he left on the tracing paper."[103]

Right after the stripping of the East Wall, Kahn visits La Jolla to observe the results of the patterns and profiles. The photographs also show Salk in front of the monumental test. The pencil drawings made on the wall

45-46. Salk Institute for Biological Studies, La Jolla, California, 1959-65.

45. "Mechanical Wing Looking South," 12 February 1963 (LIK, 030.IV.A.540.8.38).

46. "Mechanical Wing Looking Southwest," 1 March 1963 (LIK, 030.IV.A.540.8.35).

94 Smith-Emery Company, report, 21 May 1964, LIK, 030.II.A.37.6.

95 Marshall D. Meyers, letter to Thad E. Harden, Jr., 21 August 1969, LIK, 030.II.A.37.6.

96 Louis I. Kahn, *Address*, 5 April 1966, now in, Alessandra Latour, ed., *Louis I. Kahn: writings, lectures, interviews*, New York, Rizzoli, 1991, (pp. 208-220), p. 215.

97 Louis I. Kahn, letter to Jury of Fellows of The American Institute of Architects, 1 November 1973, published in, August E. Komendant, *18 years with architect Louis I.Kahn*, Englewood, Aloray, 1975, (pp. 1-3), p. 2.

98 See worksite photograph, 14 January 1963 (August E. Komendant Archive, AAUP, Philadelphia, 029.III.28.7).

99 See worksite photograph, 12 February 1963, LIK, 030.IV.A.540.1.

100 Dale L. Harvey, letter to Louis I. Kahn, 18 February 1963, LIK, 030.II.A.107.36.

101 *Ibid.*

102 Report on a meeting held on 20 February 1963, LIK, 030.II.A.108.46.

103 Marshall D. Meyers, *Interview with Alessandra Latour, [...], March 19, 1983*, in, Latour, *Louis I. Kahn, l'uomo, il maestro* cit., (pp. 71-88), p. 81.

47-48. Salk Institute for Biological Studies, La Jolla, California, 1959-65. Photographs of the Mechanical Room east wall with Jonas Edward Salk (Salk Institute Archive, La Jolla).

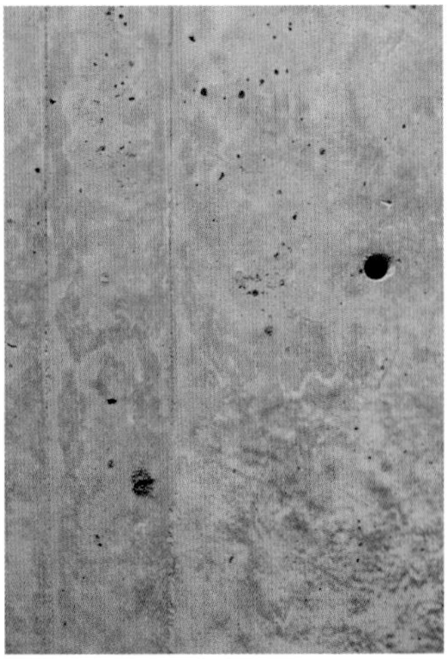
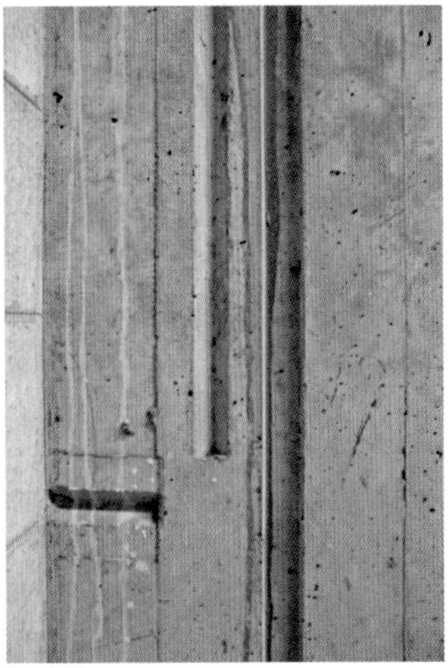
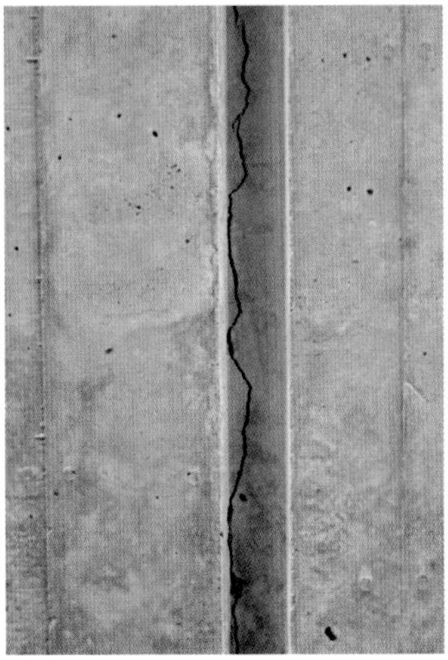

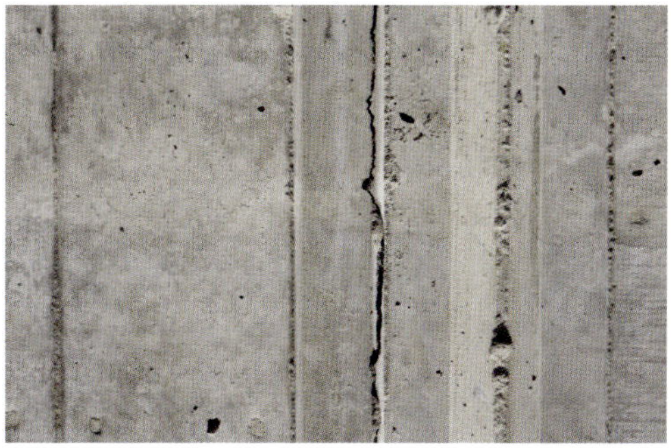

49-60. Salk Institute for Biological Studies, La Jolla, California, 1959-65.

49-58. Details of the "fossils" of the panels and reliefs in the Mechanical Room east wall (photographs by the author, 2010).

59-60. Sketches made on the Mechanical Room east wall (photographs by the author, 2010).

61-64. Salk Institute for Biological Studies, La Jolla, California, 1959-65.

61-62. Sketches from Kahn notebook (LIK, Kahn notebook).

63. "Retaining Wall at the Service Tower," LA-347, March 1963 (LIK, 030.I.C.540.31).

64. Patio perspective with Retaining Wall and Service Towers (LIK, 030.I.C.540.018a).

document the first discussions on the effects of the design of the joints and possible modifications. Some sketches address the way to protect the edges of the plywood panels with L-shaped pieces, probably in metal, and the variation for a strip with a pointed V-shaped hollow, more slender and with a narrower base than the various pointed V elements cut during the trials.

Some of the most telling drawings from the visit are contained in the three-page sequence of Kahn's notebook, opened by a view of the coastal landscape. The play in the rocks of cavities in shade or protruding parts in light reflects what Kahn is trying to achieve with the joints of the formwork panels. Kahn is interested in refining the design of the profile to imprint on the concrete, both in relief and in hollows. In the sketches, he re-examines the system to align the strip against the two panels and the profile to sculpt in the same strip to obtain the imprint of a molding on the concrete wall. The profiles of the molding that seem to interest him, after having seen the trials on the worksite, are those with a V and diamond shape[104].

The study of the variants of the profiles of the trials of the East Wall induces Kahn to reconsider the pattern of the imprints for the walls of the Studies, now to be decorated, as shown in his notebook sketches, with protruding moldings with a rectangular section, forming a cross and compartments. There is, instead, no sign of awareness that the definitive drawings prepared for the East Wall did not cover all the imprints that are now revealed by the worksite test, where the joints between the strips and panels have produced visible signs on the concrete.

The V joint and the bleed-out

The many sketches and the sets of technical drawings following the East Wall trials document the desire of Kahn and his staff to try out every possible variation of the plywood panel formwork system to obtain a new expression of mural composition.

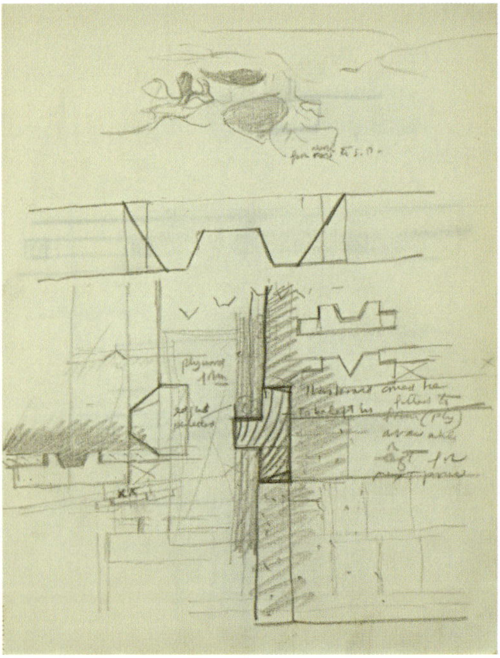

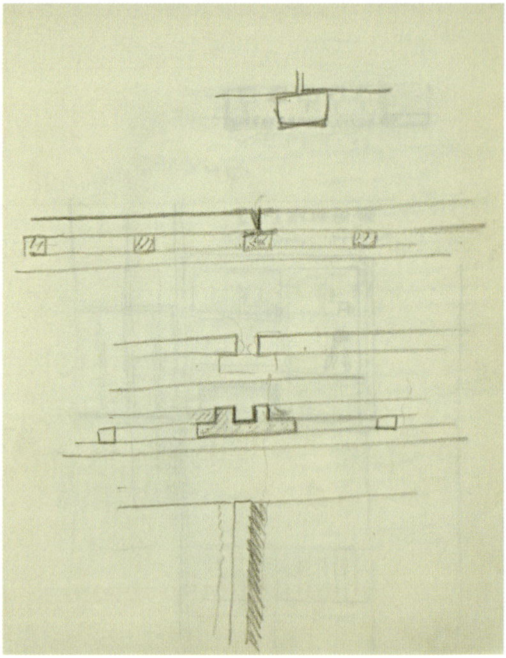

The drawings of March 1963 document the new patterns for the Retaining Wall, the Service Tower, the Laboratories and the Studies. While the first two continue to have patterns of panels framed by moldings, as in the tests on the East Wall, the Laboratories and Studies use a different system, confirming that Kahn continues to pursue the goal of a diversified character for the various parts of the Salk Institute.

After the East Wall trials, Kahn develops a variation for the Retaining Wall, with a pattern of horizontal panels that combines the vertical positioning of "shrinkage joints" (abbreviated with an "s") and "pour joints" (abbreviated with a "p") – the former are larger than the latter. Thus the formwork system is a set of grooves, reliefs and imprints that take the assembly of the panels, the pour phases and the dilations of the structure into account. In the design for the Retaining Wall, the pattern of the spreaders creates a quadrangular quilting of 24 × 26 inches, which nevertheless is not matched between the wall and the parapet, due to the staggering of the panels of the formwork.[105] It is worth noting that in the drawing, for the first time in the series of those for the formwork of the Salk Institute, there are indications of the joints of the vertical strips, to obtain at staggered levels in the various wall sectors, to prevent their imprints from creating a regular composition. Instead, there are still no signs of the joints between the strips in the corners of the framing bands.

Drawings like those for the facades of the Service Tower, also dating back to March,[106] document the application of reliefs and grooves as done

[104] Louis I. Kahn, notebook 1901-1974, LIK, 030.VIII.7. ("This board could be fitted to form (ply) available or left for next pour").

[105] See the drawing "Retaining Wall at Service Tower," LA-347, March 1963, LIK, 030.I.C.540.031. Drawing LA-348, with the plans of positioning of the formwork for the "Retaining Wall at Service Tower," is dated 10 March 1963 (LIK, 030.I.C.540.031). A perspective view documents the study of the pattern of the Retaining Wall as compared to that of the Service Towers (LIK, 030.I.C.540.018).

[106] See the drawing "Retaining Wall at Service Tower," LA-321, 10 March 1963, LIK, 030.I.C.540.031.

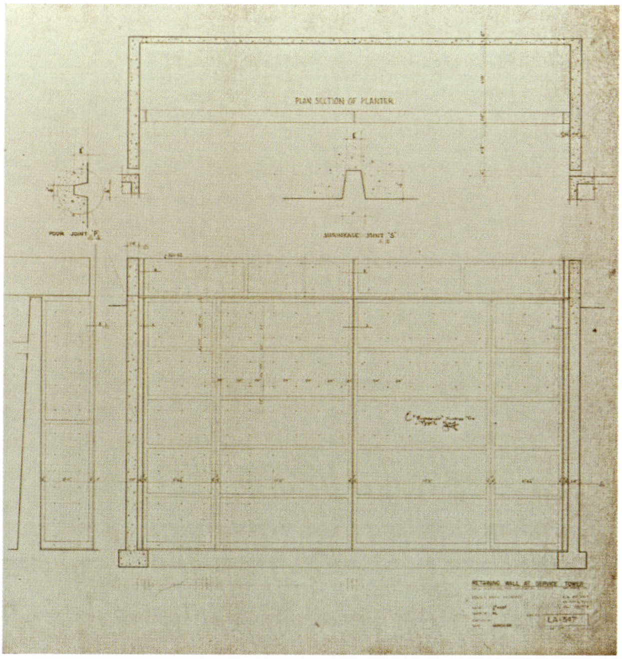

65-66. Salk Institute for Biological Studies, La Jolla, California, 1959-65.

65. Section and elevation of patio with Service Tower and Retaining Wall, LA-321, March 1963 (LIK, 030.I.C.540.31).

66. Service Tower formwork study (LIK, 030.I.C.540.22).

on the East Wall to obtain framed compartments (the selected motif belongs to the solution B, that of the V-shaped relief). The accentuated reliefs of the frames and grooves break up the volume of the towers into a sequence of large compartments corresponding to the levels, in turn constituted by series of framed panels.

The vertical or horizontal positioning of the rectangular plywood panels continues to be examined, as demonstrated by the sketch for the composition of the formwork for the inclined wall sector of the Studies – the panels are placed horizontally. Kahn opts for the T-shaped wooden strip, but instead of sculpting the head, he goes back to reconsider the technique of beveling of the edges. In several documents, the T strip is called a "Coupler Block," indicating its role as a joint between two panels. In a sketch, Kahn envisions beveling the two edges of the post of the T and those of the contiguous plywood panels, so that the combination of the three elements will produce two V joints. That fact that the T strip is inserted between the panels produces two V-shaped imprints separated by about 2 inches.[107] It is precisely this V imprint, though further modified in the construction system and therefore in its measurements, that becomes one of the particular characteristics of the design of the concrete of the Salk Institute.[108]

One of the first important studies for the new design of the imprints of the formwork on the facade of the Laboratories also dates back to March 1963.[109] The band formed by the wall enclosing the spaces for physical plant systems with the Vierendeel trusses – the wall is perforated by vertical loopholes – and the parapet of the gallery at the upper level, the crowning, is subdivided into compartments by means of the imprints of the formwork panels and grooves, while the holes of the spreaders create the usual graphic quilting. The imprints are designed with a single line, with the exception of the double line under the dripstone. At this date, in the documents available, the solution of the plywood panel with beveled edges to obtain a slim V like that of the Carpenter Center has never been used – solution G for the East Wall test had been developed to produce a marked relief. Therefore it is not possible to establish if the pattern of lines documented in the drawing in March still refers to the imprints to be obtained through the simple adjoining placement of the panels, or if it documents an initial application of the solution that is destined to become definitive. The fact that the double line traced below the dripstone depicts a band protruding from the flat face, whose technical making will be described in drawings on 15 April which also show the definitive solution of the panels with slightly shaped edges, would seem to favor the hypothesis that the drawing documents the final version of the V joints. In the set of drawings we also see drawn the projection of the imprints of the rectangular panels onto the ceiling of the galleries, proving that all the surfaces are subject to the same desire to control the pattern. All the lines generated by the formwork, beyond the configuration of their section, correspond to the ones that are actually built.

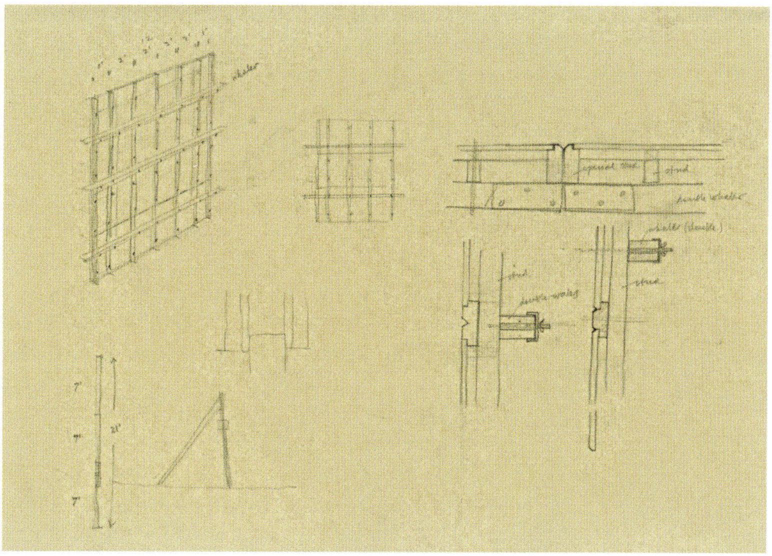

[107] See the study sketches, n.d., LIK, 030.I.C.540.022.
[108] A telling factor regarding the logic followed by Kahn and his team is that in the study for a pattern of panels and strips, two solutions are included, correlated in terms of position and molding of the strip: the strip recessed with respect to the panels, leaving an imprint on the wall of a band in relief, corresponds to a V-shaped groove, while the strip flush with the panels corresponds to a V-shaped relief. See the study sketches, n.d., LIK, 030.I.C.540.022.
[109] The drawing is dated March 1963, but it is revised on 23 January, 17 February and 9 September 1964, (drawing, "Balcony Elevation-Service Tower Side," LA-377a, LIK, 030.I.C.540.031).

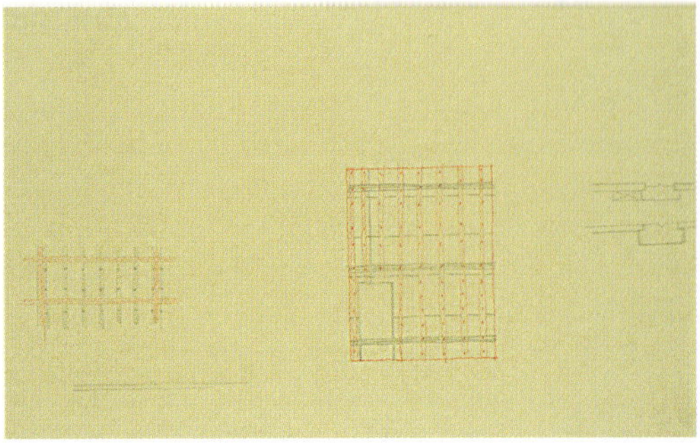

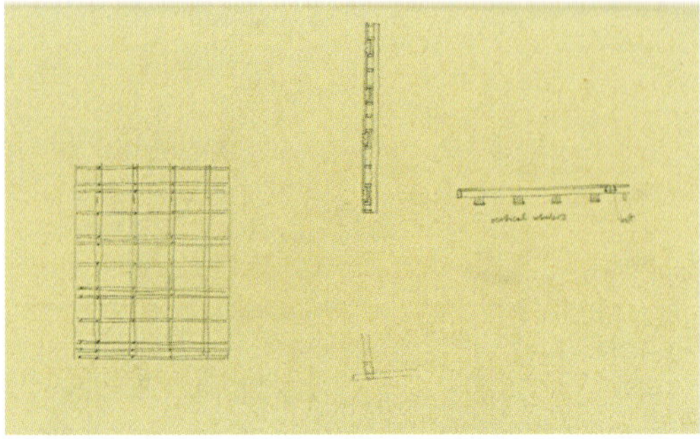

67-70. Salk Institute for Biological Studies, La Jolla, California, 1959-65.

67-68. Service Tower formwork studies (LIK, 030.I.C.540.22).

69. "Balcony Elevation — Service Tower Side," LA-377a, March 1963 (LIK, 030.I.C.540.31).

70. "Studies — Elevation Section," LA-366, April 1963 (LIK, 030.I.C.540.31).

The set of drawings "Studies. Elevation/Section," prepared in April for the staircases between the Laboratory and the Studies, confirms the architectural concrete of the previous drawings.[110]

Kahn is guided towards the choice of the V joint by reflection on the behavior of concrete during pouring, in its phase as "liquid stone," against the joint between the panels. The joint ceases to be a sort of molding that comments on the assembly between panels, and takes on, beyond the form itself, a different meaning with respect to the V profile used in the Carpenter Center. Now, for Kahn, the use of the V joint means intervention on a critical technical detail in the construction of formwork, where the pour may bleed through the joint, creating flaws on the concrete surface. Kahn's choice is to take advantage of that seepage, giving it a form. In this technical sense, the V joint of the Salk Institute has its forebears in the imperceptible imprints the poured concrete left in the cylindrical wall of the Yale Art Gallery, oozing into the joints of the boards of the formwork.

"In pouring concrete I use a wood form, which actually becomes better the more you use it," Kahn explains. "Where the one form meets the other, I have allowed the concrete to bleed out by making a cut in the plywood which doesn't quite penetrate the first layer of plywood. When the concrete pours, there isn't a resistance to the flow outward at the joint. Trying to make it stronger and stronger by keeping the wood together makes insufficient accuracy in the alignment, and I came to the realization that if you could allow it to bleed out some, you would then have no trauma at the joint, no honeycombing. And this proved to be right."[111]

It is worth re-reading the recollections of Meyers on this detail that determines the character of the concrete of the Salk Institute, because they help us grasp the particular technical importance of the V joint of Kahn:

"At Salk he first developed this language of formwork marks that represented events in the pour and the joining of plywood contact surfaces or the meeting of the larger, reusable form panels. These details were not merely subjective patterns he desired, but grew out of the fundamental and practical problems of the craft. The 'vee' joint which he devised at the butting of two sheets of plywood came from an awareness that water tended to leak from these joints carrying sand and cement with it, leaving the concrete disfigured with the scars called 'honeycombing.' Realizing that the pressure of the water would always seek these 'cracks' in the formwork, he devised a way to control the mark made by this leak rather than try to erase it. He accepted nature's way by chamfering each of the meeting edges of the two plywood sheets with a 45° angle, 1/4" long so that they formed a 'vee' when they butted. The resulting convex vee-joints tend to panelize and control the random, uncontrollable discolorations in the concrete surface, especially when these linear protrusions are struck by rays of the sun."[112]

Pattern studies for the formwork of the Service Towers

The technical development of the formwork system of the Salk Institute is only the starting point of another type of research on formwork, which now involves the formal quality of what is shaped on the concrete. The passage from boards to plywood leads Kahn to an accurate design of the whole formwork system, so that the imprints of the joints can be tuned to the various parts of the building, in a coordinated plot of essential lines. The imprint of the formwork, once boards are no longer used, stops having the purpose of distracting the gaze thanks to its dense pattern, but becomes a fundamental line for composition, to coordinate with the lines of the openings, the corners of the walls, the parapets, the floor slabs, the landings of the staircases, the electrical sockets, the drips. The entire process of definition of the formwork is rendered complex precisely by the need to foresee, in its design, the installation of physical plant systems, including sockets and switches.[113] The temporal sequence with which Kahn reaches the definition of his own formwork system, when construction is already under way and the design has already been set, is not without consequences on the

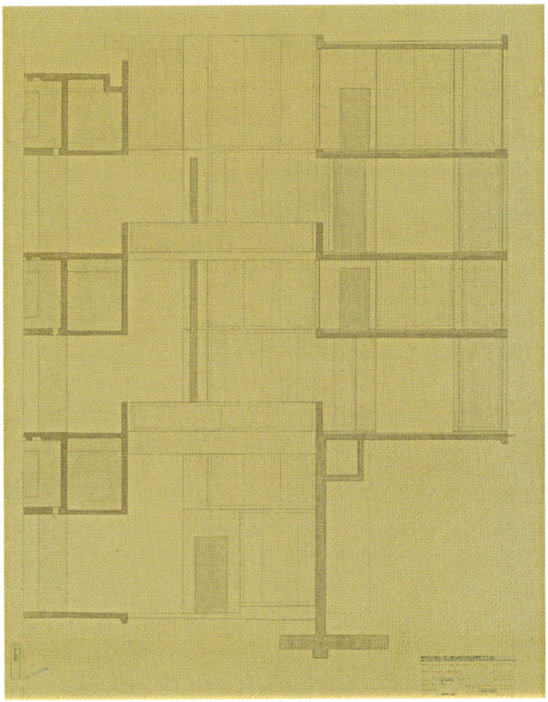

[110] See the drawing "Studies Elevation/Section," LA-366, April 1963, LIK, 030.I.C.540.031.
[111] Louis I. Kahn, conversation with William H. Jordy, in, Richard S. Wurman, *What will be has always been, The words of Louis I. Kahn*, New York, Rizzoli, 1986, p. 241.
[112] Marshall D. Meyers, *Interview with Alessandra Latour, [...], March 19, 1983*, in, Latour, *Louis I. Kahn, l'uomo, il maestro* cit., (pp. 71-88), p. 81.
[113] "Architect's formwork drawings should be reviewed by Electrical and Mechanical sub-contractors so that sleeves, blockouts, boxes, inserts, etc., be indicated on drawings before submission to Builder," is stated by Kahn's staff (Frederick Langford, "Report on concrete work," 23 March 1964, LIK, 030.II.A.107.7).

pursuit of agreement between the assembly of the panels and the fundamental lines of the building.

The first significant studies to tune the formwork system to the various parts of the Salk Institute begin in April 1963 and have to do with the Service Towers. Kahn proposes modification of the formwork system of the Service Towers, sacrificing solution B. At first he applies the solution D used in the East Wall, that of the flat strips, leaning towards a design of the concrete wall based on the imprints of joints, but without the horizontal protrusions and grooves that form a shadow corresponding to the floors.[114] In this solution the panel, though it may seem like the protagonist of the pattern in the drawing, becomes a subordinate component with respect to the concrete surface. Over time, weathering would further diminish the visibility of the framing imprints. To make the imprints of the flat strip coincide with the opening for the door, Kahn alters the width of the plywood panels, creating off-axis effects with the vertical imprints in the levels above. The imprints that form the frame of the door are thicker than those between the panels – they have the width of the vertical imprints along the edges of the volume of the towers. It is worth noting that in the drawings of the concrete facades of the Service Towers there are indications, with dotted lines, of the thicknesses of walls and floor slabs, as if to try to keep under control the correspondence between the compartments and the patterns defined by the framed panels, and the load-bearing structures. The drawing bears witness to the difficulties of the formwork system of Kahn, which are not limited to technical questions but also extend to control of the composition of the panels, their arrangement and sizing, in relation to the structures and the basic lines of the Salk Institute.

One variant of the Service Towers, also dating back to April, documents an evolution of the pattern and design of the formwork imprints. The size of the door opening determines a new pacing of the panels, to align all the vertical frames of the panels on the various levels. It is clear that Kahn wants to find a weave of lines to imprint on the concrete that will be coordinated with the openings and the walls, and therefore a formwork system as suitable as possible for the measurements of the designed building. The fundamental change in the new version has to do with the design of the strips

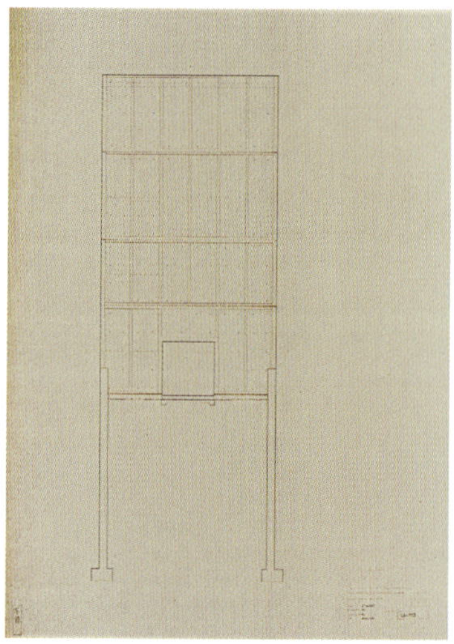

71-73. Salk Institute for Biological Studies, La Jolla, California, 1959-65.

71. "Service Tower Elevation," LA-326, April 1963 (LIK, 030.I.C.540.31).

72. "Service Tower Elevation," LA-327, April 1963 (LIK, 030.I.C.540.31).

73. "Service Tower El.[evation] North," LA-325, April 1963 (LIK, 030.I.C.540.31).

that frame the panels.[115] Also after the inspection of the East Wall trials, which demonstrated the fundamental role of all the imprints, now all the lines that go into producing those imprints are drawn, including those of the technical joint between the strips. In the corners of the volume a vertical strip is depicted, between two horizontal grooves that indicate the presence of the floor slabs; the two horizontal strips that frame a large compartment (aligned against the corner strips) become recognizable; the frame of the door itself is composed of a horizontal strip resting, at its ends, on two vertical strips.

Having designed all the imprints of the parts of the formwork in contact with the pour is the first decisive step towards the definitive development of the system. As in the case of the set of drawings of the Laboratories in March, again in this drawing of the Service Towers it is not possible to establish whether the imprints are simply thin lines resulting from the coplanar assembly of strips and panels, or if they have a more accentuated relief generated by the beveling of the parts – the V joint.

The two versions for the Service Towers demonstrate that Kahn intends to match the openings with the design of the formwork, in the case of the strips that produce thick framing bands. They also indicate Kahn's intention to create, on all the concrete of the Salk Institute, a weave of imprints no longer having the marked reliefs tested in the East Wall. We might even hypothesize that the solution of the drawing of March 1963 for the Laboratories influences the variant for the Service Tower with all the imprints of the formwork elements.

A series of drawings demonstrates that the positioning of the formwork for all the walls is studied in plan and elevation, in order to proportion the rhythm of the spreaders and grooves based on the length of the walls and the size of the openings. Kahn follows a modular logic in the layout of the parts, in order to have the most symmetrical, regular sequence possible in every place in the complex.

The philosophy of the panel: Langford's letter

During March 1963 the Salk Institute worksite progresses. Evidence from Komendant points to some misunderstandings that arise between the Fuller Company and Kahn's staff regarding the making and deployment of the formwork, in these early phases of the work – "the Fuller superintendent did not like the form design, considered it too expensive and impractical," and rejects any "help" and any "interference from Kahn's local office."[116] Kahn himself does not make linear worksite progress any easier, with his many variations as the work proceeds regarding the formwork and the quality of the concrete. The first concrete poured with the new system of imprints does not meet with the approval of Kahn and his staff, who cannot oversee the assembly of the formwork, due to the veto of the Fuller Company.[117]

One crucial document on the construction of the Salk Institute, and to understand Kahn's idea of concrete and the meaning he is seeking in the imprints of all the traces of the formwork and the pour, is the letter entitled "Design of formwork for concrete," written by Langford on 5 April 1963, precisely in the wake of the first difficulties that arise on the worksite.[118]

[114] See the drawing "Service Tower Elevation," LA-323, April 1963, LIK, 030.I.C.540.031, and the drawing, "Service Tower Elevation," LA-326, April 1963, LIK, 030.I.C.540.031.

[115] See the drawing "Service Tower Elevation," LA-325, April 1963, LIK, 030.I.C.540.031, and the drawing "Service Tower Elevation," LA-327, April 1963, LIK, 030.I.C.540.031.

[116] Komendant, *18 years with architect Louis I.Kahn* cit., p. 58.

[117] *Ibid.*, p. 60.

[118] Frederick Langford, letter to Greer Ferver, 5 April 1963, LIK, 030.II.A.89.3.

The letter is not a mere document of worksite chronicles; it contains a true apologia of the process of construction in exposed concrete. In some passages, the concrete itself becomes the narrator of the process of formation of the traces of construction on its own skin.

The letter is written in a particular moment of the construction, when work is starting on the parts in exposed concrete, and based on indications from Salk himself that Langford should clearly explain Kahn's thinking on the meaning of the material to the engineers and builders, so that all the persons involved in the construction, including carpenters and masons, can base their work on the same philosophy. To explain "what Mr. Kahn wants to express in the concrete work on this building" is the stated objective of Langford's letter.

"Design of formwork for concrete" serves to indicate, with force and authority, the expertise to which Kahn lays claim, and which until now has usually, on worksites, in the hands of the contractor: that of the design of the formwork. Together with the great quantity of drawings of imprint patterns on exposed concrete walls of the Salk Institute produced by Kahn's team, the letter represents the profession of Kahn's belief in formwork as the ultimate tool in the hands of the architect for the control of exposed concrete and of the construction reduced to pure form. All the formwork materials, from the plywood panels to the plastic sheets applied to the surface in contact with the pour, or the spreaders, are explained in detail in terms of their impact on the final result, at the moment of stripping, in the design of the concrete surfaces. It is clear that the letter sets out to instruct the builders on a different use of those materials usually deployed in an ordinary way, in order to avoid further technical misunderstandings on the Salk Institute worksite that might threaten the results Kahn seeks, because the concrete must truly be untouchable after stripping.

Langford's document explains how the imperfect joint line between plywood panels, or the "bleed line," can become, once corrected, a "hydrostatic relief," namely the V joint; how the relief lines permit reduction of the perceptibility of possible pour defects; how the polyurethane film to place on the plywood is fundamental for the goal of expression of the true nature of concrete; and how the entire conception of the formwork system represents a veritable "philosophy," in which every line has to rigorously correspond to the construction process, becoming its visible "recording" of the phases and the technical characteristics, in stated pursuit of a "construction process" that must be "expressed honestly." "[...] no false pour lines" must exist on the concrete surfaces of Kahn.

"The purpose of this report: 1. Acknowledge that contention exists between Builder and Architect in the area of formwork design. 2. Attempt to explain Mr. Kahn's reasons for the formwork design as it now exists. 3. Attempt to explain the construction process as an architectural expression. 4. List of factors involved in the design of formwork and suggestion of a new division of responsibility between the Architect and Builder. 5. What we hope to achieve from a better collaboration between Architect and Builder.

1. There have been many arguments between the Fuller Company staff and myself concerning the design of formwork for concrete since the beginning of this job. The results of these arguments have in no way helped the progress of the job or led to any better understanding between the Builder and the Architect with regard to this problem. I can well understand the Builder's attitude 'Show us where you want the lines and we will give them

to you.' This is the conventional approach to form design and any change in this apparently presents a jurisdictional problem. Conventional attitudes will produce a conventional building. However, since we still believe that the Fuller Company grew into the great corporation it is today by using imagination along with convention, and pride in excellent workmanship rather than exaggerated insistence in procedure and protocol, we still hope that this problem can be resolved.

2. In the evolution of this building design, Mr. Kahn has made and still must make thousands of decisions. It is impossible to state all the reasons that go into making these decisions, and yet, somehow, these all relate. A decision on any part of this building must always recognize the basic promises that are a part of the total and as long as Art is part of Architecture, some decisions must be based on pure intuition. However, a few of the reasons for design can be included in the realm of understandable logic, and these I will try and put forth in part as having an effect on the development of the formwork design. *Plywood* Plywood was selected as a suitable form material because of its excellent strength characteristics, its ease of workability, its durability for re-use and the fact that is available in large sheet sizes. *Plastic coating* Polyurethane plastic was chosen as a form coating because it produced a surface color that was the same as the color of the concrete within the wall. Plastic also produced a smooth surface that did not leave dust for a prolonged period, as many form oils do. *Floating Increment* Because this building is not a warehouse or a factory that has many repetitive parts, there are relatively few wall panels that can be used more than three or four times. To take a wall panel apart and re-assemble the plywood sheets into another wall panel is very difficult, due to the variation in wall heights and wall tie spacings. It was decided to consider each plywood sheet with its wall tie holes as a floating increment separated and framed by plywood strips that would facilitate the remaking of panels of different heights and widths. *Hydrostatic Relief Valve* The normal procedure of butting two large pieces of plywood together allows for a bleed line, but we were not satisfied with the variation in the concrete surface produced by this conventional system. We recognized the need for hydrostatic relief, but felt we could control it more easily if we used a 4" interlocking strip that would allow two lines of relief instead of one. *Expendable Parting Strip* The expendable part of the form becomes the small strip instead of the large sheet. The introduction of a new parting or juncture strip next to and surrounding large sheets that have become slightly battered marbles the outside edge area of the panel to become new again. The outside edge area, which makes corners, juncture zones, and pour terminations is much more critical than the center area of the panels. *Wall Ties* Because the wall ties play such an important part in the philosophy of the common denominator panel, which in turn plays a part in establishing an overall order in the total building, the wall patterns that are caused by the wall tie spacing must be controlled by the Architect. As expendability and number of re-uses, I may not argue in dollars and cents, but I can argue that the idea of creating a common denominator of form increment shows more reverence for the material and the hands of the carpenters that did the work. These are some of the intangible things that promote better spirit and pride in craftsmanship that will in turn produce a finer building.

3. The lines in the concrete that are left by the form are a record of the construction process. These lines vary in character because the events that caused them varied. If the Architect chooses to make these events more of

a statement or less of a statement, this is his prerogative. 'Show us where you want the lines and we will give them to you.' *Bleed Lines* These are the lines which are caused by the limitation of the plywood sheets. These same lines define the parting strips which also represent the hydrostatic relief during the pouring process. We might call them bleed lines. *Panel Junctures* The plywood is assembled into large panels that are limited in size by crane handling during erection and stripping. After the large panels are lifted into place, they must be joined together in such a way that the juncture element can be removed after pouring, to facilitate stripping. Once again, a record of the construction process. *Pour Joint* The last lift of each pour should always terminate at a chamfer strip which leaves an indenture. This means the wall continues upward during the next pour from a definite line. A color or texture change will invariably take place between pours, and the pour joint accommodates this change positively. *Shrinkage Joint* The vertical shrinkage joint is created by placing a deeper chamfer into the concrete to give location and direction to a possible crack caused by normal shrinkage. Vertical cold joints should be treated the same way. All vertical joints of this nature have been governing factors in determining the width of panels, and should always occur at the edge of a panel. This means that panels are joined in only two ways. Either the flush surfaced juncture, or the surface breaking (shrinkage chamfer) juncture. *No False Pour Lines* Indented lines that state the beginning or ending of a pour, should not be just inserted in the form and allow the concrete to continue past them. Pour line chamfers must be indicated by the Architect, and every effort made by the Builder to work to them. We want to produce an architecture where the construction process is expressed honestly. To extend a panel past a pre-ordained juncture and then cause a fake juncture in the face of the plywood to be printed in the concrete so as to appear as a panel juncture, would not be in the religion of thoughtful architecture, nor in the spirit of proud artisans of any craft."

Langford's letter also contains a list of all the elements of the formwork system, or "factors involved in form design," and the respective responsibilities of the builder (Fuller Company) and the architect. "I have listed," Langford writes, "the factors that are especially critical to the Architect in producing a fine building. The Builder should be able to detail the forms around the Architect's requirements by varying the factors that are his (the Builder's) responsibility."

The formwork system itself is subdivided into load-bearing structure and parts in contact with the pour, each assigned as the responsibility of different figures on the worksite. While the builder chooses the components of the load-bearing structure of the formwork system ("studs" and "walers") and the degree of strength of the "wall ties," the architect selects all the components that leave imprints on the concrete (the pattern of the panels, the type and arrangement of the "wall ties" and "chamfer strips"). The panel itself is subjected to a splitting of responsibilities, always in terms of structure and imprint. For the plywood, the architect establishes the size of the surface and its coating, while the builder sets the thickness and the type of wood. The architect determines the levels for the "pour joints" (while the "rate of pour" is set by the contractor).

The explanation of the formwork system cannot help but contain a persuasive rehabilitation of the role of the craftsman, as in medieval construction, going so far as to harshly attack architects who pay no attention to materials and the men who craft them.

"In the history of architecture," Langford writes, "many fine structures have developed without the influence of Architects. Men have used their ingenuity, imagination, the knowledge of their environment, limited tools, and limited materials to produce thoughtful architecture. Architects have never produced fine buildings unless they showed a respect for the materials of building and a respect for the hands that worked these materials. We sincerely want all suggestions and criticisms the Builder and his craftsmen have to offer. We must make the concrete in this building say 'I am expressive of the hands and forms that held me in place until I could grasp my inner steel and gain the strength and power that I now possess; that I must possess to hold back the tremendous earth pressures with my Retaining Walls, to span the laboratories in a single leap, to lift the studies to the grand view of the sea, and still be friendly enough to touch with human hands.' All this should contribute to the total architecture that we believe will encourage and provide for productive research and discovery at the Salk Institute."

Juncture strip, pour joint, cold joint and shrinkage joint

Langford's letter conveys the value, or the philosophy, of certain technical details of the formwork system developed by Kahn for the Salk Institute. By now that system has its own syntax, which has no forerunners in the manuals, based on the ways in which Kahn invents and combines the details of assembly of the panels. It is no coincidence, if we think about the theoretical formulations put forth by Kahn starting with *Monumentality* all the way to the texts and lectures of the early 1950s, that all the characteristics of his formwork system orbit around the design and meaning of the joint. "Juncture strip" (later also called "coupler block"), "pour joint," always horizontal, "cold joint," always vertical, and "shrinkage joint," it too always vertical, are the various types of joints that leave traces on Kahn's concrete. The fundamental corollary of the system remains the V joint, which is confirmed as the form of the cut of the edges between the panels in the formwork, to allow the "concrete to bleed out," as Kahn states.

One of the inventions that makes the formwork system of the Salk Institute so exceptional with respect to normal construction practice, in America and elsewhere, is the joint between the panels done by introducing a segment of plywood that is now defined as the "parting or juncture strip."[119] The origin of this strip can be found in the T-shaped solid wood strip that appears in the early studies and first trials done on the East Wall. This strip has a multiple technical function.

The juncture strip gives an unprecedented meaning to the panel. It makes it possible to reuse the same panel to construct formwork in wall sectors of different sizes. The different measurements can be compensated for precisely by the strip, varying its length case by case. The panel is used for multiple pours, reassembled in a system with different measurements, precisely thanks to the insertion, in the joints between the panels, of a special piece that makes up for the different size and also for certain defects caused by wear. Thanks to the juncture strip the panel becomes what Kahn and his team call the "floating increment," using a term chosen to underline its role in the organic growth of the construction, because by now in Kahn's system that panel, precisely because it is bound into a coherent whole by the juncture strip, though it may have different measurements, permits the

[119] *Ibid.*

construction of wall sectors of different sizes. "Nature, in making things, records every step of its making," Kahn states.[120]

The juncture strip also has other functions in Kahn's system. It is designed as a sacrificial part of the panel, a surrogate of those edges that are the parts of the panel most subject to wear, because it is constructed every time the panels are reassembled in a different system, making it possible to obtain, at least in the expectations of Kahn and his staff, perfect joint edges – the V joints. Due to the fact that it is the always perfect part of the V joint of the panels, replaced at each assembly operation, it is also called the "expendable part of the form."[121]

The juncture strip has another crucial function in the joint between the panels, to obtain a concrete surface free of visible defects. During the operations of pouring in formwork composed of panels, if the assembly is slightly skewed or if there is even a very small shift of one panel, the ideal continuity of the concrete surface is interrupted, producing what Kahn's team calls a "bleed line," since the compound tends to emerge through the crack.[122] It is precisely the juncture strip that reduces perception of these possible defects of the pour. So the design of the joint between the panels not only makes it possible to reuse them in a system with different measurements and always perfect edges; it also prevents, thanks to the imprint of two protruding lines, the perception of every eventual shift between the contiguous panels caused by the pressure of the pouring – a "hydrostatic relief valve," in the terminology of Kahn's team.[123]

The V lines of the Carpenter Center are imprints generated by the sharp edges of the panels directly butted together, and they do not have the value assigned to those same lines by Kahn in the perception of eventual defects of fabrication of the concrete of the Salk Institute.

Due to its complex meaning, the design of the juncture strip goes through many variations, and Kahn wants sample walls made to test all the effects of his formwork system in practice.

Kahn and his team do not limit the study of imprints to those made by the formwork system. The entire concrete construction process is analyzed in order to control every phase and therefore absorb it in an organic design, as demonstrated by the design process of the formwork system. To this end the pouring operations, the pauses between one pour and the next, and the possible openings, are seen as ways of generating other more or less visible imprints to be included in the system of those of the formwork.

So alongside the lines of the "parting or juncture strip," with their values as a "hydrostatic relief valve" and "floating increment," Kahn and his team identify another catalogue of lines, this time related to the phases of the pouring. The "pour joint," "shrinkage joint" and "cold joint" are the main elements of this catalogue, as outlined in Langford's letter.[124]

The pour joint is generated by a pour that according to the worksite logic of Kahn already outlined in the specifications of 1962, should halt when it reaches the chamfer specially inserted in the formwork to leave a groove that has the same optical role as the relief lines of the juncture strip: to prevent perception of possible shifts of color between one pour and the next.

[120] See Louis I. Kahn, *A Statement*, 1962, now in, Latour, *Louis I. Kahn. Writings, Lectures, Interviews* cit., (pp. 145-152), p. 146.
[121] Frederick Langford, letter to Greer Ferver, 5 April 1963, cit.
[122] *Ibid.*
[123] *Ibid.*
[124] *Ibid.*

The cold joint is a vertical groove obtained with a strip placed at the vertical ends inserted in the formwork; the imprint of this strip makes the pouring faces for the construction of a wall become visible.

The shrinkage joint is a deep crevice made vertically on the concrete surface, again through the application of a strip on the formwork, in order to channel, during the setting phase, possible tensions that could cause cracking – a sort of pre-cracking that safeguards the continuity of the surface. In this catalogue of lines produced by the fabrication of concrete, we should also add the traditional expansion joints introduced to divide up the overall structure of the building.

All the pour grooves and imprint lines have to be correlated with the design of the formwork system, so that the shadow of the grooves does not randomly cut through the pattern of lines. Kahn even designs the partitions of the formwork and the measurements of the panels based precisely on the expansion joints established on the basis of engineering calculations, the cold joints.

It is clear that the same focus on expedients useful to reduce the perception of defects in the fabrication of exposed concrete already demonstrated by Kahn in the works of the 1950s continues to guide him in the development of his sophisticated and ideal formwork system.

Coating of the plywood panel: the erasure of the wood imprint

The treatment of the surface of the plywood panel, alongside the joint between the panels and the various other types of joints, is another essential aspect of Kahn's formwork system developed for the Salk Institute. The focus on the color and surface finish of the concrete has reached the point of rejecting any treatment of the surface of the plywood panel with oils, to avoid the risk of leaving imprints of a color not intrinsic to the pour itself on the concrete. Instead of oils, Kahn decides to sand the panel and cover the surface with a layer of polyurethane.[125] This confirms the desire to think of the imprint to leave on the concrete only as the one generated by the construction system of the formwork and the pouring process. Precisely this polyurethane coating reveals how, through technical choices, Kahn is outlining a clear distinction between system and material, formwork and concrete, so that only the system, not the materials of its parts, will produce the imprint on the only material to be ennobled and of which to discover the true nature through control of the pouring process: liquid stone or concrete. For the final result, it is fundamental that the goal of reduction of the formwork to an immaterial system be pursued at the same time as the research on the combination of the ingredients of the concrete. It is worth noting that in none of his constructions did Le Corbusier make an in-depth investigation of the composition of the concrete, instead always concentrating, like a sculptor and *artiste décorateur*, on the design of the imprints of the formwork. It is clear that the concept of the imprint has taken on completely different connotations in the visions of concrete formulated by Le Corbusier and Kahn: an instrument for designing surfaces, in the hands of Le Corbusier, and a witness to the process of fabrication, in those of Kahn.

[125] The formwork panels, with a thickness of 3/4 inch, are treated with sanding and a layer of "catalyzed polyurethane resin" (*Laboratory for Life Science Designed to Defy Time*, in, "Engineering News-Record," 27 January 1966, (pp. 78-80, 83-84), p. 80).

Only with the sanding and the polyurethane coating of the plywood panel can the concrete take on its own color and texture, without any conceptual distinction between skin and core, as if the *Kunst* and *Kern Form* of Carl Boetticher had again found, in Kahn's concrete, that identity of values discovered in the isodomic wall in blocks of marble of the Greek temple. In any case, we should emphasize that in every pour inside formwork a sort of film is always produced, composed of the finer ingredients of the concrete without the gravel, in contact with the formwork, generating almost a coating intrinsic to the process of fabrication, including the vibration of the compound; this film could be eliminated only by immediate intervention after stripping, involving the washing of the concrete surface – an operation that cannot be done without betraying the theoretical principles of Kahn, derived from *béton brut*.

The polyurethane coating of the plywood finally permits Kahn to scrape away any vestige of the support material of the formwork, that grain that also in the case of plywood panels remains visible on the concrete. It is significant that the Carpenter Center bears evident traces of the grain of the plywood panels in contact with the pour. With the coating of the formwork, Kahn manages to direct the decorativist evolutions of international *béton brut* towards a degree zero of the material, towards the difficult goal of pure expression of the nature of concrete. To reach degree zero, the exposed concrete has had to pass, starting with its reinvention in the 1700s, through the condition of "artificial stone," a surrogate for natural stone; then it freed itself from having to imitate that natural material with stone powders added to the mix, and work done after stripping with the same tools used for dressing stone; it finally gained independence from stone, but not from the formwork with which it was tattooed by imprints to obtain decorative effects or patterns, as if an invisible coating had been spread on its skin. It is at this extremity of the evolution of concrete in pursuit of its own identity that we can place the Salk Institute, if we want to fully grasp the epochal meaning of this work, and also the huge gap that exists between the "artificial stone" discovered back in the 1700s, and the "liquid" or "molten stone" of Kahn.

After Lodoli, Ruskin, Semper, Viollet-le-Duc or Loos, in the series of statements of principle, that one should "see" the brick in brick, the stone in stone, the wood in wood and the iron in iron,[126] it is Kahn who manages to add to the list of construction materials another truth, that of the contemporary material par excellence: concrete should be concrete.

Having reduced the formwork to a panel no longer bearing the recognizable materic texture of wood is the great intuition of Kahn that makes the concrete of the Salk Institute the material no longer marked by imprints other than those produced by the system of the formwork itself, not by the materials of that system. Only an American artist sensitive to the phenomena of abstraction could be able to approach the pursuit of truth, of French origin, with that silence of form that crosses the American works of the era after World War II, those of Rothko, Newman, Pollock, where the paint is applied for its quality as paint, scraping away any figure from the canvas's surface. The absence of grain imprinted on the surface, which is a general characteristic of the concrete of the Salk Institute, intentionally eliminates

[126] On this theme see Giovanni Fanelli, Roberto Gargiani, *Il principio del rivestimento. Prolegomena a una storia dell'architettura contemporanea*, Roma-Bari, Laterza, 1994, in particular p. 15.

any mystery, any gratuitous decoration, and corresponds to the absent figure in American painting. In this perspective, the concept of liquid or molten stone expressed by Kahn, with the methods of "pouring" for a concrete without the figurative imprints of the formwork, is comparable to the concept of "dripping" of Pollock, for paint to be thrown on the canvas, in its fluid state, without the mediation of brushes and knives that rub on the canvas, thus even going beyond the abstract figuration.

So on Kahn's exposed concrete a veritable plot of "lines" can now appear, each with its own truthful connection to the construction process, so that the phases of organic growth can be glimpsed on every part. Never has an architect, prior to Kahn, broken down the whole process of fabrication of concrete in such an analytical way, to identify each single phase capable of leaving perceptible signs on a surface lacking in any type of coating, and to transform those very signs into a new form of wall of the contemporary era. By now it is clear that Kahn's concrete has evolved into something completely different from the *béton brut* of Le Corbusier, from which it had begun its journey back at the Yale Art Gallery.

With the studies in 1963 for the Salk Institute, at this point Kahn's concrete is arriving at the goal of discovery of its own essence, its degree zero, beyond any other possible imprint on its surface, from the *sculptures moulées* to the axes of the measures of the Modulor, to the grain of the wood, because Kahn has definitively set aside the idea of the shaping of the material as it is understood by Le Corbusier and his imitators. The passage from Le Corbusier's famous definition of concrete as a "natural material of the same rank as stone, wood or terra cotta," in the passages on the discovery of *béton brut* in Marseille, to Kahn's definition of concrete as "molten stone" or "liquid stone," as it is deployed in the work at La Jolla, marks a decisive turning point in 20th-century research on the nature of concrete. Those two definitions, with their apparently affinities in evoking the image of stone, actually reference a profound transformation in the meaning of concrete itself: its continued essence as a surrogate of stone, in the 18th-19th-century sense of "artificial stone," and the discovery of its geological nature celebrated on the worksite in the ritual of transubstantiation of the components, when in the span of a few hours the "liquid" becomes "solid," replicating, precisely with that act, the miracle of the age-old creation of stone; and, then, the way concrete is conceived as a sculptural material to shape in formwork, with a mold similar to those used to cast bronze sculptures, and the discovery of a specific artistic potential of concrete, intrinsic to the very process of fabrication of the materials, impacting the very meaning of the formwork. While it is undoubtedly Le Corbusier who lays the theoretical groundwork for the discovery of the nature of concrete, it is also true that only after the rise of Kahn's vision of a concrete as the result of its own specific process will it be possible, for American artists like Robert Smithson, to proceed towards experiences that stage this liquid stone, no longer a prisoner of formwork, no longer forced to seek routes of escape in the narrow openings of the V joints.

Development of the system

Langford's letter is written following the first concrete works of the Salk Institute, done by Fuller Company under the supervision of Katz. Those works completely failed to satisfy Kahn and his worksite representatives, prompting him to consider the contracting of the "concrete work" to another company, as seen in "Role of the Managing Builder" signed on 17 May

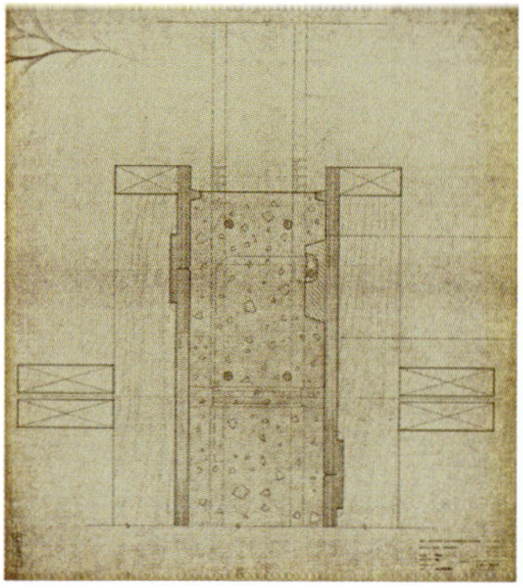

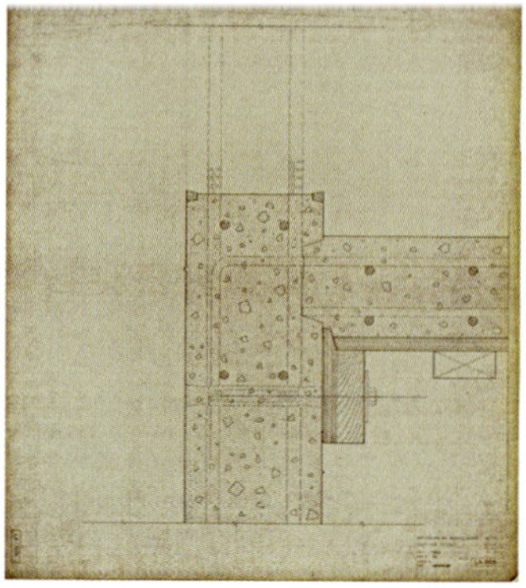

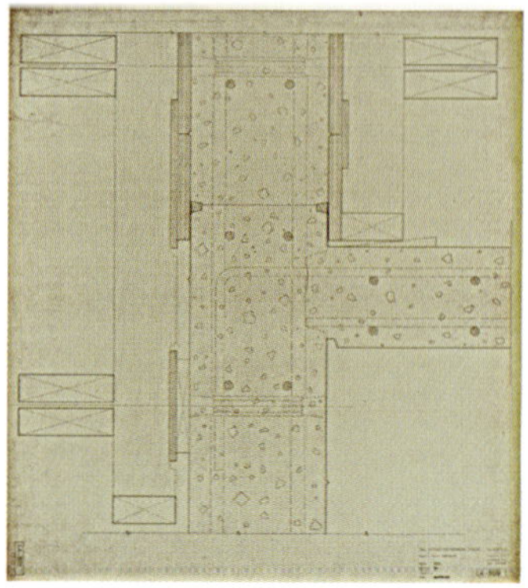

74-80. Salk Institute for Biological Studies, La Jolla, California, 1959-65.

74. Formwork details, LA-307, 15 April 1963 (LIK, 030.I.C.540.31).

75. Formwork details, LA-308, 15 April 1963 (LIK, 030.I.C.540.31).

76. Formwork details, LA-309, 15 April 1963 (LIK, 030.I.C.540.31).

77. Formwork details, LA-310a, 15 April 1963 (LIK, 030.I.C.540.31).

78. Formwork details, LA-311, 15 April 1963 (LIK, 030.I.C.540.31).

1963.[127] The member of Kahn's staff charged with overseeing the quality of the concrete at the worksite becomes Jack MacAllister, while Langford remains responsible for the design and construction of the formwork. Katz is replaced, towards September, by Paul Matt as worksite representative of the Fuller Company in charge of questions related to concrete, including those of the formwork, in collaboration with MacAllister and Langford.

After Langford's letter, Kahn and his staff reach the point of defining the formwork system for the Salk Institute, by now in a version very close to its definitive design. The letter also serves to specify the ideas that guide the development of that system.

The first working drawings of the new formwork system are dated 15 April 1963 and are even more precise than those previously made by Kahn's office. The drawings show the aggregate of the pour, the metal reinforcement bars, the spreaders and their caps, the precise arrangement of the panels in various special parts, and every other detail that might be of use for comprehension of the system.[128] The fact that the drawings have been made after the disputes addressed in Langford's letter demonstrates that the series documents the implementation process, according to the sequence to apply on the worksite, with the aim of illustrating to the technicians of the Fuller Company and the craftsmen how the phases of assembly of the formwork and the pour have to be cadenced according to Kahn's project, so that controlled imprints can be obtained in every case. By now it is clear that the logic of the worksite and that of the formwork system have to be integrated in a rigorous project, in a way never previously foreseen, to achieve Kahn's vision of exposed concrete.

In place of the solid wood strips, Kahn and his staff confirm the narrow strips cut from plywood panels 1 inch in thickness, and aligned with the large panels of the same plywood and the same thickness. The beveling does not affect the whole side of the two types of panels, as the aim is to obtain a thin line as an imprint. To consolidate the strip with the beveled edges, on the back another strip is placed, still in plywood but wider, so that it is in contact with the large panels. The cut edges between the butted panels also produce a V joint. Kahn considers the case, already examined with the solid wood strips, of doubling the strip with a third element inserted between the two, again in plywood but inserted like a blade, with the head set back slighting from the faces in contact with the pour, so as to shape a relief with a flat face on the wall. With this same solution, applying a wooden strip at the end of the element between the two strips, a deep groove is obtained, namely

79. Formwork details, LA-313, 21 April 1963 (LIK, 030.I.C.540.31).

80. Formwork details, LA-314, 4 May 1963 (LIK, 030.I.C.540.31).

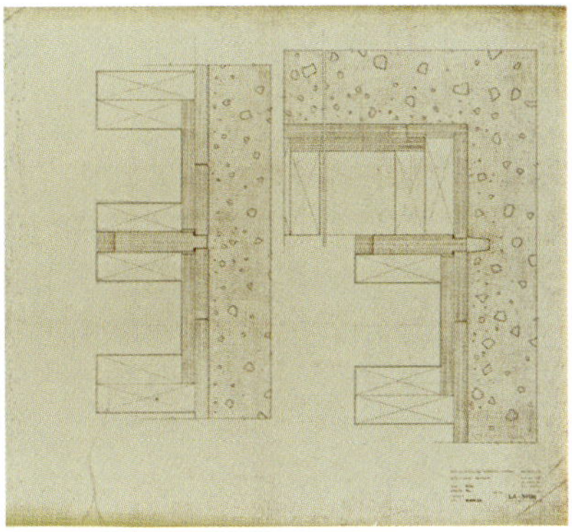

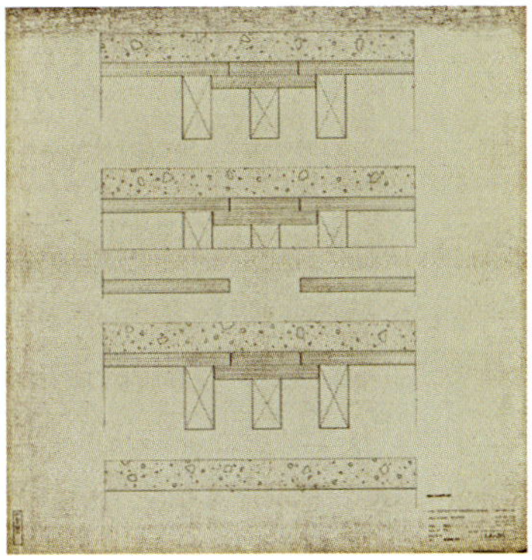

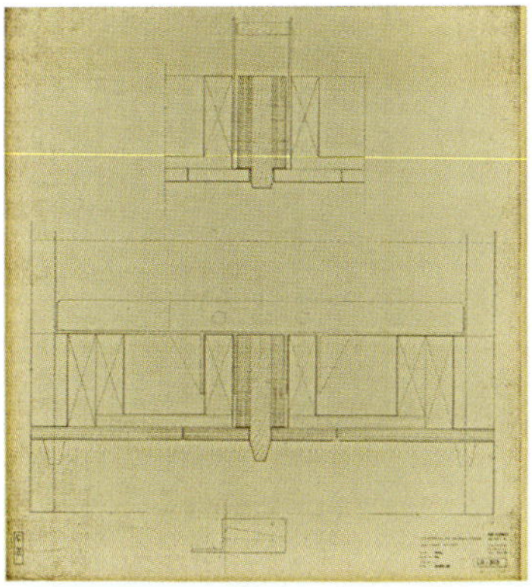

the shrinkage joint.[129] In the corners of the rectangles the formwork panel itself is sharpened and made to protrude, inserting itself in the thickness of the wall to shape a similar groove in the edge. This is a detail that demonstrates the pursuit of alternatives to the chamfer in one of the critical points of any formwork system. The other groove, less wide and deep, that of the pour joint, is obtained with thin strips mounted at the heads of the plywood panels, at the end of the formwork, to shape two narrow grooves.

The system documented by the drawings of 15 April is also destined to undergo further refinements. At the start of May the pieces to wedge between the plywood strips go back to being in solid wood.[130]

Still in May, Kahn and his team study the application of the system to the design of the Service Towers. One of the most significant variants has to do with the width of the two bands to put in the corners, narrow,[131] medium[132] or wide,[133] and the presence or absence of the two vertical imprints framing the large compartment corresponding to one level of the tower. In the three cases the original solution remains, of the door bordered by panels that form the imprints of a frame. Kahn and his staff also work, through section drawings, on the complex patterns of the formwork that produce the inner face of the perimeter walls – the panels are matched to the various structures, producing a patchwork of lines.[134] At the end of May, when the drawings for the Service Towers of the South Laboratory are prepared, a pattern appears of the main facade with the door without the framing of the previous versions, set off by two vertical panels (about half as large as the others) that pretend to carry an even larger panel placed horizontally, representing a powerful trilithic system.[135]

Kahn's office readies a series of sets of drawings with the plans of the Service Towers, covering the positions of all the parts of the formwork

[127] G.S.C., "Role of the Managing Builder," 17 May 1963, LIK, 030.II.A.89.50.

[128] See the drawings LA-307 and LA-308, dated 15 April 1963, LIK, 030.I.C.540.031.

[129] In earlier studies a single solid wood element is envisioned, wedged to create a wide groove. For variants of the "shrinkage joint" see the drawing LA-313, 21 April 1963, LIK, 030.I.C.540.031.

[130] See the drawing LA-314, 4 May 1963, LIK, 030.I.C.540.031.

[131] See the drawing "Service Tower," LA-334, May 1963, LIK, 030.I.C.540.031.

[132] See the drawing "Service Tower el. North side," LA-330, May 1963, LIK, 030.I.C.540.031.

[133] See the drawing "Service Tower el.," LA-329, May 1963, LIK, 030.I.C.540.031.

[134] See the drawing "Service Tower – Section," LA-328, May 1963, LIK, 030.I.C.540.031.

[135] See the drawing "Service Tower – Elevation of entrance building," LA-337, 28 May 1963, LIK, 030.I.C.540.031.

81-91. Salk Institute for Biological Studies, La Jolla, California, 1959-65.

81. "Service Tower Elevation," LA-334, May 1963 (LIK, 030.I.C.540.31).

82. "Service Tower El.[evation] North Side," LA-333, May 1963 (LIK, 030.I.C.540.31).

83. "Service Tower El.[evation] North Side," LA-329, May 1963 (LIK, 030.I.C.540.31).

84. "Service Tower Section," LA-328, May 1963 (LIK, 030.I.C.540.31).

85. Service Tower, "Elevation North Side," LA-337, 28 May 1963 (LIK, 030.I.C.540.31).

86. "Service Tower Plans," LA-336, 27 May 1963 (LIK, 030.I.C.540.31).

87. "Service Tower Panel Study," LA-333, May 1963 (LIK, 030.I.C.540.31).

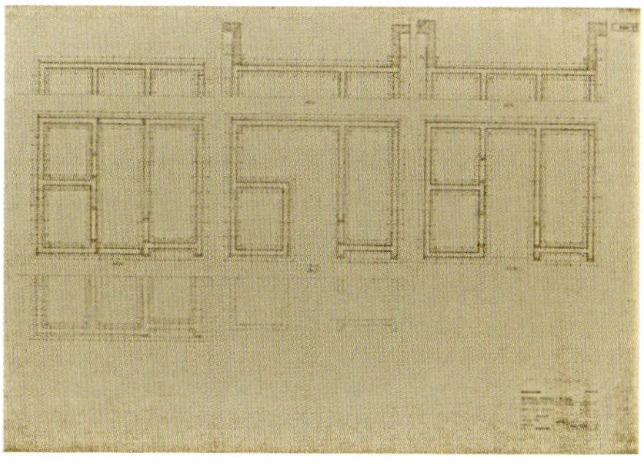

88. "South Laboratory Retaining Walls, Service Tower Side," 29 May 1963 (LIK, 030.IV.A.540.8.14).

89. "North Laboratory Retaining Walls, Service Side Tower Looking North," 29 May 1963 (LIK, 030.IV.A.540.8.13).

90-91. Retaining Wall (photographs by the author, 2010).

destined to leave traces on the pour, from the spreaders to the joints between the panels and the strips for the grooves, to the composition of the compartments of the formwork in the pour sequence.[136]

In the meantime construction continues, in May, with the Retaining Walls, in keeping with the pattern defined in the design from March, but with important variations that depend on the new formwork system: the horizontal T strips between the panels are eliminated, and the panels are now installed directly one over the other, with the V joint, while the previous vertical T strips are replaced by plywood strips of the same width, with V joints and with the intermediate element to produce the groove. The vertical strips are assembled according to a pattern that corresponds to that of the drawing from March, with joints placed at different heights; these too are emphasized with V joints.[137] So all the lines imprinted on the concrete of the Retaining Walls have a very delicate V profile. The later V joints will be more pronounced, confirming the fact that the worksite, in nearly every phase, has an experimental character. The panels of the formwork are not protected by any material, leaving the imprint of their grain on the concrete. This is a detail Kahn will change in the other parts of the Salk Institute.

On 29 May, in the North Laboratory, three Retaining Walls have already been built; the formwork are already being assembled for the fourth, and for the first two of the South Laboratory. On 28 June all the Retaining Walls of the two laboratories have been completed.

Cones and snap ties

The forms of the holes left on the concrete surface by the cones of the spreaders, in terms of diameter and depth, and the type of filler, are examined in May 1963, when on the Salk Institute worksite samples of industrial products are applied in three variants, of which two are selected for further study, namely a screw tie and a snap tie. The two systems have cones of different forms, as well as different methods of deployment and stripping. The detail of the ties and cones had been discussed at length during the course of 1962, leading to different solutions in the preparation of the various specifications. The solution chosen by Kahn to treat the imprint left by the cone in the concrete wall is the sealing of the hole by means of a lead disk, or "lead caulking," which already appears as an alternative at least from October 1962, and had also been drawn in the specifications of 6 December 1962. The seal prevents corrosion of the steel rod and thus eventual rust marks on the concrete surface, concealing the unfinished bottom of the conical hole from view. The choice of lead can be seen as a clue regarding Kahn's desire to tune the color to that of the concrete surface. With its dark gray color the caulking ensures visibility of the pattern of the holes, in the complex play of imprints that enliven the surface of the masonry sectors, reinforcing the shadow effect. For the insertion of the lead disks to seal the holes, a special steel bar is prepared with the end "ground to conform to the slope on the walls of the hole."[138]

[136] See the drawing "South Tower," LA-336, 27 May 1963, LIK, 030.I.C.540.031. Another drawing prepared for the count of the panels of the formwork of the Service Towers is the one that includes a plan with the profiles of the panels (see the drawing "Service Tower – panel study," LA-333, May 1963, LIK, 030.I.C.540.031).

[137] For the working drawing see "Service Tower – Sections," LA-335, May 1963, LIK, 030.I.C.540.031.

[138] Marshall D. Meyers, letter to Frank Sherwood, Preston M. Geren, Jr., 14 May 1970, LIK, 030.II.A.37.6.

It is worth noting that the decision to emphasize the tie holes of the formwork with lead disks simply confirms the conceptual and technical similarities between the truth of concrete pursued by Kahn and the search for truth in cladding pursued by Wagner, since what is involved here is a logical and symbolic assembly of the panels, which in the Salk Institute leave imprints, and in the architecture of Wagner remain in the work. In both cases, the goal is the truthfulness of the panel, be it a formwork panel or a cladding panel.

The choice of the type of spreaders is fundamental for the effect on the exposed concrete surface. The system has to guarantee that the operation of detachment of the ties will not damage the surface.

Kahn has an information brochure on the snap ties produced by Burke Concrete Accessories Company with wooden cones and steel rods with flat sections.[139] The innovation of these snap ties consists in the possibility of breaking them before stripping, through rotation done with a special wrench patented in 1955, to grip the head outside the formwork (the flat section prevents rotation of the rods). The snap ties make it possible to set the point of breakage, also chosen on the basis of the extraction of the cones. "A quarter to half turn is usually sufficient to break ties at the break-off position," the brochure in Kahn's possession explains. The removal of the cones can be done with a special gimlet, threaded to enter the wood. In the case of positioning of the break point "inside of conical bulge," the "action of break and removal of wood cone is simultaneous." Unlike what Kahn has in mind, the Burke Concrete Accessories Company takes for granted that the holes left by the cones "are to be grouted with cement mortar and smooth trowelled flush to wall surface to provide watertight construction."[140]

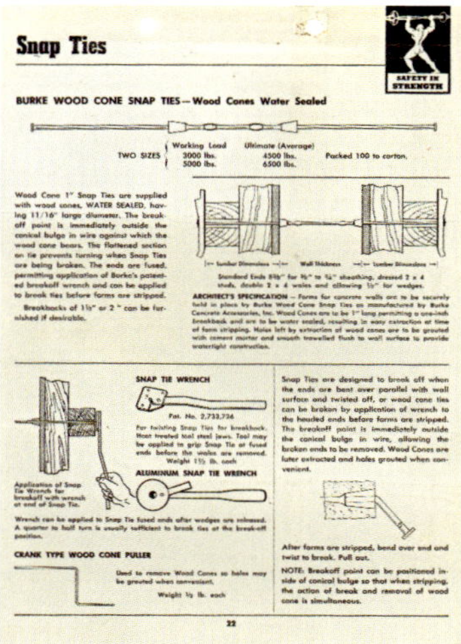
Snap Ties

BURKE WOOD CONE SNAP TIES—Wood Cones Water Sealed

TWO SIZES | Working Load 3000 lbs. / 5000 lbs. | Ultimate (Average) 4500 lbs. / 6500 lbs. | Packed 100 to carton.

Wood Cone 1" Snap Ties are supplied with wood cones, WATER SEALED, having 11/16" large diameter. The break-off point is immediately outside the conical bulge in wire against which the wood cone bears. The flattened section on tie prevents turning when Snap Ties are being broken. The ends are fused, permitting application of Burke's patented breakoff wrench and can be applied to break ties before forms are stripped.

Breakbacks of 1½" or 2" can be furnished if desirable.

ARCHITECT'S SPECIFICATION — Forms for concrete walls are to be securely held in place by Burke Wood Cone Snap Ties as manufactured by Burke Concrete Accessories, Inc. Wood Cones are to be 1" long permitting a one-inch breakback and are to be water sealed, resulting in easy extraction at time of form stripping. Holes left by extraction of wood cones are to be grouted with cement mortar and smooth trowelled flush to wall surface to provide watertight construction.

SNAP TIE WRENCH

Pat. No. 2,733,726

ALUMINUM SNAP TIE WRENCH

Application of Snap Tie Wrench for breakoff with wrench at end of Snap Tie.

Wrench can be applied to Snap Tie fused ends after wedges are released. A quarter to half turn is usually sufficient to break ties at the break-off position.

CRANK TYPE WOOD CONE PULLER

Used to remove Wood Cones so holes may be grouted when convenient.

Snap Ties are designed to break off when the ends are bent over parallel with wall surface and twisted off, or wood cone ties can be broken by application of wrench to the headed ends before forms are stripped. The breakoff point is immediately outside the conical bulge in wire, allowing the broken ends to be removed. Wood Cones are later extracted and holes grouted when convenient.

After forms are stripped, bend over end and twist to break. Pull out.

NOTE: Breakoff point can be positioned inside of conical bulge so that when stripping, the action of break and removal of wood cone is simultaneous.

22

92. *Snap Ties*, booklet (LIK, 030.II.A.107.19).

A series of sketches, dated 29 May 1963, documents the various cone samples and their different holes as they are examined on the worksite. Now the depth of the hollow is calculated, and thus the more or less accentuated play of shadows produced by the sequence of holes on the surface of the walls. The snap tie variant shows a very deep hole, but with a smaller diameter, always in order to calculate the effects of the punctuation of shadows on the surface. "These drawings made from samples now in office, not necessarily those being used," is written on the sheet of sketches – "screw tie and snap tie/cone hole caulkings in Retaining Walls."[141] The document confirms Kahn's care in the choice of every single technical detail of the formwork, aware that his new wall *opus* will be the result of the exact casting of a limited, selected number of essential constructive details, which therefore cannot be left up to the contractor's technicians.

During June other studies are conducted on the formwork system for the Retaining Walls along the Studies and for the Studies themselves, while work proceeds to define the patterns for the Service Towers. It is in this phase that Kahn and his staff examine the possibility of truncated conical caps in wood to apply to the bolts that fasten the formwork structure against the panels, to shape a particular cavity in the concrete.[142] Kahn also experiments, in the walls of the Service Towers, with the snap ties with semispherical plastic

[139] Choosing the snap ties of the Burke Concrete Accessories, Inc., Kahn and his team are considering precisely the question outlined in Hurd's manual regarding the "Effect of Tie on Finished Concrete Surface" (Hurd, *op. cit.*, pp. 52-53).

[140] *Snap Ties*, booklet conserved in the Kahn archives, LIK, 030.II.A.37.6

[141] See the sheet of sketches dated 29 May 1963, LIK, 030.I.C.540.047.

[142] See the drawing "Panel Juncture Details," LA-312, 28 June 1963, LIK, 030.I.C.540.031.

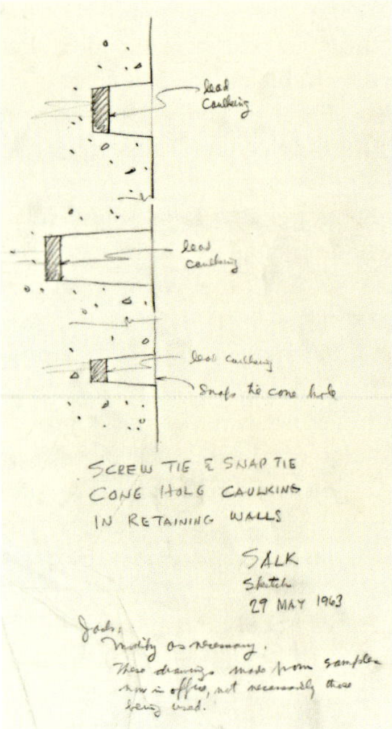

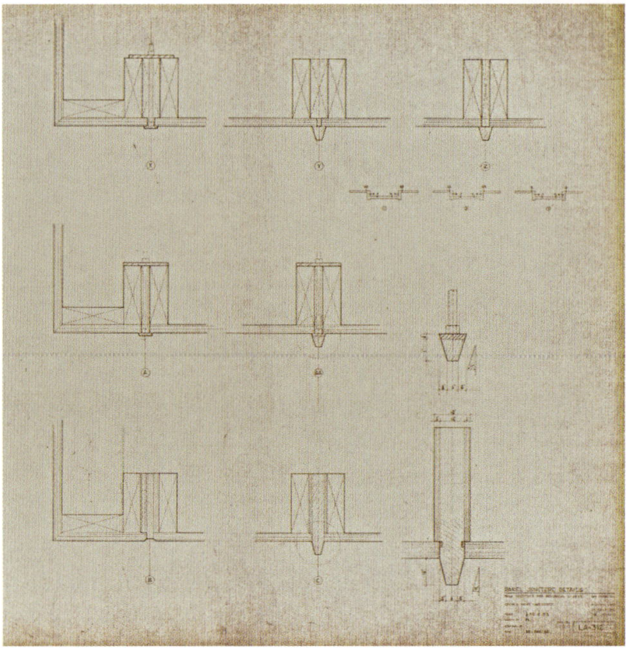

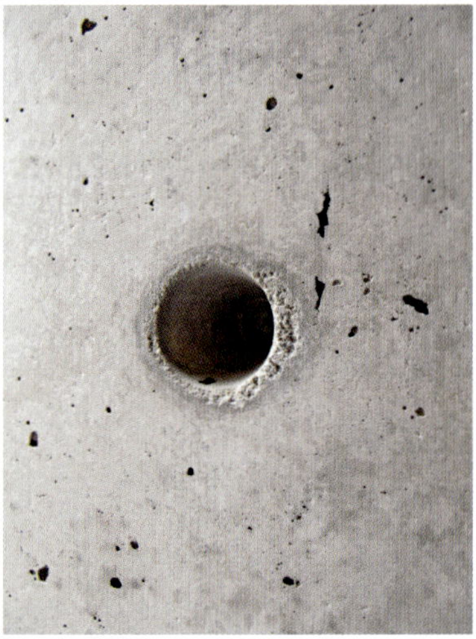

[143] This type of solution is visible in the completed building.
[144] See the drawing, n.d., LIK, 030.I.C.540.024.
[145] Each large plywood panel (7 feet -4 inches x 29 inches) contains a pattern of spreaders (24 x 22 inches) and is framed by four bands: two lateral (38 1/2 x 4 inches), one upper (8 feet x 4 1/4 inches) and one lower (8 feet x 9 1/2 inches).
[146] See the drawing, "Service Tower – Elevation," LA-337A, 25 June 1963, LIK, 030.I.C.540.031; and the drawing "Service Tower – Elevation," LA-342, 8 July 1963, LIK, 030.I.C.540.031.
[147] See the drawing "Service Tower – Elevation," LA-338A, 25 June 1963, LIK, 030.I.C.540.031, and the drawing "Service Tower – Elevation," LA-340, 25 June 1963, LIK, 030.I.C.540.031.
[148] See the drawing "Studies Bird's Eye Perspective," LA-351, 20 May 1964, LIK, 030.I.C.540.031.

cones, and imprints not sealed flush with the wall.[143] As it gradually grows, the building continues to take on visible traces of the experiments conducted during construction to test the individual details of the fabrication of the concrete, following the hidden "fossils" in the East Wall of the Mechanical Building. Thus the building itself becomes the most important documentation of the process art pursued by Kahn to achieve the goal of an ideal concrete.

The wall of the parking area, definitive test of the surface shaped by the formwork

After the construction of the trials in the East Wall, the studies in April and the construction of the Retaining Walls, and after the first walls of the Service Towers, Kahn does not stop adding new refinements to his formwork system, with the goal of obtaining increasingly subtle and complex imprints. It is clear that the Salk Institute is becoming a true masterpiece of concrete, of all eras.

An ulterior, decisive and final modification of the system has to do with the face of the recessed intermediate piece that shapes a flat, protruding band on the concrete. With the beveling of the two edges of that piece, no longer set back in its position, Kahn can shape two thin, parallel juxtaposed V reliefs. To test this modification a low wall is built bordering the parking area, outside the enclosure of the Salk Institute. The design of the formwork for this wall comes from the studies of 15 April 1963 and calls for the use of panels and strips only in plywood, with edges cut to form the V joint.[144] The drawing indicates the sizing of all the parts, their assembly to obtain a regular pattern of imprints, and the structure of the formwork.[145] In the assembly between one formwork and the next, a new, slender special piece is inserted, again with cut edges, that shapes the two vertical V-shaped lines side by side on the concrete (the width of the head of this piece is 5/8 inch). Instead, where the cold joint is to be made, a "Rubber Insert" is used. The piece of the head with the beveled edges that shapes the two neighboring V-shaped lines had not appeared in the previous drawings and tests. That piece is derived from the one that in the drawings of 15 April was inserted between two narrow plywood strips to shape a slender, flat relief on the concrete. That piece takes the place of the narrow band with beveled edges derived from the solid wood strip. This is the process behind the birth, through a long design process, of the final technical detail that typifies the concrete of the Salk Institute. The formwork system of the wall of the parking area becomes the formwork and the fundamental pattern Kahn will imprint on all the other concrete surfaces of the complex.

93-97. Salk Institute for Biological Studies, La Jolla, California, 1959-65.

93. "Screw Tie & Snap Tie," study sketches, 29 May 1963 (LIK, 030.I.C.540.47).

94. "Panel Juncture Details," LA-312, 28 June 1963 (LIK, 030.I.C.540.31).

95-97. Spacer holes in a Service Tower, North Laboratory Wing (photographs by the author, 2010).

In late June and early July 1963 the working drawings for the Service Towers are prepared.[146] The drawing of 25 June shows the composition of the panels effectively used in the construction of the main facade (five for each level, with two of different alternating lengths). But there are still differences with respect to the constructed work, in terms of the design of the lines. The door is still indicated by panels arranged in the trilith figure, later destined to vanish; but along the vertical panels that frame it, a double imprint line appears, the same one found near the corners, where it creates the image of a corner pillar. The double lines near the door stop at the first level (in the construction they continue as far as the top of the tower, and the panel of the door, not coincidentally, is changed).[147]

Confirmation of the particular meaning taken on by the formwork system in the design of the Salk Institute can be seen in the fact that Kahn's office draws the bird's-eye view of the construction phases of a Service Tower with the various panels and special pieces for the joints and grooves.[148]

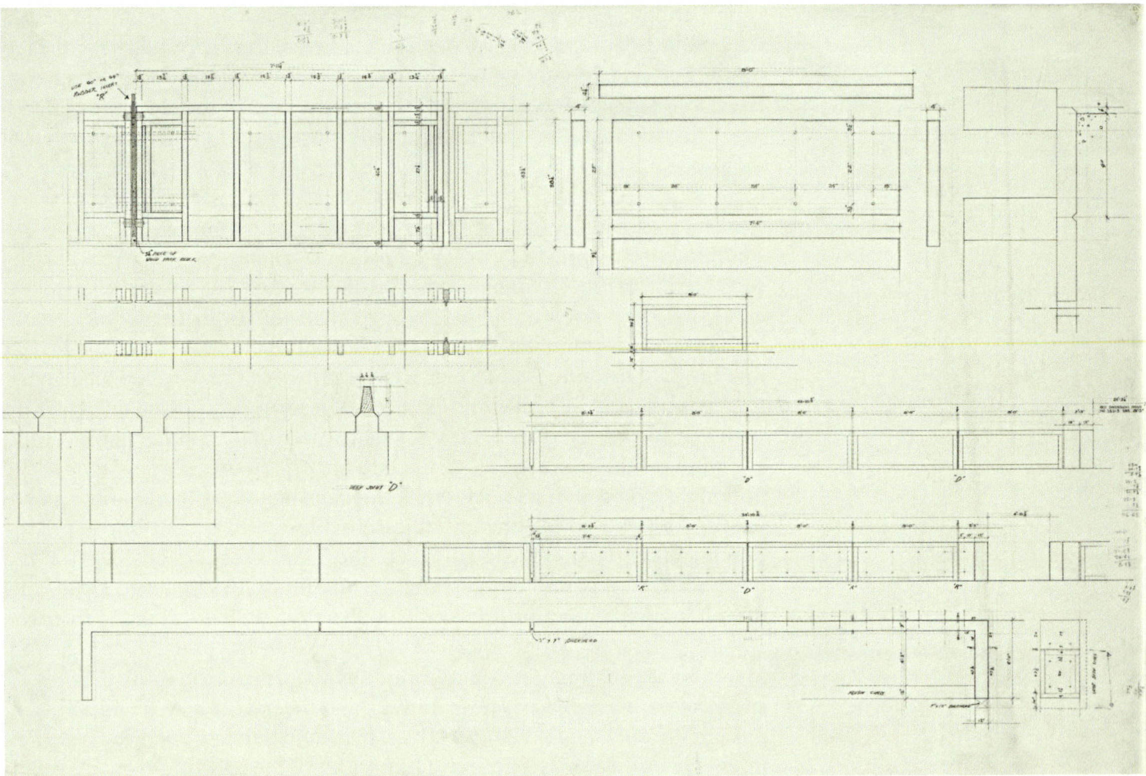

98-101. Salk Institute for Biological Studies, La Jolla, California, 1959-65.

98. Formwork drawing for the experimental parking wall (LIK, 030.I.C.540.24).

99. Parking wall (photograph by the author, 2010).

Worksite chronicles

On no other previous worksite of Kahn, as on that of La Jolla, have the initial phases of the work already in progress been marked by ongoing refinement of the details of the formwork, with the construction of trials left in the building to reach the definitive solution. During the course of June 1963 Kahn's office completes the drawings of the formwork for the "Retaining Walls," "Service Towers," study stairs," "studies," "mechanical wings," "laboratory," "west office wing."[149] At the end of July all the formwork of the Retaining Wall below the Studies of the wing of the North Laboratory Building have been completed. After one month, the Retaining Walls of the North Laboratory Building and the South Laboratory Building have already been built. In that same period, the formwork of the first level of a Service Tower of the North Laboratory Building is assembled. On 28 September the Retaining Walls below the Studies of the South Laboratory Building have been built, and the formwork is ready for the first level of two Service Towers. On that same date, the Retaining Walls below the studies of the North Laboratory Building are completed, as well as the first level of one Service Tower, and the formwork for two other towers is already assembled. On 31 October construction is under way on four Service Towers of the South and North Laboratories – the first level of the two wings is already finished. The Service Towers are completed during the course of 1964.

In the drawings of the Studies and of the connections with the North Laboratory Building prepared in June 1963, Kahn demonstrates that by now he is able to weave the various lines and grooves into a pattern composition of remarkable freedom, where the individual masonry sectors are framed, but the whole comes from the juxtaposition of the parts, which at times are not correlated with each other. In one of the most complex sectors, that of

100. "Service Towers Elevation," LA-337a, 26 June 1963 (LIK, 030.I.C.540.31).

101. "Service Towers – Elevation," LA-342, 8 July 1963 (LIK, 030.I.C.540.31).

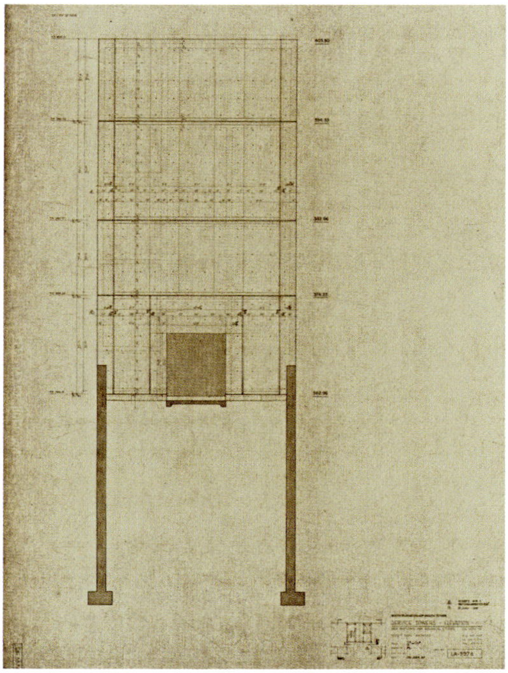

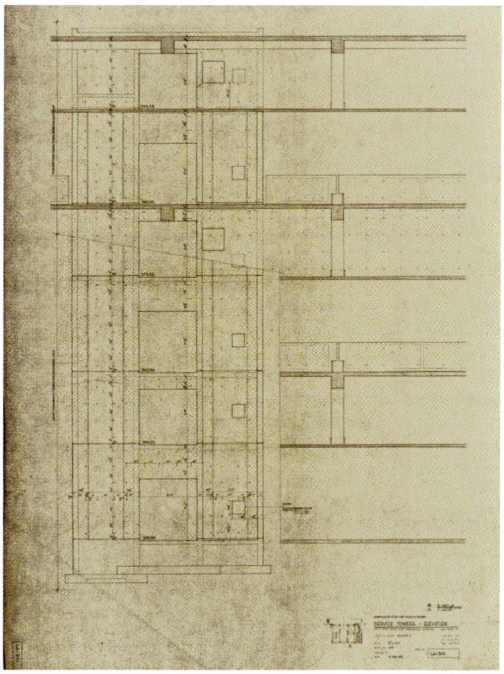

[149] See the drawing "Salk Institute for Biological Studies," 11 June 1963, LIK, 030.II.A.108.26.

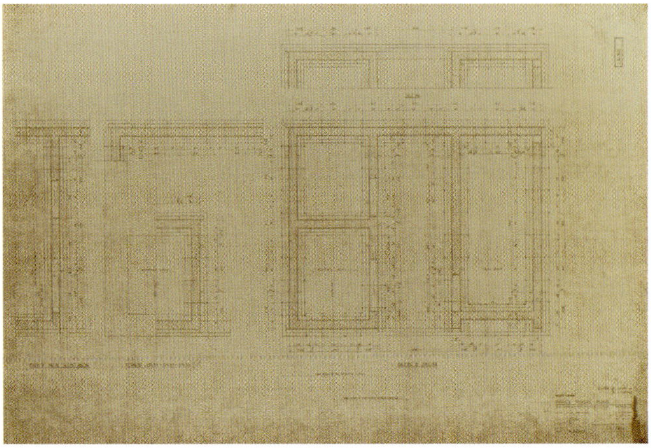

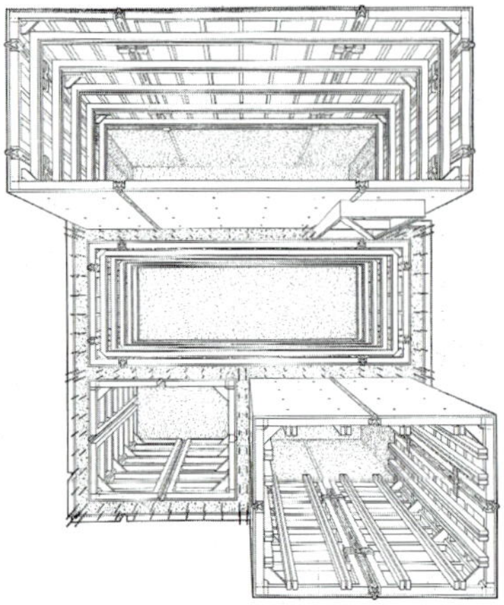

102-105. Salk Institute for Biological Studies, La Jolla, California, 1959-65.

102. "Service Towers – Plans," 8 July 1963 (LIK, 030.I.C.540.31).

103. Perspective view of the formwork system for the Service Towers (owned by Frederick Langford).

104. "North Laboratory Looking West," 29 July 1963 (LIK, 030.IV.A.540.9.35).

105. South Laboratory, construction site photograph with the walls of the patios (LIK, 030.V.D.18.4).

the staircases with the connection walkways between the Studies and the Laboratory Building, where there are large passages with parapets, Kahn uses the assembly of the formwork to design ideal posts at the extremities and a slender sill, but then the symmetrical design vanishes from one part, where the figure of the post is absorbed in the imprint of a larger rectangular panel, cut out as if the passage were a breach of an original pattern.

Fragments indicating a composition of symmetrical lines and panels clash with cut-out segments in the walls and series of irregular imprints, almost as if Kahn has found himself suspended between the discovery, in his craft, of an artistic way of proceeding by juxtaposition, and the urgent needs of a worksite that cannot allow pauses for reflection. It is in this free game of imprints that Kahn reaches the point of cutting out doors in walls without taking the pattern of lines into any account. The awareness of the artistic potential of this way of proceeding can be seen when Kahn uses the same solution of the opening cut freely between the panels for the entrance doors of the Service Towers as well, in that case following a more twisted process, but one that leads him to free himself of the charm of the framing and the reference to the trilith.

By now the image of the frame and that of the trilith have been definitively erased in the process of development of the patterns of the panels. Now the pairs of V joints that run near the door and continue on all the levels are simply the frames of panels which seem to have been cut out to make the opening of the door – over the door, panels with the simple vertical V joint have been mounted, precisely to create the effect of a breach. This choice demonstrates Kahn's arduous conquest of relative independence of the system of formwork panels with respect to the openings, as if the latter were breaches in the walls. Here again, we cannot help but notice another

106-109. Salk Institute for Biological Studies, La Jolla, California, 1959-65.

106. "South Laboratory Looking West," 27 December 1963 (LIK, 030.IV.A.540.9.3).

107. "North Laboratory Service Side. Retaining Walls Looking North," 27 December 1963 (LIK, 030.IV.A.540.9.4).

108. "Study Stair, Section," LA-365, 24 October 1963 (LIK, 030.I.C.540.31).

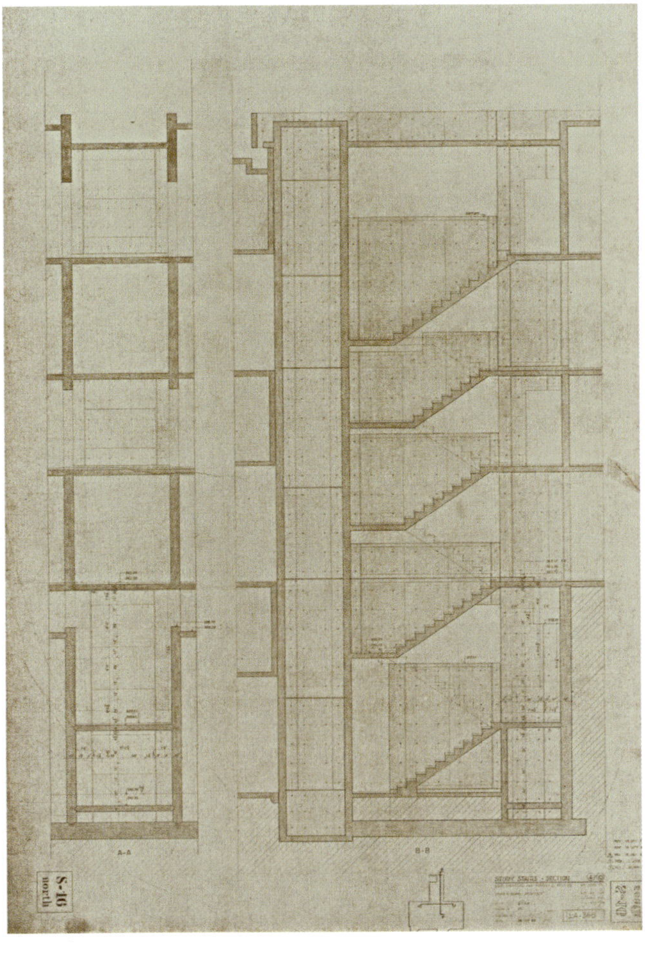

109. Formwork study, "North Office Core, Plan 342.96", LA-440, 14 November 1963 (LIK, 030.I.C.540.32).

significant similarity to the work of Wagner, and his cladding breached by the arch of the Serlian window in the Steinhof church.

In June 1963 the question of the mix design is still being examined by Ferver, Dorland & Associates, to make a concrete of the Riverside Cement Company with a strength of 6000 lb/sq.in – to achieve this, less sand is used.[150] MacAllister asks Komendant for information on the "sequence of operation for pouring and stressing of laboratory columns and Vierendeel trusses," on the exact position of "plates, joints, etc. at the juncture of the columns and the trusses," and if during the pouring, between the concrete of the columns and that of the trusses, a "paper separation" must be inserted.[151]

Only in the case of the pouring of the structural system of columns, trusses and floor slabs of the Laboratories does Kahn rely solely on the indications of the engineer. A collaborator of Komendant explains the construction phases of the structure: pour the columns, introduce the "bearing plates" in metal to permit movement of the trusses in case of seismic activity, assemble the formwork of beams and floor slabs, the reinforcement, and pour the concrete – first the "bottom chord" of the Vierendeel and the contiguous slab, then the vertical posts of the same Vierendeel together with the upper floor slab, to obtain a cold joint. After setting, when the concrete has reached a strength of 4000 lb/sq.in, post-tensioning operations are done for beams and columns, under the personal supervision of Komendant.[152] Komendant himself indicates the mix of the cement mortar to use for possible cracks caused in the beams by the post-tensioning.

[150] See Anon., letter dated 21 June 1963, LIK, 030.II.A.107.13.
[151] John E. MacAllister, letter to August E. Komendant, 12 June 1963, LIK, 030.II.A.107.13.
[152] August E. Komendant, letter to Louis I. Kahn, [21 June 1963], received on 22 June 1963, LIK, 030.II.A.89.3. Misunderstandings and discussions arise between Komendant, on the one hand, and the Fuller Company and Kahn on the other, regarding responsibilities for the positioning of the cold joints and the pour sequence for the sectors of the floor slabs and roof of the Mechanical Building (John MacAllister, letter to August E. Komendant, 1 November 1963, LIK, 030.II.A.107.13). "I am sure you would agree that the locations of cold joints in structural slabs or beams are of concern to the structural engineer," writes MacAllister to Komendant (John MacAllister, letter to August E. Komendant, 1 November 1963, cit.). For the position of the "pour line or construction joint" in the Mechanical Building, see also Charles H. Gruber, letter to Louis I. Kahn, 5 November 1963, LIK, 030.II.A.108.31. On 12 December 1963 the working drawings are made of the formwork of the "Mechanical Blg. Corridor Wall" (see the drawing "Mechanical BLDG. West Corridor Wall," 12 December 1963, LA-410, LIK, 030.I.C.540.032).

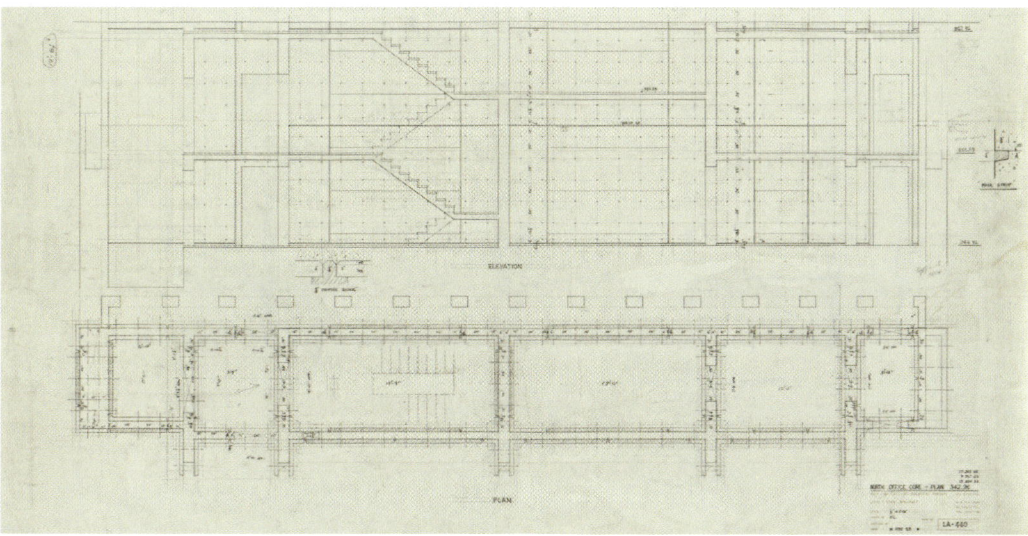

Rugged Travertine

The pursuit of concrete tattooed by signs of the construction process is not separated from research on other stone materials that become part of the catalogue of claddings of the Salk Institute. In the choice of these materials, Kahn demonstrates a sensibility similar to that of Le Corbusier, selecting materials that can be tuned to the qualities of the concrete surface.

Kahn's choice of the material for the paving of the porticos and passages of the Studies and that of the garden paths is Travertine stone, to be imported from Italy. This choice confirms his preference for stones with an imperfect surface, like the Cordova Shell that was to form the cladding of the Salk Institute in the first versions of the project. Kahn postpones decisions on the direction of the grain of the Travertine until the slabs are being installed.[153] Of the different varieties of Travertine, he discards those with a polished surface, seeing them as not typical reflections of the surface of that stone. For this reason, the choice goes to "Antique Travertine," with sawn face."[154] "All other Travertines I have looked at," MacAllister writes, "must be polished to bring out their character. Also, I think that only holes large enough to catch women's heels need be filled."[155] The character of the Antique Travertine is described by the company consulted by Kahn, Walker & Zanger of Glendale: "A dark Tuscan type Travertine with definitive bands of colors ranging from medium to dark tan and also occasional bands of gray."[156] To select the definitive type, besides examining the samples supplied by Walker & Zanger, Kahn makes a special trip to Italy, visiting Rome and Carrara.[157] In Rome he sees the Travertine of the Villa Aurelia at Porta San Pancrazio, home of the American Academy, and he is probably interested in the "Classic Roman Travertine" from the Barco quarry at Tivoli,[158] but for the Salk Institute he chooses the Travertine he sees at a marble warehouse in Carrara, the "Travertine Sis,"[159] a "Sicilian Travertine in a medium beige color with definite band markings."[160] To choose the type of working of the stone, Walker & Zanger sends Kahn six samples, of which five are in a "natural rough sawn state" and one is "slightly honed."[161]

The Travertine Sis is foreseen for "yard paving," "arcade," "portico" and "bridges" of the Studies, "west office wing," and with "saw cut unfilled" finish.[162] The canals of the gardens, originally envisioned in tile, are also indicated for cladding with Travertine.[163]

Kahn wants large slabs, although cost estimates indicate "smaller pieces of Travertine would be less expensive."[164] His explanation of the formal relationship between Travertine and concrete has to do with their similar surfaces, which do not require – as in the case of other materials to place beside concrete – the action of weathering to eventually blend in: "the Travertine and the concrete are so much together that it now looks to people as though that building were there thousands of years ago, because the decision of its harmony, which takes many years to attain, was there already, right? – as a new building it already had the harmony of something which was aged."[165]

[153] Report of the meeting of 1 July 1963, LIK, 030.II.A.107.18.
[154] John E. MacAllister, letter to Louis I. Kahn, 2 July 1963, LIK, 030.II.A.108.26.
[155] *Ibid.*
[156] Lou Bastone, letter to Louis I. Kahn, 12 July 1963, LIK, 030.II.A.108.26.
[157] See John Iberti, letter to Louis I. Kahn, 8 July 1963, LIK, 030.II.A.108.26, and Lou Bastone, letter to Louis I. Kahn, 12 July 1963, LIK, 030.II.A.108.26.
[158] Louis I. Kahn, notes attached to letter to John Iberti, n.d., LIK, 030.II.A.108.26.
[159] John Iberti, letter to Louis I. Kahn, 12 August 1963, LIK, 030.II.A.108.26.
[160] Lou Bastone, letter to Louis I. Kahn, 12 July 1963, LIK, 030.II.A.108.26.
[161] John Iberti, letter to Louis I. Kahn, 12 August 1963, cit.
[162] J. L. Wills, Job Runner, letter to Louis I. Kahn, LIK, 030.II.A.108.31.
[163] John E. MacAllister, letter

Discoloration of concrete and other defects

The cement of the Riverside Cement Company with pozzolan additive used for the Salk Institute, and manipulated by Kahn to obtain a particular color, starts to raise the first technical issues in the initial construction phases. In the second half of September 1963, when all the retaining walls of the two wings, the basement level of the Mechanical Building and the lower levels of two Service Towers, one of the South and one of the North Laboratory Building, have already been built, discoloring is observed on the concrete surface. The study of this phenomenon occupies Kahn and the technicians involved in the construction of the Salk Institute until 1964.

The discoloring involves the "5000 psi lightweight concrete"[166] used in the Service Towers, and seems to be caused by phenomena of deposited salts – "efflorescence or laitenance."[167] The concrete surfaces are also scarred by cracks and air bubbles.

The first interventions, dating back to the end of September, attempt to remove what is considered efflorescence "by rubbing with damp sand," with "some effect in improving the appearance."[168] Nevertheless, to eliminate the cause of the phenomenon, study begins on special treatments of the surface of the formwork in contact with the pour, though without achieving results.[169]

The Fuller Company introduces corrections to avoid defects on the concrete surface during the pour. To prevent staining, the corners of the forms are reinforced: "Panel pins are being used to tighten all form corners. If this is not successful, all corners will be let in."[170] To keep the cones of the spreaders from being visible, they are to be sealed with concrete, and for this reason the cones "are being pulled up tighter." To prevent discolored "wall surfaces," the formwork is kept "in contact with concrete" for one more day. Finally, to avoid the production of air bubbles, more homogeneous aggregates are selected.

The results of the interventions of the Fuller Company produce more or less important improvements, produced by the reinforced corners and by keeping the formwork in position one day longer, while the efficacy of the more homogeneous aggregates is not proven. The solution of the spreaders has led only to partially satisfying results: "The supplier is trying to obtain more accurately formed tie assemblies." "Keeping forms tight one extra day has helped."[171] Other tests on the same technical particulars, but also with different solutions, are conducted at the "Service Towers at col. 1 north and col. 13 south."[172] "Corner forms for the above towers were not let-in, but the gasketing used was fairly effective in preventing leaks. To insure perfect corners all remaining corners will be let in." Where the "cones covered with concrete" are concerned, it is reported that the Burke Concrete Accessories, Inc., "has taken sample cones and has asked the manufacturer to supply all additional cones to exact dimensions."[173]

At the start of October the Riverside Cement Company is asked to perform chemical analyses. The efflorescence turns out to be calcium hydroxide, usually formed precisely following evaporation of the water of the

to William P. Roberts, 6 November 1963, LIK, 030.II.A.107.36.

[164] *Ibid.*

[165] Louis I. Kahn, "The Wonder of the Natural Thing. Interview with Marsahll D. Meyers, Philadelphia, Pennsylvania, August 11, 1972," in, Latour, *Louis I. Kahn, l'uomo, il maestro* cit., (pp. 399-404), p. 401.

[166] John E. MacAllister, meeting report 28 October 1963, LIK, 030.II.A.107.18.

[167] Charles H. Gruber, letter to Louis I. Kahn, 2 April 1964, LIK, 030.II.A.108.13.

[168] *Ibid.*

[169] *Ibid.*

[170] John E. MacAllister, report of meeting of 21 October 1963, LIK, 030.II.A.107.18.

[171] John E. MacAllister, report of meeting of 28 October 1963, LIK, 030.II.A.107.18.

[172] John E. MacAllister, report of meeting of 4 November 1963, LIK, 030.II.A.107.18.

[173] *Ibid.*

conglomerate. "In view of this report, the significance of the discoloration was relegated in importance to a mere matter of cleaning."[174] Also in October, the National Lead Company, a firm specializing in coatings for formwork, is asked for information on surface treatments of the plywood panels and the "possible reaction with the concrete,"[175] while an attempt is also made, still in October, to keep the formwork in position for a longer period after the pour in order to reduce the evaporation time of the water, and thus to limit eventual efflorescence. The report written by the National Lead Company "offered no satisfactory explanation for the discoloration."[176]

The expedient of leaving the formwork in place to prevent the production of a "discolored wall surface" continues to fail to lead to the desired results: "less discoloration was noticed, but it is still a problem." A report is prepared by the Riverside Cement Company on this matter.[177] The reduction of air bubbles, on the other hand, has been achieved by means of an "external vibration."

The search for concrete of maximum strength for the columns of the Laboratories produces some results. From the trials conducted on different types of cement, including Colton Cement, the compounds made with Riverside Cement are the only ones that have difficulty in reaching 6000 lb/sq.in.[178] Therefore it is decided to conduct trials with samples made with Riverside Cement, pea gravel of different origins and additives to control setting time (Pozzalith or Plastiment). If these combinations still produce unsatisfactory results, use of Colton Cement is proposed. Also in October, testing is done on the strength of the compound made with Riverside Cement and Pozzalith for the columns of the Laboratories.[179]

The documents show that during a worksite visit on 4-5 November, Komendant agrees on the need to conduct trials to measure the heat generated during the setting phase of the concrete (Cement Heat Test), and he requests information on the tests already done and the types of cement involved (Riverside Cement and Pozzolan Cement), in order to understand the reasons behind the phenomena of "discoloration and crazing of the concrete walls in the Service Towers."[180] "He suggested," it is his opinion, "that perhaps the cement was unstable and that 'aluminum was separating out'"[181]. Later, Komendant was to write: "too much pozzolan and improper vibration. [...] the vibration has been too close to the surface; water, dirt, and pozzolan have been drawn to the surface. Water evaporated very quickly, which is only natural in such a dry and hot climate, causing dry shrinkage and resulting in severe crack formation and dusting of surfaces."[182]

Formwork and mix design testing confirm the ongoing pursuit of an ideal type of concrete, even when construction is already in progress. These tests significantly boost construction costs, and Salk asks for an estimate of the total cost of his building. MacAllister writes to him about this matter: "As you know, we agree with your opinion that the estimated cost of concrete,

[174] Charles H. Gruber, letter to Louis I. Kahn, 2 April 1964, LIK, 030.II.A.108.13.
[175] *Ibid.*
[176] John E. MacAllister, report of meeting of 28 October 1963, cit.
[177] John E. MacAllister, report of meeting of 4 November 1963, cit.
[178] John E. MacAllister, report of meeting of 18 September 1963, LIK, 030.II.A.107.18. See also, John E. MacAllister, report of meeting of 3 September 1963, LIK, 030.II.A.107.18.
[179] John E. MacAllister, report of meeting of 14 October 1963, LIK, 030.II.A.107.18.
[180] John E. MacAllister, letter to the George A. Fuller Company, 7 November 1963, LIK, 030.II.A.108.31.
[181] Charles H. Gruber, letter to Louis I. Kahn, 2 April 1964, cit.
[182] Komendant, *18 Years with Architect Louis I. Kahn* cit., p. 61.
[183] John E. MacAllister, letter to William P. Roberts, 6 November 1963, LIK, 030.II.A.107.36.
[184] *Ibid.* The data are supplied on 14 November 1963 and have to do with Kahn's buildings in Philadelphia – the Richards Medical Research Laboratories,

in place, is excessively high. Mr. Kahn has designed several exposed concrete buildings which may be a source of cost information."[183] Then he asks Kahn's studio for "cost data on poured-in-place concrete" including costs for "formwork, stripping, reinforcing steel, etc." of Kahn's works.[184]

Following other tests on the concrete compound, it is found that the change of the aggregates from Crestlite to Ridgelight improves the strength of the concrete.[185] MacAllister thinks this change can reduce the defect still observed in the "lightweight concrete in the Service Towers": it "is still showing a blotchy surface."[186]

Modifications to the specifications, December 1963

Certain corrections to the formwork system lead to changes in the arrangement of the parts that leave imprints on the concrete of the Salk Institute. The fact that the Fuller Company, at this point, has to always consult Kahn and his staff for every modification with respect to what is indicted in the drawings of the wall patterns can be seen in an episode in September 1963, when the company proposes a different positioning of the "construction joints," which is not approved by Kahn.[187]

After the problems encountered in the first construction phases, in October the responsibilities and procedures are revised for the approval of the designs of the formwork, done by Matt, for the Fuller Company, and by Langford for Kahn. "Mr. Matt will prepare preliminary sketches of the formwork and will review them with Mr. Langford."[188] Charles H. Gruber of the Fuller Company is also involved in the development of the formwork. The documents reveal Kahn's constant control over all the modifications proposed by the technicians of the Fuller Company. "Mr. Gruber agreed to revise the present formwork details for panel joints. The Architect is not satisfied with the present system because of the excessive leakage, at this point."[189] The complex process of design of the formwork is documented by the epistolary exchanges, for example those of the month of October, between Kahn's studio, Fuller Company and Komendant. Kahn's office indicates the designs of the walls of the staircases of the Studies and sends all "formwork pattern information" to Komendant, who proceeds to the "reinforcing for the formwork joints."[190] Once Kahn and Komendant have defined the system, the Fuller Company begins "to detail formwork."[191] Kahn's team (MacAllister in this case) also intervenes to make other changes to the design for the system of the staircases, in order to achieve a "simplification of formwork."[192]

The drawings, tests, prototypes and progress of the worksite require an updated version of the specifications for the chapters on concrete and formwork. In the second version of the section on "Forms" dating back to 23 December 1963, decisive modifications are introduced for the design of "forms for exposed vertical surfaces,"[193] impacting the formal quality of the concrete surfaces. Only in this second version of the specifications does

American Federation of Labor Medical Service Building and Eleanor Donnelley Erdman Hall, at Bryn Mawr.

[185] John E. MacAllister, report of meeting of 16 December 1963, LIK, 030.II.A.107.18.

[186] *Ibid.*

[187] Charles H. Gruber, letter to Louis I. Kahn, 1 October 1963, LIK, 030.II.A.108.31.

[188] John E. MacAllister, report of meeting of 3 October 1963, LIK, 030.II.A.107.18.

[189] John E. MacAllister, report of meeting of 7 October 1963, LIK, 030.II.A.107.18.

[190] Marshall D. Meyers, letter to August E. Komendant, 11 October 1963, LIK, 030.II.A.107.13.

[191] John E. MacAllister, report of meeting of 14 October 1963, cit.

[192] John E. MacAllister, letter to William P. Roberts, 6 November 1963, cit.

[193] "Specifications," n.d., modified on 23 December 1963, LIK, 030.II.A.107.19.

Kahn eliminate from the fabrication of his exposed concrete the panels made with wooden boards that were part of his initial interpretation of *béton brut*. In place of the "Utility Grade boards or planes" confirmed in the first version of the specifications of 1963, he now introduces plywood. "[...] Exterior type concrete form plywood, in 12 ft. lengths generally, and Utility Grade lumber of any western softwood," we read in the new version. Precisely the working drawings of the mechanical spaces and those of the retaining walls must have convinced him that large plywood panels have potential for the creation of large compartments on concrete walls, to replace the motif of the imprints of slender boards.

It is worth emphasizing the progression with which Kahn shifts from the elimination of stone slabs to that of the wooden boards to reach and bring out the use of the plywood panel, because it is precisely in this sequence that the latter assumes the value of a surrogate for the stone slabs conceived as incorporated parts of the formwork. The size of the plywood panel will leave an imprint on the concrete that evokes the image of the stone slabs that had accompanied all the phases of the design, and which Kahn had struggled to make a logical part of the construction process of the precast and poured in place concrete. Only in the light of the complex evolutions of the design of the concrete of the Salk Institute is it possible to grasp the secret that lies behind the particular nature of a concrete made, ideally, to incorporate the stone slabs as desired by Kahn.

110-111. Salk Institute for Biological Studies, La Jolla, California, 1959-65.

110. "North Laboratory Floor Slab Looking East," 28 February 1964 (LIK, 030.IV.A.540.10.9).

111. "Mechanical Wing Looking North," 28 February 1964 (LIK, 030.IV.A.540.10.11).

The point on the finish of the formwork, already modified in the first version of the new specifications, undergoes further modification in the second version precisely due to the introduction of the plywood panels. To obtain perfect corners, in this phase of the design of the formwork Kahn envisions the insertion between two consecutive panels of a strip, imprinting a cavity on the concrete that would transform the surfaces of a pillar or a wall into ideal juxtaposed planes. "In general," we read in the second version of the specifications, "forms shall be made up so that plywood does not abut plywood edge to edge, but is separated by wood strips as indicated, with form ties where required arranged in panels in a neat, regular pattern in accordance with approved shop drawings. Forms must be inspected and approved by Architect before concrete is poured."[194]

Another important consequence is produced by the change of the formwork system, once the stone slabs and wooden boards have been set aside: all the surfaces, and not just those that form the facades of the Laboratory Building, take on a similar value, since the concrete will be poured in formwork of the same material and the same measurements.

After the Service Towers, Studies and Laboratories, the definition of the patterns of the formwork continues, in late 1963 and early 1964, with the drawings of the North and South Mechanical Wings, which form the ends of the two Laboratories facing towards the main entrance. Kahn modifies the arrangement of the plywood panels, which are usually mounted vertically in the other parts of the complex. Now the narrow panels arranged vertically in the edges of the walls create the image of a colossal structural order, which unlike what was designed in the Service Towers, appears to be filled by masonry sectors designed by the imprints of the horizontally assembled panels.[195]

In the working drawings, indications are made on the composition of the formwork elements of the already constructed parts. "Tie holes must line up with existing noice rod holes in Service Tower," is written, for example, on 20 May 1964 regarding the panels of the formwork in the Laboratory.[196]

Worksite chronicles, January-March 1964: the pursuit of perfection

At the start of 1964, in preparation for the work to be done starting from the ground floor and the most representative parts of the complex, further indications are made regarding the responsibilities of the various representatives on the worksite. Langford's role of supervision of formwork and the quality of the concrete is confirmed.[197] The components of the walls made with concrete blocks are also subjected to careful study, again with the aim of obtaining a particular surface texture. In January and February the concrete block sample is indicated, for "color and texture,"[198] as well as the type of working of the joints: "all joints in concrete block walls will be struck with a 'grapevine' tool."[199]

The problem of discoloration of concrete that arose in September 1963 continues to keep the technicians busy. To eliminate the efflorescence, three solutions are tested on one of the Service Towers, "at certain areas designated by the Architect":[200] cleaning with a product formed by water mixed with 30% hydrochloric acid, rubbing by hand with sea sand, and rubbing by hand with sandpaper.[201] The poor results prompt trials with two other methods: light sanding and brushing with a detergent that does "not spot glassware."[202]

[194] *Ibid.*
[195] See the drawing "North Mechanical Wing – North Elevation," LA-405A, 10 February 1964, LIK, 030.I.C.540.032.
[196] See the drawing "Service corridor interior," LA-375a, 25 January 1964, LIK, 030.I.C.540.031.
[197] "It is our recommendation that Mr. Langford be made responsible for daily inspection of the formwork, concrete placing and form removal. His direct responsibility would be to determine if all necessary precautions were being taken to ensure the highest quality 'architectural' concrete. Mr. Langford has been inspecting the concrete work on a part-time basis and will soon be available (after completation of form drawings) on a full-time basis" (John E. MacAllister, letter to William P. Roberts, 9 January 1964, LIK, 030.II.A.107.36). In February the same text is modified, also with the addition on the reuse of panels, again to be supervised by Langford. "Economy, realized by re-use of form panels, may be made possible by adjustment of presently detailed formwork. Where these adjustments effect the Architectural requirements of the concrete, Mr. Langford will cooperate with you to take advantage of any suggestions leading to cost savings to you, and in turn to the Owner" (John E. MacAllister, letter to Charles H. Gruber, 5 February 1964, LIK, 030.II.A.108.31).
[198] John E. MacAllister, letter to William P. Roberts, 29 January 1964, LIK, 030.II.A.107.36.
[199] John E. MacAllister, letter to J. L. Wills, 5 February 1964, LIK, 030.II.A.108.31.
[200] Charles H. Gruber, letter to Louis I. Kahn, 2 April 1964, cit.
[201] Report of the meeting of 19 February 1964, LIK, 030.II.A.108.46.
[202] *Ibid.*

112-114. Salk Institute for Biological Studies, La Jolla, California, 1959-65.

112. "South Laboratory Floor Slab Looking West," 31 March 1964 (LIK, 030.IV.A.540.10.32).

113. Photograph of the construction of the North Laboratory Wing floor slab (LIK, 030.IV.A.540.4.59).

114. "1st Laboratory Floor North Wing," 30 July 1964 (LIK, 030.IV.A.540.11.9).

But the concrete discoloration is not completely eliminated, according to the Fuller Company: "Some of the above methods were effectual in removing the blotches, but disturbed the desired finish – the cure was worse than the disease – others proved unsatisfactory entirely."[203] So the recommendation is to try still other solutions.[204]

The Portland Cement Association and all the suppliers of components of the concrete of the Salk Institute are informed of the problem of the blotches and asked to provide information to the Riverside Cement Company.[205] Given the fact that the question of the stains continues to remain without a solution, MacAllister threatens to suspend pouring operations if measures are not taken to prevent the poor results.[206] The Fuller Company, on 13 March, in collaboration with the chemists of the Riverside Cement Company, conducts trials on the concrete to find out if the cause should be sought in the compound itself.[207]

Komendant is at the worksite to check "form work, reinforcing, post-tensioning tendons and placing of concrete for the first section (5 sections total) of the laboratory utility floor including the bottom chord of the two Vierendeel trusses."[208] After this visit, he indicates the modes for pouring the concrete in the floor slab and the truss, to avoid the production of cavities.[209] The Fuller Company suggests stripping of the Vierendeel trusses in three

203 Charles H. Gruber, letter to Louis I. Kahn, 2 April 1964, cit.
204 J. L. Wills, report of meeting of 26 February 1964, LIK, 030.II.A.107.18.
205 Charles H. Gruber, letter to Louis I. Kahn, 2 April 1964, cit.
206 See John E. MacAllister, "Architect's inspection report No.2," 9 March 1964, LIK, 030.II.A.108.44.
207 Charles H. Gruber, letter to Louis I. Kahn, 2 April 1964, cit.
208 August E. Komendant, letter to Louis I. Kahn, 11 February 1964, LIK, 030.II.A.108.18.
209 "Placing of concrete for bottom chord of truss must be done after in the floor slab, on both sides of the chord, at least a width of two feet is poured to avoid cavities in bottom chord. The ends of the bottom chord around and behind anchor plates must be re-vibrated (1/2 hour later) to reduce internal stresses in concrete and increase strength of concrete. In bottom chord the bars must be thoroughly after-packed at construction joint (very high shear) in addition to at least two shear keys" (August E. Komendant, "Special Instructions for Salk Institute Construction," 11 February 1964, LIK, 030.II.A.108.18). On the operations of vibration corresponding to the

days, and the use of "membrane curing": "This will obviate a continuous mess on the working deck."[210] A recipe is then supplied, probably by Komendant, for filling the cracks in the floor slabs of the laboratories.[211]

Still after the pouring, MacAllister notices cracks that have formed in the floor slab and proposes changes to the operations: "While nothing can be done to remedy this situation all future slabs should be poured in smaller increments of alternate pours";[212] he orders the Fuller Company to remove "a dark brown stain inside the Service Tower at col. ref. #7, El. 343.00" and to report on the causes of its formation; he criticizes the imperfect deployment of the neoprene joint, that has generated "very rough corners"; he recommends proper sealing of the formwork joints, saying that "more care should be taken in form stripping"; finally, he notices "excessive leakage" in the "east wall in the Medical suite" which "had scarred the base along its entire length."[213]

The biggest problems during pouring happen in the openings left in the floor slab by the "ceiling slots" in aluminium, for the passage of physical plant systems, in the North Wing of the laboratories.[214] At the beginning of March, at the position of the ceiling slots, "hardwood edge strips designed to eliminate grout pockets" are mounted, and the concrete is poured.[215]

MacAllister's final assessment of the latest pouring operations offers a glimpse of improvement in the making of the concrete: "The general level of

Vierendeel trusses see, David P. Wisdom, report of meeting of 28 February 1964, 2 March 1964, LIK, 030.II.A.108.18.

[210] Charles H. Gurber, letter to Louis I. Kahn, 24 February 1964, LIK, 030.II.A.108.18.

[211] See David P. Wisdom, report of meeting of 28 February 1964, 2 March 1964, cit.

[212] John E. MacAllister, "Architect's inspection report No.1," 24 February 1964, LIK, 030.II.A.108.44. On this point, Komendant's opinion does not jibe with that of MacAllister: "In case the base is smooth, reinforcement at proper place, the curing adequate and the joints cut at the earliest possible time, there will be no need to pour the slabs in small increments alternately. Pouring of the floor slab in sections is possible only if premolded joints are used, which is very difficult to perform. Up to date practice prefers cut joints" (August E. Komendant, letter to Louis I. Kahn, 5 March 1964, LIK, 030.II.A.107.13).

[213] John MacAllister, "Architect's inspection report No.1," 24 February 1964, LIK, 030.II.A.108.44.

[214] *Ibid.*

[215] John E. MacAllister, "Architect's inspection report No.2," 9 March 1964, LIK, 030.II.A.108.44.

115-117. Salk Institute for Biological Studies, La Jolla, California, 1959-65.

115. Formwork with reinforcing in a laboratory wall (LIK, 030.IV.A.540.5.45).

116. Formwork of a laboratory wall (LIK, 030.IV.A.540.5.94).

117. Detail of a laboratory wall (LIK, 030.IV.A.540.3.4).

concrete construction is excellent, especially in the most recent pours in the Service Towers. This same level of excellence, however, is not evidenced in the Mechanical Wings. Also, constant attention must be paid to all details to insure a consistent level of quality throughout this construction."[216]

The pouring operations completed at the beginning of March reveal the defects of the formwork system with the "parting or juncture strips" and V joints. The edges of the panels, when reassembled after their initial use, are worn, causing imperfect shaping of the V joints. "Further inspection of form panels indicates that the 'V' joints are not adequately moisture sealed. The raw edges of the plywood exposed from re-use and erosion of the coating appears to be the main cause of loss detail on the concrete. These edges should be thoroughly cleared and re-sealed before forms are re-used. Also, additional sealing before the first use would help to retain this detail. The West wall forms for the corridor at El. 342 between laboratory buildings need all 'V' joints re-sealed."[217]

Probably to prevent wear on the V joints, in May a variant of the "coupler block" is developed, with a shaped head to apply to the support.[218]

The stripping of certain Vierendeel trusses reveals imperfect finishes ("the trusses are badly spalled in several instances"[219]). To prevent this problem, it is recommended to proceed with "judicious removal of formwork in these areas."[220] In general, MacAllister wants to obtain a finish quality of the concrete that is better than what has been produced to date in the service spaces between the Vierendeel trusses: "Although the finish of the interior of the Pipe Spaces does not have to be of equal quality to the public

[216] John E. MacAllister, "Architect's inspection report No.1," 24 February 1964, LIK, 030.II.A.108.44.
[217] John E. MacAllister, "Architect's inspection report No.2," cit.
[218] See the drawing "Studies - panel details," LA-381, 20 May 1964, LIK, 030.I.C.540.031.
[219] John E. MacAllister, "Architect's inspection report No.2," 9 March 1964, LIK, 030.II.A.108.44.
[220] *Ibid.*
[221] *Ibid.*
[222] John E. MacAllister, "Supplement to Architect's inspection report No. 2," 10 March 1964, LIK, 030.II.A.108.44.
[223] See John E. MacAllister, letter to Charles H. Gruber, 6 April 1964, LIK, 030.II.A.108.31; Louis I. Kahn, letter to August E. Komendant, 10 April 1964, LIK, 030.II.A.107.13; August E. Komendant, "Special Instructions for Post-tensioning of Structural Framing," 11 May 1964, LIK, 030.II.A.107.13; John E. MacAllister, letter to Charles H. Gruber, 20 May 1964, LIK, 030.II.A.108.31. Actually, steps are taken to increase the strength of the

areas in the building, the formwork should be more carefully done than that which is in place at the time. Used, oiled plywood is mixed with plastic coated plywood in an unplanned pattern."[221] MacAllister complains that the columns of the upper laboratory in the South Mechanical Wing have rock pockets at the bases and the corners. "[...] form drawings were never submitted for these columns," he adds.[222] Compression testing on the lower chords of the three Vierendeel trusses, numbers 6, 7 and 8 in the North Laboratory, shows that the concrete is not sufficiently strong, inducing Dorland and Komendant to call for its demolition.[223]

The upper part of the external walls is "unacceptably rough," threatening the formal result of the joint with the next pour of the floor slab: "More care must be taken to ensure a neat finish at the intersection of all wall and slab planes."[224] MacAllister observes improper use of "oil coated formwork" for the walls of the vestibule "outside the Incinerator room" in the South Mechanical Wing: "The finish schedule calls for plastic coated formwork in this area."[225] The corner joints of the formwork in certain masonry sectors of the Studies "are not correctly located."[226]

The complexity of Kahn's formwork system generates the initial misunderstandings regarding the assembly of the panels.

The concrete of the north wall in the South Mechanical Wing, between the ground and first floors (Elevations 363 and 374), has been poured in formwork whose design does not match that of Kahn. "The plywood joints and wall ties are not placed correctly," MacAllister warns,[227] also pointing out that "patched plywood was used for those forms,"[228] while indicating errors in the treatment of the holes: "form tie holes are grouted above grade where they should be packed with lead."[229]

The formwork of the concrete in certain walls of the North and South Office Wings have not been assembled according to Kahn's design[230]: "1. Pour strips were not used to top out pours. 2. Intermediate walers were not used at all door openings. 3. Pour strips continue into door openings. 4. Rubber gaskets were not used. 5. Pour strips were not used on vertical cold joints."[231]

Careful supervision of every detail in order to make a perfect "exposed Architectural concrete" leads MacAllister to challenge the fabrication of the ends of the walls of the North Laboratory Building, poured on 10 April. "Please indicate to us in writing what measures you will take to ensure that all future construction is done in accordance with approved drawings," he writes to Gruber of the Fuller Company.[232]

In all ten of the Service Towers of the two wings of the laboratories the concrete of the masonry sectors of the main facades with the doors has been poured in formwork not perfectly corresponding to the drawings. What has been altered is the system of juncture strips straddling the first two horizontal grooves. "The 4th edge strip top and bottom was left off completely on the outside face of all Service Towers," MacAllister writes.[233]

already made trusses, without demolishing them.
[224] John E. MacAllister, "Architect's inspection report No.2," 9 March 1964, cit.
[225] John MacAllister, "Supplement to Architect's inspection report No. 2," 10 March 1964, LIK, 030.II.A.108.44.
[226] *Ibid.* This is the "Study Stair Retaining Wall at col. ref.#13."
[227] John E. MacAllister, letter to Charles H. Gruber, 10 April 1964, LIK, 030.II.A.108.31.
[228] John E. MacAllister, letter to Charles H. Gruber, 14 April 1964, (re: "Salk Institute for Biological Studies Concrete at Receiving Entrance"), LIK, 030.II.A.108.31.
[229] *Ibid.*
[230] John E. MacAllister, letter to Charles H. Gruber, 14 April 1964, (re: "Salk Institute for Biological Studies West Office Wing"), LIK, 030.II.A.108.31.
[231] *Ibid.*
[232] John E. MacAllister, letter to Charles H. Gruber, 10 April 1964, cit.
[233] John E. MacAllister, "Supplement to Architect's inspection report No. 2," 10 March 1964, LIK, 030.II.A.108.44.

Furthermore, the same mistake has been made in the South Mechanical Wing, regarding the sealing of the holes: "All form tie holes at the second lift, west wall, of Service Tower No. 1 South have been packed with grout. This is an exposed wall and the form tie holes are to be packed with lead."[234] On the other hand, all the imprints on the sides of the same towers comply with Kahn's design.

Precisely the two series of the Service Towers of the two wings of laboratories are cited as examples of the different methods applied by the carpenters in the assembly of the formwork, and progress is emphasized, also due to the uniformity of the patterns, which permits improvement in the implementation during the various phases. "The South Towers (Foreman - Bud Clibourne) and the North Towers (Foreman - Tony Kischassey)," Langford writes, "represent the best progress in wall-forming on the job. These Foremen have kept almost the same crews from the beginning of the job and there is evidence of a healthy competitive spirit between the crews of the North and South Towers. Excellent concrete is the result."[235] Instead, MacAllister is critical of the concrete work in the Office Wings, both North and South: "The general level of workmanship and resultant quality for these areas is far below that of the Service Towers."[236]

The construction of the border wall along the garden at the entrance, to build over the East Wall with the tests, and of the low wall bordering the garden towards the esplanade (over the East Wall of the Mechanical Building), confirms the constant supervision conducted by Kahn's team to achieve a completely designed architectural concrete. On 27 March, during the inspection of the formwork for the two walls, MacAllister and Langford halt the work, having noticed that the Fuller Company is assembling panels with tie holes without yet having the layout, since Kahn has not yet supplied the drawing of the pattern. In spite of the fact that on 30 March, MacAllister and Langford emphasize the need to halt the work, the company goes on assembling the formwork.[237] The drawing of the pattern Kahn wants for these walls, with indications of all the positions of the ties and the relationships with the holes of the different contiguous parts, is prepared on 1 April and serves as a guide for new formwork – the other formwork are completely dismantled.[238]

"Report on Concrete Work": progress, failures and utopias of the "construction process"

When the Salk Institute worksite has reached a point in which all the various aspects of the formwork and concrete have been tested, a critical assessment is made to orient the work on the other parts, "to suggest modifications or review of certain work" and to achieve the goal of "better quality architectural concrete."[239] It is clear that the work is rising from the ground marked not only by the imprints of the formwork and the pours, but also by all the attempts to create the ideal concrete desired by Kahn, which have left traces on the construction of the foundation parts: "The Mechanical Building and Wings can be excused on the lower levels for the many variations in techniques because this was the Builder's laboratory for concrete experimentation."

This is the context in which the "Report on Concrete Work" appears, prepared on 23 March 1964 by Langford for Kahn and organized in chapters: "Formwork Drawings"; "Formwork Make-up"; "Formwork Setting"; "Placing of Concrete"; "Formwork Stripping"; "Curing of Concrete"; "Cleaning of Concrete." After Langford's letter in April 1963, the "Report" is another fundamental assessment on the construction of the concrete and the progress of the worksite, written in an impassioned tone both to recognize positive progress and to criticize technical misunderstandings. This document underlines the fact that the whole play of lines staged by Kahn is aimed at the depiction of the "construction process." After all, Langford's letter and the "Report on Concrete Work" are simply other visible signs, just like the grooves or relief lines left on the concrete, of that "process" that interests Kahn because it leaves trails of the phases of growth, as if the building were a natural organism, like those studied by D'Arcy Wentworth Thompson, and perhaps because the process is also an expression of a more general American artistic form in which the value of the final work lies precisely in its fabrication, documented through films, photographs and interviews, as well as the traces left on the work itself.

"The form setting process," Langford asserts, "has been the phase that most effects the Architect's decisions on sheet layout and panel sizes. From the beginning of the job this phase has served to instruct Mr. Kahn in his desire to make an Architectural expression of the construction process. Many experiments have led the Builder along a path of constant improvement."[240]

The pages of the "Report" contain praise for the technicians of the Fuller Company who have executed the working drawings of the formwork (Carleton's name is mentioned), and the workmen who built them. "[...] the form yard has produced over 200,000 square feet of formwork. The men who work in the form yard should be congratulated for their excellent

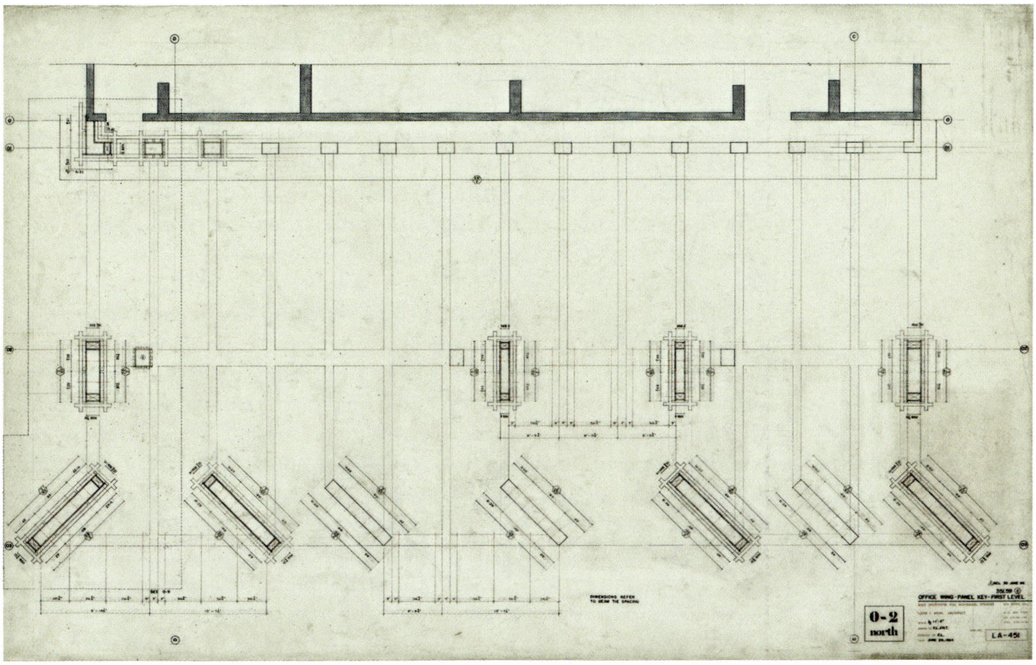

118. Salk Institute for Biological Studies, La Jolla, California, 1959-65. "Office Wing – Panel Key – First Level," LA-451, 26 June 1964 (LIK, 030.I.C.540.32).

[234] John E. MacAllister, letter to Charles H. Gruber, 14 April 1964, ("Salk Institute for Biological Form Tie Holes"), LIK, 030.II.A.108.31.

[235] Frederick Langford, "Report on concrete work," 23 March 1964, LIK, 030.II.A.107.7.

[236] John E. MacAllister, letter to Charles H. Gruber, 14 April 1964, ("Salk Institute for Biological Studies West Office Wing"), LIK, 030.II.A.108.31.

[237] Charles H. Gruber, "Weekly Job Meeting – April 15, 1964," LIK, 030.II.A.108.46.

[238] "Mechanical Building - Elevation - E. Wall Roof Garden Parapet - Lightwell," LA-416, 1 April 1964, LIK, 030.I.C.540.032.

[239] Frederick Langford, "Report on concrete work," 23 March 1964, cit.

[240] *Ibid.*

119-120. Salk Institute for Biological Studies, La Jolla, California, 1959-65.

119. "North Office Wing," 30 July 1964 (LIK, 030.IV.A.540.11.5).

120. "North Mechanical Wing – North Elevation," LA-405A, 10 February 1964 (LIK, 030.I.C.540.32).

workmanship, and overall efficiency of this operation, in spite of the challenging demands the Architect has imposed upon them. Much credit is due to the imagination, enthusiasm and teaching ability of Merle Prichard, Foreman of this phase of the work."[241]

After the experiences of the construction in progress, the various phases of assembly of the parts of the formwork and their finishing are again planned, to optimize the work and its results. The previous organization of the fabrication of the formwork with two work levels for the operations of cutting, beveling and cladding of the panels is subjected to revision. Now Langford proposes making the working of the slabs of the first layer of the formwork independent, bringing the pieces to the assembly level only after the beveling of the edges. The slabs would be stored, depending on size, in specific stacks to await assembly. The plastic film to apply to these slabs where they come into contact with the pour, namely the "plastic coating," is to be done only when the holes for the ties have already been made. "Under the present system, the sealer is applied on the assembly table and the plastic is applied in the field."[242] The fabrication of the "edge-coating" as performed until now, can be revised, in Langford's view, as can the phases of assembly and stocking of the formwork. Langford also suggests assembly of panels and support structures prior to deployment, and storage of formwork panels standing vertically instead of horizontally, as done at present, to save space on the worksite and to avoid the accumulation of dust on their surfaces.

The "Report on Concrete Work" contains an accurate description of all the defects that have appeared in the experimental phase of the formwork and pouring of the first two levels of the two wings of the Mechanical Building. Langford touches on all the main errors, from failure to mount all the necessary horizontal reinforcements, or walers, to the imperfect passages between pours of walls and slabs, while he underlines the sectors of the complex in which, together with the Service Towers, it is possible to observe the difficult process of evolution in the control of the concrete quality.

"Much of the work in the Mechanical Wings above 343 has been very disappointing. The Mechanical Wings have been plagued with the same basic form setting problems as the Service Towers. The Service Tower Foremen have shown great progress in overcoming these problems, partly due to the advantage of repetition. The Mechanical Wings with their many sleeves and blockouts, and lack of repetition make progress more difficult. The Mechanical Wings have many corners that are ragged, a large percentage of V-joint projection has been broken off, door openings in many cases show excessive bleeding because intermediate walers were not used. Block-outs for grilles show careless workmanship. Segregation has occurred along 95% of all walls that used sill-on-slab type forms, wall to ceiling junctures in some cases show a lack of care, and North and South Mechanical stair walls represent some of the worst Architectural Concrete on the entire job. The problems of form setting for the Office Cores are very similar in principle to the problems of Service Towers. Each Office Core lift with its elevator shaft, lobby, stairwell, and toilet room, is like an elongated version of a Service Tower. The exterior concrete downspouts and the expansion joint haunch are the only additional features. The entire East and West exterior walls are shaftways and require no special finish. Yet, with all this similarity, and with the opportunity of using the knowledge gained from four months' experience on the Service Towers, the office Cores are like a brand new growth on the site stumbling through their own evolutionary mishaps."[243]

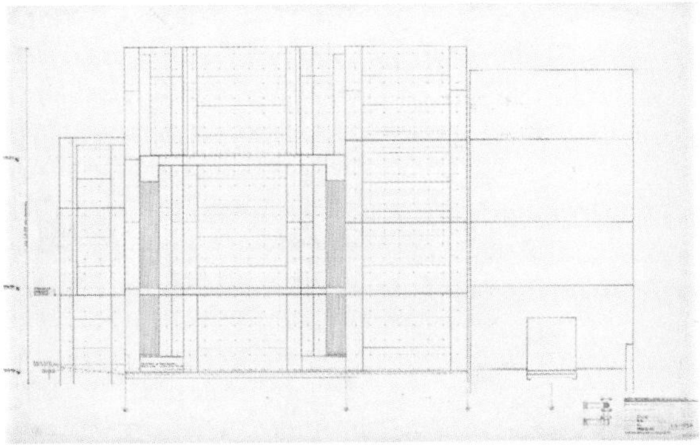

So, point by point, the "Report on Concrete Work" lists all the constructive details related to the imprints that have been made without correct interpretation of the drawings and indications of Kahn and his staff. It is clear that the complex system of lines envisioned by Kahn has generated misunderstandings among the craftsmen of the Fuller Company. In no American manual on formwork, not even in *Formwork for Concrete* by Hurd, are the making of formwork and the pouring of concrete taken to the level of a true philosophy of the construction process of exposed concrete, along the lines of what is being proposed by Kahn and his team.

The litany of errors is written in the form of a list of rhetorical questions: "Where are the pour strips that top out every pour? Where are the intermediate set of walers that are used on all door openings? Why is so much V-joint detail broken off? Why is there excessive bleeding at certain coupling blocks? Why does the pour strip continue into the door openings past the adjacent coupling blocks? If door lintels had to be a separate pour, why wasn't this detail worked out with the Architect's office? Why wasn't the rubber gasket that is used on every lift of the Service Towers used on the second lift of these Service Cores? The forms for the third lift are going up on the North Office Core and still no rubber gasket is in sight. Why weren't pour strips or at least rubber gaskets used on vertical cold joints and intermediate horizontal cold joints?"[244]

In the "Report on Concrete Work" a different organization of the work of carpenters and masons is hypothesized, in which prior to the construction or assembly of the parts, the philosophy of Kahn's formwork system would be explained, helping the individual workers to understand the cultural reasons behind that system, and to be able to take part, as protagonists, with their own proposals for improvement. The "Report on Concrete Work" makes it clear that the concrete of the Salk Institute cannot achieve the goal of absolute truth sought by Kahn unless all the workers are aware of being actors in that very system. Precisely in the light of the sections of the "Report" on the craftsmen and workers, Kahn's concrete reveals an intention to be the vital casting of all the energies that drive the worksite, crystallized in eternal form in its bulk, like those fossils captured in the magma of the Cordova stone. Only by transforming the worksite can his concrete become vital process-generated material, made eternal: an organic sculptural concretion facing the ocean, born of the efforts of a working community.

[241] *Ibid.*
[242] *Ibid.*
[243] *Ibid.*
[244] *Ibid.*

"Every Foreman on this job seems to be conscientious," writes Langford, "with a sincere desire for quality work. The reason the Mechanical Wings and Office Cores are below par in form setting workmanship is because of an inexcusable lack of educational communication between the various Foremen of each area of the job. Each Foreman who is in charge of form-setting, in addition to being completely familiar with his particular area of the buildings, should also be made aware of all forming problems, and educated to the respective solutions that have occurred in every area of this job. Perhaps each Foreman should be encouraged to submit sketches and ideas to the Superintendent for consideration. The large number of men on the job necessitates the formulation of a kind of sign language, which can be interpreted from one crew to another. For instance, the pour strip which tops out each pour should occur everywhere regardless of whether the wall is exposed or not. In the same way the carpenters should indicate on the forms as they are buttoned up, which forms should be removed first in the stripping process."[245]

After the "Report on Concrete Work," in any case, other episodes of lack of coordination between the contractor and the architects still occur, in the making of the formwork and the pouring. MacAllister and Langford often have to intervene, to halt the assembly of formwork without the patterns approved by Kahn or to indicate execution that does not comply with the drawings.[246]

Towards a new concrete surface

Following many unsuccessful attempts on the La Jolla worksite to prevent the formation of marks on the concrete surface, on the day after the preparation of the "Report on Concrete Work," 24 March 1964, the decision is made to discuss the question of the "discoloration of concrete" in a special meeting held on 2 April, where Komendant, MacAllister and the representatives of the Riverside Cement Company are informed of the Fuller Company's decision to assign the study of the phenomenon and the identification of possible remedies to the Smith-Emery Company of Los Angeles, a firm specializing in analysis of the technical problems of structures, especially those in reinforced concrete.[247]

The technicians of the Smith-Emery Company make careful worksite inspections to survey the most problematic discolored areas; they study the

121-122. Salk Institute for Biological Studies, La Jolla, California, 1959-65.

121. Photograph of the construction of the North Mechanical Wing (LIK, 030.IV.A.540.4.17).

122. "South Study Towers Looking Southwest," 31 December 1964 (LIK, 030.IV.A.540.12.10).

formwork system and the characteristics of the plywood panel surfaces; they analyze the individual components of the concrete compound, including the water used on the worksite, their proportions, and the methods of fabrication and deployment of the compound; they take samples of the concrete surfaces for spectrographic analysis, conducting tests at the worksite and in their own laboratories.

The immersion tests in distilled water conducted by the Smith-Emery Company on samples of pure pozzolan, cement with pozzolan additive and Portland cement alone, particles form when the pozzolan is present. Laboratory analysis to discern the possible presence, in the concrete aggregates, of calcium oxide (usually the cause of common efflorescence on concrete surfaces), shows that the "Calcium Oxide content of the aggregates, even if liberated, would not be the primary cause of the discoloration." The analysis of the Pozzolith and the substances contained in the water used on the worksite to form the compound also tend to exclude any effect, on their part, in the discoloring of the surfaces.

To find the water-soluble substances and colorants contained in the plywood of the formwork and the plastic coating film in contact with the concrete with pozzolan additive, the Smith-Emery Company subjects the two materials to the tests indicated by Method A of "Cold Water Solubility" (calculation of the percentage of tannins, gums, sugars and coloring matter contained in the wood and the plastic film). The tests show that sugars, which might cause the discoloration phenomenon, are present in the plastic film solution, but with too low a percentage "to be a significant factor in the form discoloration," while the sugars found in the plywood can have no influence in terms of discoloring if the panels are coated with the plastic film. The Smith-Emery Company does the same testing on a plywood panel with veneered face, then covered with a plastic membrane (but the percentage of soluble substances is greater than in the plywood with the plastic film alone).

Also to test the effects of the surfaces of the panels on the pour, the Smith-Emery Company makes concrete wall samples using formwork of different materials: metal, natural plywood and plywood with plastic coating. In some of these samples the concrete is subjected to vibration, both inside and outside the formwork, to test the possible effect of the work on the discoloring.

"Initially," the report reads, "all samples in the saturated state showed a slight deposit of material, especially around the 'bug' holes usually present in all concrete. Three samples of this concrete plus a 'blank' sample using lightweight aggregate and plain Portland Cement were subjected to a leaching action using distilled water. All three Portland Cement-Pozzolan samples displayed an appreciable amount of an encrustation of material which was then dried and analyzed. [...]. The 'blank' sample of plain cement concrete did not leach any material."

The Smith-Emery Company has some sample masonry sectors made on the worksite, "two test panels along with a portion of tower concrete," with the goal of identifying working processes that can eliminate the discoloration. The concrete components are mixed to make the compound

[245] *Ibid.*
[246] "We would expect that your awareness of our desire for the highest quality Architectural Concrete would be more acute" (John E. MacAllister, letter to Charles H. Gruber, 15 April 1964, LIK, 030.II.A.108.31).
[247] Charles H. Gruber, letter to Louis I. Kahn, 2 April 1964, cit.

123. Salk Institute for Biological Studies, La Jolla, California, 1959-65. Photograph of the construction of a Study Tower (LIK, 030.V.D.18.6).

more workable. "A revision was made on Job Mix 64-C-6 using 3/4-inch Crestlite Aggregate and Caudell-Johnson sand. The sand fraction was increased, along with the total water and cement-pozzolan combination, maintaining the water-cement ratio." Given the fact that vibration has also been indicated as an operation that influences the discoloring, to test its effects one of the panels is subjected to "extreme over-vibration," and the other to "under-vibration." "Both test panels exhibited evidence of the same discoloration present on other work on the project."

At the end of April the Smith-Emery Company is already capable of informing the Fuller Company of certain conclusions of its research, explained to Gruber, MacAllister and P. William Roberts, the engineer representing the Salk Institute, at a meeting on 1 May. But the official report, signed by the engineer George Battey and entitled "Report of special investigation," is dated 21 May.

The Smith-Emery Company ascertains that the efflorescence on the surfaces of the concrete of the Salk Institute is composed of calcium and silica with percentages of aluminium, magnesium, iron, sodium, potassium and small quantities of other elements. These are the chemical components of pozzolan. "These spectrographic analyses," the conclusion reads, "indicate that the chemical composition is a combination of the Pozzolan and the result of Portland Cement hydration, which produces an abundance of CaO." The Fuller Company had already summed up the findings of the Smith-Emery Company at the meeting of 1 May: "the blotching is caused by Pozzolan cement which was deposited on the surface."[248] Komandant claims that he was the one to discover the cause of the discoloration. "The problem is rather simple too much pozzolan and improper vibration," he recalls having stated at one of the meetings held in 1963.[249]

The paragraph on "Construction Practices" of the "Report of special investigation" is a fundamental document for comprehension of the technical processes of fabrication of concrete on the Salk Institute worksite, and the formal effects for the surfaces.

"An inspection of the job concrete surfaces shows localized but widespread incidence of discoloration. These blemishes appear to be in areas of greatest restraint of the forms. The restraint consists of studs, walers, tie rods, and strongbacks as required to hold the forms in place until the concrete hardens. During usual placing operations it is desirable to have the water-cement paste be forced to the form face to produce the desired surface texture. Physically, this involves the movement of water through the plastic concrete to the surface. This water tends to collect and to form 'bug' holes in the areas of the greatest form restraint. This would indicate that in the normal mechanical handling of the concrete in the forms the discoloration is showing up where water is accumulating and cannot be worked back into the concrete. These form areas do not have the same harmonic characteristics as those areas of less restraint during vibration. Where less restraint is present, some or all of the water is worked back into the plastic concrete and reduces the incidence of discoloration. The test panels cast with a more workable mix showed the majority of discoloration around the periphery, which was the location of greatest restraint."

[248] John E. MacAllister, "Minutes of Meeting – May 1, 1964," LIK, 030.II.A.107.18.

[249] Komendant, *18 Years with Architect Louis I. Kahn* cit., p. 61.

124. Salk Institute for Biological Studies, La Jolla, California, 1959-65. Photograph of the construction of North Office Wing (LIK, 030.V.D.18.11).

"Based on our study of this problem," we read in the conclusions of the Smith-Emery Company, "we believe the discoloration is a result of a combination of conditions naturally occurring in and the result of the Portland Cement-Pozzolan combination, the normal movement of moisture through the concrete, the energy required to compact the concrete, coupled with the use of almost totally impermeable forms. None of the materials used, in our experience, has singly caused this type of a discoloration. However, the conditions listed have compounded themselves to adversely cause this discoloration. The phenomenon may be peculiar to the San Diego area due to materials and atmospheric conditions."[250]

The recommendations of the Smith-Emery Company to prevent discoloring have to do with the mode of preparation of the compound, mixing the pozzolan and the cement shortly before use, and a possible alternative form coating. Komendant recommends trying the use of panels that have been "steamed to see if this treatment might prevent the blotching."[251]

"Weekly report on concrete work"

After the pouring of the basement levels, with the Retaining Walls, the start of the construction of the walls of the Service Towers, and after all the tests and analyses to achieve a perfect surface, it seems increasingly clear that the worksite of the Salk Institute is destined to become an experimental place for the making of a type of formwork and a type of concrete that are utterly unusual in contemporary construction. Also for this reason, Kahn's office develops a system capable of recording all worksite operations, to offer an overview of the evolution of the quality of the concrete from the viewpoint of the results relating to its surfaces. Once again it is Langford who takes charge of organizing and managing this form of control over the concrete work, aided in the task by Wisdom.

Starting in March 1964 weekly tables are compiled on the operations of deployment of the formwork ("first side" and "closing side"), the pouring and stripping done in the various parts of the Salk Institute. Service Towers, Mechanical Wing, Laboratories, Office Cores, Office Wings, Studies and Studies Stairs are the "locations" reported in the tables, with indications of the levels and sections into which the plan is subdivided (from 0.3 to 16), to precisely identify the sectors where the work is being done. The information also includes how many times the formwork have already been used ("previous uses") and the duration of the protection with tarps of the wet concrete after stripping ("curing, pouring-day + 7"). The "remarks" area contains certain considerations on the concrete made in the various parts, including the quality of the pour, the operations of surface cleaning, the judgment of the results, improvements or worsening of the execution.[252]

Titled "Weekly report on concrete work," the tables are another method invented by Kahn's studio to control the progress of the making of the concrete and to complete the picture of the true "philosophy" of the formwork, outlined by Langford back in 1963 in "Design of form work for concrete"[253] and confirmed in the *Report on Concrete Work* of which the tables are the logical complement.[254]

[250] George Battey, engineer of the Smith-Emery Company, "Report of special investigation," 21 May 1964, LIK, 030.II.A.108.31.

[251] MacAllister, "Minutes of Meeting – May 1, 1964," cit.

[252] See, for example, "Weekly report on concrete work," 20 March 1964, LIK, 030.II.A.108.35.

[253] Frederick Langford, letter to Greer Ferver, 5 April 1963, LIK, 030.II.A.89.3.

[254] Frederick Langford, "Report on concrete work," 23 March 1964, LIK, 030.II.A.107.7.

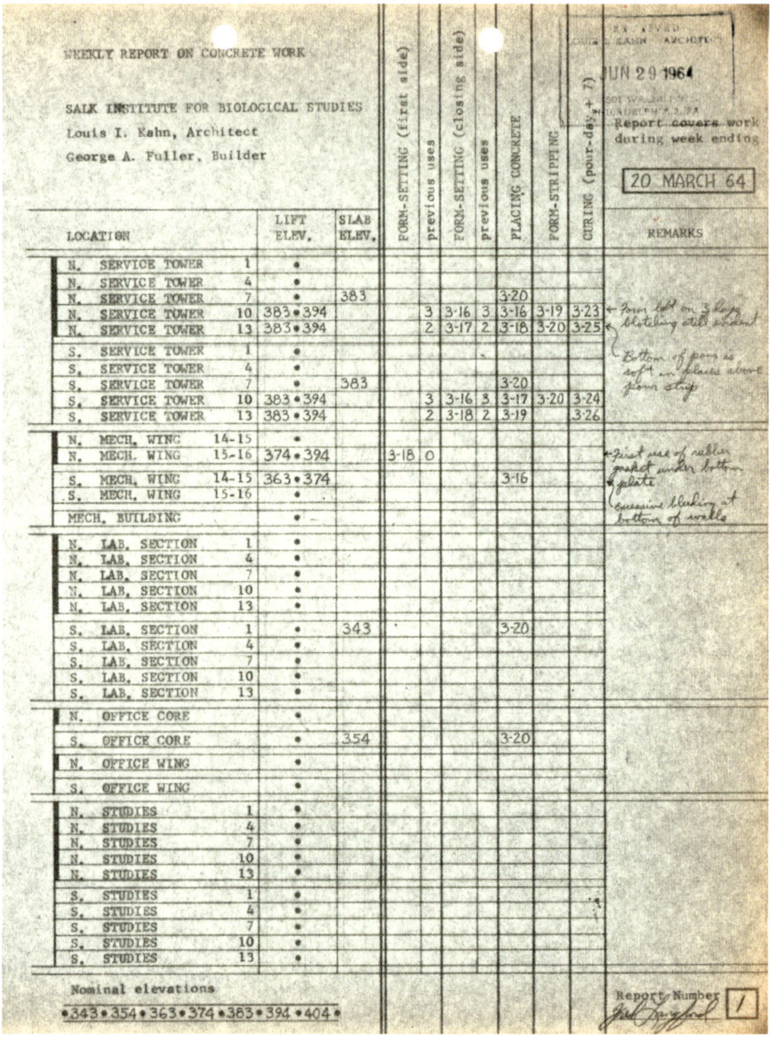
WEEKLY REPORT ON CONCRETE WORK

SALK INSTITUTE FOR BIOLOGICAL STUDIES
Louis I. Kahn, Architect
George A. Fuller, Builder

Report covers work during week ending 20 MARCH 64

RECEIVED LOUIS I. KAHN ARCHITECT JUN 29 1964

LOCATION		LIFT ELEV.	SLAB ELEV.	FORM-SETTING (first side)	previous uses	FORM-SETTING (closing side)	previous uses	PLACING CONCRETE	FORM-STRIPPING	CURING (pour-day + 7)	REMARKS
N. SERVICE TOWER	1	•									
N. SERVICE TOWER	4	•									
N. SERVICE TOWER	7	•	383					3-20			
N. SERVICE TOWER	10	383•394			3	3-16	3	3-16	3-19	3-23	← Form left on 3 days
N. SERVICE TOWER	13	383•394			2	3-17	2	3-18	3-20	3-25	← blotching still evident
S. SERVICE TOWER	1	•									Bottom of pour is
S. SERVICE TOWER	4	•									soft in places above
S. SERVICE TOWER	7	•	383					3-20			form strip
S. SERVICE TOWER	10	383•394			3	3-16	3	3-17	3-20	3-24	
S. SERVICE TOWER	13	383•394			2	3-18	2	3-19		3-26	
N. MECH. WING	14-15	•									
N. MECH. WING	15-16	374•394		3-18	0						← First use of rubber gasket under bottom
S. MECH. WING	14-15	363•374						3-16			← plate
S. MECH. WING	15-16	•									Excessive bleeding at
MECH. BUILDING		•									bottom of walls
N. LAB. SECTION	1	•									
N. LAB. SECTION	4	•									
N. LAB. SECTION	7	•									
N. LAB. SECTION	10	•									
N. LAB. SECTION	13	•									
S. LAB. SECTION	1	•	343					3-20			
S. LAB. SECTION	4	•									
S. LAB. SECTION	7	•									
S. LAB. SECTION	10	•									
S. LAB. SECTION	13	•									
N. OFFICE CORE		•									
S. OFFICE CORE		•	354					3-20			
N. OFFICE WING		•									
S. OFFICE WING		•									
N. STUDIES	1	•									
N. STUDIES	4	•									
N. STUDIES	7	•									
N. STUDIES	10	•									
N. STUDIES	13	•									
S. STUDIES	1	•									
S. STUDIES	4	•									
S. STUDIES	7	•									
S. STUDIES	10	•									
S. STUDIES	13	•									

Nominal elevations
•343•354•363•374•383•394•404•

Report Number 1

125. Salk Institute for Biological Studies, La Jolla, California, 1959-65. "Weekly Report on Concrete Work," 20 March 1964 (LIK, 030.II.A.108.35).

The first table reports on the work done from Monday 16 to Friday 20 March 1964, confirming that the idea of making a "weekly report on concrete work" depends on the strategy outlined that same month with the "Report on Concrete Work." The last table is that of the week ending on 12 February 1965, when the last pours are done, those for the South Office Wing, which are judged to be "excellent."[255]

The question of the efflorescence on the surfaces of the concrete of the Service Towers, a central concern of the technicians, can still be seen from the notes on the work done in March and April 1964.[256] On the walls of lev-

[255] See "Weekly report on concrete work," 20 March 1964, cit., and "Weekly report on concrete work," 19 February 1965, LIK, 030.II.A.108.35.

[256] See "Weekly report[s] on concrete work," of 20 March 1964, 27 March 1964, 3 April 1964, 10 April 1964, 17 April 1964 and 24 April 1964, LIK, 030.II.A.108.35.

[257] "Weekly report on concrete work," 3 April 1964, LIK, 030.II.A.108.35.

[258] *Ibid.*

[259] "Weekly report on concrete work,"

el 383-394 in the North Service Tower of section 10, blotching is noted, in spite of the fact that in order to prevent it the formwork has been kept in place for three days after the pouring (the assembly of the formwork, already used three times, terminates on 16 March; that same day, the concrete is poured; stripping happens on 19 March). Some "excessive blotching at bottom of walls" is seen in the South Mechanical Wing, in section 14-15 at level 363-374 (again, the pour was on 16 March). Certain solutions for intervention on the blotched surfaces, discussed at meetings and in correspondence, are recorded in Langford's reports. "Blotches that were 3 or 4 months old were removed from a portion of this wall using beach sand and water," we read in the report of 3 April 1964 regarding the Retaining Wall in the South Studies, section 10, level 343-363.[257] On 3 April, in the North Service Tower, section 4, level 383-394, work is done on several flawed surfaces, seven days after the pour: "blotches were successfully removed from a portion of the south wall by rubbing with beach sand and water."[258] The pour in the South Office Core, at level 354-363, displays after stripping, on 16 April, "excessive blotching at coupling blocks" and "oil plywood material of plastic around door openings 3 very bad rocks pockets."[259]

Kahn's complex formwork system continues to not always be implemented to the letter in the various construction phases. In the case of the North Service Tower, in section 13 at level 383-394, the pour done on 18 March does not exactly comply with the indicated level of the pour joint ("bottom of pour is soft in place above pour strip"[260]). In the pouring operations, the efficacy of Kahn's system of reuse of formwork through the juncture strip is put to the test. Often the results of the pour are underlined in relation to the number of times the formwork has already been used. "Very good in spite of much used plywood" is the comment on the pour of 15 April (stripped on 17 April), in formwork already used five times, at the North Service Tower, section 7, level 383-394.[261] The same judgment is made for the pour of the South Service Tower, section 1, again at level 383-394 (here too the formwork have been used five times).

The notes also report different kinds of flaws even when the pour is generally considered good. For the work done in the South Mechanical Wing, section 15-16, level 374, the following remarks are recorded in April: "top of loading dock above lintel, concrete finish is very bad under pour strips near windows."[262] The finish of the upper part of the wall, built in the roof of the Mechanical Building (formwork assembled on 6-7 April; pour done on 8 April, stripped on 10 April), is criticized because the formwork "could have been planted with more care."[263] "Difficult pouring due to close spacing of steel," is the comment on the North Laboratory, section 13, level 363 (poured on 10 April[264]): "'topping out' of balcony, starter walls and slopes of the gutter were finished in a *very unsatisfactory* way." "Exterior walls in pipe space are excellent; interior walls show some rocks pockets and bulges," referring to the North Laboratory, section 13, level 354-363 (poured on 10 April, stripped on the 14th[265]). "In general a very good wall except for

16 April 1964, LIK, 030.II.A.108.35.
[260] "Weekly report on concrete work," 20 March 1964, cit.
[261] "Weekly report on concrete work," 17 April 1964, LIK, 030.II.A.108.35.
[262] "Weekly report on concrete work," 3 April 1964, cit.
[263] "Weekly report on concrete work," 10 April 1964, LIK, 030.II.A.108.35.
[264] *Ibid.*
[265] *Ibid.*

126. Salk Institute for Biological Studies, La Jolla, California, 1959-65. Photograph of the concrete blocks in the construction site (LIK, 030.IV.A.540.6.47).

patch in plywood on col 0 line," is reported on the Retaining Wall of the Studies of the North Wing, section 1, level 343-363.[266] The stripping of the North Service Tower, section 10, level 349-404, on 23 April, turns out to be "very good except for leakage at one corner."[267] Also in the South Service Tower, section 7, level 383-394, where stripping is done on 20 April, the concrete is "very good except for horizontal cold joint."[268] The stripping on 20 April of the South Mechanical Wing, section 15-16, level 374-394, shows "improvement over previous lift": "No *tremies* used. Why? Some walls show excessive loss of detail."[269] At the end of the operations performed after stripping on 23 April, in section 7 of the slab at level 354 of the South Laboratory, the report indicates "negligence in clean-up after pour" – "clean-up talus 5 times is much labor after concrete hardness."[270] Upon stripping, on 21 April, of the Retaining Wall of the South Studies, section 1, level 343-363, the walls are assessed as "good," "but not as good as adjacent Retaining Wall, which was poured 8 months ago."[271]

Starting in the month of May the observations in the "Weekly report on concrete work" offer a glimpse of gradual improvement in the implementation, destined to produce even better results, already during the course of the summer.

"Good walls" emerge from the pouring, on 1 and 5 May, of various sections of four of the North and South Service Towers, at level 394-404.[272] The same approval is expressed for the pouring, from 5 to 15 May, of the walls in three other Service Towers, on the same levels. The walls in these three Service Towers, stripped from 12 to 15 May, are even considered "especially good considering the number of reuses," confirming the efficacy of the formwork system invented by Kahn (the formwork has already been used from 5 to 8 times). "Very good" is also the assessment of the surfaces of the South Mechanical Wing, section 14-15, level 374-383, stripped on 28 April. "Excellent walls in pipe space" are seen in the North Laboratory, section 10, level 354-363 (stripping 30 April). Similar praise is made of the "pipe space corridor" – "walls are very good" (North Laboratory, section 4, level 354-363; stripping on 13 May).[273]

The treatment of the formwork surfaces with oil produces certain marks on the concrete walls. "V-Joints are good, but form was not worked down thoroughly, resulting in oil coat appearance; incorrect tie holes that are plugged are not acceptable," are the comments written following the stripping, on 30 April and 4 May, of section 13 in the North Mechanical Wing, level 363-366 ("Air Intake").[274] In the same point of the North Laboratory, on 4 May the parapet is stripped: "Hand rail is excellent inside but has the oil coat look on the outside same as air intake."[275]

"Uneven pour stage in stairwell was pointed out to Paul, Max, and Bill Harvey. It was not corrected. In general the concrete is better than lifts 1 and

[266] "Weekly report on concrete work," 17 April 1964, cit.
[267] "Weekly report on concrete work," 24 April 1964, LIK, 030.II.A.108.35.
[268] *Ibid.*
[269] *Ibid.*
[270] *Ibid.*
[271] *Ibid.* In sectors 15-16 of the North Mechanical Wing, the first use is made of the "rubber gasket under bottom plate" (here the formwork is assembled on 18 March, and have never been used before). From the report of 10 April 1964 the North Service Tower, section 1, and the South Service Towers, section 4, at level 383-394, stripped respectively on 10 and 6 April, have "corners, details, etc. very good." Also regarding the Service Towers, the possibility of saving time in stripping operations is considered ("Stripping time and crane time would be less if all snap ties were removed").
[272] "Weekly report on concrete work," 1 May and 8 May 1964, LIK, 030.II.A.108.35.
[273] "Weekly report on concrete work," 15 May 1964, LIK, 030.II.A.108.35.
[274] *Ibid.*
[275] *Ibid.* "Plugged tie-holes are very obvious as they were on North air Intake," is the comment on the South Mechanical Laboratory,

2 but it is still not good enough," are comments on level 374-383 of the North Office Core. "Dirt should be removed at col 0 before it stains concrete," for the North and South Studies, in the report of 15 May.[276]

Pour defects continue to remain in the concrete of the ends of the walls and the openings for the elevators. "Starter walls are still not satisfactory," is the comment on the North Laboratory, sections 1 and 7, level 354-363 (stripping done from 18 to 21 May).[277] "Elevator door opening is terrible," in reference to the South Office Core, level 363-374, stripped on 20 May - "often door openings show leakage. Elevator call box is bad for the 7th time. No pour strip half way up on door into toilet." "Good walls, some bottom leakage," is the assessment of the North Mechanical Wing, sector 14-15 (poured on 25 May, stripped on the 27th).[278]

In the month of July Kahn's office prepares other specifications regarding the organization of the area of the Salk Institute, including pavements and pools of water. Other notes on concrete pouring methods contained in these same specifications reflect the techniques applied on the whole worksite and Kahn's formwork system, with indications of the various joints, including the pour joints:

"Place concrete in places of final deposit as rapidly as possible to avoid flawing by methods that will prevent segregation of aggregates. Do not drop concrete more than 6 feet without an approved means of controlling fall. Never place partially hardened concrete. Carry on concrete as continuous operation until section or panel between predetermined construction joints is completed and so cold joints are not formed except at construction joints. Thoroughly compact concrete by pubbling or spading with suitable tools and vibrators during placing. Work around reinforcement, embedded fixtures and into corners of forms."[279]

The preparation of the working drawings of the various parts of the Salk Institute continues to proceed in keeping with the pace of the worksite, section after section. It is at the moment of the pouring of the Office Wing of the North Wing that the final pattern drawings of the forms are made, in May and June.[280] On 7 July Langford informs Komendant that he has just "completed the formwork drawing for the Office Wing."[281]

The formwork system invented by Kahn, including the pour phases and breaking up the pouring operations depending on the design of the joints, also to obtain the imprint motif, comes into conflict with the structural system designed by Komendant, also because of the location in a zone of seismic activity, with the aim of obtaining monolithic connections at certain points.[282] Upon seeing the drawing of the "pour sequence" envisioned by Kahn for the Studies, sent to him by Langford on 7 July,[283] Komendant insists that the proposed system is "structurally unsound," and that it will compromise, due to the deep grooves, the durability of the metal bars, which can

section 15-16, level 374-394.
[276] "Weekly report on concrete work," 15 May 1964, LIK, 030.II.A.108.35.
[277] "Weekly report on concrete work," 22 May 1964, LIK, 030.II.A.108.35.
[278] *Ibid.*
[279] "Site Development Work. Salk Institute for Biological Studies, San Diego, California," July 1964, LIK, 030.II.A.107.5.
[280] See "Office Wing - Panel Key - First level," LA-451, 26 June 1964, LIK, 030.I.C.540.032. On 30 July the formwork is assembled for the pillars of the portico (see worksite photograph dated 30 July 1964, LIK, 030.IV.A.540.11).
[281] Frederick Langford, letter to August E. Komendant, 7 July 1964, LIK, 030.II.A.107.13.
[282] On the question of the antiseismic structure see, Louis I. Kahn, "Treatise on earthquake forces and their effect on structures," n.d., LIK, 030.II.A.67.21b.
[283] "Studies Plans Walls-Slab," LA-388, 7 July 1964, LIK, 030.I.C.540.032.

be reached by humidity in a capillary way: "The static system of the studies is changed, if floor slabs and columns are not poured monolithically […]."[284]

The recommendations of Komendant and the decisions of Langford

On the question of the discoloring of the concrete of the Salk Institute, after the results of the chemical analyses and tests conducted by the Smith-Emery Company, and after the adjustments and remedies tried on the worksite and documented since March 1964 in the "Weekly report on concrete work," Komendant too intervenes, after receiving a copy, at the start of August, of the letter sent by Gruber to Langford, containing new instructions regarding the formwork. The letter from Komendant to Gruber on 8 August is the only document among those connected with the Salk Institute from which it is possible to deduce something regarding his contribution to the design of the formwork. After having spoken of his many "suggestions to avoid or at least reduce such discoloration," made during worksite inspections, Komendant discusses what he believes are the causes of the phenomenon, partially accepting the conclusions of the tests of the Smith-Emery Company: high percentage of pozzolan additive in the compound; type of coating of the formwork panels; method of vibration of the concrete.

On the quantity of the pozzolan added to the compound and the effects of its percentage on the discoloring, Komendant simply cites the conclusions of the Smith-Emery Company: "there is no doubt that the added Pozzolan into the mix is one of the causes. To the best of my knowledge, there is no study made about the maximum allowable percentage of Pozzolan the cement can take under normal conditions without losing its stability during hydration." Only later does he state that the percentage of pozzolan had been increased with respect to what had been established during the mix trials, in order to obtain a given chromatic tone (Komendant does not seem to have been informed that the brown color was chosen by Salk, on the model of Pei's concrete, and not by Kahn). "[…] later it came out that Kahn wanted a little bit more brownish color for the concrete, so somebody wanted to please him and ordered the change of mix so that even Fuller was not aware of it."[285]

Relying on his worksite experience, Komendant addresses the construction processes that produce what in his view, as already indicated by the Smith-Emery Company, is the main cause of the discoloration: the movements of water inside the pour, generated by the degree and point of application of the vibration. According to him, precisely these movements produce what he identifies as two types of discoloring: the normal vibration, along

[284] August E. Komendant, letter to Frederick Langford, 15 July 1964, LIK, 030.II.A.107.13.

[285] Komendant, *18 Years with Architect Louis I. Kahn* cit., p. 62.

[286] August E. Komendant, letter to Charles H. Gruber, 8 August 1964, LIK, 030.II.A.107.13. "But most of all, the water movement in concrete is caused by vibration during placing (pumping action and disuniformity in hydration resulting in chemical heat generation). At present (a normal practice), the vibration takes place at the center of the wall resulting in the movement of water toward the center. During hydration, to obtain equilibrium, the excess water moves from the center of the wall toward surface carrying free colloidal size particles disuniformly to the surface (discoloration). Also, it is common practice to vibrate the deposited concrete layers to the maximum density. Due to this and considering the difference between surface and atmospheric temperatures the water is forced to the top horizontal surfaces of each deposited layer. This procedure causes horizontal discoloration. The counteractive measure would be to reverse the water movement in concrete by use of 'plate-vibrators' which can be applied between form and reinforcing (space only approx. one inch). In this case the excess water is drawn to the margins of the wall during placing and pulled toward center during hydration to establish equilibrium. Thus the colloidal size free solids remain inside concrete. Furthermore, not to vibrate the individual layers independently to the maximum density but the underlying layers using the so-called deep vibration. By doing so the disuniformity of concrete between different layers is destroyed and more efficiency with less vibration is obtained (vibration under over-load). There is no doubt

the middle zone of the pour, already produces water deposits along that same zone; in the resulting process of capillary spread towards the edges, the water drags all the soluble particles to the vertical surfaces; when taken to a maximum, to obtain better compacting of the pour, that same vibration also brings water with its soluble particles to the upper surface of the pour, causing a horizontal band of discoloring in that zone (with every intermediate pour accompanied by vibration such bands are produced, all the way to the completion of the pour at the level of the pour joint indicated by Kahn).

To prevent or reduce discoloring, Komendant suggests giving the water of the compound an inverse movement with respect to the one used to date, i.e. from the edges toward the middle zone of the pour, achieving this by means of small plate vibrators to apply between the reinforcement and the formwork, and he recommends the elimination of the intermediate vibrations, replacing them with a single vibration applied in depth, when the whole pour up to the pour joint has been completed.[286] These practical recommendations, which to some extent translate what has been reported by the Smith-Emery Company on the causes of the discoloration, were not contained in the report of the Smith-Emery Company.

In the fluid mechanics of the pour that causes the discoloration, Komendant also attributes a non-negligible role to the nature of the waterproof surface of the formwork, again in this case agreeing with the conclusions of the Smith-Emery Company, which had suggested changing the panels.

"The special characteristics (physical & chemical) of the forms," Komendant writes, "also influence the movement of water. In the present case plastic coating is used, which in accordance with my experience (testing and observations) influences water movement and especially attracts solid (colloidal size) particles. Due to these characteristics I recommended from the very beginning of the construction to use 'form armor'[287] instead of plastic coating and also to use punch rubber in joints of form panels to avoid leakage."

According to Komendant, under his supervision, the Fuller Company decides to make a wall sector using the vibration method he has proposed. "I [...] showed them how to vibrate properly, and explained to the fellows why it has to be done this way. I also showed how to keep walls moist at least for three days. [...] after three days the forms were stripped. [...] no pouring tracks and no honeycombs. The later inspection revealed no dust on the surface and so the problem was solved."[288]

From the documents, it appears that it is Kahn's office, and Langford in particular, that proposes, from 11 August, a way of assembling the

in my mind that the abnormal discoloration in the Salk case could be largely controlled by using the above discussed practice and method. The use of external vibration does not give the acceptable results, it may even increase the general discoloration (wave, resonance pattern). Besides, even a powerful external vibration in case of large wooden forms is inefficient and causes tremendous damage to the forms (leakage, etc.). [...]. The problem is rather simple: too much pozzolan and improper vibration. The proper vibration is determined by the movement of water and light particles during placing and hardening of concrete. In placing, the movement of water and so-called dirt should be toward the center of the item vibrated. When drying out and hardening starts, only the excess water will be drawn to the surface to balance the evaporation and to reduce the drying out. In this case the vibration has been too close to the surface; water, dirt, and pozzolan have been drawn to the surface. Water evaporated very quickly, which is only natural in such a dry and hot climate, causing dry shrinkage and resulting in severe crack formation and dusting of surfaces" (Komendant, *18 Years with Architect Louis I. Kahn* cit., p. 61).

287 With the term "form armor," Komendant probably intends to indicate a panel to be treated with "multi-purpose epoxy protective coating," normally used to protect surfaces of various materials (Ernest W. Flick, *Epoxy Resins, Curing Agents, Compounds, and Modifiers. An Industrial Guide*, New Jersey, 1993, p. 404).

288 Komendant, *18 Years with Architect Louis I. Kahn* cit., p. 62. Nevertheless, in the documents conserved in the Kahn archives there is no trace of this test done by Komendant.

formwork to avoid the problems of water concentration in critical points indicated by the report of the Smith-Emery Company. Langford suggests a device to install at the top of the individual formwork blocks, to transmit the vibrations along the edge of the pour, taking advantage of the reinforcement bars. In this way it is possible to dispense with the plate vibrators recommended by Komendant. The purpose of Langford's device is to attract the water of the pour along the edges, but at the same time to keep it distant from the plastic surfaces of the formwork. The tests to be conducted involve the wing of the Studies under construction.[289]

Triumph in concrete

The progress of the worksite at La Jolla in the construction of the concrete, from the summer of 1964 to the last pour, in February 1965, continues to be recorded in the "Weekly report on concrete work."

The making of the V joints, destined to determine the character of the Salk Institute, is a particular focus in the various worksite sectors. On 9 June 1964, when the pour of section 14-15, level 383-394 of the South Mechanical Wing is stripped, matters have taken a negative turn: "door openings, corners and V-joint detail have gone down in quality." The cause of the marred fabrication of certain V joints in sections 1, 4 and 13 of the South Study Stairs, level 343-351, is identified as the stripping time (the pour is done on 14 July, but not stripped until the 21st): "Walls are good but there would have been less V-joint breakage if the forms were stripped 24/48 hours after pouring, not 7 and 8 days after." Leaving the formwork for so long after the pour is considered a defective procedure. "The inside form was left on 10 days. This is too long," is written regarding section 13 of the South Study Stairs, level 363-374, poured on 3 November and stripped on the 13th.[290]

Only in rare cases does the reuse of formwork for multiple pours lead to unsatisfactory results, as in the railing of sections 4, 7 and 10 of the North Laboratory (poured on 26 May and stripped on the 27th and 28th): "Inside of handrails facing laboratory is excellent. Outside of handrails is very good except finish is dull in many places due to length of time form was in use."[291]

In general, where the reuse of formwork is concerned, the comments noted down in the "Weekly report on concrete work" indicate the success of Kahn's method. "Very good in spite of much used plywood," is the comment on the "bridges" in the South Study Stairs, level 361-363, stripped from 21 to 25 September.[292] "Portico columns have been excellent even after 8 uses of some forms," we read in the report of 6 November, regarding the work done in the North Studies. The use of plastic coating facilitates perfect utilization and correct reuse of the formwork: "Some column forms were completely recoated with plastic. Concrete produced was excellent. In general almost all Laboratory columns have been very good, even after much form use."[293]

"These starter walls look good except for 'top out' in some places," is the note regarding the North and South Stairs in sections 7, 10 and 13, poured from 2 to 5 June. The columns of the North Laboratory, section 10,

[289] See Frederick Langford, letter to Charles H. Gruber, of the George A. Fuller Company, 11 August 1964, LIK, 030.II.A.108.31.
[290] "Weekly report on concrete work," 13 November 1964, LIK, 030.II.A.108.35.
[291] "Weekly report on concrete work," 29 May 1964, LIK, 030.II.A.108.35.
[292] "Weekly report on concrete work," 25 September 1964, LIK, 030.II.A.108.35.
[293] "Weekly report on concrete work," 6 November 1964, LIK, 030.II.A.108.35.
[294] "Weekly report on concrete work," 5 June 1964, LIK, 030.II.A.108.35.
[295] "Weekly report on concrete work," 19 June 1964, LIK, 030.II.A.108.35.
[296] "Weekly report on concrete work,"

level 363-374, stripped on 5 June, are also "good except for some leakage at bottom."[294]

Certain observations in the "Weekly report on concrete work" have to do with the work of teams of workmen in the various sectors of the worksite. Langford had already underlined the importance of using teams of well-instructed masons and carpenters. Their varying skills often have a clear influence on the quality of the concrete. "9 lifts, all identical in plan and still little improvement. Show lack of form setting *education* and *supervision*," is noted regarding the concrete stripped on 15 June in the South Office Core, level 374-383 (poured on 10 June).[295] At times the craftsmen produce damages during the stripping phase, caused by incorrect handling of the formwork, as in the "very good walls" of section 10 of the North Study Stairs, level 343-351: "one corner broken off by stripping crew" (the stripping is on 13 July, the pour on the 10th).[296] Also in section 4 of the South Study Stairs, level 354-363, poured on 20 August and stripped on the 24th, in spite of "excellent formwork, excellent pouring," errors are made in the stripping phase: "careless stripping [...] causing many chips on wall top."[297]

In some cases the carpenters continue to make mistakes in the assembly of the formwork with respect to the patterns indicated by Kahn. One wall sector, or column, has been "poured with plywood layout reversed," in sections 7 and 10 of the North Studies, level 363-374, stripped on 5 August, though this part comes in for praise for its "excellent corners."[298] There are other cases of irregular assembly, as in section 10 of the South Laboratory, level 374-383, stripped on 5 October: "Outside walls are very good. Some form patchings are evident." Sometimes the level strips ordered by Kahn in the forms fail to be inserted, as in the North Office Wing, where the "waterproofing strip is completely out of the concrete in places."[299] Other errors in the positioning of the strips of Kahn's system are found during the course of the work, as in the sector of the North Office Wing, level 383-391, stripped on 4 December: "In general the walls are very good. But in one area the pour strip was left out, and on the North a form strip was placed where none should have occurred."[300]

But the most significant note on the work of the squads, contained in the "Weekly report on concrete work," is written at the time of the stripping, on 31 August, of the pour done on 27 August for level 394-407 of the South Office Core: "The Office Core crew and their foreman are not in the same league as the other wonderful craftsmen on this job. It's frightening to think that this same group of men are going to produce the *Office Wings*."[301]

After the analyses of the blotching and after having hypothesized the effects of the vibration of the pour in the discoloration of the concrete surfaces, the vibration operation is observed to record its results. At the time of stripping, on 10 July, of the pour done on the previous day in section 15-16, level 383-407 of the North Mechanical Wing, the work is considered "good for carpentry, [but] bad vibration."[302] Also in the sector of the Mechanical Building poured on 7 July and stripped on the 9th, the vibration

17 July 1964, LIK, 030.II.A.108.35.
[297] "Weekly report on concrete work," 28 August 1964, LIK, 030.II.A.108.35.
[298] "Weekly report on concrete work," 7 August 1964, LIK, 030.II.A.108.35.
[299] "Weekly report on concrete work," 24 July 1964, LIK, 030.II.A.108.35.
[300] "Weekly report on concrete work," 4 December 1964, LIK, 030.II.A.108.35.
[301] "Weekly report on concrete work," 4 September 1964, LIK, 030.II.A.108.35.
[302] "Weekly report on concrete work," 10 July 1964, LIK, 030.II.A.108.35.
[303] *Ibid.*

127-128. Salk Institute for Biological Studies, La Jolla, California, 1959-65.

127. View of North Mechanical Wing (photograph by the author, 2010).

128. North Laboratory Wing and a Service Tower (photograph by the author, 2010).

is flawed: "south east corner shows lack of vibration."[303] The vibration operations can also damage the formwork, as seen after the stripping on 2 October of section 1 of the North Laboratory, level 394-404: "Excellent walls. However many vibration bursts are showing up on the forms."[304] Instead, in section 13 of the same laboratory, at the same level, the vibration generates flaws: "Very good outside walls. A few spots where vibration could have been better."[305]

Other problems arise in the pouring phases, producing alterations of the concrete surface of a different nature than those analyzed with reference to the discoloration. Worksite supervision by Kahn's staff quickly finds these unexpected results, preventing further repetition of their causes. "Jacket was not used on the bottom of forms which allowed some blotching, but in general walls were very good," is the comment at the time of stripping, on 29 June, of sections 4, 7 and 13 of the North Study Stairs, level 343-351 (poured on the 25th, with forms held in position for 4 days, never previously used).[306] On 15 July, when section 15-16 of the North Mechanical Wing, level 383-407, is stripped, the report states, "1 1/2 panel produced much darker surface."[307] Still in July, in the South Laboratory, sections 1 and 7, level 363-374, other stains are found on the columns ("columns have good corners, tight bottoms, but green patches on form transferred to concrete"),[308] while "dull-dark areas on concrete" are observed in section 15-16 of the North Mechanical Wing, "Air shaft,"[309] (stripped on 31 July), and similar flaws are found in section 13 of the North Laboratory, level 374-383 (stripped on 3 August): "very good transfer of carpentry detail, surface is slightly 'dull-dark.'"[310] Marks of other types are caused by impurities left on the formwork, in section 13 of the North Studies, level 363-374, stripped on 22 September: "Everything excellent except sawdust was not cleaned out of bottom of form before pour."[311]

Following the sighting in July and August of random color variations on the concrete surface, some duller and darker than others, in the North Mechanical Wing and the North Laboratory, Langford intervenes to ask the Fuller Company to take steps to eliminate the phenomenon. Langford's hypothesis is that the cause of these different tones is deposits on the surfaces of the formwork prior to pouring of substances present in the atmosphere: a natural phenomenon like the dew that descends during the night long the coast of La Jolla could have, in his view, an impact on the color of the concrete surface. The fact that any agent that could generate surface alterations is avoided demonstrates the desire of Kahn and his staff to strive for a concrete that is only expressive of its own nature.

"The existing procedure," Langford writes to Matt on 24 August, "for setting the outside forms of the cantilevered box [of the North Laboratory] is no longer acceptable. The carpentry work on these forms, both 'make up' and 'setting,' has been excellent, however the resulting concrete finish is very unsatisfactory. The concrete surface that these forms have produced has shown variations in light to dark from panel to panel. We feel the major cause for these variations is the length of time that expires between when the forms are set and when the concrete is placed in these forms. Our construction location near the ocean may have something to do with this prob-

304 "Weekly report on concrete work," 6 November 1964, cit.
305 "Weekly report on concrete work," 23 October 1964, LIK, 030.II.A.108.35.
306 "Weekly report on concrete work," 3 July 1964, LIK, 030.II.A.108.35.
307 "Weekly report on concrete work," 17 July 1964, cit.
308 "Weekly report on concrete work," 24 July 1964, LIK,

lem. The heavy dew that is prevalent here at night may cause deposits on the exposed forms that eventually affect the concrete surface. The deposit on the forms may be a mystery, but the fact remains that the longer the forms remain exposed to the elements the darker and duller the finish concrete becomes. We realize this means a certain amount of time lost for form modifications and for re-bar placing procedure, but the change must be made in the interest of more consistent concrete surface tones and texture."[312]

In the last months of 1964, when the worksite is fully activated and the teams are well-trained for all operations, the comments written by Langford in the "Weekly report on concrete work" generally contain praise for the quality of the concrete, often defined as "excellent" or "very good." He also adds enthusiastic remarks about the workers: "study stair concrete continues to reflect superb workmanship," he writes, for example, following the stripping on 14 December of the wall sectors in the North and South Studies Stairs. Certain expressions included until February 1965 demonstrate the growing enthusiasm of Langford for the quality of the concrete that is rising on the worksite: "outstanding"; "exterior sections of cantilevered box that have been stripped are excellent"; "every study column poured to date has been excellent"; "study arcade columns continue to show excellent formwork panel very good concrete placement"; "excellent formwork, excellent pouring"; "carpentry work on bridges has been very good"; "clean up after pour was very good"; "undersides of North bridges were stripped last week and the results are very fine"; "beautiful concrete."

030.II.A.108.35.
[309] "Weekly report on concrete work," 31 July 1964, LIK, 030.II.A.108.35.
[310] "Weekly report on concrete work," 7 August 1964, cit.
[311] "Weekly report on concrete work," 25 September 1964, cit.
[312] Frederick Langford, letter to Paul Matt, 24 August 1964, LIK, 030.II.A.108.31.

129-130. Salk Institute for Biological Studies, La Jolla, California, 1959-65. North Laboratory Wing and a Service Tower (photographs by the author, 2010).

The last pour is done in the South Office Wing, on 19 February 1965. The other concrete parts have to do with the landscaping of the surrounding area of the complex, and are completed at the end of April.[313]

It is no coincidence that the concrete surfaces of the Studies and the Offices, among all those of the Salk Institute, are the cleanest, brightest and also the palest, similar in effect to the surfaces of the Travertine, as if after the tests and laboratory analyses, and the various attempts in the different phases of the construction, Kahn, his team and the craftsmen had managed to make concrete into a completely different material from the one used until now in all the works of architecture in *béton brut* around the world. Pozzolan, used since the time of the Romans as a fundamental ingredient of concrete, returns to a decisive role in the contemporary era precisely with the concrete of the Salk Institute, not to obtain setting under water or particular strength, but to achieve a desired surface quality. The adjustments of the vibration methods and the coating of the formwork in contact with the pour eliminate the discoloration defects, contributing to make Kahn's concrete in the final phase of the construction absolutely exceptional.

"This office sent Langford to La Jolla to direct the construction of all concrete," Kahn would write in 1966. "Working with the contractor, this office, through Langford's comprehension of the Builder's responsibilities in regard to the concrete, proved to the Owner that the construction of all formwork for the concrete should be taken over by the Architect. Fred Langford directed the making of all the drawings, amounting to about 200. These drawings were fully accepted by the contractor, though at first resisted. The concrete work, now completed, serves as an example for many other works that have followed."[314]

It is in the light of the complex design process and all the experience gained on the worksite to reach the goal of the truth of the formwork system and the nature of concrete that we can read Kahn's statements from 1966 circa, precisely against the backdrop of the construction of the Salk Institute. Nothing of the multiple manipulations of the constructive logic deployed to create an artfully coherent system can be seen in those statements, as if the design were simply the transcription of the work on paper.

"The concrete in this building could be so exciting," he declares, in reference to the Meeting House, "I would imagine the concrete as very, very smooth, by the way; I'm trying to make you want to put your hand over it and feel it, and I would do it simply by insisting that the corners of the buildings must be made of one piece, not be pieced together, so that all corners are very, very hard. And then another thing: where it naturally will bleed, you let it bleed. By actually allowing the concrete to come out, not trying to

[313] On 29 April, Kahn asks Komendant to make "a final site inspection of the finished concrete work," for the part related to the structures (John E. MacAllister, letter to Louis I. Kahn, 29 April 1965, LIK, 030.II.A.107.37).

[314] Louis I. Kahn, letter to State Board of Examiners of Architects, 23 August 1966, LIK, 030.II.A.10.7.

[315] Lecture, without date, Salk Institute Meeting House, and typewritten "Kahn," LIK, 030.II.A.54.25. See also what Kahn says about the concrete of the Salk Institute in, Kahn, *Address*, 5 April 1966, now in, Latour, *Louis I. Kahn: writings, lectures, interviews* cit., (pp. 208-220), p. 216. "The color of the concrete is a kind of almost silium color. It is a color of rose," Kahn asserts (*ibid.*, p. 215).

[316] Kahn, *Address* cit., p. 216.

make it hold back, it will come out in the form of a bead that peels off evenly because it is not very strong. When this small bead comes out, you can knock it off very easily. The corner then looks like a sketched line, but it's very clean around it – only because you allow the concrete to bleed out where naturally it wants to bleed instead of trying to force it into not doing so. With this technique we've gotten very smooth concrete so far, not wood. All the walkways are Travertine but the building itself is smooth. The building has a rose-like color by the use of pozzolan which is in the concrete and takes away from its green color, which does not sympathize with the landscape. The reason why most people paint their buildings in La Jolla is because the concrete is such a mean color against the sky."[315]

The arduous, tenacious struggle of Kahn and his faithful team to tame the logic of concrete construction in the Salk Institute is summed up in a phrase pronounced by Kahn himself in a lecture: "Every bit of the formwork was designed; nothing was left to the contractor."[316] The glowing, clean image of the surfaces and the perfection of the edges of the concrete walls make the Salk Institute look as if it had been built with the gigantic sculpted stones of the craftsmen of bygone days. But it is only a mnemonic effect, because that form of concrete is the objective of Kahn's discovery of the degree zero of that material.

In the summer of 1966, Kahn is interviewed, to explain the characteristics of the Salk Institute, by Joe Bell on behalf of Richard S. Huhta, the editor of Concrete Construction Publications, Inc. It is not surprising that in the interview Kahn describes the concrete of the work, and that the title of the interview is *Triumph in Concrete.*[317] The questions addressed by Kahn

range from the potential of concrete for the configuration of any type of structure, to the portrait of the La Jolla worksite, with the craftsmen operating as in a medieval construction, in the production of the finest concrete.

"Concrete makes it possible for an architect to build an unconcerned structure around the mechanical necessities of a building. [...] the integration of the mechanical, electrical and architectural – this is the beginning of new shapes [...]. Concrete is amenable to this style of design because it is the most plastic of all structural materials. We are most able to shape it to our own needs. Concrete makes it possible for an architect to build an unconcerned structure around the mechanical necessities. [...] concrete is excellent for this type of use because it is also rigid. [...] in the Salk Institute, we wanted a concrete that was finished when the forms came off. This meant we had to control pockets. The forming to accomplish this sounded like cabinet work to the builder, and that's almost what it was. But it paid off handsomely – in appearance, and finally in economy, too. We used some forms as many as 12 times. When it was impossible to make a joint absolutely tight, we made a virtue of it, letting the concrete bleed out. We looked at it as a release of concrete's inhibitions – a recognition on our part that you can't make a perfect joint. The joint is the beginning of ornament. That's a longtime rule of architecture that is too often ignored in working with concrete. Builders always try to eliminate or cover joints. That's wrong. [...] it takes as much care to make formwork as it does to make the mix. Good concrete should have carefully supervised operations from beginning to end, and this includes high concern over the forms. At Salk, we designed all the formwork, but it required much persuasion to take it out of the hands of the builder. We resisted his prerogative to do the forms.

131. Salk Institute for Biological Studies, La Jolla, California, 1959-65. View from the west of the two Laboratories Wings with Offices and Study Towers (photograph by the author, 2010).

We had to. He simply couldn't see what was in our minds. You see, concrete had to be the marvel of this building. It had to be the finish, the end-all. Most builders believe in repairing concrete after it is built. We said, 'No, it has to be perfect when the forms come off.' This insistence had a remarkable effect on the workers. The construction people at Salk really looked forward to coming to work. They felt like sculptors. They would come voluntarily on Saturday or come to the job site early so they could get started. Almost every carpenter on this job came individually and left his name with us when the job was over, so if ever we did another concrete building, he could work on it. They all felt real pride in this building. Before, they always worked on the theory that everything they do is covered up anyway, and if it's going to be covered, then why bother? At Salk, it was different. I feel that respect for the effect of producing the design is reflected in the organization of the building and in the performance of every man on the job. The builder can get twice as much work out of each man if this respect is felt. It adds up to the saving of money in the willingness to work. The architect hopes for this tone and always tries to get it. But he can't make it himself. At Salk, everything worked together to produce it. [...]. The most expensive thing to do is the *wrong* thing. If you compare this with what you want, it is always cheaper to do it right. Actually, the Salk Institute formwork turned out to be much less costly than the contractor said it would be when we proposed this type of forms. On the entire job, we saved 750,000 – or $25 to $30 per cubic yard – under their estimate. Why? The cost of concrete didn't change, but the cost of form make-up was much less because it wasn't nearly as difficult as expected. There were also considerable savings on the erection of wall panels. Everything was done with cranes, which made possible massive shapes. The potentialities of the crane were extended considerably on this job. And, of course, the larger the panels, the fewer joint and corner problems we have. [...] a craftsman loves to expose what he does. And this is always possible when designing in concrete. Steel is exciting where it can be exposed – in a greenhouse, bridge or factory free of fireproofing. But steel must be protected and its configuration is usually hidden. When covered up, a material's fascination to the architect ceases. The architect also loses interest when a budget becomes the predominant factor in a structure. But exposed concrete should be regarded as a fine modern material, adaptable to any type of structure. It is molten stone to the architect, and I gravitate to concrete because the imprint of the artist plainly visible."[318]

[317] Richard S. Huhta, Editor of Concrete Construction Publications Inc., letter to Louis I. Kahn, 6 October 1966, LIK, Box 71a. See also, Richard S. Huhta, Editor of Concrete Construction Publications Inc., letter to Louis I. Kahn, 17 November 1966, LIK, Box 71a. Huhta sends Kahn a text written by Joe Bell after an encounter with Kahn in summer 1966. Kahn's office sends back the text corrected by Kahn himself and by MacAllister at the start of December 1966 (Barbara Schlosser, Richard S. Huhta, "The Salk Institute/slug. Triumph in Concrete," 9 December 1966, LIK, Box 71a).

[318] During a lecture, also in 1966, Kahn goes back to explaining the nature of the concrete of the Salk Institute: "There are Travertine walks, which with concrete, look extremely well. I had, in fact, intended to make it slate because it would look well as a contrast." [...]. "This slide shows the character of concrete which has been untouched. All the forms are made out of four by twelve plywood, and at the joints the concrete bleeds out. This gives the opportunity for the concrete to be relaxed in its forming itself; not to be restrained in any way at the points where it cannot be restrained. Allowing it to go through actually perfected the concrete at the joints, and all corners were made specially, so that they made all three sides removable. The corners were never raw, they were always straight. [...]. All these holes were plugged with lead. The lead and the concrete and the Travertine are about the only materials really used. All, I think, sympathetic to each other. Every bit of the form work was designed; nothing left to the contractor. Here it shows what the effect of the bleeding of the concrete can do. In a certain light, these are all accentuated. The corners are quite true" (Kahn, *Address* cit., pp. 215-216).

132-138. Salk Institute for Biological Studies, La Jolla, California, 1959-65.

132-135. North Office Wing details (photographs by the author, 2010).

136. Study Towers of the North Laboratory Wing (photograph by the author, 2010).

137. Study Tower of the South Laboratory Wing (photograph by the author, 2010).

138. Study Towers of the North Laboratory (photograph by the author, 2010).

139-146. Salk Institute for Biological Studies, La Jolla, California, 1959-65. Study Towers details (photographs by the author, 2010).

147-153. Salk Institute for Biological Studies, La Jolla, California, 1959-65.

147-148. Study Towers details (photographs by the author, 2010).

149. View of the North Laboratory Wing (photograph by the author, 2010).

150. View of the South Laboratory Wing (photograph by the author, 2010).

151-153. Details of the spacer hole plugged with lead, of the V joint and the double V joint produced by the juncture strip (photographs by the author, 2010).

chapter two
confirmations of the smooth finish: the worksite at bryn mawr

Choice of the geometric design of the exposed concrete floor slab

The experimentation with exposed concrete, poured in place or precast, done during construction of the Richards Laboratories, the Biology Building and the Salk Institute, marks a particular phase in Kahn's technical research, leading to another fundamental step: the design and construction from 1960 to 1965 of the Eleanor Donnelley Erdman Hall, the Dormitory of Bryn Mawr College in Bryn Mawr, Pennsylvania, directed by Katharine McBride. The development of the formwork system for the Dormitory confirms that in the period around 1962 Kahn is outlining a new concept for the surfaces of his concrete. In this perspective, the experience of the precast concrete of the Richards Laboratories appears to be a crucial stage in the comprehension of the theoretical value and formal quality of a perfectly smooth surface, definitively freed of any imprints of the wood grain, as in traditional formwork, in pursuit of the true nature of concrete, beyond any traces left by other materials in the construction process. So it should come as no surprise that on the worksite for the Dormitory, Kahn and his staff strive to control all the phases of the construction process, altering the usual techniques of American construction and contractors to specify every detail of the exposed concrete, with the goal of obtaining the same finish quality typical of precast concrete even when using poured in place concrete. The role assigned to concrete blocks in the definition of the formal qualities of the works is changing significantly, precisely in the light of this evolution of Kahn's concept of the "smooth finish." It is no coincidence that the rugged surfaces of the walls built with those blocks now tend to vanish under a layer of plaster, in the main spaces of the Dormitory.

The chronicle of the Bryn Mawr worksite demonstrates that the control of the construction process on the part of Kahn's office continues to happen, as on the La Jolla worksite, when the various portions to be built are imminent, supplying working drawings sector after sector, often in response to pressure from the contractor for information on how to assemble the formwork. In the Dormitory, as in the Salk Institute, Kahn exploits the construction of the underground levels of the building to conduct tests on the formwork system, allowing him to evaluate the effects of technical details on the concrete surface, which could not be predicted through drawings alone.

The experimental worksite process is facilitated by the configuration of the Dormitory in three volumes with a square plan, known as Squares A, B and C, joined at the corners and arranged diagonally. The definitive design, developed in and around December 1962 with the collaboration of David C.S. Polk, reflects the principle of grouping of rooms around common spaces, pursued by Tyng in certain preliminary studies, still applying the system of octagonal and square cells developed in the projects for the Jewish Community Center Trenton.

Each block of the Dormitory has a large, two-story octagonal common area at its center, which is supposed to have the character of a domestic place (the one with the staircase is called the Entrance Hall; the others are the Dining Hall and Living Hall, and feature fireplaces). In each of these Halls light enters from four enormous lanterns placed on the short sides of the octagon, known as "clerestory towers." The origin of the design in polyhedral geometry determines the arrangement of the rooms, the windows and even the measurement of the narrow vertical cavities that shape the perimeter of each block, or the "undulating wall."[1]

In the choice of materials Kahn interprets the symbolic value of the composition, but he also has to take construction costs into account, as well as certain indications from McBride. Starting in 1962, the mode of construction of the perimeter walls of the three blocks is discussed, in reinforced concrete or "concrete block with precast concrete units,"[2] though Kahn's decision that exposed concrete should form the heart of the complex, as in the Yale Art Gallery or the church in Rochester, is not questioned.

The floor system, thanks to its geometric design, becomes the most characteristic structure of the exposed concrete nucleus contained in each Square. But after the construction of the floor system of the Yale Art Gallery and the many attempts to apply tetrahedra in various other cases, including the skyscraper, for the Halls of the Dormitory Kahn chooses not to translate again that geometry into reinforced concrete. And simplifies it in the motif of the square coffers, to the point of losing many of the spatial implications contained in the Tetrahedron Floor System, including the insertion of conduits for physical plant systems. Precisely the considerations regarding the wooden formwork for the construction of the tetrahedron floor prompt Kahn to find an elementary geometric design that makes the latest, unrecognizable version of the floor system of the Yale Art Gallery seem like a compositional exercise on the floor slabs with square coffers of traditional carpentry. Furthermore, Kahn's floor slab seems to be consistent with the abandoning of the polyhedral geometry of the projects of Tyng for Bryn Mawr.

[1] Precisely the profile of the perimeter walls of the three blocks is the subject of discussion at the meeting on 26 April 1962 - "undulating wall vs straight wall," LIK, 030.II.A.28.16. Kahn is against the "undulating wall" with the octagonal rooms ("Meeting June 7 1962," LIK, 030.II.A.28.16).

[2] "Dormitory for Bryn Mawr College," 20 August 1962, LIK, 030.II.A.28.1.

[3] Louis I. Kahn, *The Architect and the Building*, in, "Bryn Mawr Alumnae Bulletin," 43, summer 1962, p. 2 (cit. in, Brownlee, De Long, *op. cit.*, p. 100).

[4] "Suggested acoustical specification," 15 January 1963, LIK, 030.II.A.28.4.

[5] *Ibid.*

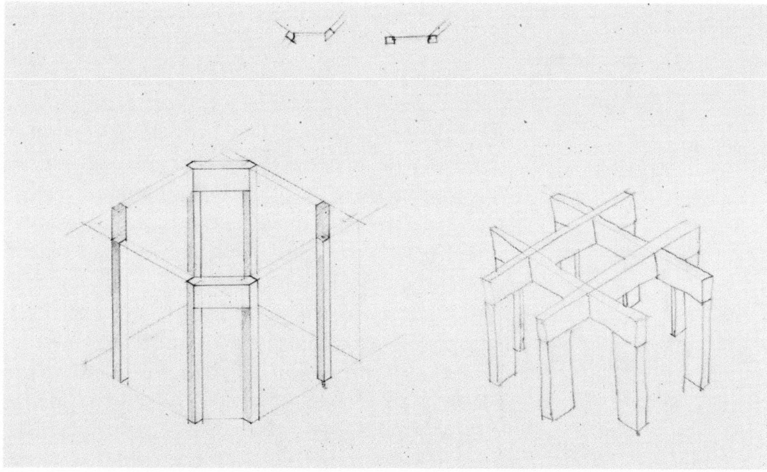

154. Eleanor Donnelley Erdman Hall, Bryn Mawr College, Bryn Mawr, Pennsylvania, 1960-65. Axonometric study for the structure (LIK, 030.I.C.565.1).

It is worth recalling that Henry A. Pfisterer had proposed modifications to the Tetrahedron Floor System of the Yale Art Gallery, transforming it into a floor system with coffers.

In 20th-century reinforced concrete construction, the coffered floor slab selected for the Halls of the Dormitory is quite widespread, under the name of waffle slab, and it is often associated with patents for different systems of prefabrication. But for Kahn, the geometry of the waffle slab serves above all to give the space the character of an antique living area, and the concrete is poured in place, in wooden formwork, confirming the pursuit of a domestic character for the Halls, or the "presence of a house" as he puts it.[3]

At the start of 1963, square acoustic panels of the Tectum type, from the Tectum Corporation, the same ones used as permanent formwork in the floor system of the Yale Art Gallery, are envisioned for insertion in the square coffers. But the recommendations for their use indicate that all the concrete of the floor slab should have hardened at least three months prior to installation, and that the structure should be heated first to banish all humidity.[4] The investigations for the application of the panels reveal that Kahn would like to create a formwork system closer to that of the Yale Art Gallery than to the convention formwork with pieces stripped after the pour. This is why the panels are not supposed to seem to be suspended from the coffers, but should look like they have been inserted as integral parts of the construction of the floor slab (special anchors are designed to install in the coffers prior to the pouring of the concrete). It is significant that Kahn specifies the "final installation shall provide a completely concealed method of suspension."[5] Even the color chosen for the panels to be deployed is a gray that matches that of the concrete when seen from below. Everything has to convey the impression of a construction in concrete with all the parts incorporated, without claddings.

Working processes for concrete according to the specifications

At the start of 1963, Kahn envisions building the Dormitory in exposed concrete to pour in formwork with vertically assembled boards for the walls and plywood panels for the slabs. The board formwork makes the Dormitory, in this design phase, a variation on the Yale Art Gallery, the First Unitarian Church and those parts of the Richards Laboratories with concrete poured in place. The specifications in February confirm the importance of the various Kahn worksites in 1962 and 1963 for the passage from boards to plywood as the formwork material, to produce the graphic pattern of the exposed concrete.[6] The fact that in the Dormitory Kahn calls for board formwork indicates the extent to which the construction of the non-secondary walls or the foundations with plywood panels begun with the Biology Building still represents an initial stage of his experimentation with this type of formwork. After all, in the Salk Institute the boards are abandoned in favor of the plywood panels precisely in the second half of 1962 (in September, Kahn is still studying walls with vertical boards), and it is only in the first months of 1963 that Kahn orders worksite tests for the design of the formwork panels, with a degree of complexity that did not exist in the Biology Building. The project and the evolution of the formwork system for the concrete at Bryn Mawr undergo the influence of what is happening at the La Jolla worksite.

With respect to the typical entries in Kahn's specifications for his works in concrete, in the case of the Dormitory emphasis is placed on the "appearance of form contact surfaces exposed to view" of the concrete and on the ways to ensure "uniformity for all concrete for entire project." Kahn distinguishes between the two surfaces of the exposed concrete, the walls and the slabs, indicating the difference in the presence or absence of an arrangement of the panels – the pattern and "texture" on the one hand, and the "smooth finish" on the other.

The pattern and texture have to do with the imprints and grain obtained by using wooden boards. "In general, vertical surfaces shall have a definite texture and pattern imparted by the forms, and soffits and bottoms of slabs shall have a plain, smooth finish of the type achieved with well built plywood forms." Therefore the differentiation of the surfaces of the exposed concrete happens thanks to the materials of the formwork, as was the case in the Tower X of the Richards Laboratories. In the specifications for the Dormitory it is stated that the drawings of the formwork, for the exposed concrete, must indicate "size and arrangement of panels, direction of boards, location of form ties and any other features that will in any way leave a mark in the finished concrete." It is clear that by now Kahn intends to control, through the drawings, every single imprint left on his concrete, seeking a type of graphic representation of the formwork that is not found in the specialized technical manuals.

Again, in the wake of the experience of the Salk Institute, Kahn introduces the obligation of construction on the worksite of "sample panels of concrete," using the same phrases written in the specifications for La Jolla.

For the composition of the concrete, the use of Allentown Portland Cement is suggested, made by the company of the same name (the compound should have additives of Darex AEA, Vinsol Resin or similar products, as well as Plastiment or Pozzolith). The phrases on the workability of the concrete, its curing, on "placing concrete," "design of forms" or the ban on the use of "chamfers" are taken from the specifications of the Richards Laboratories.

In the paragraph "Interior Core Walls," Kahn explains, with phrases lifted from the specifications of the Richards Laboratories, the type of wooden boards ("tongue & groove boards"), their basic measurements (3/4 inch thick, 2 inches wide), and the decorative motif with the "vertical patterns with wider horizontal boards." The phrase written in the specifications of the Richards Laboratories on the optical role of the board imprints – "the imprint of these boards is used to conceal imperfections in the surface of the concrete" – is modified in the following, more general form: "The imprint of these boards is used to control the surface of the concrete."

For the retaining walls and soffits, the formwork are to be in plywood panels, "with proper coating." The possibility is left open of a design of the concrete surface without joints for a length of more than 12 feet. In this case, to make the joints between the panels invisible, Kahn accepts the solution of a slab of veneer placed on the two panels, "where necessary to eliminate joints," as he specifies. All the forms, in boards or plywood, even in the case of "precoated plywood," must be covered with special plastic coatings of the Formfilm type produced by the A.C. Horn Company, as was also the case for the Richards Laboratories.

By now the choice of spreaders and their ends has become decisive for Kahn's formwork system. While the specifications of the Richards Laboratories indicated the "Richmond Tyscru system" with "wooden Tycones," those of the Dormitory recommend "Cone Fast Coil Ties with threaded Coil Cones, using 1-3/8" long Coil Cones for 1/2" Coil Bolts," which are also the ones called for in the specifications of the Salk Institute of 1962. Again, as in the Salk Institute, the holes left by the cones have to be sealed only in the case of "exterior walls." For the sealing grout, the indications in the specifications of the Richards Laboratories are repeated to the letter (the same is true for the "cleaning" of the concrete surfaces).

For the coffered floor slabs of the Halls, which are the most characteristic features of those spaces, the specifications simply indicate that the waffle slabs should be made with plywood, with "sharp, accurately mitered corners."

For the cladding elements in precast concrete, Kahn seeks a match with the slate slabs, though limiting the indication to a generic "sympathy": "The color of the precast concrete shall be sympathetic to the slate [...]." He also emphasizes that the "surface texture" of the elements must be obtained directly with the molds, without any subsequent working. For certain precast parts the question is left open as to whether they are to be made in the workshop or at the worksite.

Apologia of plywood

The construction of the Dormitory at Bryn Mawr is assigned to the contractor Nason & Culler, Inc., based near Philadelphia.[7]

The material of the formwork of the main walls is modified on 3 May 1963, when the decision is approved to eliminate the wooden boards: "deletion of [...] form boards imprinted finish on exposed core walls."[8] Considering the fact that beside this modification an indication of the resulting savings is shown, estimated at 3000 dollars, it would appear that the elimination of the boards has been determined only by economic issues. The Dormitory confirms, with the Salk Institute, that Kahn is proceeding towards an evolution of the concept of *béton brut* quite in tune with the direction indicated by the Carpenter Center.

Having abandoned the formwork boards for the walls of the Dormitory symbolically marks the end of the first heroic era of *béton brut*, where Kahn

[6] "Construction Specifications," 25 February 1963, LIK, 030.II.A.30.1.
[7] Still in April 1963, there is discussion of the alternative of a "column system" or a load-bearing wall ("Meeting with Nason & Culler," 12 April 1963, LIK, 030.II.A.28.17).
[8] Nason & Culler, Inc., letter to Louis I. Kahn, 3 May 1963, LIK, 030.II.A.29.21.

has been the great American exponent. The work at Bryn Mawr confirms the personal path he has taken with the construction of the Salk Institute in the definition of an autochthonous architectural concrete, with the banishing of the slender boards that instead continue to resurface even in the cleanest *béton brut* of Le Corbusier, to attenuate what Le Corbusier himself calls the "hardness" of concrete, generated by the use of just one kind of formwork.

The "hardness" of Kahn's new concrete surface is officially sanctioned in the worksite documents of the Dormitory, when on 21 May the entry "Form contact surfaces" of the specifications is changed as follows: "Form for core walls shall be plywood [...]. Edges shall be beveled as approved."[9] In June it is decided to also eliminate the boards from the walls of the skylights: "Omit finish on concrete clerestory walls divested of floor boards by showing 'V' type vertical fins at 2' intervals."[10]

With the choice of plywood panels, Kahn wants to boost the degree of abstraction of the concrete walls, eliminating any possible trace of wood texture. Now that the boards have been abandoned, the imprints of wood grain have to vanish from the finished concrete, and with them the pattern and texture mentioned in the previous specifications. Now Kahn wants an absolute "absence of plywood grain pattern."[11] Back in 1961, Le Corbusier warned Sert about the "danger of using plywood with artistic veins to make pretentious decorations," since he desired, for the Carpenter Center, a concrete surface "without artistic grain."[12]

Precast concrete or cast stone, and slate slabs

After the initial phases of the design of the Dormitory of Bryn Mawr College, slate becomes the material never questioned by Kahn, because it gives the building a powerful black color against which luminous lines stand out, reflecting and accentuating the chromatic relations between walls in dark gray stones with a mutable appearance, and the frames and posts of the windows in light stones of the Rockefeller Hall or the Pembroke Hall, along Merion Avenue, other facilities of Bryn Mawr College built by Walter Cope and John Stewardson, the first in 1892-94, the second in 1900-04. At the time of the writing of the specifications, in February 1963, the slate is the "blue-black Buckingham" type sold by the Buckingham-Virginia Slate Corporation of Richmond, Virginia. For the slabs to use on "exposed surfaces" a "sand rubber finish" is indicated for the steps and thresholds, a "semi-rubbed finish" for the flooring and a "natural cleft surface" for the wall claddings.[13] According to statements made in his lectures, at first Kahn would also have liked to use slate in the Salk Institute, installed next to the concrete: "I had, in fact, intended to make it slate because it would look well as a contrast."[14]

Kahn is uncertain about the cladding material to apply on the ends of the walls of the rooms, at the vertical joints of the three blocks. The specifications postpone the choice of the material, simply conveying the formula of precast concrete, to be made "sympathetic to the slate," as is specified when Kahn contacts supplier firms. On 15 March, Harlyn E. Thompson, of Kahn's

[9] "Specification changes," addendum 5, 21 May 1963, LIK, 030.II.A.20.10. Point 2-26 © of the previous specifications covered the plywood formwork.
[10] Nason & Culler, Inc., letter to Louis I. Kahn, 12 June 1963, LIK, 030.II.A.28.17.
[11] "Job meeting," 12 August 1963, LIK, 030.II.A.28.30.
[12] Cit. in, Roberto Gargiani, Anna Rosellini, *Le Corbusier. Béton Brut and Ineffable Space, 1940-1965. Surface Materials and Psychophysiology of Vision*, Lausanne, EPFL Press, 2011, p. 537.
[13] Slabs of Alberene stone are also envisioned, probably

155. W. N. Russell and Company, *Cast Stone*, booklet (LIK, 030.II.A.28.18a).

office, calls the Cut Stone Association of Indiana, based in Bloomington, to request information on Indiana Limestone, which is generally light beige in color.[15] Kahn wants to clad the ends of the walls in concrete block with that type of stone, which is quite similar to the Aquia Freestone with which he clad the reinforced concrete pillars of the Yale Art Gallery. Having decided to give up the precast concrete envisioned in the specifications and to make use of stone is but the first step of a contamination of the nature of concrete.

The search for a material "sympathetic to the slate" moves on to the hypothesis of using an artificial material similar to precast concrete, known as cast stone, when Kahn's office contacts W.N. Russell & Company, based in Westmont, New Jersey, specializing in this type of product. The booklet supplied to Kahn by the company underline the different quality of the artificial stone of Russell & Company with respect to "just ordinary precast concrete which is usually of very poor quality and unattractive in appearance." The passages from precast concrete to natural stone and then, finally, to artificial stone, modify the value assigned by Kahn to the cladding of the ends of the walls. It is no coincidence than when the work on this cladding begins, there is a certain ambiguity in the terminology used by Kahn's staff and the contractor to indicate the slabs, which at times are defined as "precast stone."[16]

It is clear that Kahn wants to obtain a prefabricated concrete of a stone-like type, with a degree of finishing and color different not only from poured in place but also from precast concrete. The fact that Russell & Company sends Kahn a profile dated June 1959 regarding a particular type of cast stone that "shall simulate Indiana Limestone and all exposed surfaces shall have a rubbed finish with aggregate exposed,"[17] demonstrates that Kahn has contacted the company precisely to get information on the fabrication of a cast stone resembling the Indiana Limestone supplied by the Cut Stone Association of Indiana. Never before, in Kahn's previous projects, had the hypothesis of using concrete to faithfully imitate stone been considered.

The use of slabs of cast stone has repercussions on the design of the pieces to be used in the Dormitory, given the fact that thanks to casting at the workshop special measurements can be produced, sizes that could not be obtained with natural stone. It is worth noting that the cladding of the pillars of the Yale Art Gallery was done with four slabs per floor, while for the Dormitory Kahn wants to reduce the number of seams to a minimum, managing to obtain, in the definitive solution, a single slab with the height of one floor, precisely thanks to the artificial fabrication of the stone.

Confirmation of the intention of reducing horizontal seams to a minimum, which would explain the use of cast stone, comes from the sizing and arrangement of the slate slabs Kahn studies precisely to avoid horizontal joints and to accentuate the vertical continuity. It is significant that the Buckingham-Virginia Slate Corporation examines the possibility of reducing the thickness and width of the slabs to install between the windows, without altering the height, to avoid any intermediate horizontal joints because they would go against Kahn's intentions. The hypothesis of subdividing the long

for the interiors.

[14] Kahn, *Address* cit., p. 215.

[15] Cut Stone Association of Indiana, letter to Louis I. Kahn, LIK, 030.II.A.29.19.

[16] Nason & Culler, Inc., letter to Louis I. Kahn, 5 September 1963, LIK, 030.II.A.29.64. For the costs of the fabrication and installation of the precast concrete parts, on 5 April Hoopes of Structural Precast Concrete is contacted ("Meeting with Nason & Culler," 12 April 1963, LIK, 030.II.A.28.17).

[17] See the page of notes dated June 1959, LIK, 030.II.A.107.9.

slabs into two or three pieces for vertical installation is put aside due precisely to Kahn's principle of reduction of horizontal joints to a minimum. To respect the vertical design sought by Kahn using, in any case, multiple stacked slabs, the Buckingham-Virginia Slate Corporation suggests reduction of the width of the slab which, in spite of the laying with horizontal joints, would produce, according to the company, the "vertical effect" desired by Kahn. The correspondence between the companies confirms that Kahn is seeking a precise design of the joints of the slate slabs. There is even mention of the "architect's system of all vertical joints and no horizontal joints." "It is our observation that the architect is striving to emphasize the vertical dimensions of the building by eliminating the horizontal joints […]."[18]

Nason & Cullen, Inc., assesses the hypothesis of using Pennsylvania slate to reduce construction costs, and contacts three suppliers in the Bangor zone for estimates. Kahn is informed that due to the quarrying methods used at Bangor, it is necessary to reduce the size of the slabs.[19]

The slate installed at the Dormitory is from Pennsylvania, "clear for exterior, ribbon for interior,"[20] supplied by the Structural Slate Company based in Pen Argyl, Pennsylvania. The question of the horizontal joints needed in order to use the sheets from Pennsylvania radically changes what had been discussed with the Buckingham-Virginia Slate Corporation. Kahn introduces not a joint but a supplementary strip, to be made in precast concrete, that runs at the height of the sills and serves to clamp the masonry of the slabs.[21]

The concrete blocks for the construction of the walls are supplied by the York Building Products Company, Inc., located in York, Pennsylvania. During the course of 1963 the question of the visibility or concealment of the blocks has not yet been resolved by Kahn, who hesitates between a neat cladding and a rugged surface finish. Shortly after the preparation of the specifications, the walls of the two staircases at the ends of the Dormitory are envisioned in exposed concrete block.[22] Kahn and Nason & Cullen, Inc., examine the possibility, later rejected, of leaving the blocks exposed also in the walls of the rooms, with shaping of the mortar in the joints – "striking the joints." In this case, the possibility of painting the walls is considered.[23] The uncertainty about whether to leave the blocks visible in the staircases and dorm rooms confirms the changes in progress in Kahn's idea of the concrete surface.[24]

Design and construction of the first formwork

One fundamental chapter in the technical definition of the design of the Dormitory of Bryn Mawr College has to do with the preparation of the drawings of construction details, and the worksite tests of what, for Kahn, is becoming a true formwork system, after the start of the work at La Jolla. Kahn's concrete shaped by plywood panels and perfectly smooth stands out clearly from that of international Brutalism precisely due to the study of every detail of the assembly and deployment of the formwork. It is no coincidence that the form ties, from the size of their pattern to the choice of the diameter of the holes and their sealing materials, play a leading role in Kahn's concrete as the only ornamental pattern generated by the construction process, taking the place of all the sophisticated decorative patterns developed by Le Corbusier and his imitators to vary the effect of "hardness" of the stereotomy of the formwork.

The drawings made from July to December 1963 document the control exerted by Kahn over details of the assembly of the panels, their arrangement to obtain particular alignments of the imprints of the joints, and the positioning of the spreaders, calculated to make the holes form a pattern based on a square grid.[25]

The Bryn Mawr worksite begins operation on 1 July 1963, with the excavations for the foundation. To experiment with the technical characteristics of Kahn's concrete, prior to the construction of the three blocks the formwork of the retaining wall are assembled, with the first pour in the low curved wall that serves as the border of the parking area, together with the northern end of the retaining wall. The contractor asks Kahn to provide the "form design for North Retaining Wall" and to approve the "plastic cone snap ties."[26]

Kahn designs the North Retaining Wall with the typical panel measuring 4 × 8 feet; but there are quite a few exceptions to this standard size, because the panels are assembled in large compartments defined by expansion joints, and the difficulty lies in balancing the pattern of the spreaders, the imprints, the expansion joints and the two opposite faces of the wall at the height of the parapet. The drawing also shows the face of the "wall model" (actually a parapet, with three 4 × 8 feet arranged horizontally), the "tie layout" and the design of the edges cut in the joint to produce a V-shaped fin imprint.[27]

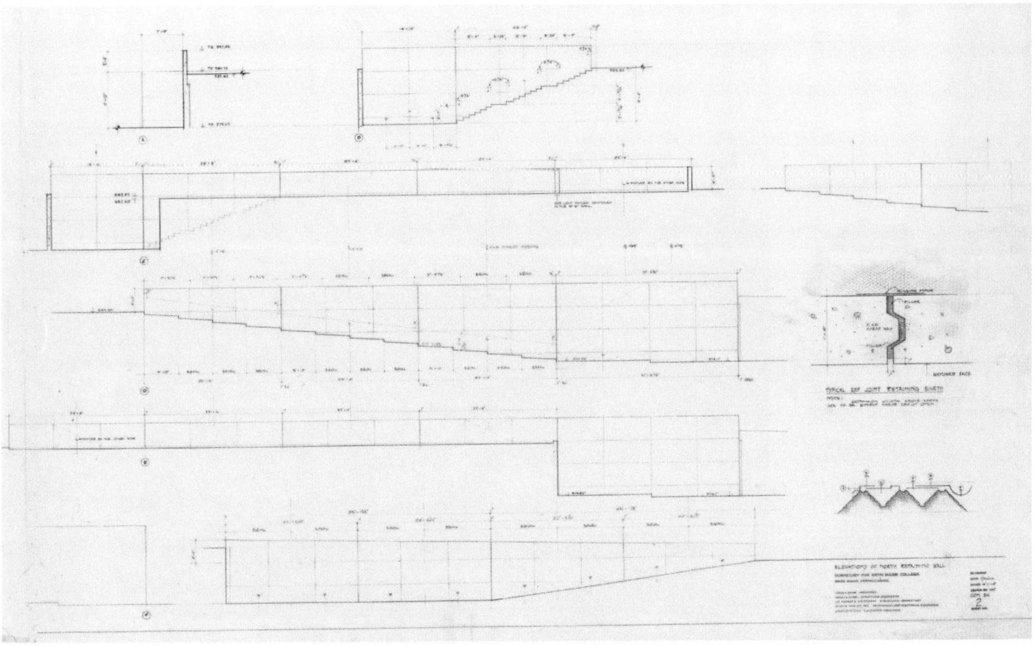

156. Eleanor Donnelley Erdman Hall, Bryn Mawr College, Bryn Mawr, Pennsylvania, 1960-65. "Elevation of North Retaining Wall," 26 July 1963 (LIK, 030.I.C.565.2).

[18] Buckingham-Virginia Slate Corporation, letter to John H. Roberts, 12 April 1963, LIK, 030.II.A.29.19.
[19] Nason & Culler, Inc., letter to Louis I. Kahn, 24 April 1963, LIK, 030.II.A.29.21.
[20] "Specification changes," addendum 5, 21 May 1963, LIK, 030.II.A.20.10.
[21] Nason & Culler, Inc., letter to Louis I. Kahn, 12 June 1963, LIK, 030.II.A.28.17. In April and May, Kahn studies the horizontal support bands of the slate slabs, to make in precast concrete (Nason & Culler, Inc., letter to Louis I. Kahn, 31 May 1963, LIK, 030.II.A.29.21). Only on 10 September does Kahn approve the samples of the precast concrete and of the "natural cleft finish" of the slate supplied by the Structural Slate Company (Harlyn E. Thompson, letter to Nason & Culler, Inc., 10 September 1963, LIK, 030.II.A.28.21).
[22] See the handwritten not added to the specifications, n.d., [1963], LIK, 030.II.A.107.19.
[23] Nason & Culler, Inc., letter to Louis I. Kahn, 24 April 1963, cit.
[24] In August, Kahn approves the samples of "concrete block for the foundations," supplied by the York Building Products Company (Harlyn E. Thompson, letter to Nason & Culler, Inc., 6 August 1963, LIK, 030.II.A.29.64).
[25] See the drawings: "Formwork. Recreation Room Walls," 10 March 1963; "Formwork. Mechanical Room Walls," 20 October 1963; "Formwork. Living Hall Walls," 17 December 1963; and "Formwork. Entrance Hall Walls," 10 January 1964 (LIK, 030.I.C.565.002).
[26] "Job meeting," 22 July 1963, LIK, 030.II.A.28.30.
[27] See the drawing "Elevations of North Retaining Wall," 26 July 1963, LIK, 030.I.C.565.002.

At the end of July, Kahn provides the contractor with the set of drawings titled "Elevations of North Retaining Wall,"[28] leading to assembly of the formwork and pouring tests.[29]

While the making of the formwork of the North Retaining Wall allows Kahn to observe the ability of the craftsmen of Nason & Cullen, Inc., to assemble panels and spreaders in keeping with a pre-set design, the curved test wall serves above all to evaluate, prior to the pour in the North Retaining Wall, the nature of the compound and the effect of the surface shaped by the plywood.

Kahn approves the "plastic tipped accessories for concrete work" in August.[30] But the assembly of the formwork of the North Retaining Wall is criticized, as is borne out by the telephone call of Thompson to Nason & Cullen, Inc., to express disappointment with the positioning of the form ties in the plywood panels – "ties are too close together" – and to advise the contractor that the "alternate form pattern for curved wall" and the "horizontal form lines" are "not acceptable."[31]

Several decisions regarding the concrete are made by Kahn on 12 August, following the test construction of the concrete of the curved wall. The report on the inspection shows that Kahn is satisfied with the quality

157-159. Eleanor Donnelley Erdman Hall, Bryn Mawr College, Bryn Mawr, Pennsylvania, 1960-65.

157. North retaining wall (photograph by the author, 2010).

158. Detail of the north retaining wall (photograph by the author, 2010).

159. Parking wall (photograph by the author, 2010).

of the surface shaped by the plywood panels and the concrete compound, which he defines as "density";[32] he even accepts the slight wrinkles produced on the concrete by the curved panels, comparing them to the wrinkles in the skin of an elephant.[33] The fact that the compound has not been formulated on the basis of color, as at La Jolla, confirms that Kahn also thinks of concrete in terms of its exposure, or lack of it, to natural light.

During the visit Kahn does not approve of the joint in the corners of the formwork panels, so before proceeding as planned with the pouring of the North Retaining Wall which is destined to remain exposed, it is decided to construct several walls in the lower levels of Square C, not visible from the outside, transforming them into other trial samples.[34]

Though he has approved the finish of the wall samples, a few weeks later Kahn, with his team, meets on the worksite with a representative of the Concrete Service Company to discuss the type of form-coat used by Nason & Cullen, Inc., which as the documents show has left Kahn dissatisfied. The recommendation of the Concrete Service Company points in the direction desired by Kahn, of erasure of any imprint of wood grain, and therefore also of those elephant skin wrinkles: "Concrete Service indicated that a two-coat application, the first coat being thin followed by a heavier coat, would leave the finished concrete surface absent of plywood graining or pattern."[35]

While the "screw type cone ties" have been approved, uncertainty remains on the "covering of the exposed metal ties in the exterior Retaining Wall."[36] In the meantime, Nason & Cullen, Inc., requests the drawing for the "formwork pattern in the Boiler Room." It is at this point that the contractor is informed that all the drawings of the forms for the exposed concrete must be approved by Kahn "prior to construction."[37] In the last days of August, three sections of the North Retaining Wall have already been poured.[38]

The design process of the formwork continues to follow the progress of the worksite. Given the fact that on 26 August Square C is ready for the construction of the "north concrete building wall" and the "concrete core walls," Kahn is asked to supply the relative "formwork drawings," with emphasis on the need for a meeting with the physical plant engineers to definitively establish the position of the openings in the walls.[39] The drawing of the formwork of the North Concrete Building Wall – the perimeter base of Square C, which will support the walls of the dorm rooms – is made by

[28] *Ibid.*
[29] "Job meeting," 5 August 1963, LIK, 030.II.A.28.30.
[30] Harlyn E. Thompson, letter to Nason & Culler, Inc., 6 August 1963, LIK, 030.II.A.29.64.
[31] Harlyn E. Thompson, letter to Nason & Cullen, Inc., 8 August 1963, LIK, 030.II.A.28.21. The pouring of the North Retaining Walls begins after 12 August ("Job meeting," 12 August 1963, LIK, 030.II.A.28.30).
[32] "Job meeting," 12 August 1963, LIK, 030.II.A.28.30
[33] Nicholas L. Gianopulos, interview with Anna Rosellini, February 2010.
[34] "The Architect has approved the sample concrete wall for color and finish – finish meaning the density and the absence of plywood grain pattern. The projecting 'fin' at all form edges was not approved. Mr. Nason, Jr. did not want to hold up the progress of the formwork construction in order to pour another sample for 'fin' approval. Mr. Kahn agreed to allow the concrete walls to be poured without 'fin' approval only after Mr. Nason, Jr. agreed to remove the first pour of a finished wall if the 'fin' detail was not to the Architect's satisfaction. Mr. Nason, Jr. stated at this meeting that they would pour concrete walls in the Boiler Room for 'fin' approval prior to the exterior Retaining Walls" ("Job meeting," 12 August 1963, cit.).
[35] Harlyn E. Thompson, letter to Nason & Culler, Inc., 29 August 1963, LIK, 030.II.A.29.64.
[36] "Job meeting," 12 August 1963, cit.
[37] *Ibid.*
[38] "Job meeting," 26 August 1963, LIK, 030.II.A.28.30
[39] "Formwork drawings for north building wall will be sent out Tuesday of this week," is the commitment made by Kahn's office ("Job meeting," 26 August 1963, LIK, 030.II.A.28.30).

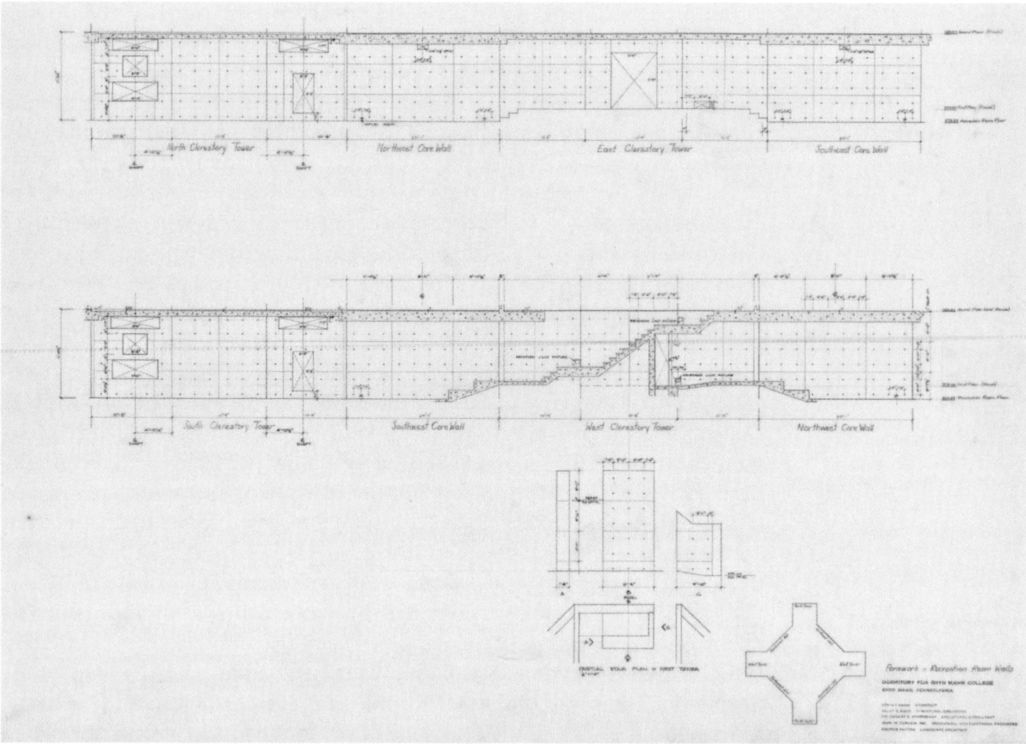
North Clerestory Tower
Northwest Core Wall
East Clerestory Tower
Southeast Core Wall
South Clerestory Tower
Southwest Core Wall
West Clerestory Tower
Northwest Core Wall
Formwork – Recreation Room Walls
DORMITORY FOR BRYN MAWR COLLEGE
BRYN MAWR, PENNSYLVANIA

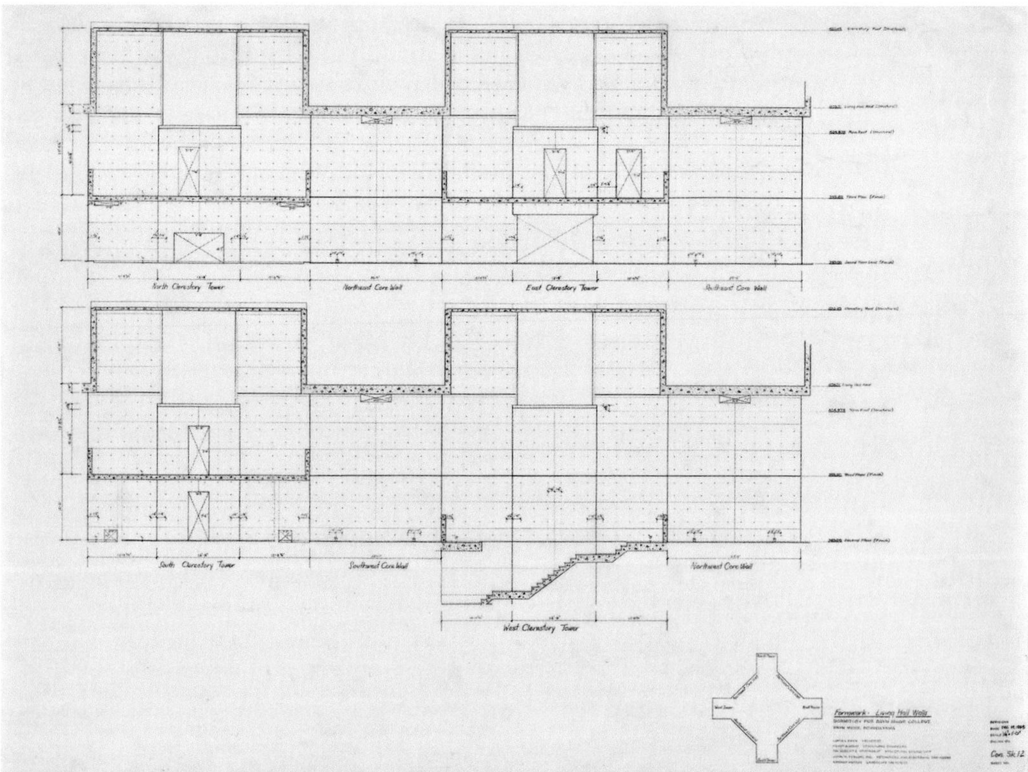
North Clerestory Tower
Northeast Core Wall
East Clerestory Tower
Southeast Core Wall
South Clerestory Tower
Southwest Core Wall
Northwest Core Wall
West Clerestory Tower
Formwork – Living Hall Walls
Con. Sk 12

160-161. Eleanor Donnelley Erdman Hall, Bryn Mawr College, Bryn Mawr, Pennsylvania, 1960-65.

160. "Formwork – Recreation Room Walls," 3 October 1963 (LIK, 030.I.C.565.2).

161. "Formwork – Living Hall Walls," 19 December 1963 (LIK, 030.I.C.565.2).

Kahn's office on 27 August.[40] All the measurements of the panels in the three masonry compartments are different. Kahn favors the cross joint between the four horizontally mounted panels in the two large lateral compartments so that the pattern of the spreaders will remain as regular as possible, though in the left compartment the vertical imprint of the cross is not located at the center of the wall. Only in the small central compartment does Kahn use two vertically assembled panels, with the imprint of the joint studded by a row of spreaders, as already happened in the construction of the Biology Building. The drawing confirms that Kahn is still playing with the optical perception effects of the imprints, just as when he assembled the wooden boards to disguise the imperfections of the pour. In the case of the formwork with panels of different sizes, he continues to rely on the potential of pattern regularity of the pattern of holes left by the spreaders.

On 3 September, Kahn's office produces the drawings of the "Formwork. Boiler Room Walls" of Square C. Those walls form the foundation nucleus of the "core walls" with the four "clerestory towers" of what will become the Dining Hall, on the level above. Thanks to the drawing of the "Boiler Room Walls" a graphic pattern is taking form on the concrete walls of the Dormitory, though not all the details are yet definitively in place.

The construction drawings by Kahn of the sections of the Dormitory, precisely starting with that of the Boiler Room Walls, depict the development of the walls, designed as rectangular surfaces extended on one floor, parts of a continuous linear sequence where each wall is recognizable thanks to the edges (the vertical lines are more marked than the others). Kahn's graphic system of the sections is developed specially to control all the assembly of the formwork panels with the drawing, from one wall to the next, seamlessly. With this type of section drawing, Kahn can see the continuous development of the pattern of the spreaders and provide the contractor with a fundamental document for the assembly of the formwork, including the panels and spreaders. The idea of a developing depiction of the walls of the Dormitory could be derived from the many exercises of representation of tetrahedra and other polyhedrons done by Kahn with Tyng, precisely because the "core walls" are part of a geometry that had originally been conceived as tetrahedral. It is significant, in this sense, that this type of development of the walls does not correspond to those used in the design of the Salk Institute, though the latter work also engages Kahn's office in the development of a complex formwork system, depicted down to the smallest details (the walls of the Salk Institute are perpendicular, with the exception of the sloped segments in the Studies of the researchers).

In the "Formwork. Boiler Room Walls" each wall sector is treated as a composition of panels to assemble according to a symmetrical order, no longer using the expedients implemented in the North Concrete Building Wall. The 4 × 5 feet panel is the module that orders the measurements of the various wall sectors, also with imperceptible extensions that range from 1 to 3 inches, to 4 1/2, to 7 inches, depending on the wall sector. The width of the large walls of the "core" generates an arrangement of three panels of 4 × 5 feet - 4 1/2 inches with one 4 × 8 feet panel at the center (but this arrangement is destined to be changed). The juxtaposition of the panels produces a weave of slender V lines in slight relief; instead, to identify the two levels depicted in the section, a continuous shadow groove is specially created by

[40] "Formwork. North Concrete Building Wall," 27 August 1963, LIK, 030.I.C.565.002. Several modifications are made on Thursday 29 August.

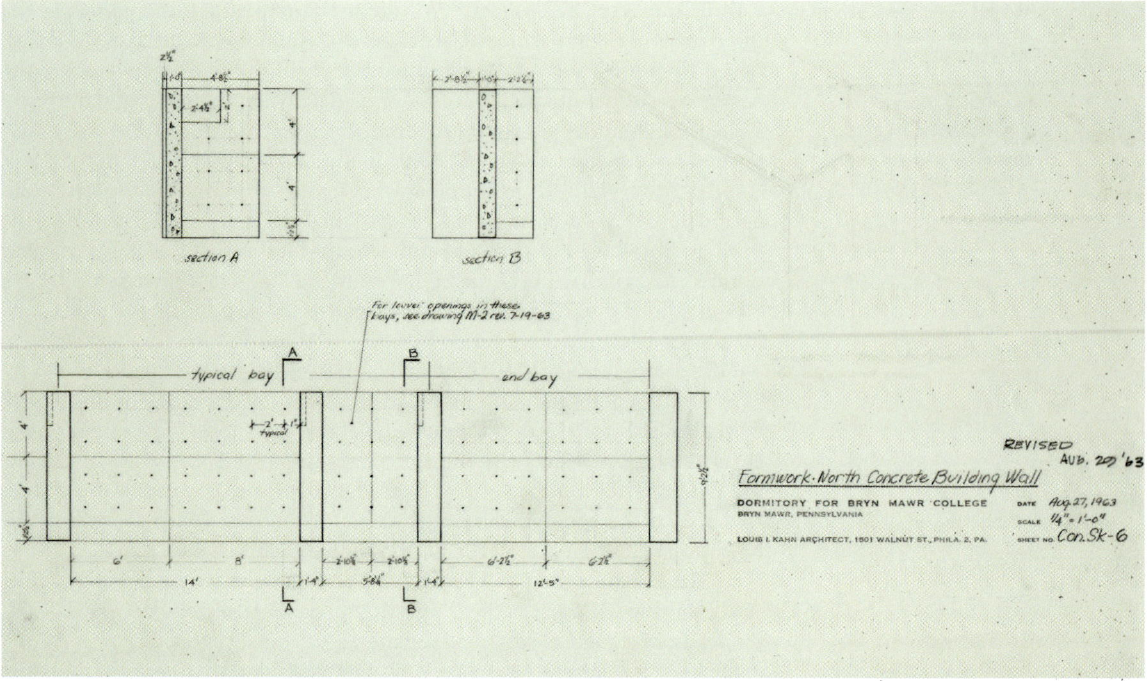
section A
section B
For louver openings in these bays, see drawing M-2 rev. 7-19-63
typical bay
end bay
REVISED AUG. 20 '63
Formwork · North Concrete Building Wall
DORMITORY FOR BRYN MAWR COLLEGE
BRYN MAWR, PENNSYLVANIA
LOUIS I. KAHN ARCHITECT, 1501 WALNUT ST., PHILA. 2, PA.
DATE Aug. 27, 1963
SCALE 1/4" = 1'-0"
SHEET NO. Con. Sk-6

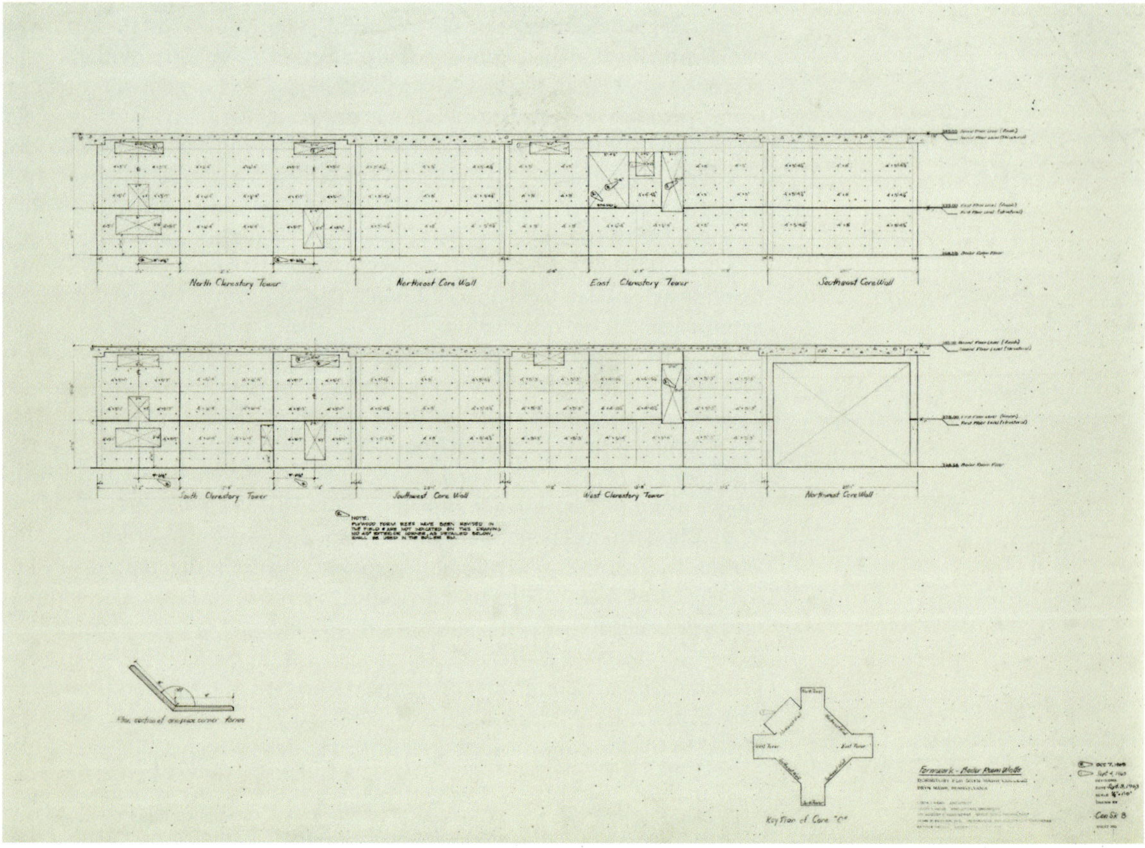
North Clerestory Tower
Northeast Core Wall
East Clerestory Tower
Southeast Core Wall
South Clerestory Tower
Southwest Core Wall
West Clerestory Tower
Northwest Core Wall
Key Plan of Core "C"
Formwork - Boiler Room Walls
Con Sk 8

162-163. Eleanor Donnelley Erdman Hall, Bryn Mawr College, Bryn Mawr, Pennsylvania, 1960-65.

162. "Formwork – North Concrete Building Wall," 27 August 1963 (LIK, 030.I.C.565.2).

163. "Formwork – Boiler Room Walls," 3 September 1963 (LIK, 030.I.C.565.2).

means of a horizontal strip inserted between the panels corresponding to the pour of the floor slab. As in other constructions by Kahn, this strip does not reveal the thickness of the floor slab, but the level of the pour of the concrete of the wall at the position of the floor slab. In all the walls the spreaders are positioned to form a regular 2 × 2 feet pattern. The exceptions to the 2 × 2 feet grid of the ties are studied for use as optical corrections – it is difficult to notice the variations in the measurements of the pattern.

Kahn's criticism of the technical finish of the corners between the panels of the formwork – "The projecting 'fin' at all form edges was not approved" – gives rise to studies, again based on the experience of the worksite at La Jolla, to resolve this joint detail. Precisely in the design of the "Formwork. Boiler Room Walls," Kahn envisions the insertion, only along the four corners formed by the walls between the central core and the towers, of special vertical pieces consisting of two wooden boards joined at an angle of 135 degrees. Each board is 8 inches wide and has the free edge cut to create the V-shaped relief line on the concrete, when butted against another panel with a similar beveled edge.

From Monday 9 to Friday 13 September, Nason & Cullen, Inc., begins construction of the formwork of the North Concrete Building Wall and the Concrete Core Walls of Square C, as well as excavation for the foundation of Square A.[41] The pouring of the concrete of the Boiler Room Walls up to the first level, marked by the groove, starts at the end of September and continues into the first days of October.[42] The lower part of Square C is the only one in the Dormitory where the concrete of the "core walls" is shaped with the special corners based on the formwork of the Salk Institute.

For the tie holes Kahn begins by indicating a seal based on the model of what was done on the worksite in La Jolla. While from the early days of August he had approved the "screw type cone ties," at the time the "covering of the exposed metal spreaders in the exterior Retaining Wall" remained to be specified.[43] At the start of September, Kahn is oriented towards the solution applied at La Jolla, and thus requests an estimate from Nason & Cullen, Inc., for "lead fill in exterior concrete wall cone tie holes."[44] The company communicates the cost for "furnishing and installing lead wool to form a plug around the form tie cone holes of all of the exposed exterior vertical faces of the concrete walls"; the depth of the visible hole after the sealing with the lead disk is also specified: "There will be a 3/4" recess from the face of the wall when the plug hole is filled."[45]

The depth measurement of the hole left by the spreader is a far from negligible technical detail for control of the pattern of imprints of Kahn's concrete wall. That depth generates a more or less intense shadow, giving the pattern of the holes more or less force in the overall design of the wall. The tie holes of the Dormitory are less deep than those of the Salk Institute, and the pattern of those cavities stands out over all the other joint imprints, unlike what happens in the concrete at La Jolla, and precisely as Kahn had foreseen in the development drawings for the formwork.

The construction of Squares A and B begins with that of the two respective "north concrete building walls." From 16 to 27 September the

[41] See "Job meeting," 9 September 1963, and "Job meeting," 23 September 1963 (LIK, 030.II.A.28.30).
[42] "Job meeting," 23 September 1963, LIK, 030.II.A.28.30.
[43] "Job meeting," 12 August 1963, cit.
[44] "Job meeting," 9 September 1963, cit.
[45] Nasons & Cullen, Inc., letter to Louis I. Kahn, 10 September 1963, LIK, 030.II.A.29.64.

formwork of these walls is assembled, and the first pouring phase for the wall of Square B begins.[46]

The pattern and joint of the formwork panels are modified during the course of the construction, in Squares A and B. The design of large panels leads to identification of the material best suited to this type of formwork. To pour the walls of the three Halls, at the end of September an order is placed for "extra long plywood panels for formwork, 10'-0" and 12'-0" lengths."[47] The special panels come from the West Coast and reach the worksite towards the end of October.[48] Certain notes on the drawings document the research on the ideal measurements for the panels: "plywood form sizes have been revised," we read, for example, in the margin of the drawing titled "Formwork. Boiler Room Walls" of 3 September 1963 – the revision is done on 7 October.[49]

Nason & Cullen, Inc., on 23 September, asks Kahn for the drawings of the formwork for the "core walls" of Square A,[50] while on Monday 30 September the firm indicates that it will assemble the formwork for that part of the Dormitory during the week. But Kahn's office prepares the definitive version of the drawings of the formwork of the first level of the "core walls" of Square A only on Thursday 3 October – this is the set titled "Formwork. Recreation Room Walls" (a modification is made on Monday the 7th, following the weekly worksite meeting). Kahn decides to directly abut the panels in the corners of the large walls of the core, to generate lines without the imprints of those special pieces that have become too typical of the Salk Institute, though already done in Square C. For the "core walls" he proposes a different pattern of panels than that already used in Square C, formed by a central panel of 4 × 8 feet, flanked by two panels measuring 4 × 6½ feet. The spreaders are still positioned to form the regular grid of 2 × 2 feet, with the usual rare and always, in any case, imperceptible exceptions.

The discussions on Monday 7 October also address the modifications to make in the openings in the concrete walls for the physical plant systems. The formwork of the "core walls" of the Recreation Room Walls are assembled in the weeks from 7 to 18 October – pouring is done starting on 21 October.[51] A note added to the drawing "Formwork. Boiler Room Walls" of Square C, precisely following the meeting of 7 October, specifies that "no 45° exterior corner, as detailed below, shall be used in the Boiler R[oo]m." Kahn has therefore decided not to continue, in the level above the groove corresponding to the "core walls" of Square C, the motif of the special corners as in the Salk Institute (the indication of 45° in the note is an error on the part of its writer, since it should have been the 135° angle of the special formwork piece). Only after the modifications made to the pattern and the corners of the formwork in Square A does Kahn decide to have the formwork in the level above the groove assembled with the new solutions. In the week from Monday 7 October to Friday the 11th the formwork of the walls of Square C are completed, up to the second level; the pouring happens in two phases, in the week from 14 to 18 October, and the week from 21 to 25 October.[52]

At the meeting on Monday 28 October, Nason & Cullen, Inc., plans to start, during the same week, construction of the formwork of the "core

[46] See "Job meeting," 16 September 1963, LIK, 030.II.A.28.30, and "Job meeting," 23 September 1963, cit. The pouring of the "north concrete building wall" of Square A is done in the week from 30 September to 4 October; that same week the northeast wall of Square A is also poured ("Job meeting," 30 September 1963, LIK, 030.II.A.28.30).
[47] "Job meeting," 23 September 1963, cit.
[48] "Job meeting," 21 October 1963, LIK, 030.II.A.28.30.
[49] "Formwork – Boiler Room Walls," 3 September 1963, LIK, 030.I.C.565.002.

walls" of Square B.[53] On the following day Kahn's office prepares the definitive drawing for these walls – the set titled "Formwork. Mechanical Room Walls." Again in this case, Kahn proposes another pattern of panels, molding the "core walls" with three horizontal panels, two measuring 4 × 6 feet, and one at the center measuring 4 × 4 feet, 1 inch. Starting on Monday the 4th, the formwork of the Mechanical Room Walls is assembled.[54] The pouring starts in the week from 18 to 22 November.[55]

At the end of the construction of the first level of Squares A, B and C, the formwork patterns in the three Squares are different. Kahn has experimented, in each one, with a different way of assembling the panels, not to create the different characteristic textures of the various blocks, but in pursuit of an ideal pattern with which to construct the main walls of the three large Halls to be placed above the first level. It is clear that the lower level of the three Squares of the Dormitory, for him, is a true experimental worksite to test the ideal pattern of his architectural concrete, so it too contains the "fossils" of the experimentation, just like the East Wall of the Mechanical Building in La Jolla.

The design drawings of the formwork system made to date demonstrate that all the measurements, both of the panels and the spreaders, are correlated to generate two modular patterns of imprints. Each panel segment that differs from the basic panel is calculated, in any case, to create a modular design of imprints. Apart from certain exceptions, it is the pattern of the spreaders that takes on a fundamental role as a unit of measure visible everywhere in the exposed concrete construction. Every opening in the walls marked by these patterns is studied to be as closely related as possible to the system of form imprints. Even the recesses made for the installation of electrical sockets are usually arranged in relation to the square pattern of the spreaders. The fact that Kahn, at this point, is calculating the measurements of the panels in keeping with the wall segments in order to control the imprints of the joints is confirmed by the formwork system developed for the staircase in the Entrance Hall, where the panels have the height of those of the walls (4 feet), but the lengths are gauged to create symmetrical figures on each face (the pattern of the ties remains that of 2 × 2 feet).

Yet none of the three patterns implemented to date in the lower sectors of Squares A, B and C is yet the definitive pattern for the three Halls.

The increase in the measurement of the panels decided on in September allows Kahn to assemble a formwork system on the scale of the large visible walls in the three Halls. The purpose, as in the case of the slate and the cast stone, is to reduce the joints to a minimum. While for the cladding of the walls in slabs Kahn focuses on obtaining a refined vertical effect, in the assembly of the horizontal panels he wants to achieve an enveloping effect for the walls of the Halls. In the research on the systematic assembly of the horizontal panels, by now Kahn again clearly distinguishes his work from the *béton brut* of Le Corbusier (the horizontal panels of the Maison des Jeunes et de la Culture at Firminy represent an exception in the stereotomy of the formwork of Le Corbusier).

[50] "Job meeting," 23 September 1963, cit.
[51] See "Job meeting," 7 October 1963, LIK, 030.II.A.28.30; "Job meeting," 14 October 1963, LIK, 030.II.A.28.30; and "Job meeting," 21 October 1963, cit.
[52] See "Job meeting," 28 October 1963, LIK, 030.II.A.28.30.
[53] *Ibid.*
[54] See "Job meeting," 4 November 1963, LIK, 030.II.A.28.30, and "Job meeting," 18 November 1963, LIK, 030.II.A.28.30.
[55] See "Job meeting," 18 November 1963, LIK, 030.II.A.28.30.

The construction of the walls in the Halls

The formwork system of the walls of the Halls is the most complete and regular of all those used in the Dormitory. Kahn has the great panels assembled horizontally for the concrete of the four two-story walls, or "core walls." While the height is still 4 feet, like that of the panels of other wall sectors of the Dormitory, the length of each panel of the Halls is obtained by dividing the length of each wall by two. In this way, the joint between the pairs of overlaid panels formwork the vertical axis of the wall, at the top of which is inserted a rectangular air duct. Each panel contains two rows of spreaders, arranged in the usual 2 × 2 feet pattern. In the juxtaposition of the panels the pattern extends from one to the next in a regular way.

For the design of the formwork of the Dining Hall, documented in the set "Formwork. Dining Hall Walls" dated 25 November 1963, Kahn uses large panels, the longest of which extend beyond 10 feet by several inches, which vary according to the walls (the height is always 4 feet). For the large inclined walls of the core, in the drawings the measurements of one panel are 4 × 10 feet 1 inch.

The assembly of the formwork and the pouring of the concrete continue at a regular pace in Squares A, B and C during the month of November.[56] In December, Kahn and Nason & Cullen, Inc., discuss a possible change in the measurements of the walls in the Dining Room and Living Room "that connect the clerestories." The contractor suggests changing from 20 feet, 1 inches to 20-0 feet to make it possible to use two plywood panels of 10'-0" instead of two measuring 12 feet, 0 inches "as had been originally agreed upon."[57] Kahn accepts the change.[58] So it is only during the course of the work that the length of the walls is changed, though just slightly, to obtain not only a regular design of the imprints but, above all, also a logical assembly of panels with the standard measurements offered by the market. The design of the formwork system for the Living Hall takes the modification of the length of the large panels into account – now the panels measure 4 × 20 feet.[59]

164. Eleanor Donnelley Erdman Hall, Bryn Mawr College, Bryn Mawr, Pennsylvania, 1960-65. "Formwork – Entrance Hall Walls," 20 January 1964 (LIK, 030.I.C.565.2).

The worksite documentation demonstrates that Kahn proceeds towards simplification of the formwork elements and their deployment, going beyond the logic of economy on the worksite. In the Dormitory he operates according to criteria different from those of the Salk Institute, where every detail of the construction process is brought out, producing a multiplicity of imprints. The design of the exposed concrete surfaces of the Dormitory achieves a degree of abstraction and essentiality comparable only to certain expressions of American abstract art of the 1950s, to paintings like those Ellsworth Kelly made by assembling rectangular panels with visible joints, or those of Frank Stella with monochrome surfaces woven with luminous lines surfacing from a black background.[60]

In the drawing of the walls of the staircases, part of the set titled "Formwork. Entrance Hall Walls" dated 20 January 1964, Kahn applies the criterion of the development of the faces, even if the walls are orthogonal to

[56] See "Job meeting," 4 November 1963, cit.; "Job meeting," 18 November 1963, cit.; and "Job meeting," 2 December 1963, LIK, 030.II.A.28.30.
[57] Harlyn E. Thompson, letter to Nason & Cullen, 13 December 1963, LIK, 030.II.A.28.21.
[58] *Ibid.*
[59] See the drawing "Formwork. Living Hall Walls," 19 December 1963, LIK, 030.I.C.565.002.
[60] See the painting by Ellsworth Kelly, *Gaza*, 1953-56 (San Francisco Museum of Modern Art), and that of Frank Stella, *The Marriage of Reason and Squalor II*, 1959 (Museum of Modern Art, New York).
[61] See "Formwork – Entrance Hall Walls," 20 January 1964, LIK, 030.I.C.565.002.

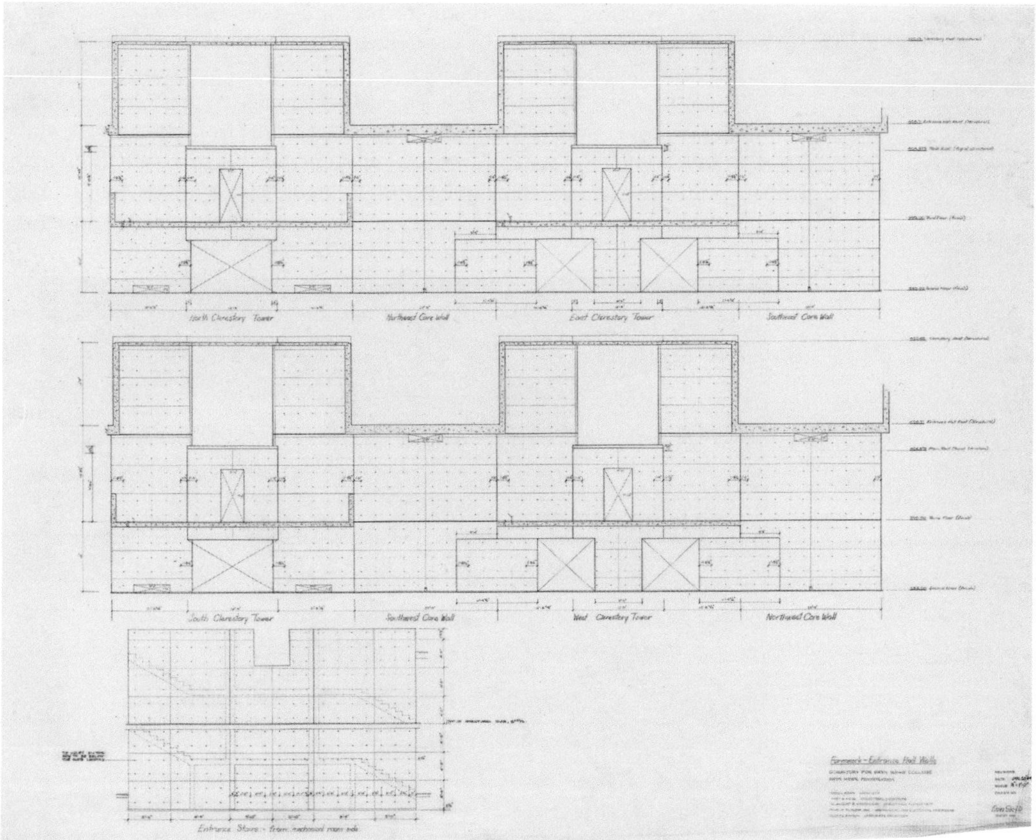

each other.[61] It is clear by now that precisely that kind of depiction permits control of the positioning and measurements of the panels and form ties of the forms. The drawing shows all the visible signs, from the spreaders to the joints between panels, to the horizontal groove at the "top of the structural slab," while dotted lines show the ramps with the steps, the walls and floor slabs, as in drawings of the Service Towers of the Salk Institute, demonstrating the complexity of the technical control to obtain an exposed concrete in the American manner, perfectly defined in every aspect of the pouring in place. Notes like "tie hole on this row to be below the slate landing" confirm the technical calculation required for the arrangement of the formwork elements on the basis of the various structures.

The design of the volume of the stairs is calculated to achieve a single, synthetic geometry of space and construction. It is no coincidence, in fact, that Kahn does not introduce any slanted lines, like that of the parapet of the final ramps, for example, to confirm the force of the pattern of imprints of the formwork system. Even the large rectangular opening for the view at the height of the landing is studied to leave a distance from the vertical rows of the spreaders that is the same as the distance left between the horizontal rows at the top of the walls. Everything is calculated so that even the smallest details of the placing of the concrete will create an ornament that is imperceptible as such, which seems like a natural consequence of the construction process, a controlled, abstract "ornament" like those of the series painted by

165-167. Eleanor Donnelley Erdman Hall, Bryn Mawr College, Bryn Mawr, Pennsylvania, 1960-65.

165. View of the Entrance Hall (photograph by the author, 2010).

166-167. Details of the surface in concrete of the Entrance Hall (photographs by the author, 2010).

Barnett Newman. This obstinate pursuit of agreement between all lines is another distinctive feature of the degree zero of Kahn's concrete. The fact that the metal railing with slender posts is installed as an extension of the concrete wall cannot help but evoke precisely the image of that same wall; that railing completes the conceptual sequence that runs from the fullness of the wall to its reduction to reinforcement without concrete, to its definitive erasure in the void above the landing. "All little things but still a whale of a difference to the uncompromising eye," Kahn writes, about his Dormitory.[62]

The design of the formwork for the waffle slab

Several development drawings for the construction of the floor slab in the Halls of the Dormitory of Bryn Mawr College demonstrate Kahn's concern with identifying a pattern of formwork with rectangular panels to assemble in tune with the square pattern of the coffers and the compositional geometry of the motif. The basic measurements of the parts of the coffers make it possible to consider a range of solutions for the formwork – width of the structural bands, 1 foot; square opening of the coffer, 2 feet. A study drawing for the floor slab, dated 14 October 1963, is emblematic of the simplification of the construction of the formwork for the coffers being pursued by Kahn, with the resulting detachment between the original compositional geometry and the most practical assembly of the formwork.

The drawing shows all the lines of the compositional geometry of the coffers, including the diagonals at the squared intersections of the structural bands. Many of those lines would have to be visible on the poured in place concrete. At first Kahn and his staff consider building as many wooden cases, or "mounds" in keeping with Alberti's definition, as there are coffers. These cases are equipped with a base frame, to be made with corners with miter joints. The frame boards, placed against each other, would leave accentuated V-shaped imprints on the pour, thanks to beveled edges to create true framing lines. The cases would rest on a deck of plywood panels. Kahn and his staff study the deck to be made with an 8 × 4 feet panel, the same one also used for certain sectors of the wall and for the perimeter band of the floor slab itself.[63] The panels are arranged along the diagonals of the coffers and assembled in a staggered way. In the same drawing, a segment traced halfway along a structural band indicates the hypothesis of arranging the deck panels in alignment with the geometry of the coffers. The whole process of assembly of the cases demonstrates that Kahn is transforming the assembly technique of the formwork into something similar to the ceiling work of carpenters.

The very next phase in the studies for the floor slab involves the elimination of the frames at the base of the coffers, so that the wooden cases are attached directly to the deck panels. It is at this point that what was a simple temporary structure for pouring, in the initial version – the deck of panels – becomes the protagonist of the definition of the pattern and imprints on the concrete. This is no mere technical detail, but a highly significant clue as to how Kahn is attempting to incorporate all the phases of construction inside the work. In the study drawing of the pattern of the panels of the deck on which to place the cases of the coffers, the possibility is considered of changing the size of the panel. The smaller side passes from 4 to 3 feet to make it possible to arrange the panel exactly straddling a series of coffers (width of one coffer, 2 feet, to which to add the two halves of the structural bands, which when summed measure 1 foot).

The definitive formwork system used for the waffle slab, namely the "Formwork Waffle Slab Decks," is designed precisely to obtain a pattern of imprints related as closely as possible to the geometry of the structure, but always based on the panel of the deck.[64] By now all the components of the formwork system are devised in such a way as to each leave an imprint

[62] Louis I. Kahn, letter to Gertrude Ely, 3 July 1964, LIK, 030.II.A.28.8.
[63] See the drawing "Formwork for Central Halls," 14 October 1963, LIK, 030.I.C.565.002.
[64] See the drawing "Formwork Waffle Slab Decks," 22 October 1963, LIK, 030.I.C.565.002. The anchoring of the Tectum panels is approved ("Job meeting," 28 October 1963, LIK, 030.II.A.28.30).

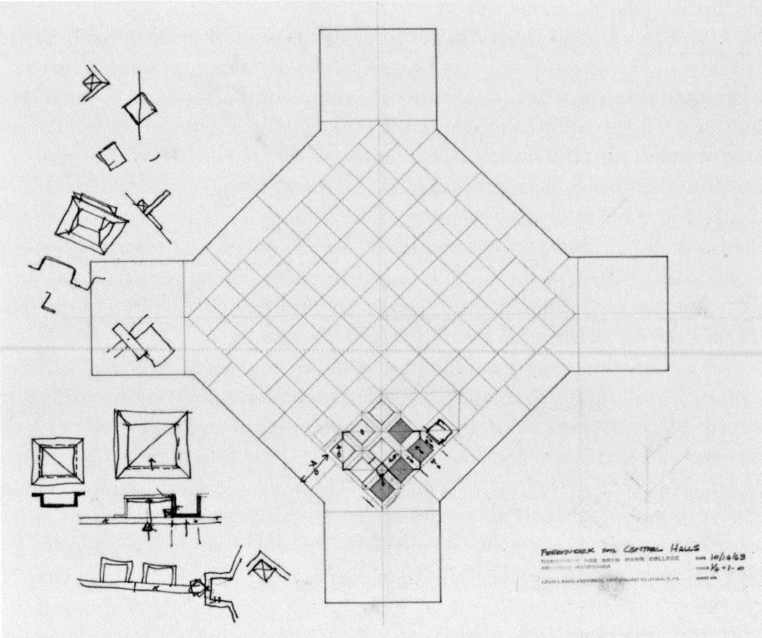

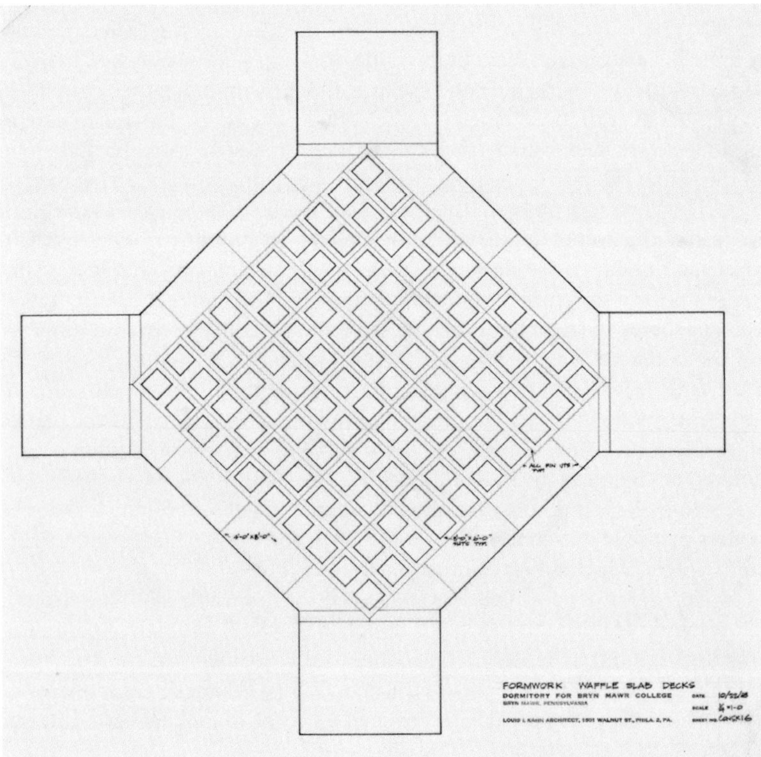

168-170. "Formwork for Central Halls," 14 October 1963 (LIK, 030.I.C.565.2).

168. "Formwork for Central Halls," 14 October 1963 (LIK, 030.I.C.565.2).

169. "Formwork Waffle Slab Decks," 22 October 1963 (LIK, 030. I.C.565.2).

170. Detail of the Waffle Slab Deck in the Entrance Hall (photograph by the author, 2010).

on the waffle slab, and all the traces of the growth on the worksite will become visible.

The surface for the pouring of the concrete of the coffers is obtained with panels measuring 3 × 6 feet, so that their combination produces a series of 6 × 6 feet squares. Thanks to this assembly, each 3 × 6 feet panel contains a pair of coffers, separated by a rib without joint imprint. The lack of this imprint is a telling clue as to the fabrication process of the concrete of the floor slab, which is poured in place and not precast. We cannot rule out the idea that this absence has been intentionally sought by Kahn precisely to reveal the nature of that concrete with non-prefabricated coffers. Had Kahn wanted to obtain a perfect unity of design between the assembly of the formwork and the design of the structure, without any intention of revealing the process of construction of the concrete, he would have had 3 × 3 feet panels assembled, thus using a measurement that would produce a pattern of imprints bordering all the ribs of each coffer typical of precast concrete, but which would have been illogical with respect to the usually rectangular panel forms of poured in place concrete. At this point the imprints between the panels can no longer be the V-shaped features of the original version, given the fact that they no longer form frames for the coffers, but are the traces of a construction process not directly related to the waffle. In the drawing of the formwork of the floor slab the type of technical implementation of the joints between the panels is indicated: "all fin j.[oin]ts," leaving slender lines as the imprints on the concrete.

Along the borders of the large coffered ceiling runs a flat band obtained with two panels of 4 × 8 feet for each side, with special pieces in the corners, also cut to produce a precise imprint, and joined with the same system. A rectangular sector of the ceiling at the second level is built in such a way as to permit removal of the sheet covering the coffers.[65]

The drawings of the formwork of the walls and slabs offer a fundamental guide for the assembly of the panels on the worksite. The contractor follows what is laid out in those drawings. After experiences such as that of the Biology Building, by now Kahn's office has refined a kind of representation through drawing of the system of assembly of the formwork that permits almost total control, also over the fabrication of the poured in place concrete.

[65] See the drawing "Detail of Inserts in Waffle Roof Slabs," 26 March 1964, LIK, 030.I.C.565.002.

Placing of the cladding in slabs of concrete and slate

The exposed reinforced concrete determines the character only of the core of each Square of the Dormitory of Bryn Mawr College. For the construction of the walls of the rooms that wrap that core, concrete blocks are used – in this part of the complex only the floor slabs are in poured in place concrete. But unlike other, previous works by Kahn, the blocks are not left exposed in any part of the Dormitory. Along the perimeter of the three volumes, the cladding of the concrete block walls is in slate and precast concrete. The slate is cut into slabs of two heights: the lower ones form the band of the windowsills; the higher ones close the wall sectors between the sills and the floor slab. The unpolished black surface of the slate forms a contrast with the smooth, pale surface of the precast concrete, in a formal play of materials that react differently to the light, similar to the one staged in the Richards Laboratories.

In the Dormitory the horizontal concrete bands have two thicknesses: a thinner one corresponding to the floor slab, and a thicker one corresponding to the sills (the same band is also installed corresponding to the flat roof). For the installation of the sill bands, special recesses are left in the concrete block walls, with the height of one row of blocks (this is why the height of the bands coincides with that of the concrete block). The bands contribute to the fastening of the slate slabs.

The ends of the divider walls between the rooms are clad with concrete slabs, with the height of one floor; where they meet, a slender horizontal band passes, corresponding to the floor.

Kahn approves the "color and texture" of the "precast concrete samples" at the start of August 1963.[66] The making of the slabs and cornices in precast concrete is assigned to the Herculite Corporation, based in Feltonville, Pennsylvania, which begins to produce the pieces for the Dormitory in

[66] See "Job meeting," 22 July 1963, LIK, 030.II.A.28.30, and "Job meeting," 12 August 1963, LIK, 030.II.A.28.30.

171-173. Eleanor Donnelley Erdman Hall, Bryn Mawr College, Bryn Mawr, Pennsylvania, 1960-65.

171. Construction site photograph showing the application of the cladding in precast concrete and slate (LIK, 030.IV.A.565.2.7).

172-173. Details (photographs by the author, 2010).

174-176. Eleanor Donnelley Erdman Hall, Bryn Mawr College, Bryn Mawr, Pennsylvania, 1960-65.

174-175. Views (photographs by the author, 2010).

176. S. R. Little, photograph of the stairs, 1966 circa (LIK, 030.II.A.55.27b).

November (the same company produced precast parts for the Mill Creek Apartments II). The Herculite Corporation defines the pieces as "cast stone." When the "first cast stone item" is removed from its mold, the presence of Kahn and of Nason & Cullen, Inc., is required, so they can "observe and inspect the stripping and the newly exposed work."[67] The installation of the slabs in "cast stone" and slate happens starting with Square C.[68] Before the start of the work, on 1 April 1964, Thompson reminds Nason & Cullan, Inc., of the precautions to be used – to discard any piece "that is chipped or damaged," and to store the pieces for possible use "later in remote areas."[69]

During June work is still in progress on the assembly of the slabs of the two types in Squares A and C, while the color of the mortar for the sealing of both types of slab is under study.[70] After completion of the installation in Square C, at the start of August, the phase of "washing down precast and slate" is followed by the operation of "caulking" – the sealing of the joints[71]. The installation of the slabs of precast concrete and slate in Square B is completed halfway through September, and that of Square A halfway through October.[72]

The concrete of the Dormitory of Bryn Mawr College looks perfect. But when the light is oblique, from particular vantage points, it too reveals those flaws of fabrication that Le Corbusier had observed in his monuments at Chandigarh, precisely through the photographs made by his staff. The photograph of the stairs of the Dormitory taken by Sam R. Little of the School of Fine Arts of Philadelphia also reveals the defects of the concrete of Kahn, if seen from an accentuated angle, in a light that reveals a surface shaped by unexpected deformations undergone by the formwork when the "liquid stone" was poured into large panels of flexible plywood.[73]

[67] R.M. Trommer of Nason & Cullen, Inc., letter to Herculite Corporation, 21 November 1963, LIK, 030.II.A.29.64.
[68] "Job meeting," 30 March 1964, LIK, 030.II.A.28.30.
[69] Harlyn E. Thompson, letter to Nason & Culllan, Inc., 1 April 1964, LIK, 030.II.A.29.63.
[70] "Job meeting," 22 June 1964, LIK, 030.II.A.28.30. For the color of the caulking mortar see Harlyn E. Thompson, letter to Nason & Cullen, Inc., 25 June 1964, LIK, 030.II.A.28.21.
[71] "Job meeting," 22 June 1964, cit.
[72] See "Job meeting," 14 September 1964, LIK, 030.II.A.29.51; and "Job meeting," 12 October 1964, LIK, 030.II.A.28.30.
[73] For the photograph sent to Kahn see Sam R. Little, letter to Louis I. Kahn, received on 14 November 1966, LIK, 030.II.A.55.26.

chapter three
design and construction in concrete at dhaka

Concrete construction practices in Dhaka

In no other Kahn worksite as in the case of the National Assembly Building is the question of the making of concrete with local methods so crucial as to even be decisive for the development of the design. The distance that separates the ways of using concrete in the Dhaka region from those found in New Haven, Philadelphia, Trenton, Rochester, Greensburg, La Jolla and Bryn Mawr is so great that the construction of the works in Dhaka takes on the features of a true foundation experience for a native system destined to influence future developments in concrete construction in the decades to come for architecture in East Pakistan (which became Bangladesh in 1971). Local architects of successive generations would continue the passionate investigation begun by Kahn with the construction of the monument of the capitol, with the goal of discovering the nature of concrete in that region. So the National Assembly Building takes on the same value as a prototype of exposed concrete possessed by the monuments of Le Corbusier at Chandigarh for the development of Indian architecture.

At the end of January 1963, Kahn is in Dhaka to discuss the program of the foundation of the capitol in the new Muslim state of East Pakistan, commissioned in August by the president Ayub Khan, on an area near Dhaka, at Sher-e-Bangla Nagar, on the Brahmaputra delta. The National Assembly Building, with its great hall, mosque, offices and rooms for ceremonies and receptions, is immediately considered the main monument, and the first to be built, of a vast capitol that includes the secretariat, the supreme court, the hospital, the residence for functionaries, schools and markets (only a few of these facilities are eventually built).

During a second trip in March 1963, Kahn, Carlos Enrique Vallhonrat – his assistant in charge of the project for Dhaka – and Komendant visit some construction sites in Dhaka to observe the local construction methods and materials, and above all to obtain information on the concrete. From their observations, reported by Komendant, it appears that the cement is of bad quality, mostly imported from China, Russia and also Romania,[1] and is mixed with fine river sand and brick fragments ("jhama chips," in the local jargon[2]). The reinforcement rods are imported in rolls, mostly from China and Japan, and stretched out by hand, using long metal mallets, by Pakistani workers.[3] The scaffolding is made of bamboo and the formwork is done with low-quality wood. The preparation of the compound and the reinforcements, and their placing, happen without machinery, again done by hand by local labor. The only machinery witnessed by Kahn, Komendant and Vallhonrat is on worksites run by foreign construction companies.[4] Apart from several works by foreign architects, like the Kamalapur Railway Station built in Dhaka in 1961-64 by the American Robert Boughey, the reinforced concrete of the structures is usually clad.

During meetings with government representatives, Kahn and Komendant propose the idea of transforming the worksite of the Capitol into a place for experimentation with new ways of making concrete, in order to bring about technical evolution in Pakistan's construction systems. This is the context of the proposal to create a workshop for the production of precast pieces in prestressed concrete. The idea does not meet with the full approval of Fala Faruqi, the Superintending Engineer in charge of relations between the Public Works Department and the group of designers, due to the consequences of reduction of jobs for Pakistani laborers on worksites.[5]

Back in Philadelphia, Komendant and one of his engineers, John Liband, sent to work in Kahn's office, and the engineers of Keast & Hood, coordinated by Nicholas L. Gianopulos, begin study of the structures and the construction systems for the work on the Capitol, and in particular the National Assembly Building. Towards April, Komendant draws up a document titled "Preliminary outline for design and construction requirements,"[6] offering a detailed illustration of the proposal for new methods of making concrete in Dhaka.

The document begins with a note that explains the orientation regarding construction, titled "Philosophy of the structural design." "The basic concept of the structural design is to integrate into modern construction the traditional crafts and materials in a manner that utilizes the natural advantages of these resources and, at the same time, is not restricted by their certain inherent limitations."[7]

The list of "local materials" for the structure includes brick in the forms of "plain and/or reinforced," concrete in the "plain, reinforced, and prestressed" versions, and "terra cotta." After the observation of the backward technical conditions on Dhaka worksites, the author of the "Preliminary outline" recommends elimination of the formwork: "Forms for horizontal and vertical structural elements are greatly reduced and, in most important areas, eliminated."

The "Preliminary outline" already reveals the plan to make use of brick, not as cladding but as veneer of an inner poured core of concrete with met-

[1] Nicholas L. Gianopulos, interview with Anna Rosellini, February 2010.
[2] The irregular and overfired bricks are crushed with rotating cylindrical devices and used as aggregate in concrete (Nicholas L. Gianopulos, interview with Anna Rosellini, February 2010).
[3] Nicholas L. Gianopulos, interview with Anna Rosellini, February 2010.
[4] Komendant, *18 Years with Architect Louis I. Kahn* cit., p. 75.
[5] Nicholas L. Gianopulos,

al reinforcement. "Reinforced" is one of the possibilities listed in the document for the brick masonry. The proposal of laying reinforced brick represents an evolution of the Roman system, transforming the brick wall into an incorporated formwork for the reinforced concrete, and it corresponds to what is stated regarding the reduction of the use of strippable wooden formwork.

In the points on the construction methods, the author of the "Preliminary outline" includes the avoidance of any "specialized heavy equipment" to take advantage of the great local resource, connecting the logic of the worksite to the overall "design" of the works: "Manual crafts are utilized to a maximum and to the advantage of the design." Then the various fundamental structural parts are listed, from the foundations to the roof, including their materials. This list covers the walls, "vertical and inclined," to build in reinforced concrete, like the retaining walls, or in brick, with possible reinforcement, or also in precast concrete and "wall panels." Already with an eye on the use of the arch figure, the author of the "Preliminary outline" specifies that certain "walls and openings are to utilize precast concrete centering." Conventional floor slabs and roofs are in precast reinforced concrete, also prestressed in some cases, while the filler elements are envisioned in terracotta, or "hollow concrete blocks" to make on the worksite with a special machine produced by Siemens, "capable of producing many various types of concrete block shapes." The large roofs are to be in "poured-in-place reinforced concrete."[8] The author of the "Preliminary outline" also lists and describes the equipment required on the worksite for the reinforced concrete parts.

The goal of transforming the Dhaka worksite into a founding experience of a constructive tradition for concrete in Pakistan becomes explicit in the description of the fabrication of the prestressed elements of the foundations, the walls ("panels"), the floor slabs and the roofs. The creation of a special factory, or "precast concrete plant," is presented as useful not just for the Capitol worksite, but more in general to launch a process of manufacture of prestressed concrete in "Dhaka and East Pakistan."

The precast and prestressed parts indicated by the author of the "Preliminary outline" belong to the K System, invented by Edward S. Klausner of the Knickerbocker Construction Corporation of New York. The K System is selected for its characteristics of fabrication and assembly, which are particularly well suited to the conditions of technical backwardness found on Pakistani worksites: "This System, with its highly efficient structural characteristics, eliminates the need for special erection equipment required by other systems. No formwork is required, and normal labor is efficiently utilized." For the reinforcement of joists and piling, the document calls for "wire strands" to be imported from Japan. The author of the "Preliminary outline" provides a series of drawings to explain the methods of installation of the beds of the factory to create in Dhaka for the production of the K System elements. The training of specialized personnel and the drawings required for the creation of the factory can be done in Philadelphia, in collaboration with the local K System, Inc.

The "Preliminary outline for design and construction requirements" also supplies a detailed description of the construction system of the dome of

interview with Anna Rosellini, February 2010.
[6] Komendant was to claim that the report had been ordered by the Ministry of Finance (Komendant, *18 Years with Architect Louis I. Kahn* cit., p. 77).
[7] [August E. Komendant], "Preliminary outline for design and construction requirements," April 1963, LIK, Second Capital – Pakistan Engineering Data – Keast & Hood.
[8] *Ibid.*

the hall of the National Assembly Building, corresponding to the design being developed until the summer of 1963 (Komendant and Gianopulos are involved in the project for this structure).

For the dome, as for the other parts listed in the outline, systems of structure and construction are studied to reduce the use of wood for scaffolding or formwork to a minimum. The dome has a double shell, like the roof of the chapel of Ronchamp, but it is to be built with a composite system specially developed to eliminate the complex wood centering otherwise required for masonry or reinforced concrete. A series of cables are placed at the upper extremity of the walls of the hall to create a curved weave on which to assemble temporary structures – a mobile platform – for the construction of the upper dome, in brick or reinforced concrete. After the construction of the dome, the cables form a shell thanks to the application of concrete sprayed through a hose with compressed air, or gunite.

The formwork of Kreier for Pakistan

The question of the technique and materials for the fabrication of concrete in Pakistan continues to be the central concern of Kahn's research for the capitol project. The evolutions of the configuration of the National Assembly Building cannot be separated from the information gathered by Kahn on concrete construction. The formwork and ways of placing the concrete determine the variations of textures of visible lines developed by Kahn for the walls of the monument.

Though the overall approach explained in "Preliminary outline for design and construction requirements" proposes construction systems marked by limited use of formwork, on 26 April 1963 Kahn meets in Philadelphia with George J. Kreier, Jr., founder of the company that produces the formwork in plastic materials used by Kahn for the residential buildings of the Mill Creek Apartments. It is clear that Kahn wants to investigate the possibilities of concrete fabrication at much higher levels of finish quality than those found on Pakistani worksites. Like the author of the "Preliminary outline," by building the main monuments of the capitol in exposed concrete Kahn sets out to transform the Dhaka worksite into a true factory for the training of local craftsmen specializing in this type of work.

The meeting with Kreier happens in order to get information about possible processes of fabrication and placing of formwork to guarantee a high finish quality for the concrete surfaces in the works to be built in Dhaka. Later, too, still in the works for Dhaka, Kahn will attempt to achieve high quality for the concrete, compensating for the technical limitations and shortcomings of materials and local craftsmen, applying different hypotheses of formwork produced using American technologies.

Kreier responds to Kahn's request of "adapting the concrete forming methods" of his "polyester fiberglass" for the projects in Dhaka, and he sends Kahn the prototype of a formwork specially developed for the worksites in

[9] George Kreier, Jr., letter to Louis I. Kahn, 29 April 1963, LIK, Second Capital – Pakistan Correspondence - Miscellaneous.

[10] *Ibid.* Kreier, Jr., writes after hearing the conference of Kahn on August 20: "I could see that I intruded on your conceptual phase when I vecame excited with formwork methods so early in the game at Daka [sic]. I don't feel, however, that from a method are inextricably involved" (George Kreier Jr., letter to Louis I. Kahn, 21 August 1963, Kreier, Jr. Archive, AAUP, Philadelphia). Kreier, Jr., sends Kahn photographs of formwork in "reinforced Polyester" used for the precast elements of the Federal center, in Hyiattsville, Maryland, Edward Durell Stone (George Kreier, Jr.,

Pakistan, where in place of the "fiberglass reinforcement" a product easily found from local suppliers is used, namely the burlap normally applied in concrete construction, also in earlier works by Kahn, to ensure proper humidity levels during the curing phases. Kreier also illustrates the ways of making the structure of the forms with bamboo. "I am certain that form systems can be developed for local manufacture which will materially expedite construction. I am confident that with reasonable direction local personnel can be trained to do the work."[9] Kreier suggests exploiting that type of formwork for prefabrication, to cut construction costs. Furthermore, after the meeting with Kahn, he indicates his willingness to train workers specialized in the making of formwork, either in Philadelphia or directly in Dhaka.

"Polyester resins and fiberglass," Kreier writes, "are available, and the quantities and costs in relationship to the overall budget of the project are insignificant. Most of the form material by the suggested technique is local. At American prices in America, material costs for the system are not over fifty cents per square foot. Costs in Pakistan remain to be developed. Ultimate forming cost per square foot of concrete placed can be as low as one cent per square foot, if precasting techniques are used. [...]. You are aware of the benefits of introducing viable plastics technologies in technically underdeveloped areas. This project represents an opportunity to introduce the most versatile resin system, and the one significantly requiring almost no capital investment into the economy of East Pakistan in a most dramatic way."[10]

After having seen the drawings of the buildings for Dhaka, Kreier suggests to Kahn the hypothesis of concrete wall construction with his formwork:

"The idea simply is to build forms for each of the rooms, the court, and the stairs, and assemble them on the prepared foundations, and then fill the cavity with concrete. Since the forms are light, they can be withdrawn and replaced in the next position by sheer manpower. The concrete can be placed by the basketful. The system minimizes supervision and maximizes quality. For the moment, I am thinking of this purely as a first-floor scheme. Masonry and partial concrete pouring might be more suitable for the upper floors."[11]

Kreier's suggestions orient Kahn's initial proposals made to the Pakistani authorities for the concrete construction system. On 16 May, Kahn communicates to the Chief Engineer of the Government of Pakistan, Kafiluddin Ahmad, the intention to use the type of formwork proposed by Kreier. In the details of the materials indicated by Kreier, Kahn calls for jute in place of the burlap. As he explains the characteristics of the plastic formwork reinforced by jute, Kahn significantly underlines the flexibility of that material, directly connecting it to the possibilities offered to the design, precisely because that flexibility permits him to easily place the curved structures of his monument. One passage of Kahn's letter is crucial in this sense: "The feasibility of plastic forms is also being studied and could lead to a technique involving the use of jute as reinforcing of the forms. Such forms are very light, tough and flexible, and imprint fine finishes. This technique, if feasible, could influence the design."[12]

letter to David P. Wisdom, 25 February 1963, LIK, Second Capital – Pakistan Correspondence - Miscellaneous).
[11] George Kreier, Jr., letter to Louis I. Kahn, 21 August 1963, Kreier, Jr., Archive, AAUP, Philadelphia.
[12] Louis I. Kahn, letter to Kafiluddin Ahmad, 16 May 1963, LIK, Second Capital – Pakistan Correspondence - Miscellaneous. "For the purpose of estimating," Kahn writes, "all walls would be considered, for the time being, as concrete. We are studying a system of construction which would combine concrete with brick, marble and tile masonry. Its development must be timed to the schedules for design and construction. The floors will be of marble slabs and marble mosaic."

177. Capitol, Sher-e-Bangla Nagar, Dhaka. First section and plan sketches, early 1963 (LIK, 030.IV.B.650.2.42, orig., MoMA, 395.1967).

178-179. National Assembly Building, Dhaka, 1962-83.

178. Plan at 48 feet level (LIK, 030.IV.C.650.2).

179. North-south and east-west sections (F. Fusaro, *Il Parlamento e la nuova capitale a Dacca di Louis I. Kahn 1962-1974*, Roma, Officina, 1985, p. 67).

Not by chance, in the project development in the summer of 1963 Kahn boosts the presence of complex curved structures for the roofing of the National Assembly Building. In certain versions of the design done in summer 1963 he transforms the motif of the *muqarnas* of Islamic architecture into a structural system in reinforced concrete with progressively staggered rings, to place on small vaults that create a pattern of Islamic character for the structures of the assembly hall and the mosque.[13] Formwork like that recommended by Kreier would facilitate the construction of such complex vaulted structures.

During the trip in May 1963, Kahn makes detailed study, together with Gianopulos and Liband, of the components and processes of concrete fabrication in Dhaka. The fact that he decides not to submit the document "Preliminary outline for design and construction requirements" to the Pakistani authorities probably also depends on these worksite visits. The crucial question posed by the "Preliminary outline," that of the precasting of prestressed parts using the K System, may have appeared hard to put into practice, to Kahn, in the context of construction in Dhaka, and a more thorough knowledge of local techniques may also have convinced him not to pursue the technical and structural orientation most favored by Komendant.

Gianopulos and Liband are busy looking for an aggregate that does not absorb humidity for the making of the compound, since the "jhama chips," in the case of exposed concrete, would trigger the surface formation of biological patinas, molds or mosses, with subsequent creation of black marks.[14] The solution is identified in the particular local geographical and atmospheric conditions, thanks to information gathered from the Public Works Department. In the zone of the Great Bends of the Brahmaputra River, during the monsoon season, the tributaries bring down from the Himalayas pieces of granite that form the gravel deposited on the riverbed.[15] The granite pebbles are transported to the worksite, selected for size and stored in mounds, ready for fragmentation.

The information of Ahmad and Gelhardt on concrete in Dhaka

Local materials and conditions for construction continue to be Kahn's main concern in the design phases. In September 1963, Kafiluddin Ahmad, sends a report to Philadelphia on the mode of construction of the government buildings in Dhaka. The report contains a detailed technical description of

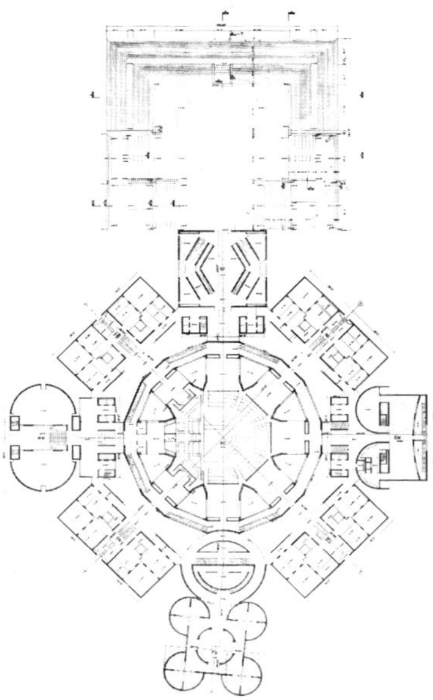

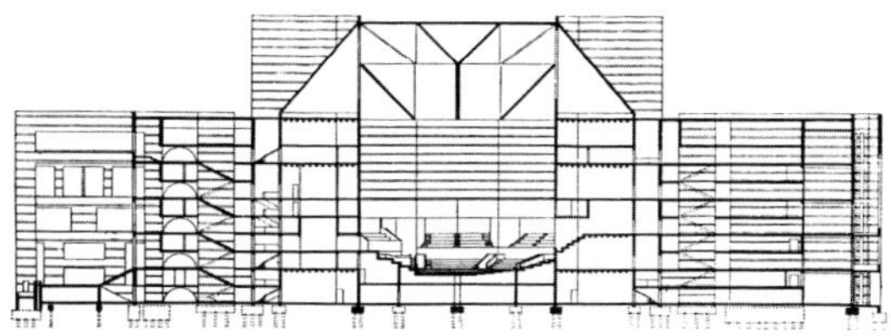

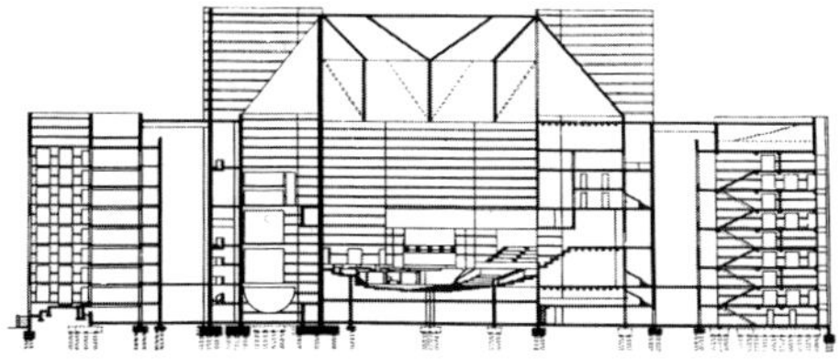

works in reinforced concrete, from the foundations to the formwork. This document indicates that the "base concrete," or that of the foundations, is composed of "cement confirming [sic] to British standard specification, produced in Pakistan or imported from abroad"; the report also indicates the characteristics of the mix design, underlining the fact that the common "coarse aggregate" for this concrete is made with brick fragments, and known as "brick Khoa."[16] For the "cement concrete," that of the floor slabs, walls and columns, use is made for "coarse aggregate" of "brick jhama chips, crushed stone, and shingles (gravel)."[17] Ahmad informs Kahn that in the Dhaka region crushed stone and shingles are rare, and that therefore for the conglomerate use is made above all of "burned (Jhama) Khoa."[18]

The section on "Shuttering & form work" contains important information to orient Kahn's choices in the formulation of his system for shaping the exposed concrete. In Pakistani terminology, as in that of India, shuttering usually indicates formwork made with metal panels. "Wooden planks" are the most common materials for the fabrication of formwork. Instead, "steel shuttering" is rarely used, due to its cost. Probably aware of Kahn's intention to create a monument in exposed concrete, the Chief Engineer points out that in Pakistan concrete poured in formwork with boards is usually clad: "Surface obtained from wooden forms is generally uneven and rough and this requires external finishing."[19] Also based on Ahmad's information, Kahn's office evaluates the costs for the composition of the concrete, with "crushed burned brick (Khana Khoa) aggregates" for reinforced concrete, and with "Jhama Khoa aggregates" for non-reinforced,[20] or with "Riverstone."[21]

Gianopulos begins to collect information on the possibilities of obtaining locally, or bringing from the United States, the machinery needed for the fabrication of the concrete – a "batching plant." Thanks to Herb Gelhardt,

[13] Louis I. Kahn, development sketches, [July 1963]. The drawing also shows a vault in reinforced concrete with hollow tubular elements for lightening.
[14] Nicholas L. Gianopulos, interview with Anna Rosellini, February 2010.
[15] *Ibid.*
[16] Kafiluddin Ahmad, letter to Louis I. Kahn, 10 September 1963, LIK, Second Capital-Pakistan. Typical Specifications (dated Sept. 10, 1963).
[17] *Ibid.*
[18] *Ibid.*
[19] *Ibid.*
[20] Cablegram, 13 November 1963, LIK, Dhaka Cooling Tower Enclosure Specifications (Exec.).
[21] Cablegram, 14 November 1963, LIK, Dhaka Cooling Tower Enclosure Specifications (Exec.).

an American who writes from Quezon City, in the Philippines, Gianopulos and Kahn's office learn of the possibility of obtaining cement mixers of the Miller type, imported by Shahnawaz and Company, based in Karachi, in West Pakistan – the company has an office in Dhaka.[22] For the concrete compound, Gelhardt recommends adding Pozzolith to achieve "some semblance of quality control and durable concrete."[23] He also provides a pithy but effective description of the construction of concrete works in Dhaka, without the use of machinery, using only local workmen:

"[...] the standard method in this part of Pakistan is to load the concrete in buckets, hand carried, and dumped by hand into the forms. This, of course, makes the process more involved. The next best thing from an overhead travel hopper would be the use of wheelbarrows, which, I might add, are seldom used. As a matter of fact, wheelbarrows would have to be imported, since they are not locally available as far as I know."[24]

The information on the modes of fabrication of concrete in the Dhaka region probably have an impact on the revising of the structures with respect to the complex geometry outlined in the fall of 1963 for the form of the roofs of the National Assembly Building. The double-shell dome also vanishes. Now the large hall is bordered by a cylindrical wall, at times cantilevered on vaults (one variation calls for a hall in the form of a truncated cone, with a spatial effect more similar to that of the parliament building at Chandigarh). The evolution of the project towards a combination of elementary volumes reaches the point of configuring a rudimentary expressive device even in the constructive details of Pakistani worksite methods.

Cladding slabs and concrete construction process

At the start of 1964 Kahn begins to work on the method for building concrete walls in Dhaka. As in the earlier studies for the Salk Institute, for the buildings in Dhaka he also envisions cladding of the concrete wall faces exposed to weathering with slabs of marble. This choice is also influenced by the desire to create a protective coating on the concrete, of symbolic meaning like what he had seen at the Taj Mahal, and able to stand up, over time, to the effects of monsoons. In a technical document for Dhaka written a few months after the preparation of the first designs of the National Assembly Building with the cladding, Kahn tellingly states: "Stone has been accepted and admired for centuries as the best building material for monumental buildings to maintain quality in appearance, use and durability over the years."[25]

In the first weeks of January, Kahn is in Karachi and Dhaka, and he informs Ahmad of the intention to build the National Assembly Building and the Hostels for the ministers and members of parliament in "marble and concrete."[26] The quantity of marble is so large that Ahmad raises the question of the costs of quarrying, working and transport to Dhaka.[27]

The fundamental lines of the texture of the cladding correspond to the construction process of the walls thanks to plywood formwork and stone cladding.[28]

The formwork are composed of rectangular panels of plywood to assemble horizontally (3 feet in height, each with 6 spreaders), that shape the two faces of the walls in different ways, depending on whether or not they will be exposed to monsoon rain. While the internal faces are smooth and the concrete remains visible, the outer surfaces are crossed by series of horizontal cornices, 4 inches high, created at the end of each pour. The height

180. National Assembly Building, Dhaka, 1962-83.
Elevation and section study drawings. Technical solution with the heights of the sheets marble of the cladding related with the heights of the formwork panels system, 31 January 1963 (LIK, 030.I.C.650.28).

of the cornice increases at the position of the floor slabs (6 inches), to make the organization of the floors of the building visible.

The cladding is connected to the process of fabrication of the concrete and the structure of the building precisely thanks to those cornices that serve as support for the marble slabs. The fact that the slabs are arranged to correspond to the pour layers constitutes another form of pursuit of a constructive logic of stone in the processes of deployment of concrete, already launched with the design of the Meeting House in La Jolla. In several variations the concrete cornices also protrude towards the rooms, to fasten the wooden panels of the *boiseries*. The fact that the cornices are always bands of exposed concrete is essential to make the material of the walls visible, even when they have cladding.

The various solutions for the formwork and arrangement of the marble slabs always reflect technical modifications to the concrete construction process. One significant variation of the positioning of the cladding slabs, with concrete cornices placed every 7 feet, has the purpose of making each pair fasten two rows of slabs – the cornices still have heights of 4 and 6 inches.[29] Again in this case, the 7 feet interval indicates the height of the pour.

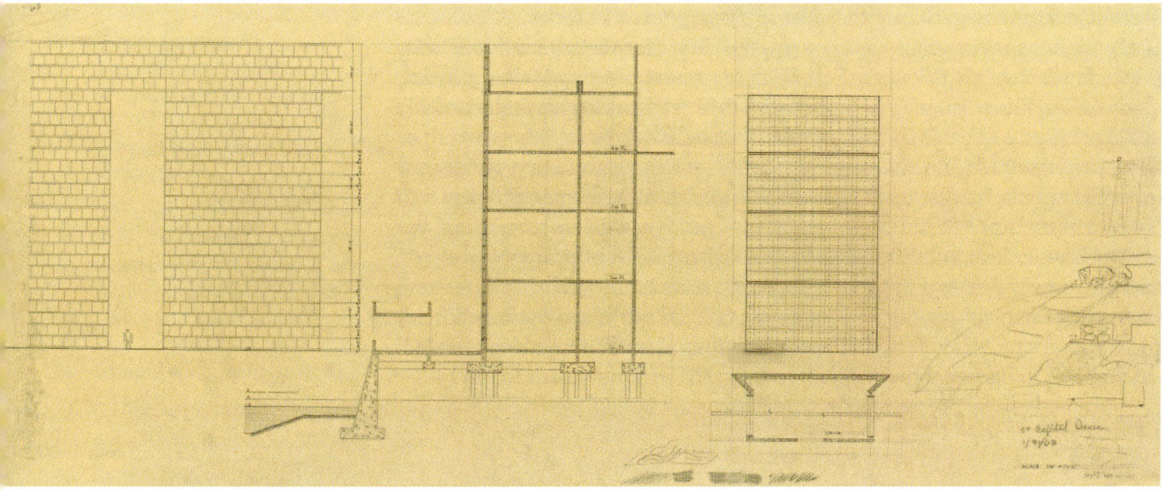

[22] Herb Gelhardt, letter to Peter Zhun, 26 September 1963, LIK, Second Capital – Pakistan Engineering Data – Keast & Hood. In the case of machinery to be imported from the United States, the choice is recommended of types that function independent of electricity, using "motor driven belts," due to the particular conditions of energy distribution in Dhaka.

[23] *Ibid.*

[24] *Ibid.*

[25] "Outline Specifications," 10 May 1964, LIK, Second Capital Special Brick Details Dated 12/7/64.

[26] Kafiluddin Ahmad, letter to Louis I. Kahn, 13 January 1964, LIK, Second Capital – Pakistan Public Works Department Correspondence - 1964.

[27] "You indicate that you will be using marble and concrete in the main building. Marble quarryng, cutting and polishing etc. will take considerable time. Marble will have to be procured from North Western province in the Western Zone of Pakistan and will have to be carried by trucks and railway wagons to Karachi, where it will have to be cut, shaped and polished before shipping to Chittagong from Karachi. All these operations will take considerable time for which it is requested that approximate quantity and shade may kindly be intimated" (Kafiluddin Ahmad, letter to Louis I. Kahn, 13 January 1964, LIK, Second Capital – Pakistan Public Works Department Correspondence - 1964).

[28] Drawing dated 31 January 1963, [sic: 1964], LIK, 030.I.C.650.028.

[29] See the drawing dated 31 January 1964, LIK, 030.I.C.650.005. A variant calls for square sheets assembled with staggered vertical joints and arranged in horizontal rows protruding from the cornices.

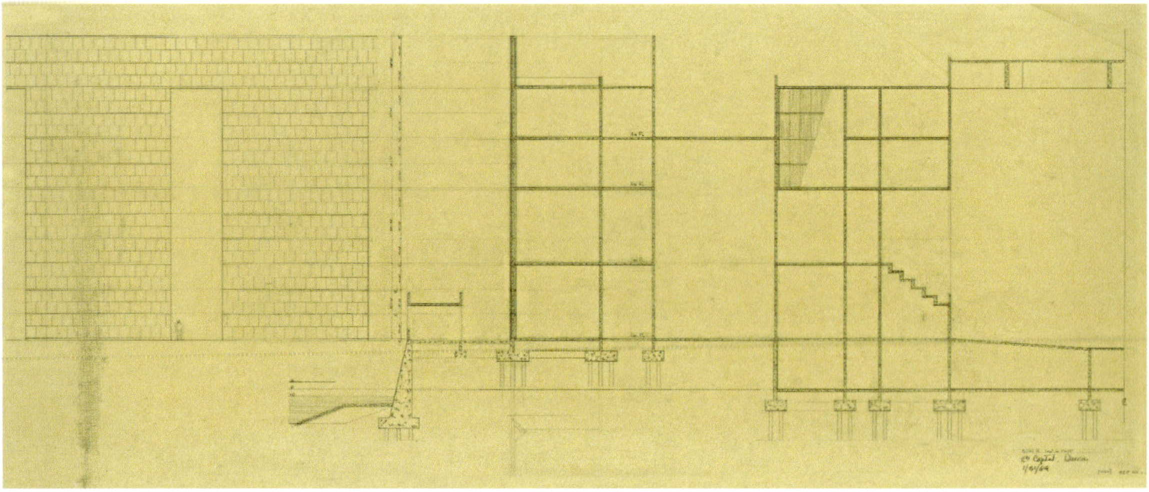

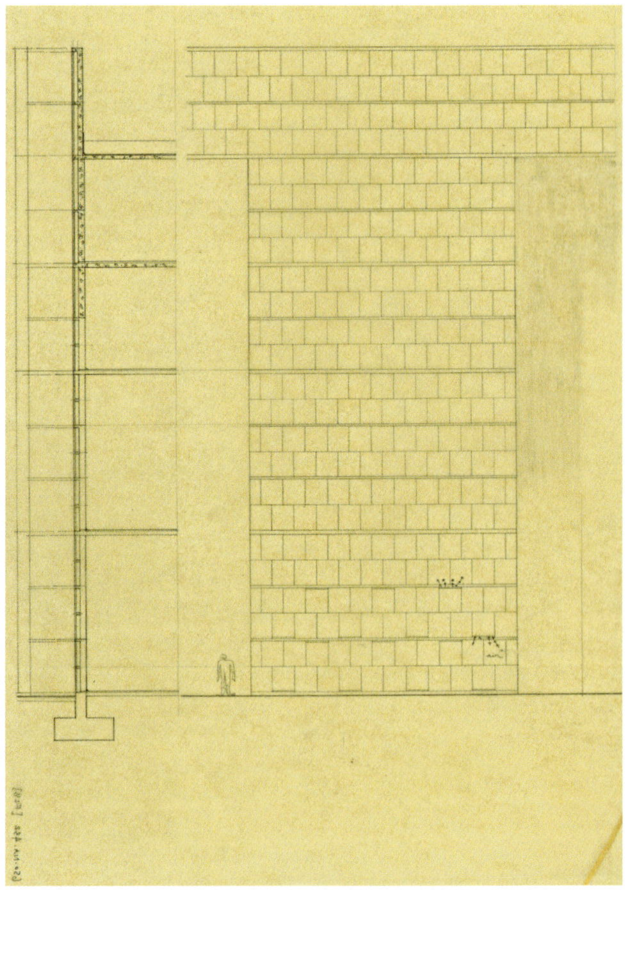

181-184. National Assembly Building, Dhaka, 1962-83.

181. Elevation and section study drawings. Technical solution with the heights of the sheets marble of the cladding related with the heights of the formwork panels system, 31 January 1963 (LIK, 030.I.C.650.25).

182. Elevation and section study drawings. Technical solution with the heights of the sheets marble of the cladding related with the heights of the formwork panels system (LIK, 030.I.C.650.28).

183. Elevation and section study drawings. Technical solution with the heights of the sheets marble of the cladding related with the heights of the formwork panels system (LIK, 030.I.C.650.28).

184. Section study drawing. Technical solution with the heights of the sheets marble of the cladding related with the heights of the formwork panels system (LIK, 030.I.C.650.28).

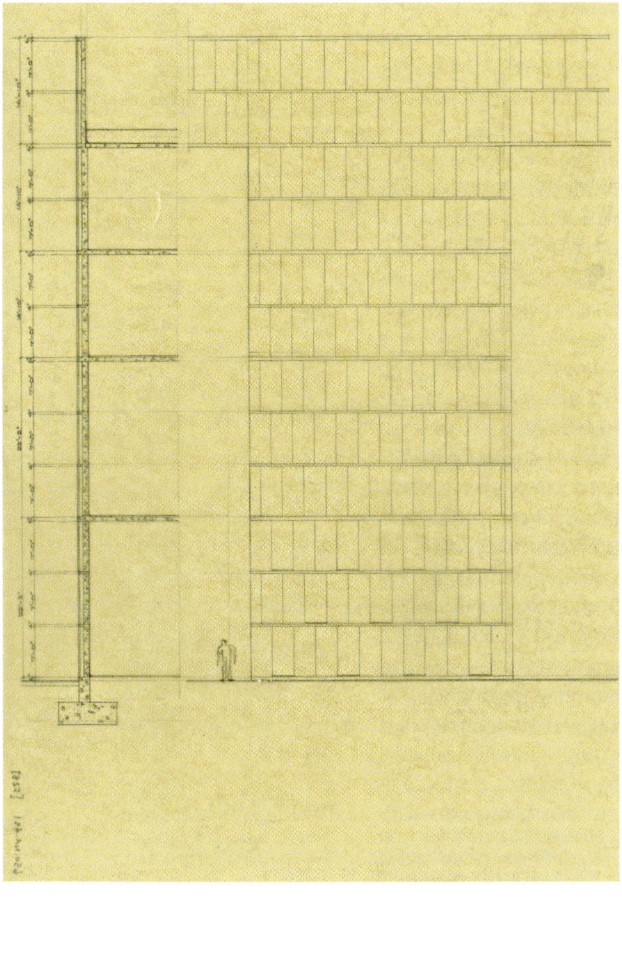

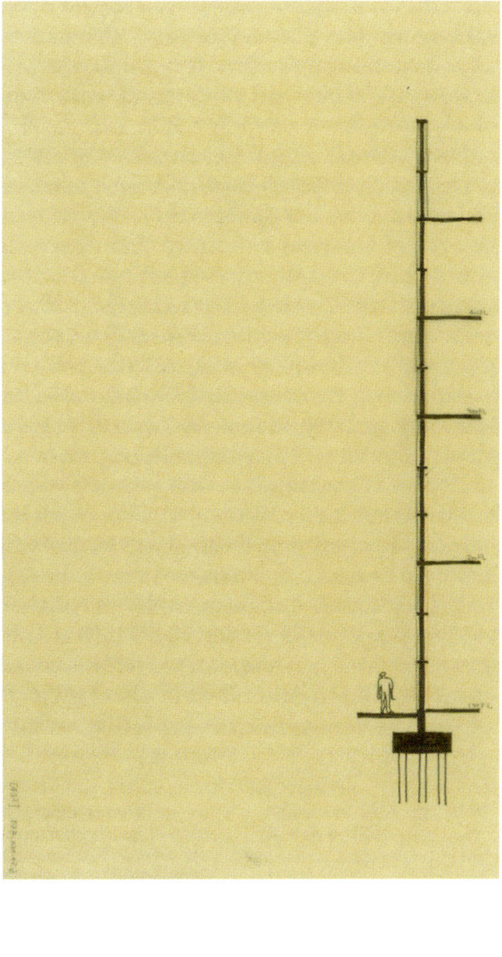

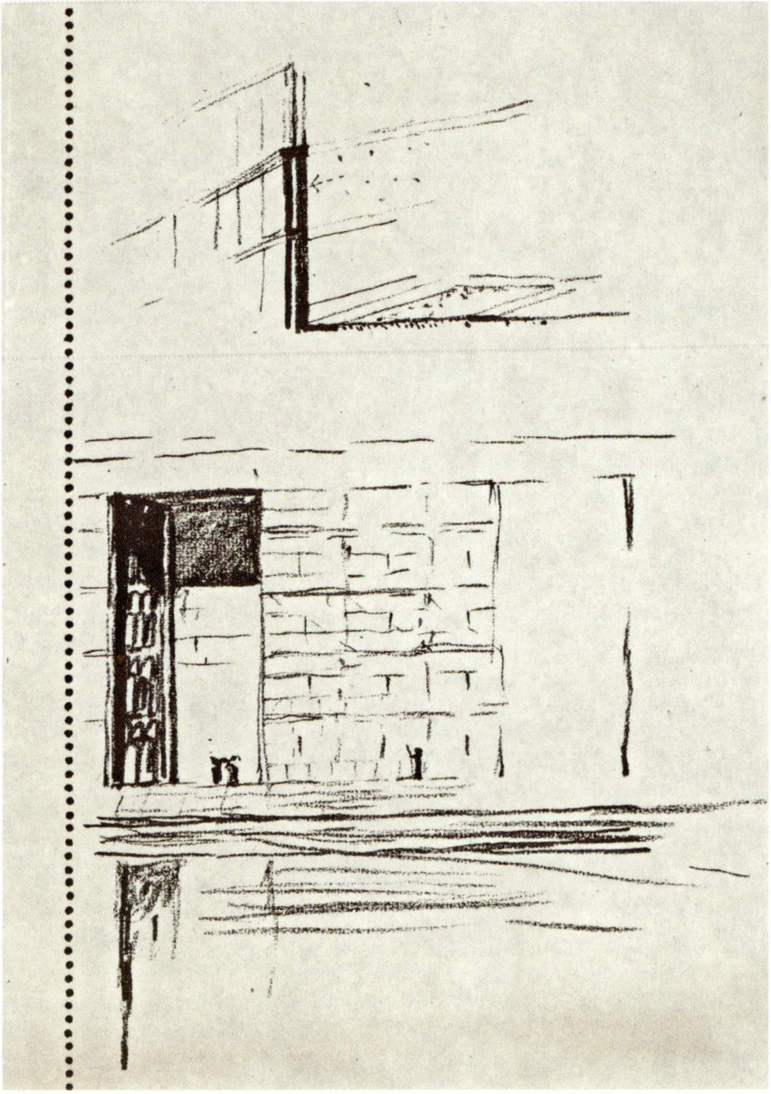

Kahn's intent to achieve a correspondence between pour height, cornice position and the height of the marble slabs can also be seen in a study that posits a height for the slabs to enable covering the distance between two cornices in the case of pouring done every 7 feet (the slabs become narrower, for reasons involving the cutting and handling of the pieces).[30] Several studies document other, more special solutions, with the cornices of the pour no longer recessed but protruding with respect to the slabs, and therefore protected with lead. In this case the marble slabs are bolted to the concrete wall with four visible fasteners, in the manner of Wagner, with the aim of accentuating the meaning of the slab as cladding – like Wagner.[31]

For the definition of the technical details of the formwork system, Kahn's office obtains information on the materials available at Dhaka, from plywood panels, for which they request the "maximum size available," to "waterproof glues" or pieces of wood for the structure.[32]

185-186. National Assembly Building, Dhaka, 1962-83.

185. Sketches for the technical solution with the heights of the sheets marble of the cladding related with the heights of the formwork panels system (LIK, 030.IV.A.650.13.32).

186. Study drawing for the bonding of the marble cladding fixed on concrete walls (LIK, 030.I.C.650.59).

The information on the cost of the marble slabs has a decisive impact on Kahn's definitive wall solution. The slabs are available in Karachi in "white & grey shade," in three main formats (24 × 12 inches, 12 × 12 inches, 8 × 8 inches, thickness 3/4 inch). "Dark grey marble slabs as shown to you at Dhaka are also available," Faruqi reports, perhaps reflecting Kahn's preference for a color similar to that of the concrete to be covered.[33] Due to the large quantity required, Faruqi asks Kahn to indicate the exact number of the slabs, and their sizes.[34]

In the design phases after January 1964, Kahn continues to envision marble cladding for the concrete walls, but after the information received from Faruqi, at the end of February he changes the construction system, though the design of the walls continues to be determined by the fabrication process of the concrete. Kahn starts to study what will become the definitive solution of the exposed concrete walls, marked by the imprints of formwork and doing away with the cladding.

In the studies at the end of February the formwork system, still in plywood, is equipped with strips to insert between the panels to obtain protruding vertical frames,[35] and strips with a trapezoidal profile to produce slender horizontal grooves to correspond to the pours (every panel also leaves the imprints of the spreaders on the concrete). The concrete fabrication process is so important for understanding of the imprints that in the drawing of the section Kahn depicts and numbers the pours required to make a wall.[36] This is the first time, in his drawings for the fabrication processes of concrete, that Kahn includes this essential information for his architectural concrete.

[30] See the drawing, n.d., LIK, 030.I.C.650.028, and the drawing, n.d., LIK, 030.I.C.650.028, with the solution of the narrower vertical slabs. For the schematic drawing of the concrete with the marble strips and claddings see the drawing, dating back to January-February 1964, LIK, 030.I.C.650.028.
[31] See the drawing, n.d., LIK, Will Broun.
[32] Duncan Buell, letter to Fala Faruqi, 4 February 1964, LIK, Second Capital – Pakistan Correspondence – Miscellaneous. The plywood is produced by two companies in Chittagong; the local industrial firms do not have products like "waterproof glues" (the measurements indicated by Buell are available). See Fala Faruqi, letter to Duncan Buell, 11 February 1964, LIK, Second Capital – Pakistan Public Works Department Correspondence – 1964.
[33] Fala Faruqi, letter to Louis I. Kahn, 11 February 1964, LIK, Second Capital – Pakistan Public Works Department Correspondence - 1964.
[34] *Ibid.*
[35] See the drawing 29 February 1964, LIK, 030.I.C.650.032. The drawing shows the system from La Jolla.
[36] See the drawing of the section, 29 February 1964, LIK, 030.I.C.650.032.

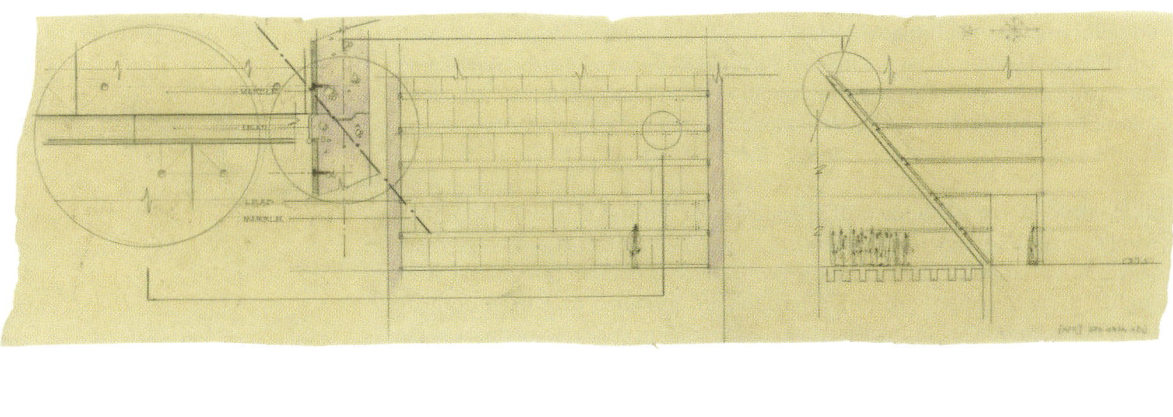

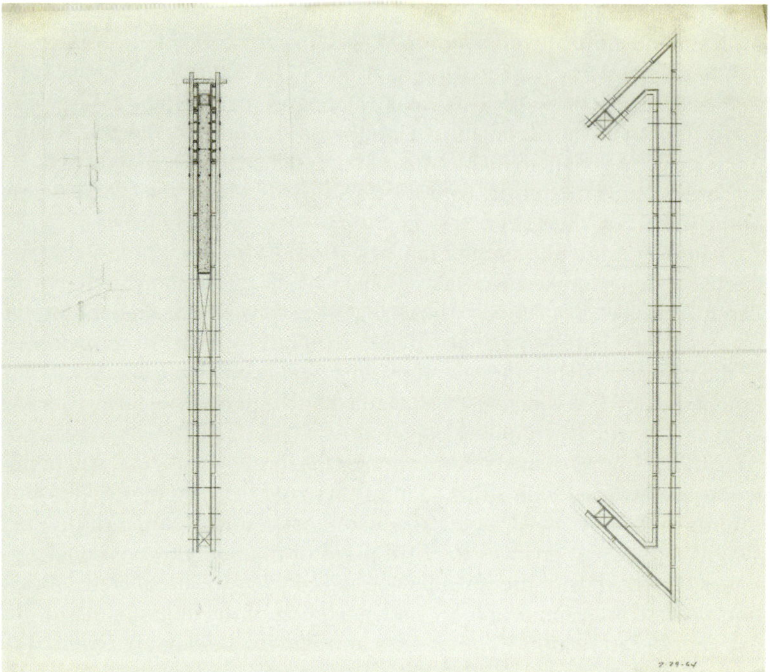

187-188. National Assembly Building, Dhaka, 1962-83.

187. Formwork study, 29 February 1964 (LIK, 030.I.C.650.32).

188. Section where is visible the assembly of the formwork panels, April 1964 (LIK, 030.I.C.650.25).

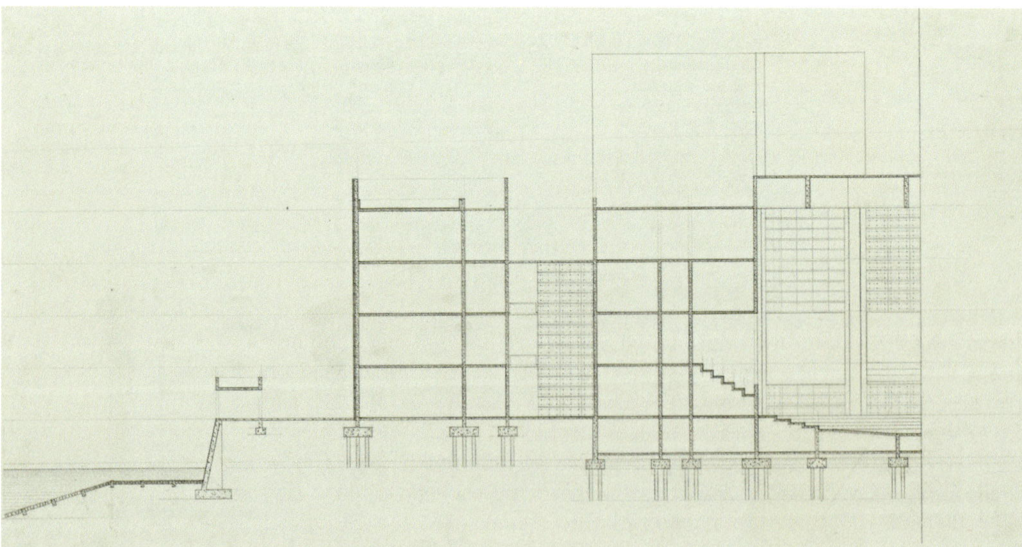

The profile of the vertical strips is designed to produce protruding frames with a rectangular section. But in the sketches on the margins of the drawing the hypothesis also appears of inserting a strip between the panels that has cut edges, like the ones used at La Jolla, to create a double-V vertical imprint. The only marble insert envisioned in this design phase is the one on the faces of the concrete walls facing inward, and it has the height of the whole first pour above the level of each floor – in the drawing it is indicated as "recess to receive marble."[37] Kahn attempts to stick to the principle of designing the walls on the basis of the fabrication process of the concrete using plywood panels and layered pours.

After having envisioned a tapestry of lines obtained with the panel imprints, alternating with high bands only at the position of the slabs, for the internal walls,[38] Kahn reaches the point of identifying a different motif of horizontal pour joints, which becomes the definitive solution. The studies in April show high horizontal recesses of separation between one row of panels and the next.[39] The marble returns as an inlay to insert in these recesses. The fact that the Pakistani engineers emphasize that the marble available in "white and grey shades," already seen by Kahn, has to be "used sparingly"[40] confirms that the high cost of importing it to East Pakistan have persuaded Kahn to eliminate the cladding called for in the initial projects.

Drafts for the specifications of the concrete and the formwork

The evolution of the design of the walls proceeds in step with the preparation by Kahn's office, on 30 March 1964, of several drafts for the specifications of the concrete, to be sent to the engineers of the Public Works Department, each with an explanatory note by Kahn attached, on 2 April. The drafts contain detailed indications on the making of the formwork panels, the preparation of the compound, the stripping phases and the treatment of the concrete surfaces, and are written on the basis of the information gathered by Kahn and his team on construction practices in Dhaka. They have the objective of defining the guidelines to refine later with other technical information, prior to the preparation of the definitive design of the construction system.[41]

The draft of the specifications on "Concrete" emphasizes, also to explain Kahn's change of views with respect to his initial proposals, that all the concrete of the walls has to be left exposed: "All walls that contribute to the structure of the building are concrete and are not to be covered with any other material, as will be true also of concrete beams and soffits of flat concrete slabs."[42]

The specifications outline methods to guarantee control of the quality of the exposed concrete: to make "trial batches" to evaluate "strength and color"

[37] *Ibid.*
[38] See the drawing April 1964, LIK, 030.I.C.650.030.
[39] See the drawing of the section, April 1964, LIK, 030.I.C.650.030.
[40] "Annexure A," document received by Kahn's office on 16 June 1964, LIK, PAC Various DWD Dec June 66.
[41] "In writing specifications for this work, which is our responsibility, we shall be in need of further information and guidance so that we may specify the workmanship and materials required for the work in a manner to suit the bidding and construction procedures that will be followed. We should also be made aware of legal requirements applying to contracts for construction work. If certain portions of the work are to be done directly by the Pakistan Public Works Department and not by contract, this too might affect the writing of specifications, since for example, the specifications for such work should not mention a contractor. We are submitting herewith a sample concrete specification in the general format which we propose for all specifications for the work. This is not intended to be the final version and should under no circumstances be used for construction, but is submitted at this time for your consideration and comments" (Kahn'office, "Notes on the concrete specification. Second Capital of Pakistan at Dhaka," 2 April 1964, LIK, Second Capital – Pakistan Specifications (2 April 1964)).
[42] Kahn' office, "Concrete," 30 March 1964, attached to, Kahn's office, "Notes on the concrete specification. Second Capital of Pakistan at Dhaka," 2 April 1964, LIK, Second Capital – Pakistan Specifications (2 April 1964).

of the concrete; to ensure "uniform quality and appearance of all concrete work exposed to view in the finished work"; to construct "large sample panels of concrete wall" for testing of formal quality before the start of construction; and to make these "sample panels" large enough to be able to judge the effect of the imprints of the formwork panels. "Samples," the draft explains, "shall be sufficiently large to include at least 1 full form panel and typical panel joints and sufficient in number to show each kind of finish required."

For the composition of the concrete, the specifications draft calls for Portland cement, "crushed stone" as coarse aggregate, "uniform light color as approved," and "Sylhet sand" as fine aggregate.[43] The formwork are described for "concrete not exposed to view" and "concrete exposed to view in finished work." In the first case, the use of seasoned wood is recommended, free of roughness that could mar the concrete surface; the joints have to be tight to prevent "leakage of cement" and similar defects; the surfaces of the formwork should be "lightly greased or oiled" to facilitate stripping. In the case of the formwork for "concrete exposed to view" the draft recommends the use of plywood panels with edges in contact with the pour "slightly beveled"; coating of the plywood with "non-staining form coating material" of the "clear urethane" type, to apply preferably "by dipping, before use"; and reuse of formwork only after cleaning and checking of their "good condition." The spreaders are "coil ties with 1 1/2 inch plastic or steel threaded cones and 1/2 inch diameter coil bolts." "Ties shall be accurately installed in pattern as shown" and the holes must remain "unfilled" as specified.

Still based on the desire to obtain a uniform concrete surface, Kahn wants the compound to be made with special machinery, namely batch mixers with "charging hopper, water storage tank and water measuring device."

The operation of pouring of the concrete is explained in all its phases, to achieve optimum results. The criteria outlined in the draft, in fact, become those used for the first construction phase. "The concrete shall be deposited as evenly as possible and shall not be dropped from a height nor handled in a manner that will cause separation. If necessary, to deposit concrete in deep, narrow forms tremies shall be used for placement as approved. Concrete shall not be deposited continuously at one spot and allowed to flow into position alongside the formwork by gravity. It shall be worked uniformly into position by rodding."

In this phase of the design of the construction systems, Kahn and his team envision manual tamping of the pour and stripping of the formwork though "mechanical vibration" is also recommended. The height of the pour is to be checked by technical personnel on the worksite: "Concrete work shall only be stopped at pre-determined position, stopping places being approved by the Engineer-in-charge. Such construction joints shall be truly vertical and horizontal as the case may be, except that in any inclined or curved members, the joints shall be strictly at right angle to the axis of the member."[44]

Concrete formwork plywood

To obtain the ideal formwork for his concrete made with plywood, in the specifications Kahn provides technical indications to improve the production of the panels normally used in Pakistan. The content of a cablegram sent to him on 13 April 1964 by Ahmad is significant: "Locally available plywood disintegrates after coming into contact with moisture. Possibility of using for shuttering ruled out. Arrangement being made for procurement of steel shuttering which will give as much smooth surface to concrete."[45]

Unlike Le Corbusier, who in Chandigarh accepts modification of his prior European *béton brut* to adapt to the typical Indian formwork made with sheet-metal panels, Kahn still does not take Pakistani steel shuttering into consideration as a system to mold his concrete, at least not in the available documentation. His vision of concrete that is totally different from that of Le Corbusier can be seen in the fact that Kahn sets out to improve Pakistani plywood, which is not normally used for formwork, attempting to change local technical conditions instead of rethinking what he now sees as the universal founding principles of his concrete. The fact that he has had a direct look at the results Le Corbusier is achieving on the Chandigarh worksite through steel shuttering, with all the defective swelling of the concrete surface caused by that type of formwork, confirms that Kahn does not accept the possibility of the "flaws in workmanship" as an aesthetic attribute of concrete. But the small size of the "steel shuttering" and the thinness of the metal sheets should also be considered, because they may have dissuaded him from experimenting with a type of formwork that would not easily permit the grandeur of the design of lines woven on the surface that is found in his American concrete, after the switch from boards to plywood panels. At the same time, the metal molds used by Kahn in the United States for poured in place or precast concrete are completely different from those available in Pakistan, since the former are made with thick sheet metal welded to achieve perfection in every detail of the surfaces in contact with the pour.

Precisely in order to shape the concrete in the American manner, on 30 March Kahn has specifications drawn up expressly on the fabrication of "concrete form plywood," for a detailed explanation of the criteria for the construction of perfect panels in Pakistan, with the methods applied at La Jolla. These specifications cover the arrangement of the "wood veneers" and of the grain from one sheet to the next, the best glue to use to assemble them in panels and the quality of the visible faces of the panel, from the orientation of the grain – "the grain of the face plies shall run lengthwise on the panel" – to the admissible size of knots, crevices or other irregularities, the type of polishing of the surface, all the way to the possible use of "plastic fillers." Certain points of the specifications have to do with the finish of the panels in contact with the pour: "Exposed veneer on both sides of panel shall be full length and width and shall have the bark or tight surface cut. [...]. Both sides of the finished plywood shall be sanded."[46]

Kahn's letter of 2 April that accompanies the sending of the specifications, "Notes on the plywood specification," significantly links the type of panel with the developments of the Dhaka project: "This specification for concrete form plywood, numbered CON PLY, is issued at this time to make possible the ordering of this material at an early date, so that it may be available when required. The quality of the concrete form plywood should meet or exceed the requirements of the specification. We have been advised that it will be possible to obtain plywood in 4 feet widths and are developing the detailed architectural design accordingly. It is highly desirable, although not mandatory, that concrete form plywood be furnished with edges sealed against moisture to prevent any likelihood of delamination."[47]

[43] *Ibid.*
[44] *Ibid.*
[45] Cablegram, 13 April 1964, in, "Materials," LIK, PAC Various DWD Dec June 66.
[46] "Concrete form plywood," 30 March 1964, LIK, Second Capital – Outline Specifications Architectural/Mechanical/Site Work May 10, 1964.
[47] Kahn's office, "Notes on the Plywood Specification," 2 April 1964, LIK, PAC – Specifications - Pending.

After the drafts of the specifications, Kahn's office begins to supply the engineers of the Public Works Department with information on the equipment required for the fabrication of the concrete. To mix the compound and cut the wood of the formwork, Buell provides specific literature on the worksite equipment for mixing, the "Rex Concrete Batching Plant" produced in West Germany, and on the "DeWalt Saw" produced in the United States.[48]

Following the draft specifications, Ahmad seriously questions the use of plywood for the formwork, advising Kahn to make use of "steel shuttering."[49] While Kahn continues to imagine the construction of the National Assembly Building and the Hostels with the same system, on 12 April Ahmad informs him of the need to avoid concrete for the Hostels: "Walls of hostels must be in brick masonry w[ith]/cement rendering on either face."[50] The reasons behind this decision made by the Pakistani technicians, destined to change the character of the complex in a decisive way, are not clear. We cannot rule out the idea that the costs for the making of the exposed concrete may have prompted them to fall back on a more traditional system, and one that in any case had already been envisioned in the report in 1963 by Komendant. To understand the logic behind decision making regarding the construction systems of Kahn's works in Dhaka, it is useful to read what is written in the second half of April regarding costs for supply of materials, when two Franki-piles are driven to determine the most reliable and economical foundation system: "gravel, sand and cement are extremely expensive in East Pakistan."[51]

The opposition of the Pakistani engineers to exposed concrete

On 20 May 1964 a meeting is held at the Dhaka worksite to examine the process of design and definition of the methods of concrete fabrication for the National Assembly Building. The minutes of the meeting point to discussion of the formwork system featuring horizontal recesses decorated with marble inserts. It is on this occasion that the efficacy is examined of the solution of using these inserts to route rainwater away from the exposed concrete. The proposal made by Ahmad in April of using metal for the formwork also returns to the fore. On this matter, the need is emphasized for study of the detail of the joint between the metal panels. At the same meeting, the question of the mix design of the concrete is also on the agenda, and all the indications converge towards solutions that guarantee uniformity of color in the final product, to obtain through control of the components, from the sand to the cement.

The fact that the sheet of notes reporting on the themes discussed at the meeting is entitled "Lou's Notes" leads to the conclusion that the opinions expressed are those of Kahn. Under the entry "concrete" the first phrase is enigmatic: "Lou now favors it." The hypothesis that this could indicate Kahn's position regarding the decision of the engineers of the Public Works Department to use brick masonry for the Hostels is borne out by the fact that the same source indicates that Kahn has approved a brick wall sample, though with certain reservations regarding the way of spreading the mortar in the joints. "But it was made of same rough brick with mortar smoothed over wide joint area and the joint scored with a tool," the notes report on the "brick wall" seen by Kahn.

[48] "The batch plant is of central importance for the quality of the concrete," Buell explains. "The proper size for the job can be determined when the rate of pouring the concrete and the method of refilling the concrete bin is known. The radial saw is a most versatile field tool, very useful in the making of wood shuttering, and recommended for your use by an American contractor skilled in the art of making reinforced concrete" (Duncan Buell, letter to Fala Faruqi, 31 March 1964, LIK, Second Capital – Pakistan Correspondence - Miscellaneous).

"a. Lou now favors it. b. Must use small lifts. Many horizontal joints. Do special detail at these joints. Develop detail to throw water away from wall. c. Perhaps use marble inserts. d. Maybe use their metal forms, but show them how to join them, etc. e. Select aggregate from one source. f. Mix *all* sand (from several sources) into *one* consistent mass for whole job. g. Select Portland cement from *one* source only. h. Color of Pakistan concrete is acceptable. It merely has too many variations."[52]

Ahmad's communication about the construction of the walls of the Hostels and the first trials for the bricks of these walls are reflected in the document where Kahn reasserts the unity of the complex, to be obtained through the choice of one single material with which to build the National Assembly Building and Hostels. This is an extraordinary confirmation, in Kahn's vision, of the role of the "form" which is independent not only from the functional program of a work, but also from the materials themselves. Kahn continues to believe it is indispensable to construct all the buildings with the same material, whatever it is, be it concrete or brick:

"[...] the Assembly Building and the Hostels must be constructed of the same material to achieve the unity of design that is essential to the project. The materials being considered are concrete or brick masonry. For either material this project requires some deviation from construction practices in Dhaka to achieve a high quality of workmanship. The differences in methods and materials are described in separate sections below. It must be noted that the choice of structural material will affect the plans somewhat in that the geometry of intersections in the structural systems would necessarily differ."[53]

The "Outline Specifications" contain the explanation of the quality required by Kahn for the two materials proposed as alternatives.

The section on "Concrete" emphasizes the fundamental characteristic for Kahn, namely the color. All the reasoning focuses on the aesthetic qualities of the concrete, so much so that the "Outline Specifications" seem utterly original with respect to similar technical documents produced by Kahn's office in the past. One statement reveals the essential characteristic of Kahn's research on the nature of concrete, to be discovered through the study of its components: "The color cannot be pre-selected. It must be chosen from among the best combinations of Portland cement, aggregate and sand." This conception of the color of concrete to be achieved by experimental means, by controlling the ingredients of the compound, reflects the experiences on the La Jolla worksite. To obtain the ideal color, the procedure involves the making of sample walls, as at La Jolla, each done with a different mix. "It is suggested that slag cement not be used in first samples because of its dark grey color," the document recommends. The "Outline Specifications" also urge that once the color of the concrete has been selected, all the ingredients be procured in sufficient quantities "to ensure uniform quality and appearance" in all phases of the construction. "Concrete walls, piers, columns, beams, slab soffits will be exposed in the final building, therefore the utmost skill in science, technology, and workmanship will be required."

[49] Cablegram, 13 April 1964, cit.
[50] Cablegram, 12 April 1964, in, "Materials," LIK, PAC Various DWD Dec June 66.
[51] Swissboring, letter to Fala Faruqi, 27 April 1964, LIK, Second Capital – Pakistan Swissboring Report – 6 additional tests dated 4/10/64 (Xtra copies).
[52] "Lou's Notes," 20 May 1964, LIK, Lou's Notes 5-20-64.
[53] "Second capital of Pakistan at Dhaka. Outline Specifications," 10 May 1964, LIK, Second Capital Special Brick Details Dated 12/7/64.

Kahn reaches the most complete theoretical formulation of the form of concrete as pursuit of the truth of the construction process, to obtain through all the imprints and traces of the work: "The imprint of forms, ties, joints and the joints between pours become the finished design in concrete surfaces. Extra care is to be exercised to produce forms that follow in every minute detail the design of forms to be furnished by the architect."

The Pakistani engineers find a description, in the "Outline Specifications," of the role of certain details of the formwork that are fundamental for the formal quality of Kahn's concrete: "Ties for formwork are to be only those approved by the architect and spacing is to be exactly as shown on the drawings. Ties are to be coil ties with steel or plastic cones, water seal coil ties in exterior walls. Holes left by tie cones are to remain unfilled to emphasize the design."[54]

The characteristics of the other material being considered for the construction in Dhaka, namely brick, are described under the heading "Masonry." Again in this case, "color range and texture" are essential factors, and sample walls are again required to evaluate "the width of joints, the color and mix of mortar and the type of tooling for the joints."[55]

In May and June 1964 the engineers of the Public Works Department prepare a document to respond to the "Outline Specifications," containing certain remarks on the construction materials, their local availability, and the costs and processes involved in their use in Dhaka. The document is sent to Kahn at the beginning of June. What is questioned, because it clearly goes against local building practices, is Kahn's idea of leaving the concrete of the walls exposed, without repairs or coverings after formwork stripping. Economic factors and the use of non-specialized labor mean that reinforced concrete is used in Dhaka as a skeleton that is then filled with brick walls and covered with plaster. As a model of this type of reinforced concrete with infill and cladding, the Pakistani engineers indicate the Intercontinental Hotel presently under construction in Dhaka, by the contractor Macdonald Layton Costain, Ltd. (MLC).[56] Working details like the regular arrangement of the spreaders to "emphasize the design," as written in the "Outline Specifications," constitute further proof, for the Pakistani engineers, of the impossibility of leaving concrete exposed in Dhaka: "The holes which will be left by ties of scaffolding in concrete will have to be treated and will produce a patchy effect on the wall."[57]

The suggestions of the Pakistani engineers regarding the type of concrete to use in Dhaka leaves no room for doubt regarding the impossibility, in their view, of implementing Kahn's proposals without cladding: "Under these circumstances practically all the high buildings in Dhaka are made of R.C.C. frame, having brick panel walls to get uniformity and smooth surface after plastering the walls with a rich mortar in cement and sand. The outside walls and columns can be ornamentally treated with different types of finishing materials and ornamental brick facing. Under these circumstances it is suggested that the idea of making reinforced cement concrete without any sort of rendering (plaster) should be abandoned altogether.

[54] *Ibid.* The "Outline Specifications" also describe the methods for making parts in precast concrete, with steel molds.

[55] *Ibid.* Stones and marbles are limited, at this point, to pavements and floors, steps, copings or benches. The "Outline Specifications" also contain a description of the project of the National Assembly Building in relation to light and wind, so the building can also function without mechanical air conditioning devices.

[56] See the photograph showing a work under construction with the sign "MLC, MacDonald Layton

[...] if such walls are unavoidable, the same should be made in cement concrete cast in situ but both faces of the walls should be plastered to give a smooth facing. The above comments are made keeping into consideration: i) the cost of the construction; and ii) the workmanship. The workmanship in these parts is very poor, with the result that the concrete surface after stripping the forms will not come smooth and generally have the following defects: 1) In small patches it comes out with honeycombs. 2) Air bubbles along the surface make the concrete spotty. 3) Reinforced steel and aggregate exposure. These will have to be made up by rubbing the surface with cement mortar of the same proportion. That is why it is imperative to use plaster on the concrete surface."[58]

The "typical wall form"

After the opinion of the engineers of the Public Works Department it is clear that the possibility of leaving exposed concrete in the monuments for Dhaka depends on Kahn's ability to obtain formwork of superior quality to what is normally used on worksites in Pakistan. Plywood continues to be the material he envisions for the panels. It is no coincidence that starting in mid-1964 Kahn's office in Philadelphia concentrates on finding various materials and fabrication systems for the formwork, since Kahn still hopes to be able to use only exposed concrete even for the Hostels, avoiding the use of brick masonry.

After the information from Kreier obtained in the preliminary project phases on the possibility of using fiberglass formwork, and after the attempt to produce plywood of good quality directly in Dhaka, in the wake of the warnings of the Pakistani engineers regarding the difficulties involved in making exposed concrete with the typical materials of local construction, Kahn gets information about the possibility of using plywood in Dhaka that has been directly shipped from the United States. The office contacts the United States Plywood Corporation, based in New York, specialized in building materials. The company sends Kahn an estimate for the supply of rectangular 4 x 8 feet plywood panels to use for the formwork, or "Weldwood Exterior Plyform - Grade BB, faces oiled and edge sealed," in two thicknesses (3/4 and 5/8 inch), and indicates the port of Portland, Oregon as the place from which to ship the panels to Dhaka.[59] The company also sends Kahn a booklet "describing all grades of Douglas Fir Plywood."

Based on the assurances offered by the United States Plywood Corporation, starting in early July Kahn develops a version of the design for the buildings of the capitol with all the working details for the fabrication of the concrete, illustrating the logic of construction of every imprint on the walls. His new formwork system, based on the use of plywood panels, is defined as the "typical wall form." Kahn's goal of integrating all the phases of the process of fabrication of the concrete in the design of the walls can be seen in the fact that his system also covers the "scaffolding & method of pour."[60] The solution is quite close to that of the definitive design, however Kahn continues to envision the panels in plywood. Though the drawing of

Costain Limited, Hotel Dhaka Intercontinental; sub contractors ConsComstock Engineers and Constructors" (LIK, 030.IV.A.650.I.52).
[57] "Annexure A," received by Kahn's office 16 June 1964, LIK, PAC Various DWD Dec June 66.
[58] *Ibid.*
[59] Arthur J. Vasquez, letter to Duncan Buell, 18 June 1964, LIK, Second Capital Correspondence - Miscellaneous.
[60] See the drawing "Wall sections and details," 6 July 1964, LIK, 030.I.C.650.020.a.

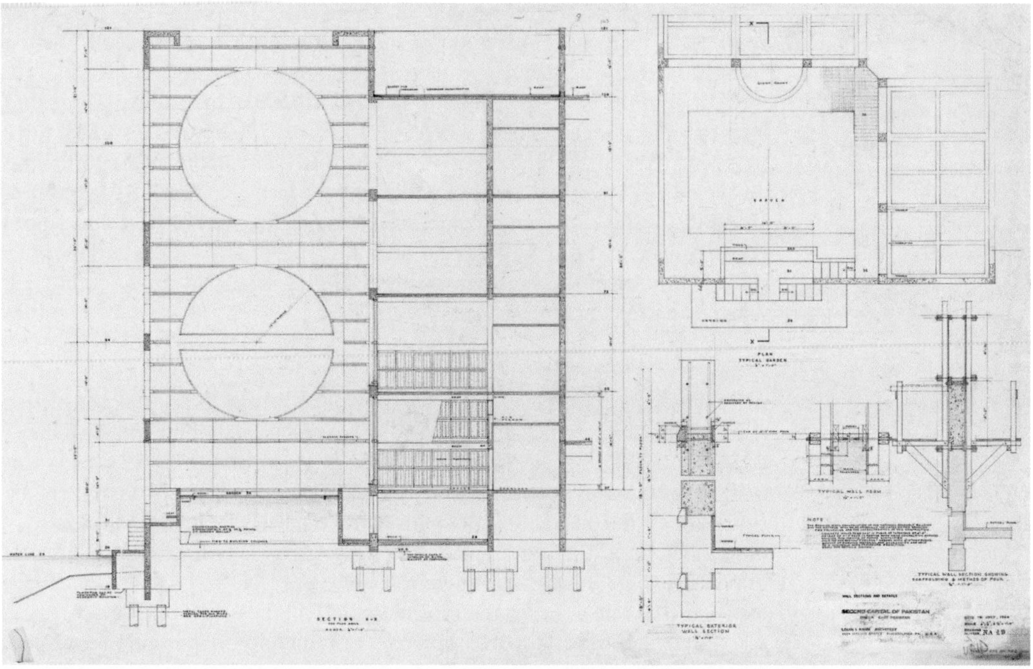

the details is for the section of the National Assembly Building with the verandas in front of the offices, the "typical wall form" also covers, as a note indicates, the construction of the Hostels. It is clear that Kahn wants to erect, with all the buildings of his complex in Dhaka, a true hymn in honor of exposed concrete.

In the design of the "typical wall form" the two scaffolding decks enable the workers to pour the concrete into the formwork using hand-operated poles. The spreaders are positioned at the lower and upper extremities of the two plywood panels, where the continuous horizontal recesses are shaped on the two faces of the wall, with a height of 6 inches and a depth of 3 inches. Each recess is to be made straddling two consecutive pours, thanks to a special strip. The recess corresponds, in negative, to the protruding concrete frame envisioned in the studies of 1963, with the slabs cladding the walls. With respect to those studies, the height of the pour has changed from 3 to 4 feet, since the measurement of the stone slabs no longer poses a limitation. The recess is then sealed with a marble cornice designed on the basis of the exposure of the surface to the weather: on the wall faces subject to rain it protrudes, with the upper surface included and with a dripstone, while on the other faces it is smooth, and flush with the wall. In the passage from the cladding in stone slabs of the first projects to the exposed concrete of the new version, the protruding cornice takes on the function of a dripstone, becoming an organ similar to the cornices inserted in the closed brick wall of the Yale Art Gallery.

A note written beside the drawing of the formwork system specifies how the walls must be built: "The bearing wall construction of the National Assembly Building and the Hostels will be as generally indicated on this drawing. The following are the characteristics of this construction. Concrete walls to be cast in place at intervals of 4'-0". Height of 4'-0" pour in keeping with hand compaction methods. Bracing and moving of forms simpli-

189. National Assembly Building, Dhaka, 1962-83. "Wall Sections and Details," 6 July 1964 (LIK, 030.I.C.650.20b).

fied. Only top and bottom 'tyscrus' needed with 4'-0" high forms. Amounts of forming material and accessories are held to a minimum with procedure resulting from the details shown."

In the case of the National Assembly Building, the cornices inserted in the recesses along the walls exposed to weather, including those facing the verandas without roofing created in front of the offices, are equipped with an overhang that makes them into dripstones. Kahn introduces a differentiated layout in the sequence of cornices, in keeping with the presence or lack of slabs abutting the walls – the cornices are doubled at the position of the floor slabs.

The extreme precision required for perfect implementation of the holes and their sealing with lead disks, as demonstrated by the experience gained on the Salk Institute worksite, probably lies behind the solution that is emerging in Dhaka, which conceals the entire arrangement of the spreaders rather than leaving their traces visible and making them a part of the formal character of the concrete. The idea of the marble cladding band to apply at the position of the horizontal rows of spreaders is paradoxically based on the desire to underscore a fundamental detail of the fabrication process of the concrete, precisely by hiding it from view. We cannot help but be reminded of the comparable detail of the joints of the drums of columns, as resolved by De l'Orme, who precisely to hide those joints from view inserts protruding decorated stone disks between one drum and the next.

"This concrete," Kahn would later explain, "is made in five-foot pours with a marble insert every five feet, and this covers the point of termination of the pour, and also acts as a wash on the outside (there are three hundred inches of rain there) every five feet on the building. This is so the building does not get damp throughout from the water that is above. Every five feet determines its own water stop. That is the general theory behind the making of the concrete. It also works with their economy and their way of pouring concrete because they have no machines. They just have a swarm of bees, people, and you don't see what the work has been during the day until they leave! For this reason they are also using sticks instead of vibrators. The idea of controlling the length of pour was part of the design."[61]

After Kahn's efforts to bring American plywood to Dhaka and after the design of the "typical wall form" the question of the material of the formwork is raised again by the United States Plywood Corporation, in the wake of Kahn's silence after the correspondence in June. On 24 August the company asks about the possibility of using its Weldwood Exterior Plyform in Dhaka.[62] Buell advises the firm to contact Ahmad directly: "You may have stiff price competition from Japan and the Philippines. We have sent them a fairly rigid specification for W.P. plywood, as a guide to their own industry, but have elected not to try to produce it in Pakistan. It would please us to know that we had good material for forming exposed concrete work."[63] So the United States Plywood Corporation writes to the Pakistani engineer to find out about local needs in the area of "concrete plyform." The letter emphasizes plywood from Japan or the Philippines can only be reused three or four times. "However, the United States Grade Plyform can be used up to eleven times, whereas if you are buying Japanese or Philippine plywood at

[61] Kahn, *Address* cit., pp. 217-218.
[62] United States Plywood Corporation, letter to Louis I. Kahn, 24 August 1964, LIK, Second Capital – Pakistan Correspondence - Miscellaneous.
[63] Duncan Buell, letter to United States Plywood Corporation, 25 August 1964, LIK, Second Capital – Pakistan Correspondence - Miscellaneous.

a lower price you are using it for half the time. This must be taken into consideration by someone who is really looking out for savings in the long run."[64] Nevertheless, all the attempts to import quality plywood for the concrete walls are in vain, and Kahn's formwork ends up being made with traditional local materials.

The information that continues to flow into the office in Philadelphia on the techniques and materials of construction in Dhaka further refine the overview of the situation, leading to modifications in Kahn's design for the buildings of the capital. In this sense, the document titled "Report on site investigation and design data for second Capital at Tejgaon Dhaka," prepared by the technicians of the Public Works Department and received by Kahn's office on 3 August, is significant, because it offers ulterior confirmation that the reasons behind Kahn's foregoing of the marble cladding lie in the costs of supply of the materials. The "Report" emphasizes the lack of quarries near Dhaka for "stone, marble & granite," while it underlines the good quality of the local brick, made both by hand and with machinery (the availability of "tiles and bricks with mat finish in different colors" is also mentioned). Regarding the ingredients of the concrete, the technicians of the Public Works Department inform Kahn about the use of two main "coarse aggregates": the "jhama chips" mixed with cement to obtain ordinary concrete, and the "shingles from the quarries of Rangapani and Seripur in Sylhet" used when high-quality concrete is required. Furthermore, based on several tests, the report indicates the possibility of using Sylhet and Durgapur sands for the compound (while the sands of Kaliapur and Savar are to be avoided).[65] Thus it is clear that in the summer of 1964 Kahn wants to make the capital complex with only exposed concrete, while the technicians of the Public Works Department continue to favor, at least for the Hostels, locally produced bricks.

Choice of cement and formwork hypotheses

From the beginning of October 1964, Kahn sends members of his staff to Dhaka, as the construction is about to begin. Roy Vollmer, Jr., is assigned the task of opening a office at the site, supervising contacts between the engineers of the Public Works Department, the construction companies, the local authorities and Kahn's office in Philadelphia, and supervising the worksite. Gianopulos and Leidigh of Keast & Hood, and Gustav Langford, the brother of Frederick and a staffer in Kahn's office, have the job of developing the fabrication of the conglomerate to be left exposed. "The primary purpose of the trip was to qualify the materials to be used," Gianopulos recalls, regarding his task of controlling the composition of the concrete in terms of strength and surface quality.[66]

[64] United States Plywood Corporation, letter to [Kafiluddin] Ahmad, 27 August 1964, LIK, Second Capital –Pakistan Public Works Department Correspondence - 1964.
[65] "Report on site investigation and design data for second Capital at Tejgaon Dhaka," LIK, Second Capital - Site Investigation and Design Data.
[66] Nicholas L. Gianopulos, interview with Anna Rosellini, February 2010.
[67] R.V., [Roy Vollmer, Jr.], "Report from Dhaka to LIK Office," 10 October 1964, LIK, PAKCAP – Correspondence to/from RoyGus October 8, 1964 thru June 30, 1965, typewritten version, LIK, PACAP Vollmer Reports.
[68] *Ibid.*
[69] "Reinforcing cage looked workmanlike but method of forming and curing could be improved for greater efficiency and accuracy. We will advise them, if possible, on better methods and also observe driving and load test. They have, to date, two rigs. One driving tower looks very fragile. They estimate three months to drive their piles. It is questionable" (R.V., [Roy Vollmer, Jr.], "Report from Dhaka to LIK Office," 14 October 1964, LIK, Roy Vollmer's Reports 10.04.'64 18 Mar. '65).
[70] *Ibid*
[71] "Wood forms (called shuttering here) are made from Gorjan Wood (local very hard wood and cheapest 11r/cu. ft.). It is subject to warping and must be reinforced with a stiffener. The supt. of a new construction (New State Bank Building, Dhaka) told me this

Kahn, his team and the engineers of the Public Works Department are also focusing on the choice of the type of cement. Vollmer emphasizes that import costs will influence the definitive choice of materials and the pace of construction. In this framework, there is also the idea of importing cement that "will be Chinese, Russian and Japanese" – the Public Works Department is already testing these three types of cement.[67]

The first application of concrete on Kahn's worksite in Dhaka has to do with the foundation piles of the National Assembly Building. In October, under the supervision of Gianopulos, load tests are being conducted on three piles. Rudy Steinberger of Swissboring states that 10 piles by 5 rows (50 piles) can be made per day.[68] Precisely the fabrication of the piles is the first phase of Kahn's team's attempt to improve the construction techniques for concrete in Pakistan.[69]

The fact that Kahn's team is aware of contributing to the evolution of concrete fabrication systems in Dhaka is confirmed by what Vollmer writes:

"Nick [Gianopulos] and Tom [Leidigh] are doing a superb job. Bit by bit we are assembling the pieces. I will try and send as much information as quickly as possible. We are all groaning under the yoke of inefficiency. We are trying to organize PWD [Public Works Department] so we can organize our work. Answers are easy, always yes. Probing is difficult. When sending sets of drawings in the future, bind them, they will stay intact long enough to be recorded. We are going to build the best buildings in Pakistan and that makes everything more than worthwhile."[70]

The first precise data on the type of formwork used in Dhaka are gathered by Vollmer, with an eye on the imminent decisions to be made prior to the construction of the National Assembly Building. Vollmer explains in detail the making of formwork with the local Gorjan wood, and their modes of deployment. This is the start of a progressive acceptance of local practices on the part of Kahn and his staff, as opposed to the insistence, throughout 1964, on the use of plywood formwork.

Usually in Dhaka four planks of Gorjan wood, 1 inch thick, are interlocked to formwork panels of 3 × 2 feet, with metal reinforcements bolted at the corners. To get a clean surface, the boards in contact with the concrete are usually covered with a metal sheet. Vollmer already proposes alterations to the technical details of the local formwork to improve the joint between the boards (he suggests the typical "tongue and groove" solutions), and the application of the metal sheet, which should be done by welding on metal reinforcements of the corners, though this technique is not widespread in Dhaka. It is Vollmer who raises the possibility of eliminating the metal covering, due to defects in its use and also due to its cost, while attempting to perfect the joint between the boards to be left in direct contact with the pour.[71]

type of form is expensive initially but economical over a long period. This is encouraging news. For a smooth finish, the shuttering is covered by sheet tin, nailed with brads to the edges of the wood. Welding and brazing is not common here, but I would suggest if a tin surface is to be used, that it should be brazed at all joints and firmly fastened continuously to the stiffener. Otherwise it will pull off. I would suggest a detail like this. I would be inclined not to cover the shutter with tin unless we spec. a heavy gauge because eventually it will tear off and might be a lot of trouble. Also the cost will be exceedingly high. If the wood joints are well made and flush, the concrete will look very good. The supt. of this job advised me that tenders for formwork and concrete should be restricted to an engineering firm who understands the need for quality work and will follow the details and specifications. Workmanship done under government tenders is of the lowest caliber because of the open bid system. Low bidders are not best usually same as USA" (R.V., [Roy Vollmer, Jr.], "Report from Dhaka to LIK Office," 16 October 1964, LIK, Roy Vollmer's Reports 10.04.'64 18 Mar. '65). For the drawings see the original manuscript, LIK, PAKCAP – Correspondence to/from RoyGus October 8, 1964 thru June 30, 1965.

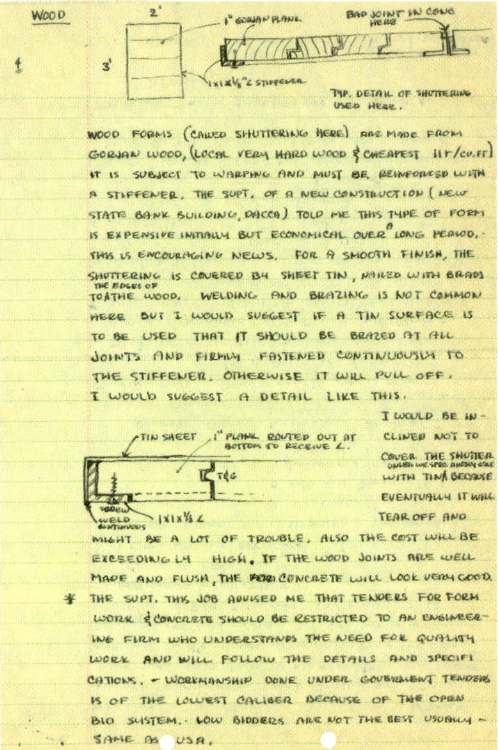

WOOD

2'

3'

1" GORJAN PLANK

BAD JOINT IN CONC HERE

1x1x1/8" ∠ STIFFENER

TYP. DETAIL OF SHUTTERING USED HERE.

WOOD FORMS (CALLED SHUTTERING HERE) ARE MADE FROM GORJAN WOOD, (LOCAL VERY HARD WOOD & CHEAPEST 11r/cu.ft) IT IS SUBJECT TO WARPING AND MUST BE REINFORCED WITH A STIFFENER. THE SUPT. OF A NEW CONSTRUCTION (NEW STATE BANK BUILDING, DACCA) TOLD ME THIS TYPE OF FORM IS EXPENSIVE INITIALLY BUT ECONOMICAL OVER A LONG PERIOD. THIS IS ENCOURAGING NEWS. FOR A SMOOTH FINISH, THE SHUTTERING IS COVERED BY SHEET TIN, NAILED WITH BRADS TO THE EDGES OF THE WOOD. WELDING AND BRAZING IS NOT COMMON HERE BUT I WOULD SUGGEST IF A TIN SURFACE IS TO BE USED THAT IT SHOULD BE BRAZED AT ALL JOINTS AND FIRMLY FASTENED CONTINUOUSLY TO THE STIFFENER. OTHERWISE IT WILL PULL OFF. I WOULD SUGGEST A DETAIL LIKE THIS.

TIN SHEET

1" PLANK ROUTED OUT AT BOTTOM TO RECEIVE ∠.

T&G

SCREW

WELD CONTINUOUS

1x1x1/8 ∠

I WOULD BE INCLINED NOT TO COVER THE SHUTTER UNLESS WE SPEC. ANODY OAK WITH TIN BECAUSE EVENTUALLY IT WILL TEAR OFF AND MIGHT BE A LOT OF TROUBLE, ALSO THE COST WILL BE EXCEEDINGLY HIGH. IF THE WOOD JOINTS ARE WELL MADE AND FLUSH, THE CONCRETE WILL LOOK VERY GOOD.

* THE SUPT. THIS JOB ADVISED ME THAT TENDERS FOR FORM WORK & CONCRETE SHOULD BE RESTRICTED TO AN ENGINEERING FIRM WHO UNDERSTANDS THE NEED FOR QUALITY WORK AND WILL FOLLOW THE DETAILS AND SPECIFICATIONS. – WORKMANSHIP DONE UNDER GOVERNMENT TENDERS IS OF THE LOWEST CALIBER BECAUSE OF THE OPEN BID SYSTEM. LOW BIDDERS ARE NOT THE BEST USUALLY – SAME AS USA.

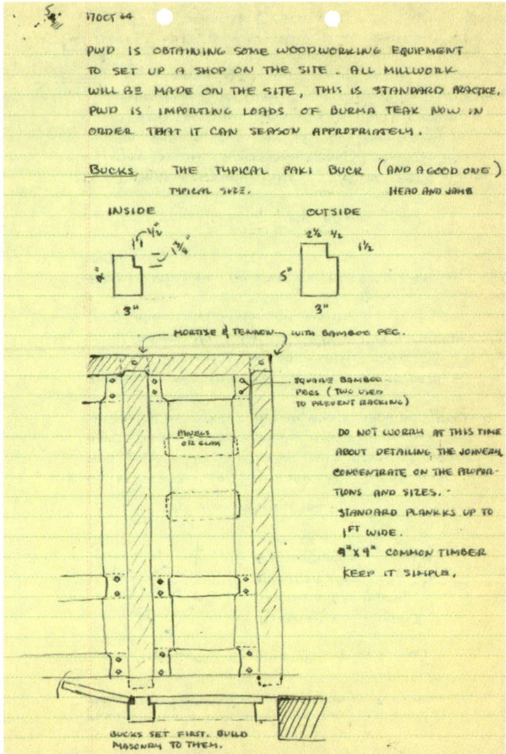

5

17 OCT 64

PWD IS OBTAINING SOME WOODWORKING EQUIPMENT TO SET UP A SHOP ON THE SITE. ALL MILLWORK WILL BE MADE ON THE SITE, THIS IS STANDARD PRACTICE. PWD IS IMPORTING LOADS OF BURMA TEAK NOW IN ORDER THAT IT CAN SEASON APPROPRIATELY.

BUCKS. THE TYPICAL PAKI BUCK (AND A GOOD ONE)

TYPICAL SIZE.

HEAD AND JAMB

INSIDE

OUTSIDE

1/2"

1 1/4"

4"

3"

2 1/2 1/2

1 1/2

5"

3"

MORTISE & TENNON WITH BAMBOO PEG.

SQUARE BAMBOO PEGS (TWO USED TO PREVENT RACKING)

PANELS OR GLASS

DO NOT WORRY AT THIS TIME ABOUT DETAILING THE JOINERY, CONCENTRATE ON THE PROPORTIONS AND SIZES. STANDARD PLANKS UP TO 1 FT WIDE. 9"x9" COMMON TIMBER KEEP IT SIMPLE.

BUCKS SET FIRST. BUILD MASONRY TO THEM.

190-192. National Assembly Building, Dhaka, 1962-83.

190-191. Sketches for formwork panels ([R. Vollmer, Jr.], "Report from Dhaka to LIK Office," 16 October1964, LIK, Roy Vollmer's Reports 10.04.'64 18 Mar. '65).

192. Equipments in the construction site, 1964 (LIK, 030.IV.A.650.1.33).

The question of the type of aggregate to mix with the sand and cement, already discussed in 1963, is not fully resolved, since Faruqi still insists on the use of the traditional "jhama chips," though recommending the method now being tested in the construction of the Intercontinental Hotel.

Kahn's worksite begins with the operation of insertion of the foundation piles, or franki piles.[72] The technical drawing made by Keast & Hood of the approximately 3000 foundation piles of the National Assembly Building is compared by Kahn to a "tapestry."[73]

For the aggregate of the concrete to use for the foundation piles the choice goes to granite fragments taken from the Brahmaputra River, but mixed with the compound only after washing, in keeping with a technique observed on the worksite of the Intercontinental Hotel. The first tests of the strength of this concrete demonstrate that the workers have not washed the granite prior to preparation of the compound, in keeping with their usual construction habits.[74]

Reinforced brick masonry

During the course of 1964, Kahn has to give up on the exposed concrete walls of the Hostels and turn to a composite system of bricks and steel reinforcements set into the concrete. This decision is influenced by the technical and economic issues of the formwork and the exposed reinforced concrete. The fact that Kahn did not initially accept the use of brick for the Hostels confirms the importance, in his view, of a unified color for the capital complex, which he always emphasizes in the documents up until 1964.

The choice of reinforcing the brick masonry with steel and concrete is based on the seismic conditions in the Dhaka region, and on several indications provided by the Geophysical Institute at Quetta, in West Pakistan.[75] This special type of masonry also has an impact on the conception of the space of the residences, at least according to what is stated by Wisdom: "Please note that there is considerable saving of masonry and space with reinforced masonry."[76] Confirmation of the fact that the reinforced brick masonry has been chosen to prevent seismic disasters is found in certain documents of the Chief Engineer, which state that the "reinforcement in masonry work" has been selected "to safeguard against seismic forces."[77]

The imminent start of work for the construction of the substructure of arches and walls of the Presidential Square, facing the site of the National Assembly Building, prompts Vollmer to organized the production of the bricks, inspecting the furnaces, selecting the manufacturers, studying

[72] Nicholas L. Gianopulos, interview with Anna Rosellini, February 2010.
[73] *Ibid.*
[74] *Ibid.* For the "test piles" and the "Swissboring" the estimate of the driving of 50 piles per day, and the indications of Gianopulos regarding the foundations, see R.V. [Roy Vollmer, Jr.], "Report from Dhaka to LIK Office," 10 October 1964, cit. Actually only 10 piles are driven per day (R.V. [Roy Vollmer, Jr.], Report from Dhaka to LIK Office, 22 November 1964, LIK, Roy Vollmer's Reports 10.04.'64 18 Mar. '65). Vollmer informs Kahn that the "concrete will be Chinese, Russian and Japanese" (R.V. [Roy Vollmer, Jr.], "Report from Dhaka to LIK Office," 10 October 1964, cit.). The excavation for the foundation of the Presidential Square begins on 14 October (R.V. [Roy Vollmer, Jr.], "Report from Dhaka to LIK Office," 14 October 1964, LIK, Roy Vollmer's Reports 10.04.'64 18 Mar. '65).
[75] David P. Wisdom, letter to [Kafiluddin] Ahmad, 6 April 1965, LIK, Pak Cap Correspondence 7 Feb 1964 May '70.
[76] *Ibid.*
[77] Chief Engineer, letter to Louis I. Kahn, 24 April 1965, LIK, Second Capital – Pakistan Public Works Department Correspondence - 1965.

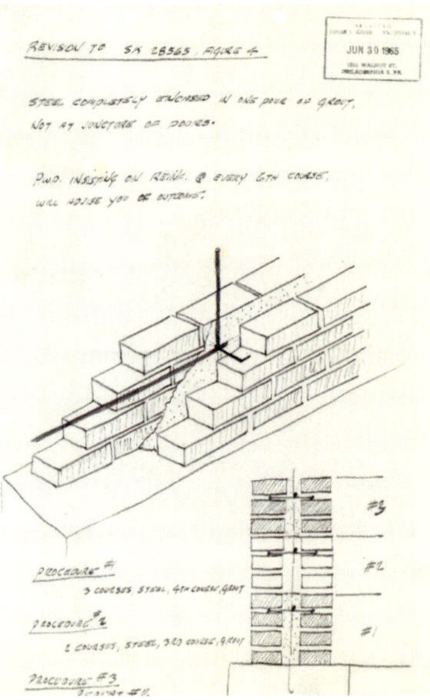

REVISON TO SK 28565, FIGURE 4

JUN 30 1965

STEEL COMPLETELY ENCASED IN ONE POUR OF GROUT,
NOT AT JUNCTURE OF COURSES.

P.W.D. INSISTING ON REINF. @ EVERY 6TH COURSE,
WILL ADVISE YOU OF OUTCOME.

PROCEDURE #1
3 COURSES, STEEL, 4TH COURSE, GROUT

PROCEDURE #2
2 COURSES, STEEL, 3RD COURSE, GROUT

PROCEDURE #3

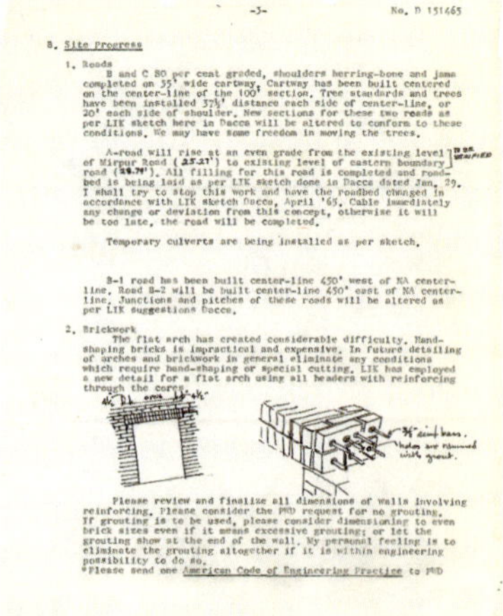

-3- No. D 151465

B. Site Progress

1. Roads

B and C 80 per cent graded, shoulders herring-bone and jama completed on 35' wide cartway. Cartway has been built centered on the center-line of the 100' section. Tree standards and trees have been installed 37½' distance each side of center-line, or 20' each side of shoulder. New sections for these two roads as per LIK sketch here in Dacca will be altered to conform to these conditions. We may have some freedom in moving the trees.

A-road will rise at an even grade from the existing level of Mirpur Road (25.27') to existing level of eastern boundary road (28.79') [TO BE VERIFIED]. All filling for this road is completed and road-bed is being laid as per LIK sketch done in Dacca dated Jan. 29. I shall try to stop this work and have the roadbed changed in accordance with LIK sketch Dacca, April '65. Cable immediately any change or deviation from this concept, otherwise it will be too late, the road will be completed.

Temporary culverts are being installed as per sketch.

B-1 road has been built center-line 450' west of NA center-line. Road B-2 will be built center-line 450' east of NA center-line. Junctions and pitches of these roads will be altered as per LIK suggestions Dacca.

2. Brickwork

The flat arch has created considerable difficulty. Hand-shaping bricks is impractical and expensive. In future detailing of arches and brickwork in general eliminate any conditions which require hand-shaping or special cutting. LIK has employed a new detail for a flat arch using all headers with reinforcing through the cores.

Please review and finalize all dimensions of walls involving reinforcing. Please consider the PWD request for no grouting. If grouting is to be used, please consider dimensioning to even brick sizes even if it means excessive grouting; or let the grouting show at the end of the wall. My personal feeling is to eliminate the grouting altogether if it is within engineering possibility to do so.

*Please send one American Code of Engineering Practice to PWD

193-198. National Assembly Building, Dhaka, 1962-83.

193. Construction of the Presidential Square, January 1965 (LIK, 030.IV.A.1.20).

194. Reinforced bricks wall in the construction site of the Presidential Square, January 1965 (LIK, 030.IV.A.650.2.36).

195. Study drawing for the reinforced bricks walls (G. Langford, Progress report 2, 22 June 1965, LIK, PAKCAP – Correspondence to/from RoyGus October 8, 1964 thru June 30, 1965).

196. Flat arch in the construction site of the Presidential Square, January 1965 (LIK, 030.IV.A.650.1.8).

197. Study sketches by R. Vollmer, Jr., for the flat arch in reinforced bricks (R. Vollmer, Jr., letter to L. I. Kahn, n.d., received by Kahn's office on 24 April 1965, LIK, PAKCAP – Correspondence to/from RoyGus October 8, 1964 thru June 30, 1965).

198. Flat arch with reinforced bricks under construction in the Presidential Square, according to the solution drawn by R. Vollmer, Jr. (LIK, 030.IV.A.650.1.9).

masonry methods and investigating possible problems for the implementation of Kahn's structural designs.

At the start of October 1964, when he is in Dhaka, Kahn discusses the technical details of connections between the reinforced bricks and the "grade beams" with the engineers of the Public Works Department,[78] and selects the bricks made by Mirpur Ceramic Works, Ltd.[79] From the various types, he chooses the largest bricks, probably considering the effect of the scale of the buildings in the landscape. He seeks a uniformity of color that nevertheless turns out to be difficult to achieve, given the artisanal manufacturing methods.[80]

Kahn's request for bricks with perfect corners prompts Mirpur Ceramic Works, Ltd (mentioned in some documents as the Mirpur Brick Company), commissioned to produce the bricks for the works of the capital, to purchase "some new dies from Germany" also capable of producing three different patterns of holes.[81]

For the color of the brick, Kahn is looking for a dark tone: "The brick color we are trying to achieve necessitates hotter firing hence more shrinkage to the brick."[82] The choice of the type of bricklaying has to take the lack of specialized workers into account. "[...] masons may have difficulty with any kind of complicated bond pattern. I advise simplicity," is Vollmer's recommendation.[83] At the end of October the masons build the first centering for the construction of the grand arches, entirely done without wood, with a temporary brick wall lightened by the inclusion of voids in the masonry, and shaped at the top with earth to shape the curvature of the arch. Among the more complex laying methods, those of the jack arches require special inclined bricks. Other special bricks are needed to make the connections between the sloped walls in the Hostels. At the start of December, Kahn's office makes the drawings for these bricks.

The most detailed explanation of the construction system of the reinforced brick masonry is contained in the section "Reinforced brick masonry" of the "Masonry Specifications" prepared on 15 January 1965.[84] The bricks are laid in two parallel faces, connected by Z brackets to place at intervals of three courses; the two faces form a narrow interspace in which the reinforcements are positioned and the concrete is poured.[85]

The construction methods of the arches are explained in detail only in the version of the "Masonry Specifications" dated 22 January. The masonry

[78] See the minutes of the meeting, 10 October 1964, "Louis I. Kahn with P.W.D," LIK, Roy Vollmer's Reports 10.04.64 18 Mar. '65. Kahn borrows a series of technical manuals, price lists and specifications belonging to the Public Works Department.

[79] See [Roy Vollmer, Jr.], manuscript, minutes of the meeting, 9 October 1964, LIK, PAKCAP – Correspondence to/from RoyGus October 8, 1964 thru June 30, 1965 (for the typewritten version see, LIK, PACAP Vollmer Reports).

[80] R.V., [Roy Vollmer, Jr.], "Report from Dhaka to LIK Office," 10 October 1964, LIK, PAKCAP – Correspondence to/from RoyGus October 8, 1964 thru June 30, 1965.

[81] "They say," Vollmer reports on the discussions with the owner of the company, "that less warping and cracking occurs with the greater number of holes, as it permits more uniform curing. The color can be the same for any of the three" (R.V., [Roy Vollmer, Jr.], "Report from Dhaka to LIK Office," 31 October 1964, LIK, Roy Vollmer's Reports 10.04.64 18 Mar. '65). For the drawings see the original manuscript, LIK, "Houses-early space studies from D. Wisdom's file."

[82] R.V., [Roy Vollmer, Jr.], "Report from Dhaka to LIK Office," 31 October 1964, LIK, Roy Vollmer's Reports 10.04.64 18 Mar. '65.

[83] *Ibid.*

[84] Louis I. Kahn, "Second Capital of Pakistan at Dhaka. Masonry Specifications," 15 January 1965, LIK, Second Capital Special Brick Details Dated 12/7/64.

[85] *Ibid.* A version of the same "Masonry Specifications" with corrections is dated 22 January 1965, LIK, Second Capital Special Brick Details Dated 12/7/64.

centering prototype is the worksite experience on which decisions are based regarding the methods of construction of the arches. According to the version of 22 January the centers should be in wood or metal, depending on the radius of the arches; in the case of centering in masonry, like that of the prototype, it is urged that the summit be covered "with building paper to permit masonry to move as the mortar shrinks." "Where arches are reinforced (see Presidential Square), pour the grout in stages as each segment is built. Grout space must be kept clear of loose mortar, and of mortar projections at all times."[86]

At the end of January the question of the manufacturing of the bricks in the color desired by Kahn on the part of the Mirpur Brick Company has yet to be resolved. The bricks with only three holes are "pinkish in color," while those with ten holes come closer to Kahn's indications, but are rejected by Faruqi (perhaps for problems of solidity, though Vollmer guarantees their strength to be equal to that of the bricks with three holes).[87]

Kahn visits the worksite on 28 January when construction is in progress of the great reinforced brick arches of the Presidential Square. "Lou looked at the brickwork and was pleased. Care is being exercised and special tools employed to keep all the joints uniform. [...] the work looks very impressive, appearing as some great ancient structure now being unearthed," Vollmer reports to the studio staff in Philadelphia following Kahn's visit[88].

Choices for the formwork

During March 1965, Kahn's office develops the first solution for the formwork without plywood panels. The system and the drawings of the "typical wall form" designed in July 1964 are repeated to the letter, without modification of any details in the formwork or in the marble inserts with which to seal the recesses made at the position of the spreaders. But now, in place of the 3/4 inch plywood, Kahn calls for panels to construct with wooden boards 2 inches thick. On the drawings we see the notation: "2-inch nom. thick hardwood formboards."[89] The fact that Kahn, in the drawing of the elevation of the wall of the Office Sector of the National Assembly Building, depicts only the horizontal cornices and the slender vertical lines corresponding to the width of the panels composed of boards conveys the idea that in this phase he imagines that only the edges of the panels will produce imprints, and that it will be possible to understand, once the concrete is poured, what was the fundamental constituent element of the formwork – namely the panel. We cannot rule out the idea, though the detail drawing suggests nothing to this effect, that the wooden boards are to be covered with a thin metal sheet, as is customary in Pakistan according to Vollmer's reports. A document of 25 March, titled "Materials schedule" and indicating the characteristics of the brick masonry – composition of the mortar and measurements for the visible and concealed bricks – and of the fabrication of the concrete, contains the description of a type of formwork that can only be the one envisioned in the drawings of 26 March. The document shows that Kahn is proposing the Pakistani formwork system, corrected in the way indicated by Vollmer, with the metal sheet welded to corner reinforcements: "The Contractor shall use smooth steel plate shuttering welded to steel angle frame in forming for all exposed concrete work."[90]

Thus, after the hypothesis of imports from America or local manufacture of plywood with American methods, Kahn turns to a completely local way of molding concrete. But when Kahn passes from the plywood panel

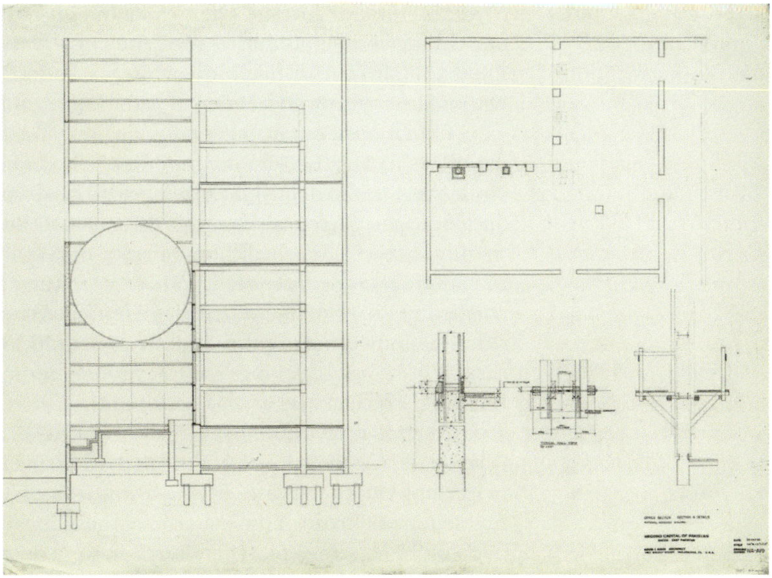

199. National Assembly Building, Dhaka, 1962-83.
"Office Sector. Section & Details," 26 March 1965 (LIK, 030.I.C.650.20a).

to the one made of boards, he is still not thinking about displaying all the parts of the new panel, and therefore the assembly of the boards; he is still considering only the visibility of the basic unit, confirming the survival of the idea of the plywood panel in the design of the imprints. This is just the first step towards the design of formwork with wooden boards, destined to undergo ulterior and decisive modifications, introduced precisely with the idea of explicitly showing all the parts of the system.

Research on concrete ingredients

Several decisions made by the representatives of technical and governmental institutions of Pakistan attempt to alter certain orientations of Kahn regarding the design and construction of the buildings of the new capital. On 1 April 1965 the Government Designs Committee states that a "traditional treatment should be given to important public buildings to bring out Islamic architectural characteristics," referring to the National Assembly Building and the residences of the ministers and functionaries. Then a telegram is sent to Kahn to inform him of the committee's decision, so that "he may reconsider facade and elevations of these building to reflect Islamic architectural characteristics."[91] The construction of the brick masonry works of the Presidential Square also meets with the first criticisms. The engineer-in-chief

[86] Louis I. Kahn, "Second Capital of Pakistan at Dhaka. Masonry Specifications," 22 January 1965, LIK, Second Capital Special Brick Details Dated 12/7/64. Concrete is also presented by Kahn as a composite system: "The concrete helped me avoid great heavy walls and achieve as much air as possible. The concrete was like the strappings on a trunk. You make it out of wood, but you use metal to reinforce the wood. That's also a composite order" (Louis I. Kahn, in, John Wesley Cook, Heinrich Klotz, *Conversations with Architects*, New York, London, Praeger and Stevens, 1973, pp. 196-197).

[87] Roy Vollmer, Jr., letter to the Louis I. Kahn office, 1 February 1965, LIK, PAKCAP – Correspondence to/from RoyGus October 8, 1964 thru June 30, 1965.

[88] "High powered meeting 19 January 65," LIK, PACAP Vollmer Reports.

[89] "Office Sector Section 8 Details," 26 March 1965, LIK, 030.I.C.650.020.a. The composition of the concrete calls for "Jhama chips" (see "Materials schedule," LIK, "Pakcap-Materials schedule (printed 3/25/65)").

[90] "Materials schedule," LIK, "Pakcap-Materials schedule (printed 3/25/65)."

[91] "Addlchief Dhaka," LIK, Second Capital – Pakistan Public Works Department Correspondence - 1965. On this occasion Kahn is reminded that in May 1964 it had been decided to give an "Islamic character" to the lighting and ventilation equipment.

Mumtaz Ahmad Qureshi, after reviewing the costs, urges Kahn to do without the machine-made bricks for the construction of the other works in Dhaka – "the use of machine-made bricks in the case of houses for the National Assembly and Hospital [should] be totally eliminated."[92]

On 12 April certain sections of the draft specifications on "Concrete" prepared on 30 March 1964 are modified.[93] It is specified that for the "mixing and testing trial batches" the reference point should be the British Standard Specifications. As coarse aggregate the materials indicated are "crushed stone," "natural gravel," "crushed Jhama Brick," or "combinations thereof." For the fine aggregate the document indicates "natural or manufactured sand or crushed graded Jhama particles, or combinations thereof." All the different combinations of aggregates and sands should be seen in terms of what is shown in Kahn's drawings for the various parts of the National Assembly Building. The changes in the specifications also have to do with certain passages in the section "Removal of forms" of the previous draft under the title "Concrete." Clarification is added on the pouring of the concrete, to be done in three phases in the case of the construction of horizontal structures, while no mention is made of the material of the formwork, which in the specifications of 30 March were in plywood for the "concrete exposed to view."[94]

Kahn's office, especially thanks to the work of Wisdom, is gathering technical information to develop the whole construction system for the exposed concrete. Expandite Limited, based in London, sends an advertising booklet for the sealing of expansion joints, both vertical and horizontal, in reinforced concrete structures,[95] while Burke Concrete Accessories, Inc., sends a catalogue of accessories for formwork, the ones already used by Kahn for the Salk Institute.[96]

After the various reports and information, the control of the quality of the concrete enters an experimental phase.

In Dhaka the order has been given to make "sample concrete walls" in concrete, to "give Lou a choice" regarding both "color & texture" and the quality of the "fine & coarse aggregate."[97] Keast & Hood, through a report written by Leidigh, provides Kahn with indications on the ways of evaluating, with tests and analyses, the type of cement and aggregate to use in the concrete, and on the efficacy of Pozzolith for the workability and formal results

[92] Mumtaz Ahmad Qureshi, letter to Louis I. Kahn, 8 April 1965, LIK, Second Capital – Pakistan Public Works Department Correspondence – 1965. "If materials are shoddy, everyone concerned in the construction shares the blame," is Kahn's reply regarding the brick: "Our present thinking about brick is that the material used in the Presidential Square is good looking in color and precision, strong, and appropriate. This material should be used in all brick buildings between Roads A and C, and also in the President's House (Louis I. Kahn, letter to Mumtaz Ahmad Qureshi, 13 May 1965, LIK, PAKCAP – Correspondence to/from RoyGus October 8, 1964 thru June 30, 1965).

[93] "Pak-Concrete specification changes," 12 April 1965, LIK, Second Capital – Pakistan Engineering Data – Keast & Hood.

[94] To the list prepared on 30 March of the concrete works, with indication of the minimum hours during which the formwork should be left, there are added the "sides of walls" – the formwork must be kept in place for 48 hours after the pour. It is specified that the concrete surfaces have to be kept damp for no less than 7 days, leaving the formwork in place (otherwise other solutions will have to be found). Also during the course of April, it is established that the concrete for the above-ground structures should be made with Pozzolith additive. "No proposed mix may be used for any exposed concrete until approved by the Architect," is the warning ("Concrete proportions," n.d., received by Kahn'office on 12 April 1965, LIK, Second Capital – Pakistan Engineering Data – Keast & Hood).

[95] *Expandite Limited*, booklet, received by Kahn's office on 14 April 1965, LIK, Second Capital – Pakistan

on the surface of the concrete. "These factors," Leidigh writes, on the subject of Pozzolith, "will tend to produce concrete of a more uniform texture, appearance, and strength."[98] Other considerations contained in Leidigh's report have to do with the position of the "construction joints," which though it is generally indicated in the drawings, also has to be reviewed during the worksite phases, case by case, depending on the static forces at work. Leidigh recommends avoiding this type of joint where the "underside of the construction will be exposed without any applied facing material [...]."

Formwork with random assembly of boards

Starting in March 1965, Kahn considers the hypothesis of appointing a staff member, to be sent to Dhaka, to supervise the "design and construction of formwork."[99] The name of the person circulates in April, when there is discussion of the possibility of sending Frederick Langford to the worksite, to be responsible for the "formwork for National Assembly." This is documented in a letter from Vollmer to Kahn: "Fred Langford to come to Dhaka for this work if PWD will pay expenses. LIK office should ship now any accessories which deems feasible to use in the concrete work and which could be manufactured locally in Dhaka. Fred is to design and supervise the making of experimental formwork. The initial drawings are to be made in the Dhaka office. This work should be taken in hand immediately."[100]

But the definition of the fundamental features of the design of the formwork takes more time than expected, and Langford is able to head for Dhaka only in the following year. From the end of April to early May Kahn's office develops another formwork system. The new project calls for panels made of wooden boards. This would become the final technical solution, though it would still be modified and revised during the phases of preparation of the formwork. All the details are assessed in terms of the chances of their being made in Dhaka, and to this end many new questions are asked by Kahn's staff to technicians expert on worksite practices in Pakistan.

The documents of Kahn's new formwork design are composed of a series of sets of drawings with legends, dated 4 May, and new draft specifications, dated 3 May. Frederick Langford is in charge of the preparation of these documents.

Next pages
200-203. National Assembly Building, Dhaka, 1962-83.

200. "Formwork Study," No. 1, 4 May 1965 (LIK, 030.I.C.650.33).

201. "Formwork Study," No. 2, 4 May 1965 (LIK, 030.I.C.650.33).

202. "Formwork Study," No. 5, 4 May 1965 (LIK, 030.I.C.650.33).

203. "Formwork Study," No. 4, 4 May 1965 (LIK, 030.I.C.650.33).

Correspondence – Michellaneous, received by Kahn's studio on 14 April 1965.

[96] "For any question about the Dhaka Project, ask Frederick Langford or Jack John E. MacAllister in our office in La Jolla," writes Wisdom (David P. Wisdom, letter to Burke Concrete Accessories, Inc., 22 April 1965, LIK, Second Capital – Pakistan Correspondence – Miscellaneous).

[97] David P. Wisdom, letter to Roy Vollmer, Jr., 22 April 1965, LIK, PAKCAP – Correspondence to/from RoyGus October 8, 1964 thru June 30, 1965.

[98] Thomas J. Leidigh, "Dhaka. Comments to accompany concrete specifications," 27 April 1965, LIK, Second Capital – Outline Specifications Architectural/Mechanical/Site Work May 10, 1964.

[99] Minutes of a meeting in Dhaka, with Kahn present, received in March 1965, LIK, Second Capital – Pakistan Pakistan Public Works Department Correspondence - 1965.

[100] Roy Vollmer, Jr., letter to Louis I. Kahn, n.d., received by Kahn's office on 24 April 1965, LIK, PAKCAP – Correspondence to/from RoyGus October 8, 1964 thru June 30, 1965.

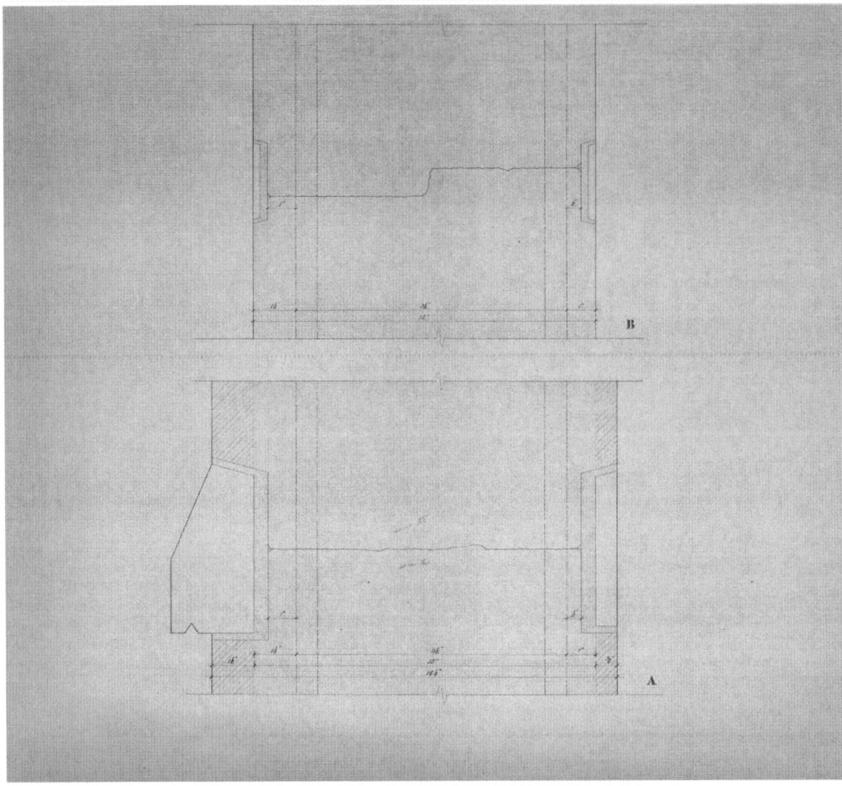

1

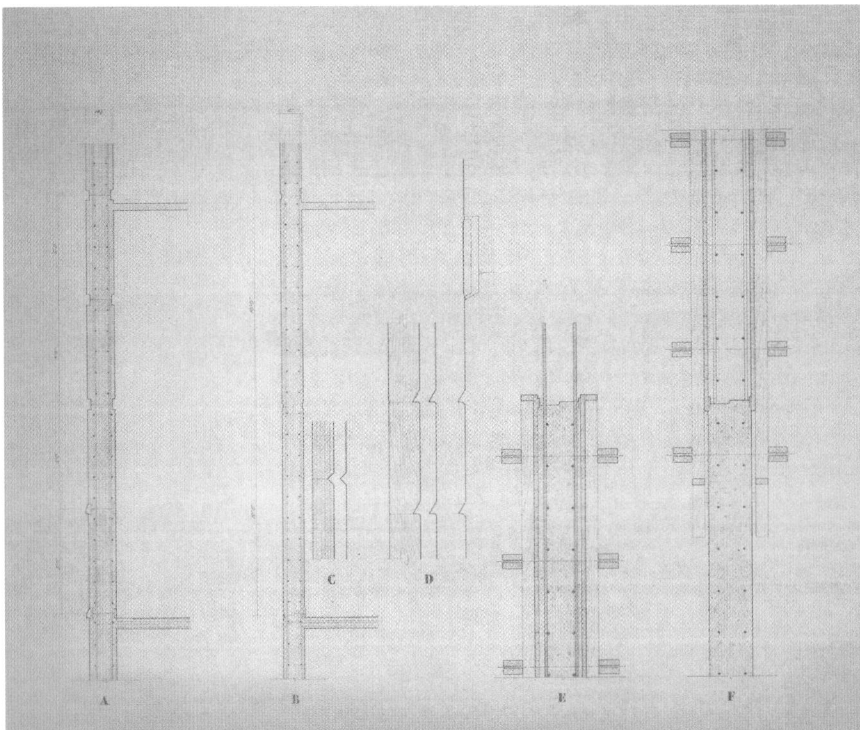

2

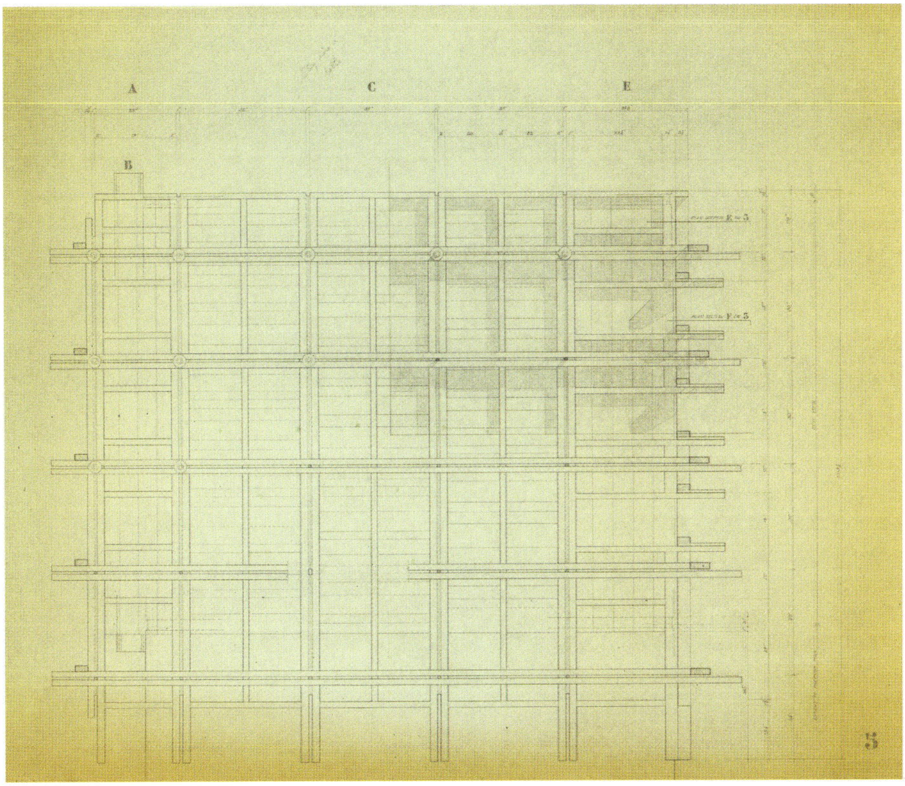
A
C
E
B
5

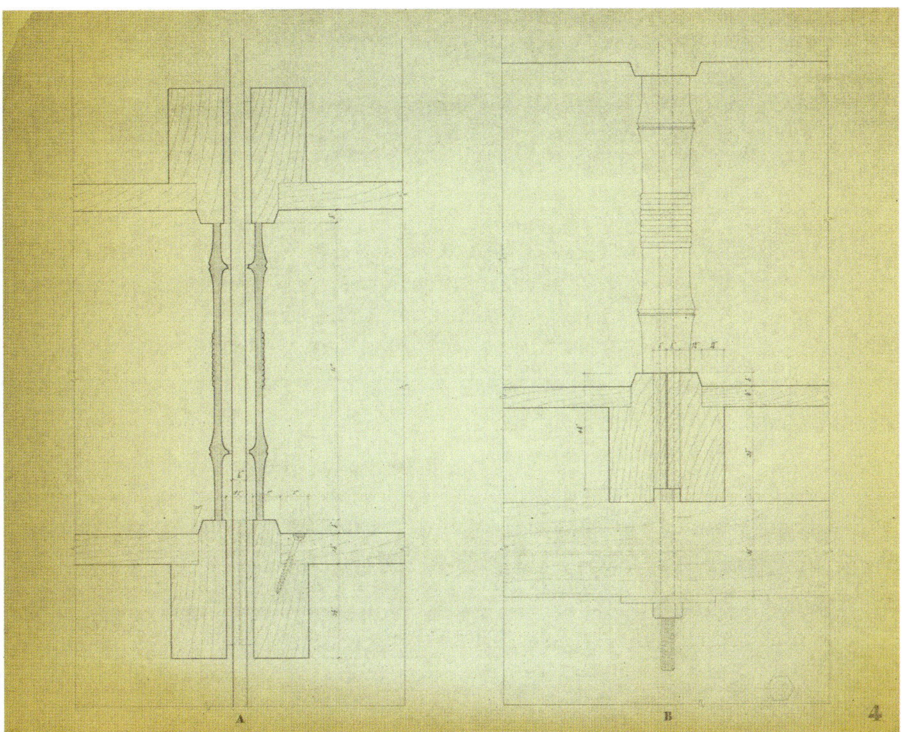
A
B
4

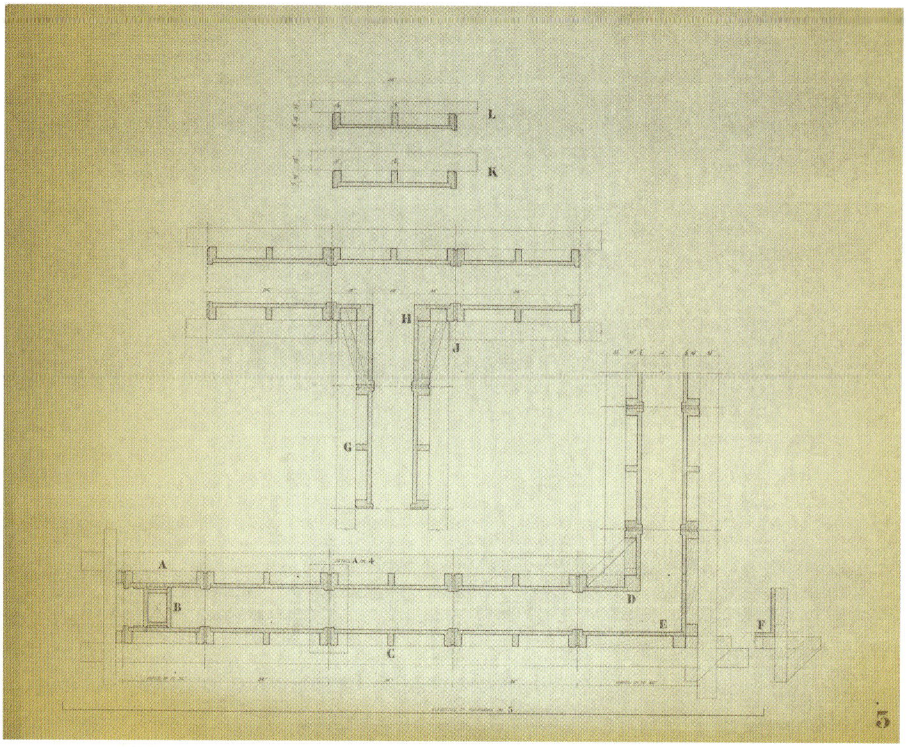

Kahn's panel system is based on a double pattern of imprints, differentiated in terms of importance in the construction process: on the one hand, the slender lines generated by the imprints of the boards; on the other, the strong chromatic bands generated by the pour phases, identified by horizontal recesses sealed by marble inserts, and by the row of spreaders, incorporated in vertical recesses and also sealed with marble. Thus the slender vertical relief frames in concrete between one panel and the next are replaced by wider white bands that now create, with the horizontal bands, a luminous grille set into the exposed concrete walls, forcefully conveying the essence of the construction.

The particular characteristic of the panel, a true exception in Kahn's designs for concrete formwork, is the size of the wooden boards, all of them different, to assemble randomly – "random width sheathing."[101]

While in the Richards Laboratories the colossal crane for the raising of the precast pieces had prompted Kahn to create precast pieces on the scale of that machinery, now for the National Assembly Building he envisions incorporating, in the design of the concrete, the typical operations of a Pakistani worksite, those done by hand according to habits that do not permit standardization of pieces all with equal measurements.

The possibility of assembling boards of different sizes, without controlling the design, makes Kahn's concrete similar to certain special results of the *béton brut* of Le Corbusier, not coincidentally those achieved in India, on worksites marked by practices not unlike those of Pakistan, when Le Corbusier decides to shape the concrete of Villa Shodan in Ahmedabad with boards arranged "without stereotomy," as he defined the random assembly of the formwork. But unlike that crucial example for the evolution

204. National Assembly Building, Dhaka, 1962-83.
Formwork study, No. 3,
4 May 1965 (LIK, 030.I.C.650.33).

of the concrete of Le Corbusier, where the boards all have the same width but vary in length, the method proposed by Kahn winds up more closely resembling the texture of different widths of the stone slabs of the floors and pavements of Le Corbusier, classified as *opus optimum*.

The possibility of using boards of different widths confirms the desire, on Kahn's part, to sum up the empirical logic of the Pakistani worksite in the design of the concrete. This marks a turning point in Kahn's research on the formwork system for Dhaka.

Kahn and his team study the possibility and the effects of diversified arrangement of the boards, in horizontal and vertical patterns. But the alternation of the boards is quite unlike the kind utilized by Le Corbusier, starting with the Unité d'Habitation of Marseille. For Kahn, in this phase of the design of the formwork, the arrangement corresponds to the intention to obtain relief imprints on the concrete that function for drainage, and to the logic of assembly of the posts of the support structure of the formwork panels, the walers, which in the corners are placed horizontally to facilitate the placing of the reinforcements at 45°, thus requiring vertical assembly of the boards. After the experience of the Salk Institute, by now Kahn is striving to make a concrete surface woven by imprints of the construction process, and as fully equipped as possible with a specific function of protection of the surfaces over time.

Drawing number 1 of the series of drawings of Kahn's formwork illustrates two variants of the reinforced concrete wall, A and B, with different thicknesses (14 1/4 and 12 inches) depending on the role of the marble cornices. In both variants the distance between the vertical reinforcement bars is the same (9 1/2 inches), while the layer of concrete from the bars to the face varies, starting with the minimum distance required to guarantee the protection of the reinforcements – 1 1/2 inch towards the outside; 1 inch towards the inside.

All the shapes of the recesses have only one side or both sides inclined to facilitate removal of the wood strip after the setting of the pour. Solution A, the one favored by Kahn and based on the studies of 1964, calls for a protruding cornice with dripstone to insert in the recesses of the face exposed to weather, and a flat piece mounted flush with the wall surface in other cases. Precisely to avoid making recesses near the bars, the thickness of the wall increases on both sides (towards the outside it is 1 1/2 inch; towards the inside 3/4 inch). The role of the two types of marble inserts is explained as follows: "It is my understanding that the exterior marble functions as a cover for the horizontal pour joint and the screw tie holes. It also acts as a water shedding device which may cut down organic growth on the surface of the wall. The interior strip covers the horizontal pour joint, tie holes, and acts as a base when the wall is intersected by a floor slab."[102]

The fact that every element of the formwork produces imprints, each one calculated to create a coherent and logical system, is demonstrated by the wall solutions generated by the two variants documented by drawing 2, which connect the height of the panels and the function of the horizontal cornices. When the marble inserts become dripstones, the height of the panels is reduced; instead, when they are simple flat frames on both faces of the wall, the panel

[101] "Formwork Study," 5 May 1965, LIK, 030.I.C.650.033.
[102] See the notes attached to the drawing of the formwork, No. 5, LIK, 030.I.C.650.033.

becomes taller. In the first case Kahn reduces the height of the panel to have cornices closer to each other, so their protrusion can offer better protection for the concrete surfaces against weathering; in the second case, the higher panel helps to speed pouring operations, eliminating the dripstones. The choice between the two solutions is based on information regarding the efficacy of the protection against rain in the monsoon climate of Dhaka, and on the availability or lack of it of machinery for the deployment of the panels, which determines the measurements of the panel and its degree of ease of handling.

Solution A calls for a wall to be made between two floor slabs separated by a distance of 20 feet, by means of a formwork with panels having a height of 5 feet, to raise 4 times during each pour operation. The defects of this system, listed in the notes attached to the drawings, lie in the long exposure of the steel bars to weather before all the pours have been completed, in the formation of four cold joints that divide the wall into horizontal segments, in the slowness of the work with respect to higher formwork and, finally, in the risk of marks left during the various pours on the already completed concrete surfaces – drying, the marks would result in a "finished wall" with a "scrubby appearance."[103]

Solution B corrects the problems caused by the low formwork. In this case, the panels with a height of 10 feet permit pouring of the wall in just two operations, with the joint between the pours again obtained through the flat marble insert, since the drainage cornice is no longer useful, due to the excessive height of the lower surface to be protected. Precisely due to the absence of the dripstones, the thickness of the wall is reduced to 12 inches and the marble inserts are smaller (as in solution B of set 1). This solution with formwork 10 feet in height is designed on the basis of information gathered by Gianopulos from the technicians in charge of the construction of the Intercontinental Hotel in Dhaka, which is being done with this system and with the use of vibrators powered by portable electrical generators, also proposed in Kahn's solution B.[104]

With solution C, Kahn seems to continue to propose, in spite of the decisions taken at this point on the materials for the formwork, the use of plywood panels joined as in the Salk Institute, with cutting of the edges to produce a V-shaped imprint – the V joint. Not by chance, the solution is defined as "Projecting concrete line (Salk project)." "The chamfer at the meeting edges of plywood sheets produced a projecting concrete line on the Salk Laboratories," reads the explanation. But the solution is included among those under study for Dhaka as an example of what was done at the Salk Institute, and the possible application of the V joint to the construction of the panels of the formwork for Dhaka is not clearly explained (in any case, it is probable that the V joints would replace the horizontal grooves with marble inserts).

The meaning of the presence of the V joint of the Salk Institute becomes clear if we consider the characteristics of solution D, in which Kahn envisions a system of dripstones to be made directly by shaping the concrete surface with the formwork. The imprints are defined as "water shedding concrete lines." In these evolutions of the dripstone from the marble cor-

[103] *Ibid.*
[104] *Ibid.*
[105] "If the problem of organic surface growth can be solved by a reduction in wall wetness, then a water shedding device that evolves from the formwork boards may be worth considering – the explanatory note reads. – Experimentation on the angle, the projection and the bottom radius would, of course, be necessary. Nick Gianopulos said that a popular size of board lumber in East Pakistan measures 1" to 1¼" in thickness, and varies

nice inserted in a special recess to a ripple on the surface of the concrete, directly shaped by the pour, the intrusion of the detail of the V joint becomes clear. In definitive terms, the lines envisioned for Dhaka would simply be a more complex function generated by the design of the joint of the formwork, not only bringing out the assembly and control of the seepage of the liquid stone, but also playing the role of protection of the surfaces from rainwater and the resulting formation of biological coatings – "horizontal sheathing flush-jointed," as indicated on the drawings. The thin protruding lines of the dripstone are obtained by cutting one edge of the wooden boards to be assembled horizontally to form the panels of the formwork. The overall effect would be that of a surface striped by protruding ripples (those of the dripstones) at irregular intervals, like wrinkles on skin.

At the position of the pour and the spreaders of the panels, Kahn continues to foresee a recess in which to inlay the cornice, which is flat at this point, no longer with a drainage function (it is indicated that the material of the insert can also be ceramic).[105]

The sections identified with the letters E and F serve to illustrate the construction process of solution D, the assembly of the formwork, the pour, the dripstone imprints left on the surface, the particular stepped profile at the top of the pour to better seal the setting joint, and the formwork repositioned higher up to proceed to the next pour, identified by the recess.

The drawings of drawing 3 illustrate the connections of the formwork required to make perpendicular walls, in a T or an L form. The details and explanations concentrate on the inner and outer corners, and on the way of installing a "corner member" or "typical edge studs" in those points subject to flaws of the pour, also with the use of "horizontal bridging" – an inclined stabilizing post. "Corner forms are extremely vulnerable to leakage which produces ragged and untrue surface," the notes warn.

The study of the formwork system also covers the detail of the spreaders to keep the panels of the formwork in position. Kahn's team's approach regarding integration in the project of the characteristics of Pakistani construction can be seen in the fact that the spreaders are to be produced locally, with the natural material most often used for scaffolding on construction sites: bamboo. The drawing 4 shows how a bamboo segment can be used as a tie thanks to binding at the middle between the two knots of the stem, at the point most subject to breakage due to compression.

The vertical structures of the formwork – the "studs" along which the ties are arranged – have a wedge-shaped head in contact with the pour, with the edges cut to facilitate stripping. In the recesses produced by the studs, other marble inserts are set. "This keyway could receive a thin marble cover strip similar to the horizontal 'pour joint' cover."[106]

The drawing 5 of the formwork structure does not illustrate all the necessary devices, besides the wood; but in the notes the need is emphasized to assess their local availability: "Form devices of this nature are usually rented items in this country. A comprehensive analysis of the materials and work methods in Pakistan should lead to creating form devices from available sources."

in width from 3" to 7". It is not Teak, but more like our Fir or Pine. Most of the drawings in this set show 1 1/8" thick form boards, or what we normally call five-quarter stock. Detail on right shows a wider radius on bottom of projection which would reduce chip-off during the form stripping process. Same detail shows a ceramic inlay as an alternate for the (¼"-3") marble pour joint cover shown on 1-B" (*ibid.*).

[106] See the notes attached to the drawing of the formwork, No. 4, LIK, 030.I.C.650.033.

The notes attached to the drawings are completed by a series of questions aimed at learning local technical information required for the development of the details of the formwork system proposed by Kahn.

The questions attached to the drawings of the marble cornices concern the reasons behind the perishable quality of concrete in Dhaka and, in particular, the possible effects of the humid climate of East Pakistan on the formation of organic substances on the concrete surfaces – "organic growth." From the questions, it can be seen that not only the dripstone cornice has been foreseen to protect Kahn's concrete; the hypothesis is also mentioned of treating the surface with special waterproofing products.[107]

The questions listed in drawings 4 and 5 cover the construction of formwork and scaffolding. Kahn and his staff request information on available thicknesses for the panels, the availability of these panels and their dimensions, the characteristics of the local lumber normally used for the building of formwork structures, the most common method for "lifting form panels," the availability of plywood, the possibility of assembling multiple panels to obtain large formwork (they also want to know about possible problems during positioning and lifting by hand in the pour phases), the possibility of installing simple machines to manufacture metal parts on the worksite, also for the formwork, and the bamboo scaffolding, including whether it is necessary to reinforce the canes[108].

At the same time as the drawings of the formwork for Dhaka are being prepared, Kahn's office staff resume work, on 3 May, on the draft of the specifications of the "Concrete" drawn up on 30 March 1964. They correct certain passages, and tone down the first part with the description of the exposed concrete, also because brick masonry has now been chosen for the Hostels – "All walls that contribute to the structure of the building" thus becomes "in general, concrete walls." At the same time, they indicate the responsibilities of the various technicians in the control of the concrete fabrication, and above all they insert the substantial modifications that have been made to the composition of the concrete, decided with the "Pak-Concrete specification changes" of 12 April 1965.

The document of 3 May is a general outline of the specifications, prior to the final preparation of the guideline text for concrete on the worksite. The other substantial change introduced in the document has to do with the material of the formwork, not mentioned in the modifications of 12 April. By now the draft of 30 March 1964 on "Concrete form plywood" is no longer valid, and Kahn's renunciation of plywood formwork has been offi-

[107] "Has anyone determined just what the 'organic growth' is that occurs on the concrete walls? Does it grow only on surfaces that get wet, or does the high humidity and dampness in East Pakistan promote this growth on all exterior concrete? Is it like the fungus that we see under old concrete overpasses in this country? Has anyone investigated the possibility of using a transparent waterproofer on the concrete?" (*ibid.*).

[108] "What is the availability of plywood? Is it true that the glue is not waterproof? What sizes and thicknesses are common? [...]. Is there any cocoon-like covering that is available for small gussets that might be made of inferior plywood? [...]. If panels were ganged (2 or more put together on the ground), could they be lifted in one piece? [...]. How much weight can an average Pakistani worker be expected to carry? How much weight can two average Pakistani workers lift?"; these are some of the questions contained in drawing 4. "What form coatings are available in Pakistan?" is the last question in the notes attached to the formwork drawing No. 5 (see the notes attached to the drawing of the formwork, No. 5, LIK, 030.I.C.650.033).

[109] See the series of notes "Formwork Study," 4 May 1965, LIK, 030.I.C.650.033.

[110] For the spreaders, the permitted types are listed, chosen from those produced by American manufacturers, indicated in the order: Burke Concrete Accessories of San Francisco; Richmond Screw Anchor Company of New York; Superior Concrete Accessories of Franklin Park, Illinois; and the Universal Form Clamp Company of Chicago. The section "Forms for concrete exposed to view in finished work" reads: "Forms for exposed concrete shall be made up as indicated on drawings, using sound, well seasoned, dressed and matched teak lumber, permitting only sound, tight knots, not over 1" in

cially decided, also on the basis of the information received from Vollmer on local formwork methods and the possibilities of their improvement to obtain high-quality exposed concrete. It is clear that the document of 3 May interprets the indications in the contemporary design drawings of the formwork made with boards.

With respect to the draft in 1964 in which plywood was indicated for use, now Kahn specifies teak as the material of the formwork panels. It is worth noting, however, that the notes to the sets of drawings of the formwork specify, based on information from Gianopulos, that the wood in Dhaka is not teak, but a wood "more like our Fir or Pine."[109] The painstaking description of the spreaders, their parts and positioning is necessary because Pakistani construction does not usually involve concrete marked by the imprints of formwork.[110]

Pattern studies for the Office Sector of the National Assembly Building

Though all the documents for the formwork are ready in May 1965, the sets of drawings, notes and questions are not sent to Dhaka until the start of the following year. As soon as the detail drawings and specifications of the formwork are done, during the course of May Kahn's staff works on the possibilities of the form of the various random textures produced by the imprints in the wall of the Office Sector of the National Assembly Building, already examined in the project for the formwork in March. Two drawings for that wall show the alternatives that can be obtained through the arrangement of the panels and boards, and through the height of the pour.

In the first drawing of the wall the position of the spreaders in vertical rows generates upright bands, corresponding to the studs with the spreaders, and horizontal bands corresponding to the pour phases, and at the height of the floor slabs. The resulting pattern creates narrow rectangular compartments arranged vertically, designed by the dense series of horizontal lines of different heights caused by the imprints of the irregular boards.[111] The horizontal arrangement of the boards means that the rows of spreaders are positioned along the vertical recesses, while the horizontal recesses mark the various pour levels. The logical distinction between the two types of recesses is so important for Kahn's formwork that special drawings are made, without the marble inlays, to show the positioning of the spreaders and the pour levels, and with the marble inlays that cover both the vertical and the horizontal recesses, making the two logics unrecognizable.

In the second drawing of the wall, the height of the pours is halved to obtain closer horizontal bands. The vertical bands are limited to the one at the

diameter. Forms shall be coated with approved non-staining form coating material on face and edges, preferably applied by dipping before use. Form coating shall be clear urethane type coating as directed. The use of oil on forms for exposed concrete is prohibited. Forms may be re-used as long as they remain in good condition, but must be thoroughly cleaned, inspected and approved before each re-use. Snap ties for walls shall be fabricated from steel wire, formed to break off in wall approximately 1" back from face of concrete, with flattened portions to prevent twisting of tie wall. Ties shall have suitable washers to space form faces required distance apart and upset ends on projecting portions to permit wedging formwork tight against washers, using slotted malleable iron wedges or other suitable devices. Use water seal snap ties for exterior walls with swaged washer in center of tie to break surface continuity and prevent water seepage. Ties shall be accurately installed in pattern as shown.
Coil ties shall be made of 2-wire struts welded to a helical wire at each end. Coil shall project approximately 3 turns beyond ends of struts to thread into plastic or steel cones, 1-1/2" long, and in turn receive special coarse thread 1/2" dia. coil bolts with steel washer under head to bear on form. Bolts and coil cones shall be removed when forms are stripped. Use water seal coil ties for exterior building walls, with swaged washer
at center of each strut" ("Concrete," 3 May 1965, LIK, Second Capital – Outline Specifications Architectural/Mechanical/Site Work May 10, 1964).

[111] See the drawing, n.d., No. 8, LIK, 030.I.C.650.033. See also the drawing, n.d., No. 7, LIK, 030.I.C.650.033. The drawing, n.d., No. 6, LIK, 030.I.C.650.033, shows the vertical grooves that contain the tie holes, prior to the insertion of the marble.

205-209. National Assembly Building, Dhaka, 1962-83.

205. Formwork study, No. 6 (LIK, 030.I.C.650.33).

206. Formwork study, No. 7 (LIK, 030.I.C.650.33).

207. Office Sector section with the study of the pattern of the formwork, No. 9 (LIK, 030.I.C.650.33).

208. Study sketches for the pieces of marble (R. Vollmer, Jr., letter to L. I. Kahn, n.d., received by Kahn's studio on 24 April 1965, LIK, PAKCAP – Correspondence to/from RoyGus October 8, 1964 thru June 30, 1965).

209. Study sketches for the strip section with drip functions (R. Vollmer, Jr., letter to L. I. Kahn, n.d., received by Kahn's studio on 24 April 1965, LIK, PAKCAP – Correspondence to/from RoyGus October 8, 1964 thru June 30, 1965).

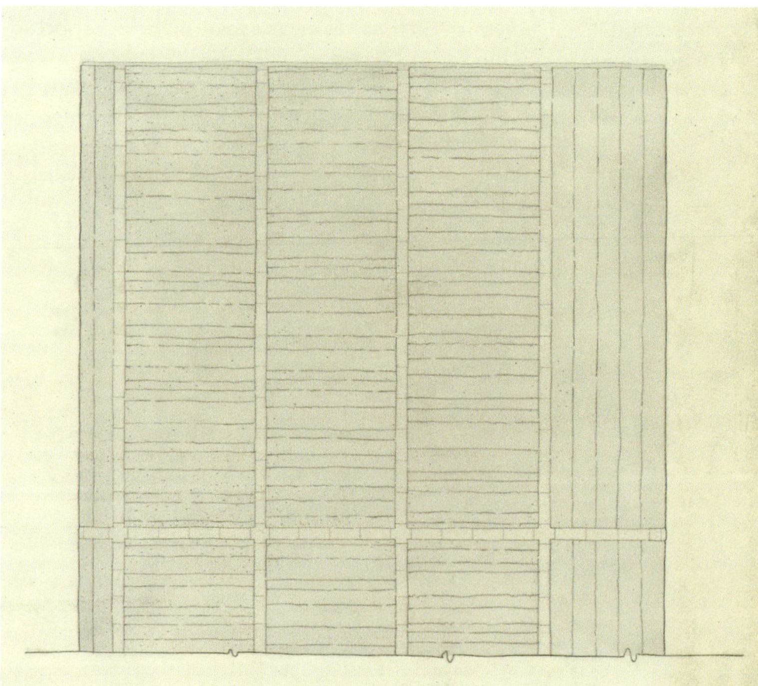

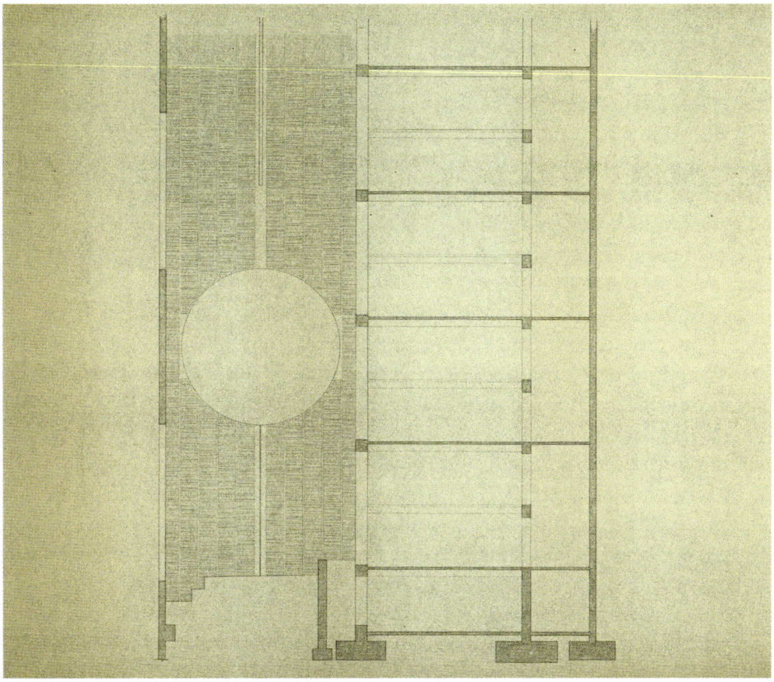

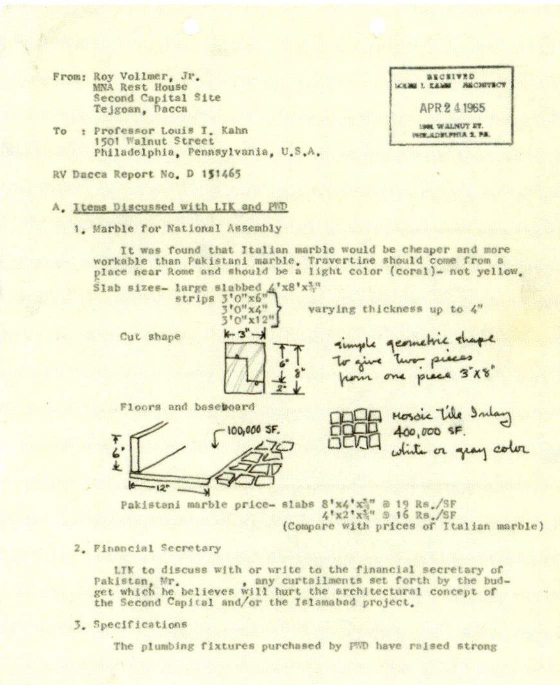

From: Roy Vollmer, Jr.
MNA Rest House
Second Capital Site
Tejgoan, Dacca

RECEIVED LOUIS I. KAHN ARCHITECT APR 24 1965 1501 WALNUT ST. PHILADELPHIA 2, PA.

To : Professor Louis I. Kahn
1501 Walnut Street
Philadelphia, Pennsylvania, U.S.A.

RV Dacca Report No. D 151465

A. Items Discussed with LIK and PWD

1. Marble for National Assembly

It was found that Italian marble would be cheaper and more workable than Pakistani marble. Travertine should come from a place near Rome and should be a light color (coral)- not yellow.

Slab sizes- large slabbed 4'x8'x¾"
strips 3'0"x6" / 3'0"x4" / 3'0"x12" } varying thickness up to 4"

Cut shape

simple geometric shape to give two pieces from one piece 3"x8"

Floors and baseboard

100,000 SF.

Mosaic Tile Inlay 400,000 SF. white or gray color

Pakistani marble price- slabs 8'x4'x¾" @ 19 Rs./SF
4'x2'x¾" @ 16 Rs./SF
(Compare with prices of Italian marble)

2. Financial Secretary

LIK to discuss with or write to the financial secretary of Pakistan, Mr.   , any curtailments set forth by the budget which he believes will hurt the architectural concept of the Second Capital and/or the Islamabad project.

3. Specifications

The plumbing fixtures purchased by PWD have raised strong

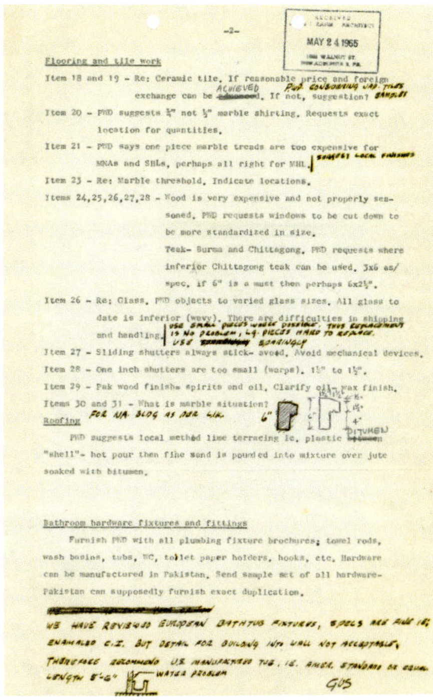

-2-

RECEIVED LOUIS I. KAHN ARCHITECT MAY 24 1965 1501 WALNUT ST. PHILADELPHIA 2, PA.

Flooring and tile work

Item 18 and 19 - Re: Ceramic tile. If reasonable price and foreign exchange can be ~~advanced~~ ACHIEVED. If not, suggestion? PWD CONSIDERING JAP. TILES SAMPLES

Item 20 - PWD suggests ¾" not ½" marble shirting. Requests exact location for quantities.

Item 21 - PWD says one piece marble treads are too expensive for MNAs and SHLs, perhaps all right for MHL. SUGGEST LOCAL FINISHES

Item 23 - Re: Marble threshold. Indicate locations.

Items 24,25,26,27,28 - Wood is very expensive and not properly seasoned. PWD requests windows to be cut down to be more standardized in size.
Teak- Burma and Chittagong. PWD requests where inferior Chittagong teak can be used. 3x6 as/spec. if 6" is a must then perhaps 6x2½".

Item 26 - Re: Glass. PWD objects to varied glass sizes. All glass to date is inferior (wevy). There are difficulties in shipping and handling. USE SMALL PIECES WHERE POSSIBLE. THUS REPLACEMENT IS NO PROBLEM, LG. PIECES HARD TO REPLACE. USE SPARINGLY

Item 27 - Sliding shutters always stick- avoid. Avoid mechanical devices.

Item 28 - One inch shutters are too small (warps). 1½" to 1½".

Item 29 - Pak wood finish- spirits and oil. Clarify oil- wax finish.

Items 30 and 31 - What is marble situation? FOR NA. BLDG AS PER LIK.

Roofing

PWD suggests local method lime terracing ie, plastic ~~bitumen~~ "shell"- hot pour then fine sand is pounded into mixture over jute soaked with bitumen.

Bathroom hardware fixtures and fittings

Furnish PWD with all plumbing fixture brochures; towel rods, wash basins, tubs, WC, toilet paper holders, hooks, etc. Hardware can be manufactured in Pakistan. Send sample set of all hardware- Pakistan can supposedly furnish exact duplication.

WE HAVE REVIEWED EUROPEAN BATHTUB FIXTURES, SPECS ARE ALL IE; ENAMELED C.I. BUT DETAIL FOR BUILDING INTO WALL NOT ACCEPTABLE, THEREFORE RECOMMEND U.S. MANUFACTURED TUB, IE. AMER. STANDARD OR EQUAL LENGTH 5'-6" WATER PROBLEM

GOS

middle, and the one near the corners between walls. The result is long rectangular panels arranged horizontally, with horizontal board imprints, and narrow vertical panels at the corners, with vertically assembled boards. In this case, all the boards have the same width, producing a regular pattern of imprints.

In both variations the vertical axis of the wall is emphasized by an imprint that generates wider bands than the others, still decorated with marble inlays. Though there are no indications in the drawings to this effect, it seems clear that to obtain that imprint special pieces are used in the formwork, of the type applied at the joints between consecutive panels in the formwork of the Salk Institute.[112]

The gradual development of the formwork design demonstrates that for Kahn the question of truth is more complex than simple issues of the nature of materials. The connection between parts does not only show a joint; it also represents a phase of growth of the structure.

The whole height of the National Assembly Building (excluding the volume of the central hall) is subdivided into 22 pour layers, each indicated by a horizontal cornice. If we consider the fact that the vertical bands also intercept juncture lines in the construction of the building, it is clear that Kahn has identified an ornament that expresses the growth of the organism on the worksite. The assumption of the construction and its phases of evolution as the basis of the expression of the strong lines of the work produces an ornament that marks the growth of the building in time, like the concentric rings of a tree trunk or the knots along stems of bamboo. The marble strips wind up taking on a value similar to the organic growth tetrahedra developed with Tyng. The tetrahedral geometry has been utterly transfigured, become unrecognizable, because what determines the growth of the National Assembly Building is the logic of its concrete construction. The pattern of grooves and imprints that covers the Salk Institute now becomes a powerful expressive form. The simple recessed groove of the Salk Institute is widened to contain a chromatic inlay that points to the construction of the

[112] The vertical bands are of different types; some are obtained with two strips of marble cladding with a slender groove in the middle that makes the concrete base visible (drawing, n.d., No. 10, LIK, 030.I.C.650.033); others are made with a single large strip that starting at the level of the floor slab splits into two pieces, separated by a groove (drawing, n.d., No. 8, LIK, 030.I.C.650.033). Two dimensioned sketches of the section of the cornice with dripstone function are found in, Gustav Langford, letter to David P. Wisdom, 30 April 1965, LIK, PAKCAP – Correspondence to/from RoyGus October 8, 1964 thru June 30, 1965. See also Public Works Department, "Specifications for a single-storied brick masonry structure - Seismic structures," n.d., received by Kahn's office on 24 May 1965, LIK, PAKCAP – Correspondence to/from RoyGus October 8, 1964 thru June 30, 1965.

[113] "Printed and sent informally to Roy Vollmer 5/24/65 by D. Wisdom for his further disposition," LIK, PAKCAP – Correspondence to/from RoyGus October 8, 1964 thru June 30, 1965.

[114] "After we have had sufficient experience with brick masonry in all its engineering and technological requirements, then it would be possible to consider concrete also in its best systems for the buildings which are planned between Roads A and B, the Hospital Building, etc." (*ibid.*).

[115] Roy Vollmer, Jr., "Report," 27 May 1965, LIK, PAKCAP – Correspondence to/from RoyGus October 8, 1964 thru June 30, 1965.

[116] Vollmer also proposes making all the joints "straight and even" (Roy Vollmer, Jr., "Report," 27 May 1965, PAKCAP – Correspondence to/from RoyGus October 8, 1964 thru June 30, 1965).

[117] Roy Vollmer, Jr., Gustav Langford, letter to Kafiluddin Ahmad, 4 June 1965, LIK, PAKCAP – Correspondence to/from RoyGus October 8, 1964 thru June 30, 1965.

[118] Gustav Langford, Progress report 2, 22 June 1965, LIK, PAKCAP – Correspondence to/from RoyGus October 8, 1964 thru June 30, 1965.

[119] *Ibid.* For the pouring of the concrete of the roof of the Presidential Square, see the worksite photograph, 8 June 1965, LIK, 030.IV.A.650.2.80. On the back of the photograph is written: "that good Chinese cement" with the date stamp 8 June 1965. The contractor of the Presidential Square is Ahmaddullah (see back of photograph, 8 June 1965, LIK, 030.IV.A.650.2.81). The pouring of the concrete on the inclined part is documented by photographs that show men with various tools, from sticks to pans, with which to make the compound penetrate the reinforcements (see, for example, the worksite photograph, 8 June 1965, LIK, 030.IV.A.650.2.76).

concrete. The conception of the joint, thanks to Kahn, takes on the meaning of a line that expresses organic growth. It is no coincidence, in fact, that in the documents regarding the work on the National Assembly Building, Kahn's staff often uses the term "growth."

The roof of the Presidential Square, first formwork for the concrete of Dhaka

It comes as no surprise that a few weeks after having completed the design of the formwork system for the walls in Dhaka, in February 1965, Kahn reconsiders exposed reinforced concrete as the material for the construction of the buildings around the National Assembly Building, in place of brick. The proposal is brought back to the fore in the context of replies to criticisms voiced by Qureshi regarding the excessive cost of machine-made brick. First Kahn explains the reasoning behind his acceptance of the change from concrete to brick, the "only material indigenous to Pakistan": "I changed the buildings from their original design in concrete to that of brick, honoring thereby the advice that brick would prove to be more economical. Though I first resisted this change, I have now discovered, in the development of the design, some beautiful shapes that are true to the order of brickwork."[113] Then he makes the proposal to continue the construction of the residence and other facilities in exposed reinforced concrete.[114]

In Dhaka Kahn's exposed concrete makes its first appearance on the worksite of the Presidential Square, under the supervision of Vollmer and Langford. At the end of May, when the construction of the monumental arched portico has reached the roof, the shuttering is ordered for the pouring of the concrete of the top slab.[115] For the intrados of the roof visible from the portico, Vollmer proposes formwork with panels measuring 3 × 6 feet, in metal, to assemble with two types of connections, "lap jointed" (interlocking overlaid) and "butt jointed" (juxtaposed). The latter solution is to be used when the horizontal posts of the temporary structures that are then used to form the seal run below the connections, "so conc.[rete] leakage is minimized."[116]

In spite of the precautions taken by Vollmer, the poured concrete of the first roof sector, completed on 2 June, is not without defects, due to the lack of expertise of the technicians of the Public Works Department and the orders given to the craftsmen. The operations prior to the pour indicated by Kahn's staff, for example, have not been performed. "The result is that the formwork was damaged by your visiting troupe and also due to insufficient preparation the concrete has spilled all over the brick surfaces of the walls below."[117] "This pour was done too hastily and consequently has many small leaks down the walls. All stains were cleaned but not 100% effective," Vollmer observes, though he does indicate improvement in the second pour, again for the roof of the Presidential Square: "Same pour stripped June 18 nice and clean with uniform texture and color."[118] A third pour is slated for 28 June.[119]

210-212. National Assembly Building, Dhaka, 1962-83.
Concrete pouring of the Presidential Square roof (LIK, 030.IV.A.650.2.80; 40; 76).

Information from Vollmer and Gustav Langford for formwork development

"The complete design is unfolding as construction work proceeds," is the curt comment of Kahn to sum up the creative process of the National Assembly Building, opening the description of the grand buildings shown in the models, structures that are slowly taking form on the worksite with the pattern of the piles and the walls emerging from the ground, harbingers of the future work.

"The initial stages of construction, shown in photographs," he writes, "are by nature devoid of any of the aesthetic character expected in the finished product. Pictures of earth being moved, leveled and compacted give no hint of a tree-lined avenue. Photographs of the tops of concrete pile foundations extending deep into the earth bear no seeming relation to the building to rise high above them. And none of this rough looking work gives visible evidence of the care given to its design and placement. But as the work rises above the ground, partly hidden as it may be by scaffolding and the machinery of construction, an interested observer may be able to sense the forms taking shape and to recognize in them the becoming of an expression of government and its institutions, a government working for the common good."[120]

Precisely in order to integrate the proceeding work of construction in Dhaka in the development of the details, Kahn's staff makes arrangements with the Public Works Department for Frederick Langford to go to Dhaka for the supervision of the technical details of the formwork and their construction. Qureshi promises to provide Langford with all necessary help, providing him with carpenters, plywood and teak, as requested by Kahn. But he also sets the technical limits of the situation: "The carpenters will be placed at the disposal of Mr. Langford who can make them work as he deems fit. It would not be possible to arrange for any special kind of implements which are not available within the country and Mr. Langford should try to get the work done with normal implements of the carpenters."[121]

For the assembly of the formwork outdoors, Vollmer recommends that Langford come to Dhaka in October, at the end of the rainy season (in case of earlier arrival, the work can in any case be done inside barns). From the technical suggestions provided in his usual worksite report sent to Kahn, it appears that Vollmer has not been updated on the studies done in Philadelphia at the start of May for the formwork with wooden boards. Also in the wake of Qureshi's comments and with the aim of communicating his own opinion to Langford, Vollmer proposes the use of "local-made panels of plywood" or the panels known as "pressboard" made in Dhaka with "jute fibers." As he had already proposed in 1964, he suggests reinforcing these panels with "steel angles" to apply in the corners, and covering their parts in contact with the pour with either a metal sheet to weld along the corner reinforcements,

213-215. National Assembly Building, Dhaka, 1962-83.

213. Concrete pouring of the Presidential Square roof (LIK, 030.IV.A.650.2.85).

214. Construction of a reinforced brick arch of the Presidential Square (LIK, 030.IV.A.650.3.14).

215. Study sketches for the construction of the arch (W. N. Thompson, letter to H. Wilcots, 27 April 1968, LIK, PAC – Correspondence To/From Gus January 1968 thru Dec 31, 1968).

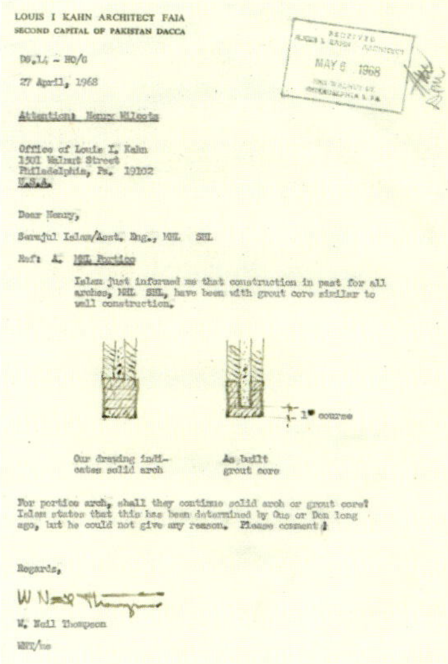

LOUIS I KAHN ARCHITECT FAIA
SECOND CAPITAL OF PAKISTAN DACCA

DO.14 – HO/G

27 April, 1968

RECEIVED LOUIS I. KAHN ARCHITECT MAY 6 1968

Attention: Henry Wilcots

Office of Louis I. Kahn
1501 Walnut Street
Philadelphia, Pa. 19102
U.S.A.

Dear Henry,

Serajul Islam/Asst. Eng., MNL SHL

Ref: A. MNL Portico

Islam just informed me that construction in past for all arches, MNL SHL, have been with grout core similar to wall construction.

1" course

Our drawing indicates solid arch

As built grout core

For portico arch, shall they continue solid arch or grout core? Islam states that this has been determined by Gus or Don long ago, but he could not give any reason. Please comment.

Regards,

W. Neil Thompson

WNT/ne

or with a special epoxy. Any type of "rubber bushing" and "gasket" can be made locally, while the scaffolding will be in bamboo tied with ropes.

During the course of 1965 two visions for Kahn's formwork are outlined, that of Langford and that of Vollmer, and the two perspectives are perceptibly different, depending on the degree of knowledge and inclusion of local construction practices. The fact that Vollmer is oriented towards taking all the characteristics of local techniques and materials as founding features of the project, following the indications of Qureshi, is confirmed by his advice regarding the formwork: "Fred must start thinking of a simple, flawless, primitive method. Forget sophistication of any sort – it just can't be done. Keep everything direct and simple and the fewer processes the better. [...] aside from the equipment Fred has named which is not available, if he has any ideas or sketches pertaining to what he wants to do he should send them to us prior to his arrival and we shall try to get things underway before he arrives."[122]

In June the driving of the foundation piles of the National Assembly Building is reaching its conclusion – on 22 June 99% of the work has been done.[123] The worksite comes to a stop in August, awaiting the working drawings for the walls. Kahn tries to make an agreement with Ahmad to have a technician of the Public Works Department work full-time in the Philadelphia office, to guarantee rapid project development.[124] As soon as he is back in Philadelphia, Kahn writes to Vollmer, in August: "The design of the Mosque is now finished and is extremely simple and beautiful. There is a play of large windowless openings in walls of marble and concrete of a different design than that of the other walls of the building. [...] the appearance of the inner walls of the office wing facing the main promenade has remained undetermined but now becomes the same as all other solutions of openings which were more tightly related to problems that had to be solved. Until recently these four walls were more subject to arbitrary decisions. Since returning from Dhaka, I have determined what these openings are and they are being recorded on the 1/16" scale drawings."[125]

The matter of the roof of the hall and its ceiling also evolves. Kahn is looking for a system that can combine the problem of acoustics with the form of the structure (technical consultants are called in).[126]

After having considered the use of Travertine quarried near Rome, marble from Pakistan is chosen, since it is more economical, and Kahn continues to study the cornice with its drainage function to insert in the grooves of the exposed concrete walls, seeking to simplify its profile to facilitate shipping of the material in strips, and the cutting of the stone itself.

[120] Anon., "1st Photograph. Opposite 1st Photo. Site Model (from south, if possible)," typewritten text sent to Roy Vollmer, Jr., on 9 June 1965, LIK, PAC Progess brochure.
[121] Mumtaz Ahmad Qureshi, letter to Roy Vollmer, Jr., n.d., [post 21 May 1965], cit. in, [Roy Vollmer, Jr.], "Dhaka Report," 19 June 1965, LIK, PAKCAP – Correspondence to/from RoyGus October 8, 1964 thru June 30, 1965.
[122] Anon., [Roy Vollmer, Jr.], "Dhaka Report," 19 June 1965, cit. The question of the reinforcement of the brick masonry continues to be discussed.
[123] Gustav Langford, "Progress report 2," 22 June 1965, cit.
[124] Roy Vollmer, Jr., letter to Louis I. Kahn, 13 August 1965, LIK, PAC Pending.
[125] Louis I. Kahn, letter to Roy Vollmer, Jr., 25 August 1965, LIK, PAKCAP – Correspondence to/from RoyGus June 30, 1965 thru December 31, 1965. The hypothesis of painting or plastering the brick walls of the houses to protect the masonry from the effects of the monsoons is forcefully challenged by Kahn: "The plastering will erase the beauty of the natural texture [...]."
[126] Louis I. Kahn, letter to Roy Vollmer, Jr., 25 August 1965, LIK, PAKCAP – Correspondence to/from RoyGus June 30, 1965 thru December 31, 1965.

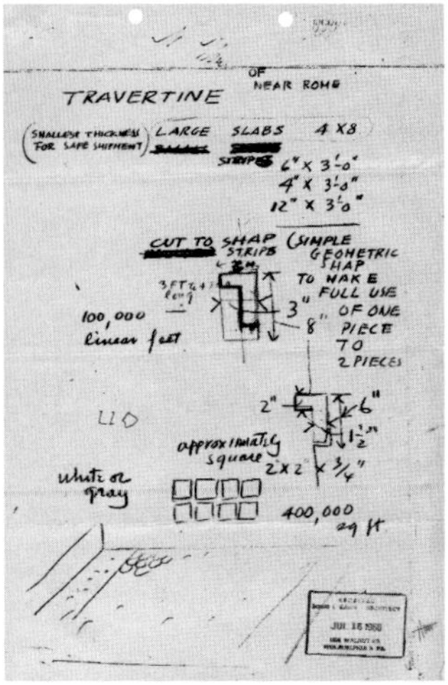

216. National Assembly Building, Dhaka, 1962-83. Study sketches and notes for the strips in travertine (page attached to, Engineer of the PWD, letter to Northern Engineers, 20 November 1965, LIK, PAC – Correspondence – RoyGus July 1966 thru December 1966).

In November he discusses the hypothesis of an L-shaped piece of 3 × 6 inches (about 3 feet long) to sculpt in one or two blocks (the variation "white or gray" also appears).[127]

In December 1965, Kahn's team in Dhaka has not yet been informed of the decisions made in Philadelphia regarding the formwork, and the arrival of Frederick Langford at the site. "In what stage is the formwork for the N[ational]A[ssembli] Building and does Louis I. Kahn intend to send someone to Dhaka to get this work underway, or should Roygus [Roy Vollmer and Gustav Langford] do it themselves?" is Vollmer's question.[128] He also urges that the working drawings should contain indications of the conduits to be installed prior to pouring, explaining that procedures in Dhaka are different from those of the United States: "It is always a cut and chip proposition with the result being patchwork."[129]

The series of working drawings of the formwork produced in May 1965 is not sent to Dhaka until January 1966. On the 27th, Gustav Langford responds to the questions listed in sets 1, 4 and 5 of those drawings. His considerations are addressed to the studio of Kahn in Philadelphia, to the attention of his brother Frederick.[130]

Regarding the organic patina that forms on the concrete surface, Gustav has found no valid explanations, not even regarding its nature. In his view, it might be an organism living in dust carried by the wind, which grows on the moldings of walls and penetrates into their cracks. Gustav notices that in brick walls this organism attacks first the mortar of the joints, and then spreads to the bricks only if they are porous (instead, it is not found on the "smooth hard burned bricks" of the Hostels of the capitol). The humid climate of monsoon season, in his view, favors the fast growth of this coating.

His concluding considerations on the growth of the organic coating, which happens in zones where the surface is not constantly washed by rainwater,[131] lead Gustav to suggest that the protruding cornices envisioned by Kahn and his team in Philadelphia might even be damaging: "I really believe that if the wall is completely free from projections, regardless of size, and the surface is relatively smooth and free from honeycombing, you will have little or no growth."[132]

"There are no stock sizes of lumber available on the local market," is Gustav's reply to the request for information sent from Philadelphia about the existence of standard wood pieces available in Dhaka – each piece is cut to order, based on the required measurements.[133] Gustav warns about the possible effects on the formation of biological coatings of "variation of lumber sizes" with the main protruding joints, and cautions that there is a "great lack of availability of any product which requires a precise result." He recommends immediate contact with the Public Works Department to obtain the machinery necessary for precise working of lumber.[134]

After having explained the total lack of precision in carpentry at Dhaka, which would be needed to make Kahn's formwork, and after having observed that the solution proposed by Philadelphia to prevent bleeding depends on solid joining of the planks to be assembled in Dhaka, Gustav writes: "Fred, if you think of these workmen in the light of butchery rather than carpentry, you will be much closer to reality. I feel most certain that all of those planks which would produce the detail would be knocked out of place and not taken out as the procedure of stripping should be done."[135]

Gustav answers the other questions sent from Philadelphia with apt observations: plywood "is difficult to obtain in large quantities and is very expensive" (panels of 4 x 8 feet, formed by six layers, are available); "glue is not waterproof"; transport of panels from the workshop to the construction site, their assembly and raising are all done by hand, carrying them on the head, though a crane is provided on the worksite. What is less certain, according to Gustav, is the possibility of manually raising the whole corner piece called for in the formwork (the detail indicated with the letter E of drawing 3[136]); in the case of a panel composed of multiple parts, the raising should be done by crane. Regarding the bamboo, for which he provides several average measurements of height and diameter of the poles, he advises reinforcement with "rattan" because it is more resistant to torsion. Finally, to facilitate stripping of the formwork, a pale oil can be used, or old motor oil, which nevertheless leaves a "charcoal gray finish."[137]

The formwork man: Frederick Langford and the Experimental Walls

All the technical information exchanged between Dhaka and Philadelphia until the start of 1966 forms the preparation for the arrival at the worksite of the expert responsible for the production of formwork in Kahn's most important projects: Frederick Langford. At the end of January, Frederick requests other information on the work force and the machinery available in Dhaka to work the wood of the formwork (how much weight one Pakistani worker can carry, what types of cranes are available, the voltage of the electricity), while Gustav sends a list of the machinery ordered by the Public Works Department to give him an idea "of the kind of woodworking shop" he can count on in Dhaka. "On one of the Salk formwork drawings you asked if certain machinery was available. I believe all will be except for the portable items, i.e. routers, drills and saws," Gustav adds, confirming what will be the initial model pursued by Frederick for the panels of the National Assembly Building.[138]

A few days after reaching Dhaka, on 16 February Frederick sends Ahmad an impassioned letter on the "Concrete and Formwork for National Assembly Building," written to lay the ideal groundwork and to convey the required enthusiasm for the major changes of logic in the construction of the formwork he sets out to introduce, in order to achieve the goal of the transformation of Kahn's design into a true monumental sculpture of exposed concrete. In a few sentences, he sums up the main phases of the design process of the formwork, always weighing his words to emphasize Kahn's intention of obtaining a sculptural concrete, for which he is

[127] Engineer of the PWD, letter to Northern Engineers, 20 November 1965, LIK, PAC – Correspondence – RoyGus July 1966 thru December 1966. The drawing attached to the letter is by Kahn. See also A.M. Musa of Northern Engineers, letter to David P. Wisdom, 11 July 1966, LIK, PAC Pending.

[128] Roy Vollmer, Jr., "Bi-Weekly Report No. 6," received 23 December 1965, LIK, PAC – Bi-Weekly Reports – Dhaka Office Sept. 29, 1965 thru Feb. 22-1967.

[129] *Ibid.*

[130] Gustav Langford, "Questions on NA Formwork Dwgs Nos. 1, 4 & 5," 27 January 1966, LIK, PAC – Correspondence – RoyGus Oct 1965 thru June 1966.

[131] "Another theory is that any projection out from a building will gather days of dust and the first few minutes of a rain will splash this muddy mixture back and up on the walls and the rest of the rain just dos'n't [sic: doesn't] seem to wash it completely clean again; this dirty stain is compounded every time it rains. Again the hot humid air adds to the spreading of this fungus. I believe that in cases where you get direct sunlite [sic: sunlight] you only have a dark dirty stain and not really the presence of green moss-type growth. So far this explanation pertains only to brickwork and plaster, I haven't seen any actually on a concrete surface, but has been suggested and warned by the PWD that it will undoublly [sic: undoubtedly] appear after a short period of time" (*ibid.*).

[132] *Ibid.*

[133] *Ibid.*

[134] *Ibid.*

[135] *Ibid.*

[136] See the drawing of the formwork details, No. 3, LIK, 030.I.C.650.033.

[137] Gustav Langford, "Questions on NA Formwork Dwgs Nos. 1, 4 & 5," 27 January 1966, cit.

[138] Gustav Langford, letter to Frederick Langford, received at Kahn's office on 31 January 1966, LIK, PAC – Correspondence – RoyGus Oct 1965 thru June 1966.

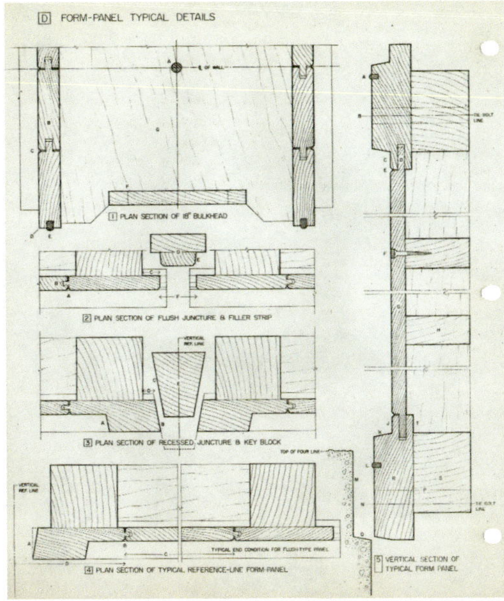

217-219. National Assembly Building, Dhaka, 1962-83.

217. Construction site of the Experimental Walls of the North Entrance, under the direction by F. Langford, 1966. Photograph showing workers, components of the formwork system and tools for transporting and pouring concrete (F. Langford, "A Report on Concrete and Formwork. The National Assembly Building, Second Capital Project," 30 June 1966, LIK).

218. Drawing of the formwork system designed by F. Langford, 1966 (F. Langford, "A Report on Concrete and Formwork. The National Assembly Building, Second Capital Project," 30 June 1966, LIK).

219. Experimental Walls construction, under the direction by F. Langford, North Entrance, 1966. Formwork panels (F. Langford, "A Report on Concrete and Formwork. The National Assembly Building, Second Capital Project," 30 June 1966, LIK).

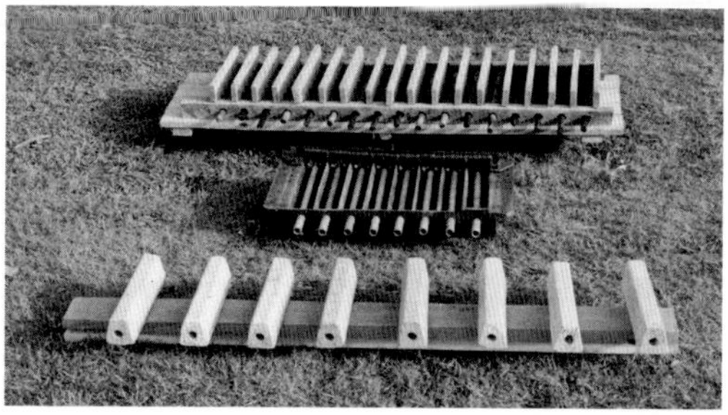

220-228. National Assembly Building, Dhaka, 1962-83.

220. Concrete spacers and their metallic molds (F. Langford, "A Report on Concrete and Formwork. The National Assembly Building, Second Capital Project," 30 June 1966, LIK).

221. Concrete sleeves and spreaders equipped with holes prefabricated by means of aluminum cylindrical molds and steel molds. (F. Langford, "A Report on Concrete and Formwork. The National Assembly Building, Second Capital Project," 30 June 1966, LIK).

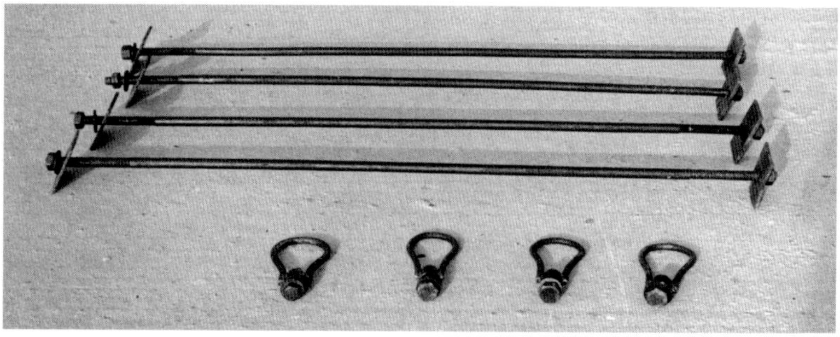

222-224. Metal tools for the formwork system (F. Langford, "A Report on Concrete and Formwork. The National Assembly Building, Second Capital Project," 30 June 1966, LIK).

225-226. Tools to measure the amount of the aggregates and transportation of the concrete (F. Langford, "A Report on Concrete and Formwork. The National Assembly Building, Second Capital Project," 30 June 1966, LIK).

227. Hopper used during the construction (F. Langford, "A Report on Concrete and Formwork. The National Assembly Building, Second Capital Project," 30 June 1966, LIK).

228. Vibrators for the concrete (F. Langford, "A Report on Concrete and Formwork. The National Assembly Building, Second Capital Project," 30 June 1966, LIK).

simply in charge of implementation, ready to train the craftsmen in its making.

"The past few days have been very enjoyable and extremely informative since my arrival in Dhaka. I am intrigued with the sense of energy and growth potential that your interesting city possesses. A disturbing but understandable fact, however, is the gap that exists in construction skills between your country and mine. I was well aware of this gap before I came, therefore, if I can play a part in teaching new methods, particularly in the field of cement work, I will return to my country realizing a great satisfaction. I can speak on behalf of Prof. Kahn in this respect because his passion for excellence in workmanship is as strong as mine. As early as April 1965, while we were completing the construction of the Salk Laboratories in California, I started making preliminary studies of formwork for the National Assembly Building. During the past three months I worked in close association with Prof. Kahn and his Structural Engineers studying the many ramifications of 'Poured in Place Concrete' for your building. Now that I am in Dhaka I feel very fortunate to be given the opportunity to continue my investigation and begin experimentation. Paradoxically as it may seem, it has been my experience in the past that the problems involved in promoting good workmanship are often minimal if the artisan is learning a new skill. The formwork methods for the National Assembly Building, as we will be proposing them, will be completely new, and will be a challenge to your carpenters, form setters, and pouring crews. However, the willingness and spirit of cooperation that I have witnessed in the past few days while touring Dhaka and the Second Capital Site assures me that it is possible to produce Architectural Concrete for the National Assembly Building that should make all of Pakistan proud.

229-234. National Assembly Building, Dhaka, 1962-83.
Experimental Walls construction, under the direction of F. Langford, North Entrance, 1966 (F. Langford, "A Report on Concrete and Formwork. The National Assembly Building, Second Capital Project," 30 June 1966, LIK).

The concrete should be so excellent that it complements the sculptural magnificence of Prof. Kahn's concept, so that all the world will look to this building as a great work of Architecture in a great new Country."[139]

Langford's arrival at the worksite marks the definitive development of the formwork system for the National Assembly Building. All the individual parts are designed to create a coherent, regular system of lines, no longer with the degrees of randomness embraced in the previous studies for panels with boards of variable width.

For the assembly of the formwork, or "experimental panels," and for the construction of the sample walls, or "experimental walls," Langford employs 4 carpenters, 16 masons, from 50 to 70 workers, as well as 14 craftsmen charged with deploying the metal shuttering and the bamboo scaffolding.[140]

The "experimental panels" are made by carpenters using manual tools. At the only sawmill equipped with some electrical machinery, that of the

[139] Frederick Langford, letter to Kafiluddin Ahmad, 16 February 1966, PAKCAP – Correspondence – PAK PWD 1966 - 1967.

[140] Frederick Langford, "A Report on Concrete and Formwork. The National Assembly Building, Second Capital Project," 30 June 1966, LIK, unarchived. This document would later be published by Langford himself (Frederick Langford, *Concrete in Dhaka*, in, "Mimar," 1982, No. 6, (pp. 50-55), p. 50).

235-238. National Assembly Building, Dhaka, 1962-83.
Experimental Walls construction, under the direction of F. Langford, North Entrance, 1966 (F. Langford, "A Report on Concrete and Formwork. The National Assembly Building, Second Capital Project," 30 June 1966, LIK).

Forest Industries Development Corporation, located about 5 miles from the worksite, Langford has certain parts of the formwork made, which nevertheless turn out to be imprecise – "the work that was produced was poor due to the lack of skilled workers."[141]

The formwork are of two kinds: those for the walls connecting the foundation piles, or "starter walls," and those for the upper walls, the "experimental walls." They differ in terms of materials and finish.

The formwork of the "starter walls" are in Gorjan wood, which is easy to work (the boards are 1 or 2 inches thick, and 6 inches wide); the panels are treated with motor oil, producing a "greenish gray surface" on the concrete.[142]

The "experimental panels" used for the formwork of the upper walls are composed of series of vertical boards, two horizontal posts shaped and attached at the ends, and a structure applied to the back. To make the parts of the panel in contact with the pour, Langford chooses "champa" wood, after having discarded teak, which though "ideal" turns out to be "too

[141] Langford, *Concrete in Dhaka* cit., p. 50.
[142] *Ibid.* During the construction of the "starter walls" Langford also experiments with special pieces, like those for the corners and the "tee sections" (*ibid.*).

expensive," as well as "chapalish," which "proved very soft and also contained fluids with concrete staining ability."[143] For the wood of the structure, he chooses "chapalish" because "it received nails well."[144]

The assembly of the formwork proves to be the fundamental ingredient of the design of Kahn's concrete, where all the parts that produce imprints must conform to a regular arrangement. It is telling, in these terms, that the screws used to attach each board to the posts of the structure, and which are the only elements whose exact position is difficult to control, are driven deep enough to seal the tops, leaving no imprints on the concrete.[145]

The boards are 6 inches wide, joined with the "tongue and groove" system and assembled vertically, also as a result of Gustav's information on the formation of an organic patina under every horizontal protrusion. The assembly of the boards is made visible by the vertical edges, cut to shape a regular sequence of V joints on the concrete. This is the same technical detail used in the formwork of the Salk Institute, but the rhythm of the imprints varies in the two works, due to the different measurements of the constituent parts of the panel – plywood in La Jolla, wooden boards in Dhaka.

The boards terminate with tongues to insert in special grooves made in the wing of the two horizontal posts of the panel. The edges at the joint between the boards and the horizontal posts are worked to leave the usual V joint imprint. The assembly of the pieces is studied to make the horizontal imprints at the ends of the panel link up with the vertical imprints, combining to form a texture of the wall surface entirely based on the assembly of the individual parts of the panel.

The detail of the assembly of the horizontal posts with the vertical boards is of the greatest importance for an understanding of the goal pursued by Kahn of the visibility of all the imprints of the constituent parts of the panel. Precisely by also making the presence of the posts visible, that wing is introduced which is the sole part that can produce, with the joint of the boards, the horizontal V joint.

In the design phases no hint appears of the intention to also give the horizontal V imprint the function of a seal of the formwork of the floor slabs abutting the walls, though mention of this is made in the worksite documents: "The projection of the concrete formed by the V joint in forms is tending to keep floor from closing tight against walls. Medium hard rubber was suggested to close gap between slab form and wall."[146]

The two horizontal posts of the panel have a protruding head so that each of them makes a horizontal recess in the concrete. The function of the recess has become complex, at this point: it defines the height of the pour, the position of the spreaders and the plane on which to place the panels for the next pour.

The formwork panel has an iron hook for raising on the worksite, attached at the center of the upper post. The horizontal posts have different holes, because their functioning differs. The upper post has simple circular holes; the spreaders are placed through them, along with the required bars. On the other hand, the lower post is placed against the recess already made in the wall, serving as a base for the panel when it is repositioned for the pouring of the next level, and it has openings instead of the holes, to facilitate passage of the metal bar through the spreaders that are already in place, even when there is a margin of error in their positioning.

The consecutive assembly of multiple panels horizontally produces a joint called the "flush juncture," to seal with a special vertical T strip, or "filler strip"[147] that once in position has its head flush with the plane of the boards. The head has cut edges to produce two close V-shaped imprints, together with the contiguous boards, marking the passage from one panel to the next. It is clear that this "filler strip" is technically derived from the "coupler block" used in the formwork of the Salk Institute.

To imprint vertical hollows in the walls, some of which Kahn envisions corresponding to the construction and pour joints, Langford develops two special panels, each with a vertical border reinforced by a protruding post with an inclined side. When the panels are placed side by side, the posts form a large vertical opening, or "recessed juncture," and they are designed to receive a third element of the shaping of the concrete, a wedge-shaped vertical strip known as the "key block," which produces a deeper, narrower groove. Thus the "recessed juncture" is organized on two planes, of which the narrow central one of the "key block" is the deepest. This system of vertical imprints is based on the studies of 1965, but it will not be visible in all its features in the finished work, because the narrow vertical groove at the middle will be buried under a single slab of marble that will block off all of the "recessed juncture."

The "filler strip," "recessed juncture" and "key block" are simply parts of the constructive grammar of the formwork system of Langford, which also includes true special pieces, those of a quarter circle, created to shape the circular openings.

Before the pour, the panels are treated with a rag soaked in a type of motor oil and "with special attention to the chamferred junctures ('V' joints) of the vertical boards." [148]

The assembly of the panels requires a series of techniques to always shape imprints that reflect the logical operations of the construction process. Following the making of the "experimental walls," Langford develops a range of solutions for the positioning of the panels, so as to create true optical corrections capable of compensating for every exception in the system; in his words, the "junctures allow for length corrections," including "fixed filler strips," "variable filler strips" and "flexible inserts." So in Dhaka, too, we can see Kahn's intention of formulating a system to control the flaws in the fabrication of the concrete.

"Length corrections in order to respect reference control lines are a problem inherent in all concrete forming systems – to preserve the integrity of the individual panels, a make-up (dimensional correction) zone must be established – the logical location for this zone is at vertical gap that occurs between form panels. Fixed filler strips were used in the vertical gaps between panels on the building experimental walls. Variable filler strips must be employed when a panel series includes a corner form – these strips will have to be custom shaped for each pour. Flexible inserts (rubber or neoprene) should be studied as a joint sealer and as a variable width gap-filler."[149]

Langford's pursuit of artisan fabrication of all parts of the formwork even extends to the devices like spreaders, usually supplied by industrial sources. After having envisioned the use of American industrial products

[143] Langford, *Concrete in Dhaka* cit., p. 52.
[144] *Ibid.*
[145] See the photograph dated 1 May 1967, LIK, 030.IV.A.650.4.40, and the drawing published in Langford, *Concrete in Dhaka* cit., p. 54.
[146] Harry M. Palmbaum, "National Assembly Building Construction Report," n.d., [July 1967], LIK, PAC Various Notes Dhaka Jan 66.
[147] Drawing published in, Langford, *Concrete in Dhaka* cit., p. 54.
[148] Langford, "A Report on Concrete and Formwork" cit.
[149] Langford, *Concrete in Dhaka* cit., p. 53.

and after the idea of using segments of bamboo, Langford studies a way to locally manufacture special spreaders, nimbly exploiting the height of the recess Kahn had calculated to accommodate the marble inserts, protruding from the wall or embedded flush with it.

Langford's spreaders are large prisms of concrete with a square section, equipped with a central hole, of a length equal to the thickness of the wall measured at the position of the horizontal recesses. These spreaders are fabricated at the worksite using a special metal mold capable of shaping 8 pieces at a time (two such molds are made). Once the compound poured into the mold has set, the individual piece is extracted by sliding out the metal pipe that served to make the hole. Sleeves and spreaders equipped with holes to keep the reinforcement away from the formwork are prefabricated in concrete by means of aluminum cylindrical molds ("donut mold") and steel molds ("spreader mold").

After having defined the basics of the formwork system for the National Assembly Building, towards the end of February Langford begins to provide Faruqi with indications on the leveling of the terrain at the position of the piles where the walls are to be erected, and on the devices to attach to the piles to assemble the formwork – "all pile caps will have to be star-drilled and wood plugged, in order to secure the bottom plate of all formwork" – as well as on the concrete samples to be made in the first days of March.[150] "Fred will send a complete report of progress on concrete formwork," the colleagues in Philadelphia are informed on 23 March.[151]

In the work organization, Langford perceives the impossibility of training the foremen to perform regular cycles of operations, though in any case he does manage to create small and more efficient teams guided by one worker: "Workers were always willing to cooperate to the best of their ability – willing to work long hours and rebuild formwork when necessary – form panels were handled and cleaned with loving care – some men became skillful in a very short time."[152]

At the end of April some working drawings arrive from Kahn's office in Philadelphia, but they are incomplete regarding the information on the positioning of the conduits of the various physical plant systems, as requested by Vollmer; indications are also lacking on the "expansion joints."[153]

The first aboveground walls of the National Assembly Building are not constructed until May. The work begins with the foundation beams up to level 26.5 of the G.E. Block (the walls of the whole block are made) and the N.E. Block (only the walls towards the lake), and inside the G.E. (the walls of the staircases, up to level 33). "[…] it is Fred's intention to extend the pour to elevation 38 on the four interior walls of the stair unit," we read in a report sent

[150] Frederick Langford, letter to Fala Faruqi, 26 February 1966, LIK, PAKCAP – Correspondence – PAK PWD 1966 – 1967.
[151] Gustav Langford, letter to Henry N. Wilcots, 23 March 1966, LIK, PAC – Correspondence – RoyGus Oct 1965 thru June 1966.
[152] Langford, *Concrete in Dhaka* cit., p. 51.
[153] Roy Vollmer, Jr., Gustav Langford, "Bi-monthly report No. 13," 26 April 1966, LIK, PAC – Bi-Weekly Reports – Dhaka Office Sept. 29, 1965 thru Feb. 22-1967.
[154] Gustav Langford, Roy Vollmer, Jr., "Bi-Monthly Report No. 14," 13 May 1966, LIK, PAC – Bi-Weekly Reports – Dhaka Office Sept. 29, 1965 thru Feb. 22-1967.
[155] The walls arranged in a rhombus are completed under the supervision of Frederick Langford up to the height of the first large circle (see the photographs dated 18 November 1966, LIK, 030.IV.A.650.3.26, and LIK, 030.IV.A.650.3.11).
[156] David P. Wisdom, letter to Gustav Langford, 18 July 1966, LIK, PAC – Correspondence – RoyGus July 1966 thru December 1966.
[157] Langford, *Concrete in Dhaka* cit., p. 50.
[158] *Ibid.*, p. 52.

239. National Assembly Building, Dhaka, 1962-83.
Experimental Walls, North Entrance, 1966 (F. Langford, "A Report on Concrete and Formwork. The National Assembly Building, Second Capital Project," 30 June 1966, LIK).

to Kahn in mid-May.[154] Unlike the worksite of the Salk Institute, where the sample walls were made in secondary spaces, for the National Assembly Building the experimental walls are those of the presidential entrance (the G.E. or North Entrance), probably also because of faith in Langford's abilities.[155]

The objective of the walls Langford has built is "finding out how to produce excellent concrete with the materials and tools *at hand* in Dhaka," Wisdom points out.[156]

To make a concrete always with the same percentages of sand and crushed stone, Langford has wooden containers made with two lateral shafts, like a stretcher. The mixing of the compound is done in cement mixers of the "two-bag" type, after having rejected the "one-bag" mixers as being unsuitable "to provide proper lift rate" and to guarantee the same quantity of water in the various mixtures.[157] From the cement mixer, the concrete is discharged onto the ground, on a sheet metal bed; then the large metal basins are filled using shovels; two men help to put the basin on the head of the worker who will carry it on a route between 50 and 300 feet to the point where the concrete is poured into the formwork. Langford notices that the time required for these operations, together with the fact that the concrete is transported in small quantities, causes the mix to already separate in the basin; after the pour, a further "separation or a lack of adhesion" happens inside the formwork.[158]

To regulate the pouring of the concrete in the formwork, Langford has special sheet-metal hoppers made with a rectangular opening, equipped with two removable parts for better guidance of the descent of the compound: an extension of the opening downward ("chute") and a flap with three sloped sides to insert above the mouth ("hopper extension"). For the construction of a wall sector of about 30 to 35 feet, two hoppers with the removable accessories are placed above the panels, at the extremities of the formwork. Langford calculates the sequence and rhythm of the pours in the two hoppers in order to obtain, in spite of the operations done by hand, layers of concrete without clusters of gravel and visible joints. The workers pour the concrete from the basin first in one hopper and then in the other, under the supervision of two workers who control the angle of the mass of concrete in the formwork and its mixing. When the layer of concrete gets high, the lower extension of the opening and the three-sided flap of the hoppers are removed, and the workers start to simultaneously pour the concrete into the two hoppers. The hoppers are definitively removed when the layer of concrete reaches about 18 inches from the top of the wall segment. At this point a third worker intervenes to control the operations to be done in the middle part of the form; he is the one who ensures that at the center of the pour the concrete mixes without forming any "vertical cool joint" (the pour of the concrete in this middle part of the form begins only after the two lateral pours have reached a length of at least 10 inches each).

The pour of the first walls is mixed in the formwork with a bamboo stick, in keeping with local methods. The two workers in charge of the hoppers perform this operation at every pour from the basin, and are joined by a third worker, responsible for the middle part of the formwork. At the same time, wooden mallets are used to strike the formwork. Following "unsatisfactory" results ("many rockpockets" are produced), Langford has the Public Works Department purchase 14 vibrators with gasoline engines. Only three of the vibrators are used, each of them operated by the workers responsible for the three pours; in the final phases of the work, Langford also experiments with an electric vibrator.[159]

Stripping turns out to be less problematic than the other operations, as Langford explains: "wooden wedges were made to assist the breakaway process – bolt threads were being damaged during removal until form-panels were handled without any benefit of hoists which presented dangerous situations at times […]."[160] For the curing, Langford makes use of jute sacks spread over the walls and kept damp for seven days. But the fabric marks the concrete surface, while more or less white stripes are produced due to the varying dampness of the sacks.

To get the smoothest possible concrete surface, Langford conducts experiments during the first pouring phases, using different products to spread on the panels. First he has the panel surfaces in contact with the pour treated with two coats of shellac – a "greenish-brown dull surface resulted"[161] – and then he has this layer removed, applying two coats of "imported waterproof varnish." With this dual treatment he obtains a "good surface" as the result of the first concrete pour. Nevertheless the varnish begins to detach after the second or third reutilization of the panel. To try to protect the varnish, he adds a layer of polyurethane, but without success – "this peeled on the first usage." Then he has the various layers removed from the panels, returning to the natural wood, and treats the surface with only polyurethane. To establish the number of coats of polyurethane required to obtain a good concrete surface, he has two groups of 8 panels prepared, respectively treated with two or three coats. After this series of tests he establishes the principle that will orient the construction of concrete in Dhaka: the panels must be covered with three layers of polyurethane ("the three-coat system proved satisfactory").[162]

Regarding the second series of walls erected by Langford with polyurethane-coated panels, Gustav and Vollmer write: "The form panels have proved to be highly successful in terms of re-use. Some uncoated forms (those used for starter walls) have been used some 16 times, and are now showing signs of wear. This is primarily due to lack of coating and mishandling, which hopefully can be corrected. The forms coated with polyurethane have had some 6 uses and are perfectly intact. No scraping [added by hand: for cleaning] is necessary and they are washed and wiped clean."[163]

For the treatment of the surface of the panels, PVC is also taken into consideration, a material that has recently become available in Dhaka, though "no details have been worked out for its placement in a form," and the first works do not seem to require formwork with such a treatment ("We have not encountered any walls in the pouring operation to date which require PVC").[164]

In spite of the difficulties in training the workers to apply the logic of concrete construction according to Kahn's methods, the results obtained by Frederick Langford are satisfactory, as reported by Roy and Gustav, who nevertheless – precisely in the light of Langford's return to Philadelphia – warn Kahn of the need to send a replacement in order to continue to supervise the workers: "Fred is continuing to experiment with the pouring operation. Each pour improves over the previous one and some excellent concrete is the result. The big difficulty is in the organization or lack of same with the P.W.D. Superintendents. They have difficulty in grasping the idea of the team-work and timing required for this kind of work. L.I.K. must send a Full-time Superintendent to attend to work only on the National Assembly Building. Spot supervision will be ineffective. Fred will give full report on his return."[165]

In June, Frederick Langford leaves Dhaka, but he continues to be informed about the work and to offer his views.[166] During his time at the worksite, Langford has taken photographs and notes for the preparation of a detailed report on the experiments conducted under his supervision: "A Report on Concrete and Formwork. The National Assembly Building, Second Capital Project," dated 30 June 1966.[167] The document is prepared along the general lines of the similar report he prepared in 1964, the "Report on concrete work" regarding the construction of the Salk Institute. But the style of the new report shifts towards the creation of a manual for the making of Kahn's architectural concrete in Dhaka. It is probable that the idea of the report reflects the initial desire, already expressed by Kahn and Komendant, to transform the worksite of the monuments for the capital into the basis of a technique and style of exposed concrete for Pakistan.

"Local Conditions Relating to Concrete Work During the Experimental Period," "Form-panel Design Criteria and Recommendations" and "Placement of Concrete - Preparation and Procedure" are the three large chapters of Langford's report, which contains the description of the materials of the compound and the formwork, the manual tools and machinery to use for the

[159] *Ibid.*
[160] *Ibid.*
[161] *Ibid.*
[162] *Ibid.*
[163] Langford, Vollmer, Jr., "Bi-Monthly Report No. 14," 13 May 1966, cit.
[164] The sheets of aluminium used for the formwork of the floor slabs in Hostels and MNAs will be replaced by PVC (*ibid.*).
[165] *Ibid.* In June, Frederick Langford's report is in Kahn's hands, as indicated by, Anon. [Roy Vollmer, Jr., Gustav Langford], "Bi-momthly Report No. 15," 9 June 1966, LIK, PAC – Bi-Weekly Reports – Dhaka Office Sept. 29, 1965 thru Feb. 22-1967. M. G. Siddiqui, "Additional Chief Engineer" of the Public Works Department, meets on multiple occasions with Vollmer, Gustav and Frederick Langford to be kept abreast of the progress of the works (M. G. Siddiqui, letter to Louis I. Kahn, 28 June 1966, LIK, PAKCAP – Correspondence – PAK PWD 1966 – 1967).
[166] See Gustav Langford, letter to Henry N. Wilcots, 4 July 1966, LIK, PAC – Correspondence – RoyGus July 1966 thru December 1966.
[167] Frederick Langford, "A Report on Concrete and Formwork" cit.

preparation and placement of the concrete, the work organization, the climate and its consequences for the exposed concrete surfaces.

Langford covers the construction of the "experimental walls" in a series of practical recommendations ranging from the type of cement (among those imported from China and Russia, and the local cement, produced at Sylhet, he indicates the Russian cement as the ideal choice because it "produced best concrete color of the three brands") to the other components of the concrete (he recommends the stone chips and sand of Sylhet), to the advice to use scaffolding, starting at the level of 38 feet (for the stripping it will be necessary "to incorporate mechanical means"[168]). The experience of the "experimental walls" prompts Langford to provide the following indications on the timing of pour operations, to get the best results from the panels with polyurethane, in terms of the formal quality of the concrete surface: "the polyurethane coated panels produce the best concrete surface when the wall is poured rapidly – a lift-rate of 5 feet per hour should be the minimum lift-rate – 10 feet per hour should be the maximum."[169] Langford provides also some rules for the positioning of the groove, like the "control-joint recess": "shrinkage cracks occur every 20 to 30 feet in long walls, therefore it is important to locate control-joints at these intervals as an attempt to contain these cracks."[170]

The chapter on "Form-panel Design Criteria and Recommendations" contains detailed analysis of the various hypotheses that emerged during the course of the design of the formwork, related to the materials of the panels. Also in the wake of the results achieved at La Jolla, plywood ranks first in Langford's preferences, seen as a "good panel material," but it is not possible to use it in Dhaka, since a waterproof quality is not locally available. Instead, he advises against "steel form-panels" because they rust in tropical climates, and are not economically feasible unless reused about 100 times. He is also against "rough wood board panels" because the imprints left on the concrete facilitate the growth of fungi ("this would be objectionable," Langford comments). The list of formwork materials also includes the fiberglass considered by Kahn in the initial phases of the project, as suggested by Kreier, but not available in Dhaka.

The question of the life of the panels is crucial for their reuse, in order to obtain wall sectors featuring the same imprints. "Five-foot lift allows high reuse factor" is the principle derived from Langford's experience, with the aim of constructing a panel capable of accompanying the entire construction of the building, so much so that it can be raised to the 22nd level called for in the construction of the walls.[171]

One section of the "Report on Concrete and Formwork", titled "Marble Recess Contains Bolt Line and Pour-joint," is on the marble cornices that are fundamental for the expression of the concept of organic growth of

[168] Langford notes, after the problems encountered during the fabrication of the formwork, the difficulty of the "manufacturer" to make the "tongue and grooved joint" due to the lack of "proper cutter head"; the difficulty of applying the filler strip in the case of the walls of insufficient thickness "to allow a man to get inside the formwork and caulk the joints with string" ("an expandable rubber or neoprene filler strip is being investigated," is the remedy searched by Langford); some defective executions of the joint detail between typical form-panel elements that have caused broken or imperfect horizontal V joints; the sealing of screw heads with glue and sawdust instead of wood ("one hour in sun and all plugs started to melt"); and certain difficulty in the opening of the formwork due to the insufficient slant on the horizontal members in relief that form the marble recess (Langford,"A Report on Concrete and Formwork," cit.).

[169] Langford, *Concrete in Dacca* cit., p. 52.

[170] Langford, "A Report on Concrete and Formwork," cit.

[171] "Potential life of each panel should be based on usage from the first pour to the top-out pour (22 lifts) without major repair – the panel should have the potential to be resurfaced and used 22 more times – 44 uses before cutbacks for secondary surfaces. Proper care exercised in stripping and cleaning procedures is a

the work, as desired by Kahn. Langford goes into the details for the construction of the panel to create the recesses for the marble slabs, warning against adoption of a type of solution that paradoxically will be precisely the one chosen by the Pakistani contractors hired for the placement of the concrete. "Recess-maker cannot be a nailed-on or glued-on shape, because of the many uses desired from each panel – it should be part of the solid edge-member, molded from one piece of timber."[172]

Langford's definition for the pouring operations in the "experimental walls" is "advancing slope method," which recurs in formwork manuals like that of Hurd regarding the construction of tunnels.[173] "After many unsatisfactory experimental walls," Langford explains, regarding the operations of pouring through movable hoppers, "this advancing slope method was the only procedure found to eliminate rockpockets. It is possible that with the add mixture Pozzolith #3 the conventional horizontal layer-type system of placement will be satisfactory."[174]

The stability of the individual panels is guaranteed by means of protection of the wood from the water of the compound and of the monsoons. To this end, Langford advises polyurethane coating not only on the surface in contact with the pour, but also on the back of the panel, "in order to minimize effects of rain and high humidity," and the construction, in any case, of the formwork in such a way as to permit expansion of the wood when exposed to rain.[175] "Minimum moisture absorption is a key factor in maintaining dimensional stability in wood form-work – polyurethane coated surfaces allow very little moisture intrusion during the pouring process."

The possibility of multiple uses of the panels depends on the stiffness of their structure, which Langford explains must be guaranteed by the "perimeter frame," "glued at corners and bolted to the heavy back-up frame" (for this purpose, the boards must also be joined with the tongue and groove method). For the "glued joints" he advises use of "urea formaldehyde" because it is a waterproof resin available in Pakistan.

Langford lists the main expedients to protect the already placed concrete during the pour phases. When the panels are assembled in the special groove of the lower wall sector the openings should be sealed "with a cotton rope" to "prevent leakage at the bottom of the new pour."[176] "String caulking," he writes, "can be very helpful in the prevention of leakage around filler strips, stair-keyways, circular block outs, etc."

He then provides a list of operations for the checking of the formwork, to be done prior to each pour,[177] and recommendations on how to proceed with the construction of the walls beyond the level of 38.42, or that reached by the "experimental walls": "A scaffolding system with a hoist system counterpart will greatly affect formwork erection procedures, pouring procedures,

prime factor in form-panel life" (Langford, *Concrete in Dacca* cit., p. 53).

[172] *Ibid.*

[173] Mary Krumboltz Hurd, *Formwork for concrete*, Detroit, American Concrete Institute, 1963, p. 270.

[174] Langford, *Concrete in Dhaka* cit., p. 54.

[175] *Ibid.*, p. 53.

[176] *Ibid.*

[177] "Corners and vertical references on control lines. Check bottom of all key blocks and filler strips for leak-holes. Fill the crack that is formed by bottom recess marker of form panel and previous casting with mortar. Check wedges behind all juncture elements (key blocks – filler strips). Check stair key-ways for inserts and string caulking for marble insert. Check reinforcing steel for proper distance from all form surface (adequate wooden spreaders and pre-cast donuts). Check for spreader loops (closed position). Check panels for excess oil (this is the time for final wipedown if wall thickness permits). All wood shavings, nails, debris, etc. should be cleaned out of the bottom of the wall. Check all bolts for equal pressure. Wet down hoppers, panels, top of previous pour. Station a man under the formwork with bucket and water to wash down any leaks that may occur. Start the mixer and holler 'Thara Thari' (Quickly!)" (Langford, *Concrete in Dhaka* cit., p. 53).

[178] *Ibid.*

and stripping procedures."[178] Furthermore, he believes it is necessary to equip the worksite with an electrical lighting system, to make it possible to continue pouring at night, in case of delays in the fabrication and placement operations of the concrete during the day.

Langford advises treating the formwork panels with polyurethane to get a smooth concrete surface that will resist biological growths. The smoothness goal, for Kahn's team, is that of the surface of terrazzo, as stated in a fundamental phrase in the report: "Polyurethane resin form-coating will produce a surface almost as smooth as terrazzo if proper care is exercised [...]."[179]

The concrete surfaces, the plaster surfaces and the mortar of the joints are exposed to humidity and drying, which favor the growth of biological patinas. Based on the recommendations provided by Imperial Chemical Industries, a well-known British varnish manufacturer, Langford advises protecting the surfaces with silicon, to be applied every five years.

The report concludes with a document titled "Outline of preparatory work to be considered," which sums up the operations to perform in order to organize the production of perfect formwork in Dhaka.

Langford's photographs show the various tools, from the vibrator to the hopper, the panels of the forms, and the methods of transport of the compound. All the pictures are taken with the purpose of documenting equipment and work phases, with angles typical of the illustrations of formwork manuals. But it is the photograph with the walls of the presidential atrium under construction that best represents the document-manifesto of the new vision of architectural concrete tenaciously pursued by Kahn and his team, starting with the construction of the Yale Art Gallery, which finds its apotheosis on the Dhaka worksite. All the parts of the formwork system designed for the National Assembly Building are visible, arranged on the ground as in a catalogue, and all the workers responsible for the assembly of the formwork and the construction of the concrete walls stand in a row. The maker of the complex machine organized to create Kahn's exposed concrete in Dhaka is at the center of the picture.

But the goal of perfection set by Langford with the wall samples and his report turns out to be far from simple to achieve on the Dhaka worksite.

The Engineers, the Inter-Dominion Construction Company and the reorganization of Kahn's office in Dhaka

After Langford's departure, at the start of July 1966 the engineers of the Public Works Department launch a series of tests on the concrete, but they complain that they have not yet received from Langford the report with "comments and suggestions" on the "experimental walls" he has built.[180] The works for the construction of the reinforced concrete of the National Assembly Building undertaken by Langford are assigned to two companies, The Engineers, directed by a "refined engineer,"[181] and the Inter-Dominion Construction Company, directed by a "hustler."[182] All the documents requested from Kahn's office in Philadelphia, from the detail drawings of the physical plant systems, of the reinforcements of the walls, beams and pillars, and of the circular openings with their radii, to the report by Langford, serve to prepare this construction phase.[183] The two companies assemble formwork panels without fully obeying the indications contained in Langford's report, or the recommendations of Kahn's staff, with the goal of achieving the greatest number of reuses for the panels.[184] Gustav Langford and Vollmer warn their colleagues in Philadelphia about the risk of having formwork built with the sole purpose of being reutilized as many times as possible:

"The Contractors feel most confident that they can produce form panels which can be used 30 or 40 times but are at a complete loss as to solving the scaffolding problem. We should have specific construction and form panel specifications or the contractors will construct panels which will suffice for the reuse factor but will not listen to reason when it comes to designing a panel which reflects the true system of its fabrication. Their attitude is; 'we'll support the corner panel with *something,*' 'we'll stuff *something* between the vertical joints to make it look *good,*' 'blow outs can be *patched'* etc. I'm working on form panels and bolt layouts for walls contained within the pile cap series, C, D, E, will send copy for comment when completed."[185]

To establish the general guidelines of the project organization and the work, and also to temporarily replace Vollmer who has to return to Philadelphia, Kahn sends another expert on concrete from his staff to Dhaka, on 5 August, MacAllister, who like Frederick Langford was involved in the construction work for the Salk Institute. MacAllister stays in Dhaka for only a few weeks.[186] The moment is critical, however, for the organization of the system of decision-making relations between Dhaka and Philadelphia, also following the termination of Gustav Langford's contract in September. MacAllister even fears the risk of closure, for lack of funds, of Kahn's office in Dhaka.[187] MacAllister's intervention during his short stay is limited to certain specifications on the control methods to apply on the worksite, on the part of the engineers of the Public Works Department.[188] To this end, he asks Kahn for "a copy of form used on Salk Project for this purpose," confirming the extent to which the experience of the La Jolla worksite has marked a turnaround in the design of the concrete.[189]

To avoid delays in the progress of the work, Siddiqui asks if it is possible to have Frederick Langford return to Dhaka, accompanied by Kahn's engineer, "to follow up work done on concrete formwork."[190] From Philadelphia the proposal arrives that Langford be hired as a consultant for the concrete of the National Assembly Building with a special professional contract drawn up with the Public Works Department, to prepare the drawings of the formwork needed for the continuation of the work. But the government representatives, at this point, see the question of the design of the formwork as a substantially closed episode.[191]

179 *Ibid.*, p. 52.
180 Superintending Engineer, letter to Roy Vollmer, Jr., 9 July 1966, LIK, PAC – Correspondence – RoyGus July 1966 thru December 1966. For the answer see David P. Wisdom, letter to Gustav Langford, 18 July 1966, LIK, PAC – Correspondence – RoyGus July 1966 thru December 1966.
181 Henry N. Wilcots, letter to David P. Wisdom, 8 June 1967, LIK, PAC – Correspondence – to/from Gus June 1967 thru December 1967.
182 *Ibid.*
183 Gustav Langford, letter to Henry N. Wilcots, 29 July 1966, LIK, PAC – Correspondence – RoyGus July 1966 thru December 1966. The Engineers is entrusted with the construction of the concrete works in the SW, PE and AC sectors, while the Inter-Dominion Construction Company is assigned those in sectors NE, SE, NW, GE, EW, WW.
184 Work resumes with the construction of the walls in sectors NW and AC up to the level of 28.42 feet; but certain technical issues immediately arise. "What is typical horizontal construction joint for 28.00' level slab, is slab keyed into walls or does it float on compacted fill? Must know immediately so formwork can be adjusted if required," Gustav writes on 4 August to Kahn's staff in Philadelphia (Gustav Langford, letter to Henry N. Wilcots, 4 August 1966, LIK, PAC – Correspondence – RoyGus July 1966 thru December 1966).
185 Gustav Langford, Roy Vollmer, Jr., "Bi-monthly Report No. 16," 8 August 1966, LIK, PAC – Bi-Weekly Reports – Dhaka Office Sept. 29, 1965 thru Feb. 22-1967.
186 Gustav Langford, letter to Henry N. Wilcots, 4 August 1966, LIK, PAC – Bi-Weekly Reports – Dhaka Office Sept. 29, 1965 thru Feb. 22-1967. Vollmer leaves the worksite.
187 See John E. MacAllister, notes to Kahn's office, n.d., [August 1966], LIK, MacAllister – Notes from Aug '66 trip to Dhaka; and MacAllister, "Contract for National Assembly, n.d., 1966, LIK, MacAllister – Notes from Aug '66 trip to Dhaka.
188 John E. MacAllister, notes to Kahn's office, n.d., [August 1966], LIK, MacAllister – Notes from Aug '66 trip to Dhaka.
189 *Ibid.*
190 *Ibid.*
191 John E. MacAllister, "Contract for National Assembly," n.d., [1966], LIK, MacAllister – Notes from Aug '66 trip to Dhaka.

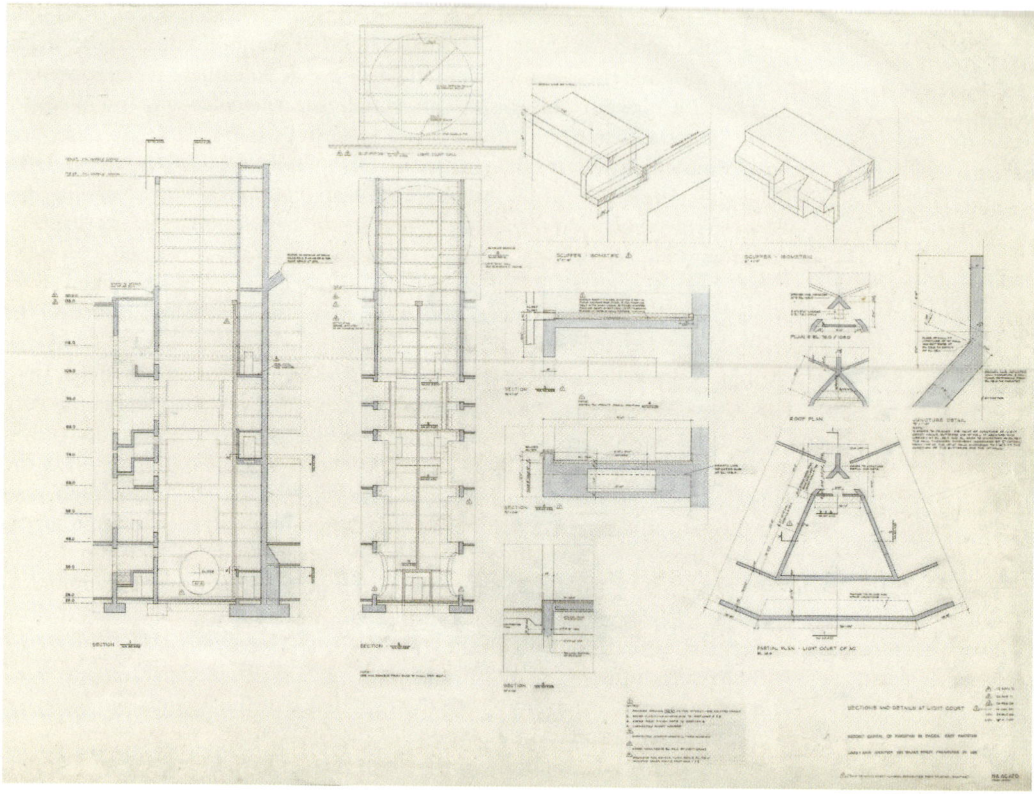

Kahn continues to receive requests from Siddiqui and MacAllister for a staff member to manage the Dhaka office, without making a contract with Frederick Langford (at this stage, the name of Henry N. Wilcots appears), in order to avoid, as MacAllister explains, leaving crucial technical aspects of the architectural concrete in the hands of the technicians of the Public Works Department and the hired contractors. "[...] Gus must have the authority to make decisions here, when necessary. It is not a question of whether LIK or Gus Langford makes decisions; it is a question of whether P[ublic]W[orks]D[epartment] or Gus Langford makes decisions."[192]

Towards the end of August, Gustav Langford remains the sole member of Kahn's staff in Dhaka. "What is Fred's progress on form panels? Is he coming back? If so, when and what can I expect him to bring?" he writes to his friends in Philadelphia.[193] A note indicates that Frederick has indeed begun to make the drawings of the formwork still needed.[194] In the correspondence between Dhaka and Philadelphia the possibility emerges of taking advantage of the presence at the worksite of John Gunther, charged with the fabrication of the metal casements, at the expense of the Public Works Department. "John made all of the form panel accessories and is quite competent, check with Fred," is Gustav's view.[195] In the meantime, samples of materials for the formwork and for the concrete arrive from the United States, to solve the problems of waterproofing of the brick masonry ("polyurethane form coating," "concrete additives," "masonry coating" and "caulking materials"[196]).

The person who joins Gustav Langford in the supervision of the worksite is the young engineer Harry Palmbaum, a former staffer of Keast &

240. National Assembly Building, Dhaka, 1962-83. "Sections and Details at Light Court," 24 October 1966 (LIK, 030.I.C.650.38).

Hood hired by Kahn on the recommendation of Gianopulos.[197] Palmbaum reaches Dhaka towards 13 September (the visa application dates back to 1 September).[198]

In the initial phases of his stay, Palmbaum takes part in the definition of certain aspects connected with the formwork, and discusses the positioning of the "control joint bulkhead"[199] with the contractors and the engineers of the Public Works Department, but above all he is studying the roof of the assembly hall, for which he proposes a structure in prestressed concrete. One of the main difficulties is that of raising the eventual precast units or of supporting the formwork of a pour in progress: "it may be possible to cast comparatively lighter sections and then post-tension them."[200] What Kahn recalls about the foundation work, begun prior to knowing the dimensioning of the roof, is fundamental to understand the developments of the structure of the hall: "The crowning roof structure combines the needs of acoustics and the sheltering of natural light. The structure is light in weight to be in keeping with the loads allowable on the piles."[201]

The plastic-covered panel of the Inter-Dominion Construction Company

The requests sent from Dhaka to Philadelphia during the months to come and throughout 1966 always have to do with the indications, in the definitive drawings of the various levels, of all the masonry elements and conduits to be foreseen prior to the pouring of the walls and floor slabs, and the various construction joints, while the question of the "vertical marble placement" also starts to appear.[202] The construction of the first walls of the building indicated as the "mosque" takes place in just 21 days, completely out of the control of Kahn's team. Certain rooms are plastered prior to the

[192] "The sample walls in place are accepted," MacAllister writes, "and will be the basis of all concrete work on the National Assembly. Mr. Kahn has done his job as far as PWD is concerned and they do not expect further concrete formwork drawings from him except for the final revisions of existing drawings. (J.E.M. [John E. MacAllister] explained the necessity for completed detailed formwork drawings and complete control of all concrete work. Mr. Langford's experience on the Salk Project was mentioned along with the resultant savings to the owner). Mr. Siddiqui said that Fred Langford's hard work while here was appreciated by everyone and the result of this work speaks for itself. The discussion with the builder and PWD of a separate contract for concrete consultation and formwork drawings is considered by PWD to be unethical and a bad reflection of the professionalism of Mr. Kahn's office. They will not answer Mr. Langford's letter and further discussion of the subject will only tarnish Mr. Kahn's reputation. (These comments by Mr. Siddiqui, however brutal, were straight off the cuff, and absolutely frank)" (*ibid.*).

[193] Gustav Langford, letter to Henry N. Wilcots, 29 August 1966, LIK, PAC – Correspondence – RoyGus July 1966 thru December 1966.

[194] See the handwritten note in the margin, Gustav Langford, letter to Henry N. Wilcots, 29 August 1966, cit.

[195] *Ibid.*

[196] John E. MacAllister, "Objectives of JEM Site Visit," n.d., [August 1966], LIK, MacAllister – Notes from Aug '66 trip to Dhaka.

[197] At the end of 1964, Keast & Hood is forced to interrupt collaboration on the project due to the delays in government payments, which are not sufficient to cover office expenses (Nicholas L. Gianopulos, interview with Anna Rosellini, February 2010). Perhaps this is also the reason behind the exit, in the same period, of Komendant from the job.

[198] See Louis I. Kahn, letter to Consulate General of Pakistan, 1 September 1966, LIK, PAC – Correspondence – RoyGus July 1966 thru December 1966, and Gustav Langford, letter to Henry N. Wilcots, 17 September 1966, LIK, PAC – Correspondence – RoyGus July 1966 thru December 1966. A series of photographs of the Assembly Building taken by Palmbaum is conserved at the University of Pennsylvania, Rare Book and Manuscript Library Collections.

[199] Anon. [Ghulam Mohammad], "Comments on Fred Langford's concrete report section Q," n.d., [November 1966], LIK, PAC – Correspondence – RoyGus July 1966 thru December 1966.

[200] M. G. Siddiqui, letter to Louis I. Kahn, 1 December 1966, LIK, PAKCAP – Correspondence – PAK PWD 1966 – 1967. Siddiqui is able to discuss the prestressed concrete structure with representatives of Freyssinet Pakistan Company, Ltd., and of Ibisi, The Hague.

[201] Louis I. Kahn, letter to M. G. Siddiqui, 19 December 1966, LIK, PAKCAP – Correspondence – PAK PWD 1966 – 1967.

[202] Gustav Langford, "Bi-monthly Report No. 17," 5 October 1966, LIK, PAC – Bi-Weekly Reports – Dhaka Office Sept. 29, 1965 thru Feb. 22-1967.

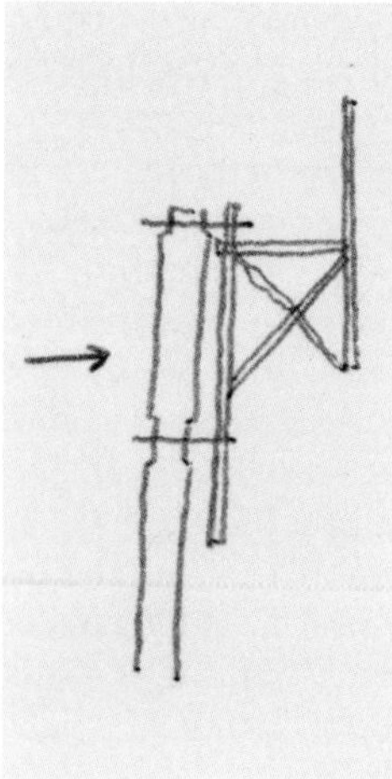

241-243. National Assembly Building, Dhaka, 1962-83.

241, 243. Sketches for the scaffolding systems. Propositions by The Engineers and by the Inter-Dominion Construction Company ("Outline of preparatory work to be considered," n.d., received by Kahn's studio on 18 November 1966, LIK, PAC – Correspondence – RoyGus July 1966 thru December 1966).

pouring of the floor slab, and Gustav writes that "the whole job was rushed and poorly constructed."[203]

In spite of Kahn's team's expectations, who hope to be able to produce formwork panels in the workshop installed at the site for carpentry and metalworking (at the start of October the workshop is almost complete), The Engineers and the Inter-Dominion Construction Company continue to use their own equipment for this type of job.[204]

To create a system that can guarantee high-quality production of panels, starting precisely with his own report on the experimental walls, Frederick Langford prepares a list of the spaces and machinery required to set up a sawmill on the worksite for the construction of formwork, and of all the materials and operations for its assembly. The document, titled "Outline of preparatory work to be considered," contains a series of chapters on the works for the formwork, and each chapter specifies the related materials and operations. Gustav gathers the information requested by Frederick from The Engineers and the Inter-Dominion Construction Company.

For the sawmill the document calls for spaces for the seasoning of the wood, the cutting and storage of the boards, as well as "work-benches," an "assembly line," a zone for "dehumidification and dust-proofing for form-coating," and cutter-heads made in Germany. For the training of the workmen, it suggests programs for the carpenters and the mechanics. But the production system described by Langford runs up against the organization of the two companies, which operate separately, each with its own subcontractors: "Each contr. has his own sub.contr. for supply & fabrication of form panels," Gustav writes in response to Frederick's inquiries.

For the boards of the panels, the type of wood used for Langford's experimental panels is confirmed, namely champa (indicated in the document as "champi teak"). For the structure, in place of the "chapalish" indicated by Frederick, the contractors use "jarul" or "gorjan." For the "scaffolding system" the two contractors consider two variants, one involving industrial production in metal, the other based on on-site fabrication with bamboo poles. In the case of the use of metal, The Engineers hypothesize assembling scaffolding attached to the walls thanks to the holes left by the spreaders and not

242. Formwork made according to the model developed by F. Langford (LIK, 030.IV.A.650.3.28).

yet plugged by the marble seals; in the case of the bamboo, a sketch of the proposal of the Inter-Dominion Construction Company shows the scaffolding assembled on the ground, while a worker is using a rudimentary device for the raising of the formwork panel (a frame with a hoist, placed at the top of the wall under construction). Gustav reports that both companies are procuring the polyurethane, while with the goal of obtaining a "consistent exterior color" the Public Works Department has already placed an order for the cement, to purchase an "enormous amount from one source."[205]

The question of Frederick's presence at the worksite is raised again when the work resumes at the start of November. Both The Engineers and the Inter-Dominion Construction Company, though they recognize Frederick's "knowledge," do not seem to urge his arrival in Dhaka. "Maybe they want to try first to prove that they do not need an American expert to solve all their difficulties," is Gustav's guess.[206] The critical level of the pour above which as yet unpredictable difficulties may still arise for the two contractors is 33.00 feet. "If Lou's idea is to have Fred here only when he is absolutely needed then the time is not now," Gustav writes. "If he wants Fred here to help prevent possible failures, and at the same time to help establish sensible construction techniques that will result in good architectural concrete, then his presence would be most helpful when they start to pour the 33.00 feet level walls of the AC which will be sometime in December."[207]

Several photographs taken from the Presidential Square document the progress of the work on the National Assembly Building in November 1966.[208] The walls erected under Langford's supervision display all their value as models of the quality of the concrete for the walls still to be made. The work concentrates on the construction of the low walls connecting the ends of the piles, below the level of 28.00 feet.[209]

The contractors operate in an empirical way, without testing modifications using drawings or samples on the worksite.[210] The Engineers tries to stick to the model of panels and working modes developed by Langford, while the Inter-Dominion Construction Company introduces modifications that tend to compromise the original system, even to a significant extent. Gustav Langford believes in the capacity of both companies to make good

[203] Gustav Langford, letter to Henry N. Wilcots, 8 October 1966, LIK, PAC – Correspondence – RoyGus July 1966 thru December 1966.

[204] Gustav Langford, "Bi-monthly Report No. 17," 5 October 1966, cit.

[205] The Publics Works Department does not believe it should have to order Pozzolith for the concrete compound to use on the worksite ("Outline of preparatory work to be considered," n.d., received by Kahn's office on 18 November 1966, LIK, PAC – Correspondence – RoyGus July 1966 thru December 1966).

[206] Gustav Langford, letter to Henry N. Wilcots, 14 November 1966, LIK, PAC – Correspondence – RoyGus July 1966 thru December 1966.

[207] *Ibid.* Gustav informs Kahn's office that the contractors have decided to move up the date for the start of work on level 33.00 feet to mid-December (Gustav Langford, letter to Henry N. Wilcots and Frederick Langford, 18 November 1966, LIK, PAC – Correspondence – RoyGus July 1966 thru December 1966). On the same occasion, he asks if Kahn has made decisions regarding the panel of the Inter-Dominion Construction Company.

[208] See the photographs dated 18 November 1966, LIK, 030.IV.A.650.3.11, and LIK, 030,IV.A.650.3.26. For the exposed concrete walls of the National Assembly Building a sample sector is made (see the photograph dated 18 November 1966, LIK, 030.IV.A.650.3.29).

[209] These walls are poured in formwork with the boards assembled horizontally, using the same V joint. The formwork are documented by a photograph taken during the shoot in November (see the photograph, LIK, 030.IV.A.650.3.28, with note "typical below 20.00' formwork"; the picture reaches Kahn's office on 18 November 1966).

[210] Mohammad specifies that "formwork is to be designed by the contractors": "They are following almost the same as shown in the sample prepared by Mr. Langford" ([Ghulam Mohammad], "Comments on Fred Langford's concrete report section Q," n.d., [November 1966], LIK, PAC – Correspondence – RoyGus July 1966 thru December 1966).

244. National Assembly Building, Dhaka, 1962-83.
Sample wall built next to the Presidential Square, according with the formwork system proposed by the Inter-Dominion Construction Company - thin plastic sheets wrap each board (LIK, 030.IV.A.650.3.29).

exposed concrete, but he fears that the critical details of the formwork system may produce defects in the pour precisely because the contractors "will not extend themselves any further i.e. in terms of experimentation."[211]

The first substantial modification to the experimental panel is made by Inter-Dominion Construction Company to make up for the impossibility of obtaining polyurethane to treat the panel surfaces, as recommended by Frederick Langford's report. The company uses a protective coating with thin plastic sheets to wrap each board. The attachment of the sheets weakens the interlock system of the boards on the posts at the ends of the panel through the wing that produced the V imprint, altering the frame of the panel and therefore also the imprints of the experimental panel patterns.

With the panel boards coated in plastic, Inter-Dominion Construction Company makes a sample wall segment next to the Presidential Square and outside the perimeter of the National Assembly Building, since it differs from those already built by Langford and those under construction by The Engineers, which has instead managed to purchase sufficient quantities of polyurethane.[212] The report of Gustav on the panel and sample wall made by the Inter-Dominion Construction Company is of maximum importance to understand how all the staff members are attuned to Kahn's vision of a concrete to be shaped with a panel built in such a way as to display all its components through the imprints:

"This experiment has resulted in the elimination of the top and bottom horizontal V joint – i.e. 2" above and below the marble recess. A sample wall was poured using this panel covering method and the result was extremely successful. All vertical fins are sharp and 99% intact, color is very consistent, the surface is very smooth and almost completely without air bubbles. The elimination of the top fin also overcomes the shrinkage cracks which always resulted immediately below the top horizontal V joint. I realize that this panel design and fabrication has deviated from Fred's which was used on the GE and that LIK will seriously object to the method as it does violate the whole idea of the finished concrete reflecting the technique of wood panel construction [...]. The impression achieved by having both top and bottom

horizontal joints is one where the wall is made up of a series of vertical fins confined within a rectangle. Whereas the elimination of the top and bottom fins produces an effect of continuous vertical lines being interrupted only at the vertical marble position."[213]

Siddiqui is impressed by the sample wall of the Inter-Dominion Construction Company and approves the construction of the walls of the NE Office Wing using the "revised form panel design."[214] Precisely with an eye on Kahn's criticism of the suppression of the two V joints, Gustav asks the colleagues in Philadelphia if it is possible to treat the panels with oil for kitchen tables in place of the polyurethane, thus doing away with the sheets of plastic that alter the panel design. "Please have LIK comment on this," is his request regarding the changes being made with respect to the experimental panels.[215]

The correspondence between Philadelphia and Dhaka offers no evidence of Kahn's reaction to the panel of the Inter-Dominion Construction Company. On the worksite, modifications are made to the panel, which would lead us to presume decisions were made to keep the plastic sheets, given the results on the surface finish, though applying the sheets to the panel boards in such a way as to also obtain the horizontal V joints. To do this, the horizontal posts are broken down into two layers; the outer layer is not coated, since it is in contact with the concrete of the recess, and it serves to attach the plastic sheet on the other layer.[216] But again in this case, as in the first panel of the Inter-Dominion Construction Company, Langford's original design of the horizontal imprints is altered, as it no longer exactly indicates a single piece of the panel.

One phrase written by Gustav refers to the modifications made by the Inter-Dominion Construction Company to the first panel with plastic coating: "V-joint elimination. This we tried but the detail of terminating the horizontal V-joint into the marble recess was very poor so we are back with top and bottom horizontal joints."[217]

Since The Engineers has polyurethane, for the AC sector the company has 36 panels "constructed as per Fred's drawing," so that several walls of the level 33.42 feet can be poured in January 1967.[218] But the quality of

[211] "The simple straight wall," he writes on 14 November, "will be poured extremely well as indicated by their 28.42' starter walls. Each contractor is very conscious of accuracy and exactitude but it is in the detailing i.e. construction joints, bulk heads, corners, blockouts, T beam, intersections and stair details, where the results may deteriorate. I have this impression from the way they say they will handle the details and also by their lack of investigation and supporting designs and sketches" (Gustav Langford, letter to Henry N. Wilcots, 14 November 1966, LIK, PAC – Correspondence – RoyGus July 1966 thru December 1966).

[212] See the photograph dated 18 November 1966, LIK, 030.IV.A.650.3.29.

[213] Gustav Langford, letter to Henry N. Wilcots, 14 November 1966, LIK, PAC – Correspondence – RoyGus July 1966 thru December 1966.

[214] *Ibid.*

[215] *Ibid.*

[216] See, for example, the photograph, LIK, 030.IV.A.650.4.54. The breakdown of the post is clearly visible in the photograph, LIK, 030, IV.A.650.2.66, which documents a panel of the Inter-Dominion Construction Company and bears the note "plastic sheet [and] + patch"; the photograph shows a sheet added to correct the coating. See also the note, "back of typical ID panel," photograph, LIK, 030.IV.A.650.2.63.

[217] Gustav Langford, letter to Henry N. Wilcots, 27 January 1967, LIK, PAC – Correspondence – RoyGus January 1967 thru May 1967.

[218] Gustav Langford, letter to Henry N. Wilcots, 15 December 1966, LIK, PAC – Bi-Weekly Reports – Dhaka Office Sept. 29, 1965 thru Feb. 22-1967.

these works is still not satisfactory: "The easternmost wall over pile cap series B has been poured to 33.42 ft. The surface is very good but the corners and V-joints are extremely poor as they bled during pouring."[219]

To avoid pouring the concrete into the formwork using the operations applied in the experimental walls and described in Frederick Langford's report, the contractors conduct tests to improve the workability of the compound. As an alternative to the Pozzolith additive recommended by Langford, which is costly to import, they try Lissapol N, but without positive results, because it generates "more air holes in surface of concrete." Another product tested is Toxement S.C., sent by Kahn to Dhaka, again as an alternative to Pozzolith.[220]

The search for a logic for the vertical strips

Kahn's imprint system sets out to be rigorous in the form and positioning of every element, and the first modifications made by the contractors to the experimental panels are evaluated precisely in terms of their conformity with the principles of truth pursued by Kahn. But while the function of the horizontal strips of marble has been clear for some time, the vertical strips pose questions for Kahn regarding the logic of their positioning inside the imprint system.

A series of drawings made by Gustav Langford are sent to Kahn to specify the position of the vertical marble strips. Several positions of the strips, like those in the Office Blocks or the GE, are based on previous drawings prepared in Philadelphia. But for the bands whose position has yet to be defined, Gustav Langford wants to find a constructive criterion that corresponds to that of the horizontal strips, so that the vertical strips will also indicate growth phases of the pours. For this reason, in his drawings the position of the vertical strips is calculated on the basis of the quantity of concrete to be poured each time, with the aim of obtaining a correspondence represented precisely by the distance between the strips, between the construction process and the marble ornament.

245-246. National Assembly Building, Dhaka, 1962-83. Formwork built by the Inter-Dominion Construction Company (LIK, 030.IV.A.650.4.57; LIK, 030.IV.A.650.2.60).

"I was trying to establish a geometry," he explains to Kahn, "into which the vertical marble strips and the ceiling construction joints would unite, however, this so far has become impossible, as there are 2 major considerations, i.e. the 40.00' +/- allowable dimension of any one pour and the logical amount of conc.[rete] which can be poured before a horizontal construction joint is required. Please have LIK [Kahn] review these prints and the prints of the proposed construction joints by the contractors – hand carried by Harry Palmbaum. If it is imperative that a geometry uniting both be established then please let me know as soon as possible, as this will effect 28.00 level form panel construction."[221]

In the technical details designed in October 1966, we see the particular of the vertical joint between one pour and the next to construct a continuous wall. At that point, on both sides of the wall, the design calls for recesses for the insertion of slender strips of marble, with a width of one foot. The note indicates that this is "recessed marble at vert.[ical] constr.[uction] joint."[222]

Even after Gustav's studies for the positioning of the vertical strips to match the pours, the question remains open, and in December the

[219] Gustav Langford, letter to Henry N. Wilcots, 9 January 1967, LIK, PAC – Correspondence – RoyGus January 1967 thru May 1967. All the formwork, including those of the ceilings with relative expansion joints, are assembled according to Kahn's approval (Gustav Langford, letter to Henry N. Wilcots, 16 January 1967, LIK, PAC – Correspondence – RoyGus January 1967 thru May 1967).

[220] [Ghulam Mohammad], "Comments on Fred Langford's concrete report section Q," n.d., [November 1966], LIK, PAC – Correspondence – RoyGus July 1966 thru December 1966.

[221] Gustav Langford, letter to Henry N. Wilcots, 6 October 1966, LIK, PAC – Correspondence – RoyGus July 1966 thru December 1966.

[222] See the drawing "Developed section thru circulatory stairs," 11 October 1966, revised on 16 January 1968, LIK, 030.I.C.650.038.

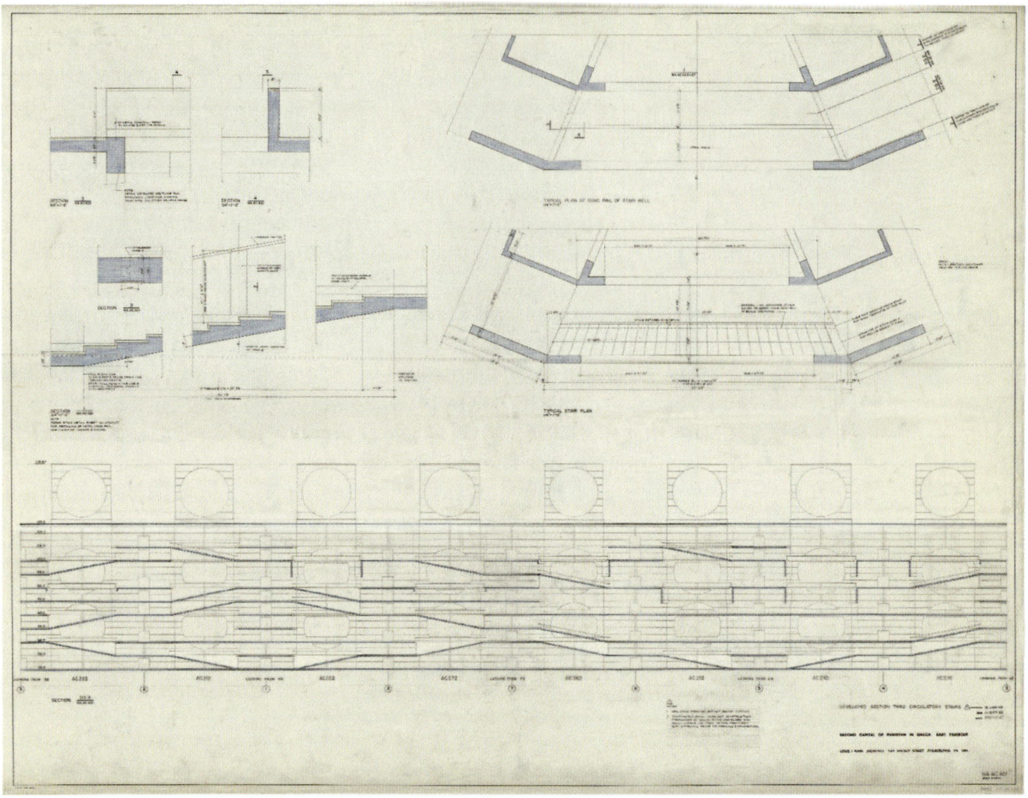

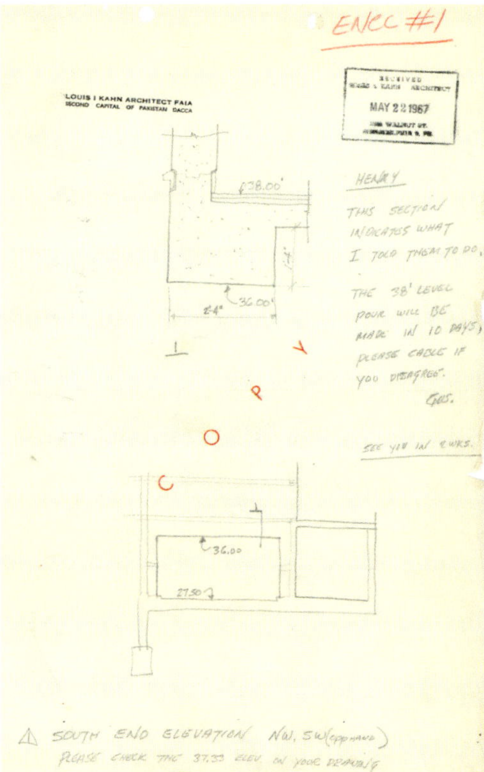

247-248. National Assembly Building, Dhaka, 1962-83.

247. "Developed Section Thru Circulatory Stairs," 11 October 1966 (LIK, 030.I.C.650.38).

248. Sketches and notes for the positioning of the vertical strips, 18 May 1967 (G. Langford, letter to D. P. Wisdom and H. N. Wilcots, 18 May 1967, LIK, PAC – Correspondence – RoyGus January 1967 thru May 1967).

Philadelphia office is urgently requested to supply the positions of these strips in the NE and SE Office Wings, given the fact that the "contractor is constructing his form panels presently for these two Office Wings […]."[223] In February 1967 another similar request is sent for drawings that indicate the "vertical marble strip positions for construction joints & at other places."[224] The fact that "other places" are mentioned confirms the lack of definition of the positions of the vertical marble strips in the overall design.

It is at this point that Kahn's team defines other rules for the vertical strips and construction joints. The order now arrives from Philadelphia to establish the position of the vertical strips according to that of the conduits of the electrical system, as well as the axes of symmetry of rooms and openings. This way of detaching the vertical strips from the logic of fabrication of the concrete demonstrates the impossibility of giving that motif the same meaning as the horizontal cornices.

The decision regarding the positioning of the vertical marble strips takes Gustav by surprise: "I see now that the vertical marble strip placement is being determined by the electrical layout and centerline of room premises. This differs greatly from my assumption that the placement would be determined structurally and by construction joint alignment. I must have electrical layout and finalized vertical marble strip locations immediately for EW and WW."[225]

After Kahn's clarification, Gustav repeatedly complains that he has not received the technical indications necessary for the positioning of the vertical strips: "I have requested several times for finalized vertical marble placement. This is a must. […] final placement for all buildings is immediately required."[226] He sends frequent requests to Philadelphia to know the positions of the "electrical for AC circulatory, Westinghouse elevator cuts," and the "finalized vertical marble locations."[227] Finally, again as a result of Kahn's silence regarding the "PE vertical marble location," he starts to propose solutions himself, shown in drawings sent to Philadelphia: "This indicates what I have gleaned from the PE section [crossed out: and what I believe to the best of my knowledge] where the vertical marble should be located. The Engineers want to start fabrication of forms immediately, therefore, your final comment or revision is required."[228]

For the four cylindrical volumes, a drawing is sent from Philadelphia to Dhaka with indication of the "vertical marble locations," showing that the strips in correspondence with the four walls, which determine the boards of each cylinder, no longer have a "complementary strip on the inside."[229]

[223] Gustav Langford, letter to Henry N. Wilcots, 5 December 1966, LIK, PAC – Correspondence – RoyGus July 1966 thru December 1966.

[224] See "Drawing for works in progress: National Assembly Building," n.d., received by Kahn's office on 15 February 1967, LIK, PAKCAP – Correspondence – PAK PWD 1966 – 1967. The same document shows that in section drawings of 19 July 1966 the position of the "vertical marble strips" in the circular walls is not clear. "Is there any vertical marble strip on the wall of East elevation other than central one? If so, its distances from ends are required."

[225] Gustav Langford, letter to Henry N. Wilcots, 6 February 1967, cit. On the marble quarries at Mullaghori, in West Pakistan, and at Ayub Nagar, in East Pakistan, see I. Ahmad, letter to the Additional Chief Engineer, 12 February 1967, LIK, PAKCAP – Correspondence – PAK PWD 1966 – 1967.

[226] Gustav Langford, letter to Henry N. Wilcots, 16 February 1967, LIK, PAC – Correspondence – RoyGus January 1967 thru May 1967.

[227] Gustav Langford, letter to Henry N. Wilcots, 22 February 1967, LIK, PAC – Bi-Weekly Reports – Dhaka Office Sept. 29, 1965 thru Feb. 22-1967.

[228] Gustav Langford, letter to David P. Wisdom and Henry N. Wilcots, 18 May 1967, LIK, PAC – Correspondence – RoyGus January 1967 thru May 1967. Attached to the letter there is a drawing for the positioning of the horizontal cornice with the dripstone, corresponding to the floor slab of level 38.00 feet.

[229] Gustav Langford, letter to Henry N. Wilcots, 12 July 1967, LIK, PAC – Correspondence – to/from Gus June 1967 thru December 1967.

249-250. National Assembly Building, Dhaka, 1962-83.
Construction site photographs (LIK, 030.IV.A.650.4.53; 42).

Unlike the experimental panels of 1966, at this point to make the vertical recesses positioned to respond to different logics the workers apply unseasoned wood, not perfectly square and even poorly attached, to the panels, resulting in imprints on the concrete of profiles that are so irregular at times as to be hard to seal, it is feared, with the marble inserts.[230]

During a worksite inspection in July 1967, Kahn, when he sees the grooves imprinted in the concrete, says he is concerned about the "relationship of vertical fins and openings."[231] This is further confirmation of the uncertainties regarding those strips, which Kahn continues to see as arbitrary with respect to their horizontal counterparts. In the meantime, the question arises of the cutting of the marble slabs to insert in the recesses of the cylindrical walls, as to whether or not they should be curved. Following the difficulties and defects in the making of the recesses, Kahn believes it is worth introducing "design modifications for the marble details": "I request that no indent for precut marble be made until a consistent accuracy of workmanship has been attained. A precision of concrete workmanship which would promise standardization of the marble details is most essential."[232]

Precisely in the light of the complicated functional and technical definition of the marble inserts, the pattern of the strips begins to even take on a symbolic connotation that never previously appears in the project documents, and is utterly alien to the pursuit of their rigorous constructive essence.

The issue of the continuity with Islamic architecture of the monuments of the new capital, already raised repeatedly during the project phases and brought back to the fore in 1965 in the form of an official request on the part of the Government Designs Committee, must still be a concern of the government authorities of Pakistan, since Siddiqui feels compelled to write, in the summer of 1967, to defend the buildings that are being erected as ex-

[230] Harry M. Palmbaum, "National Assembly Building Construction Report," n.d., [July 1967], LIK, PAC Various Notes Dhaka Jan 66.
[231] Donald Barbaree, letter to Henry N. Wilcots, 28 July 1967, LIK, PAC – Correspondence – to/from Gus June 1967 thru December 1967.
[232] Louis I. Kahn, letter to M. G. Siddiqui, 31 July 1967, LIK, 030.II.A.10.9. During the same inspection in July, after seeing certain ceilings painted white in the rooms of the National Assembly Building, Kahn intervenes to give precise orders: "Color will be introduced in upholstery and drapes using approved Maris color samples [...] color combinations for rooms being composed" ("The month of July 1967," LIK, PAC Transmittals – from Dhaka office Nov. 1 '65 thru 1/19/68). The question of color also has to do with the Hostels, for example the ceilings of the verandas, for which "a couple of LIK suggested colors" are requested (Donald Barbaree, letter to Henry N. Wilcots, 28 July 1967, LIK, PAC – Correspondence – to/from Gus June 1967 thru December 1967). For the Hostels, we find written: "Samples of material for tapestry and curtains have been taken to Philadelphia by

pressions of the new state, an interpretation of Kahn's works in Dhaka as "Islamic Architecture," where the latter is presented as the pursuit of the truth of materials and structures. The marble strips and concrete surfaces are the focus of his reasoning on the Islamic meaning of the National Assembly Building, explained in "Philosophy of Architecture of Ayus Nagar." Behind Siddiqui's phrases, we can glimpse Kahn's ideas about materials; even the critique expressed against any cladding reflects Kahn's concerns about possible inappropriate intervention on the surfaces of his exposed concrete, utterly unusual for monuments in Pakistan. The fact that the marble strips take on a crucial role in Siddiqui's text is explained precisely by the problems posed by that ornament.

"According to Kahn this is the age of 'miracle materials'," Siddiqui writes. "These can and must be used in their pure form to produce beautiful buildings in which each material exhibits its pure form and function. If the buildings are polluted with smears of paints and plasters they look ugly and impure. In the legislative enclave of the capital modern material has been honestly and truly used. The National Assembly is a marble and concrete structure. No plaster will ever pollute its face. Concrete is a man-made material and therefore imperfect. This imperfect material will be laced with marble at 5' intervals. Marble is so perfect that it will automatically hide the imperfections in concrete by attracting all attention to itself. Marble will absorb the atmosphere and will present to the viewers a kaleidoscopic shade at different periods of the day in each season. Concrete by its sturdy nature will provide strength and stability. According to Prof. Kahn marble is the 'Woman' and the concrete is the 'Man.' The combination of these 'male & female' materials, one giving strength and the other beauty and colour, symbolises creation."[233]

Professor. [...]. Professor mentioned that the walls should be of different colors and not all white as suggested by his local representative" (typewritten document, received by Kahn's office on 4 August 1967, LIK, PAKCAP – Correspondence – PAK PWD 1966 – 1967). For Kahn's definitive choice of the fabric colors see Henry N. Wilcots, letter to Donald Barbaree, 2 August 1967, LIK, PAC – Correspondence – to/from Gus June 1967 thru December 1967.

[233] M. G. Siddiqui, "Philosophy of architecture of Ayus Nagar," n.d., received by Kahn's office on 21 August 1967, LIK, Pac- Entrance Gate - Ayut Nagar. All Correspondence.

251-253. National Assembly Building, Dhaka, 1962-83.

251-252. Sketches for the formwork system (G. Langford, letter to D. P. Wisdom and F. Langford, 9 May 1967, LIK, PAC – Correspondence – RoyGus January 1967 thru May 1967).

253. Photograph of the concrete wall, with note: "Indentations not typical. Better than average" (LIK, 030.IV.A.650.4.34).

It is worth emphasizing the symbolic role take on by the marble strips in Siddiqui's interpretation, in order to fully grasp the complexity of values of that ornament, which in the design process is not always consistent with the logic of construction usually found in similar inventions of Kahn. Precisely in the light of what Siddiqui writes, the reason for the continuing existence of those strips in the various project phases becomes clearer; they are never challenged, while we can see the effort to find a function for them, any function beyond that of pure ornament, to justify their presence in Kahn's eyes. It is precisely the pattern of these strips that attenuates the force of image of the vast exposed concrete surfaces, and becomes a factor regulating the density of gray tones, just as Le Corbusier had used color to created calculated contrasts with the *béton brut*. We can also notice a similarity between Siddiqui's assertions on the optical power of "attracting all attention" of the system of strips, with respect to the imperfections of the concrete, and Kahn's statements during the construction of the Mill Creek towers, the Richards Laboratories or the Salk Institute, regarding the same optical effects of the holes left by the cones of the spreaders and the surfaces striped by the imprints of boards or V joints with respect to the defects in the making of the concrete.

The conclusion of Siddiqui's text is emblematic of the atmosphere in which the construction of a work lacking in any Islamic particulars takes place: "I leave it to the readers to judge whether a design based on the above philosophy will be Islamic or not."[234]

During the period of discussion of the Islamic character of the National Assembly Building, Kahn's office in Philadelphia receives, on 31 July, a photograph taken by Mirze A. Rahman, Section Officer of the President's House in Karachi, of the detail of cladding in marble slabs of the Taj Mahal, described in a handwritten note on the back of the print.[235] There is little doubt that the marble strips of the National Assembly Building, in the eyes of the Pakistanis, represent the link between Kahn's work and the Islamic tradition. "The image is that of a many-faceted precious stone, constructed in concrete and marble," Kahn remarks regarding his work.[236]

The panels of The Engineers and of the Inter-Dominion Construction Company

In spite of the initiatives and criticisms, Kahn's team is not able to control the modifications to the experimental panels made by the contractors, each in their own way, depending on their needs and the materials at their disposal. All the worksite documents offer glimpses of imprecise management of the decisions for the placing of the concrete, so much so that from the fall of 1966 to the spring of 1967 problems arise in the making of the details and the application of different construction methods.

The construction of the wall sectors of the National Assembly Building, from the level of the foundation beams to the horizontal groove of the first pour, is photographed in April 1967, with the aim of documenting all the types of defects produced in the fabrication of the concrete, from those caused by the panels and their imperfect assembly to the marks of various kinds produced on the surfaces. Gustav Langford takes the pictures to monitor the progress of the worksite and to document the defects. The photographs are sent to Kahn, who receives them on 1 May. "Some photos," Langford warns, "represent the average result and others the worst. I have tried to categorize the photos according to the existing problems – patching, bleeding, alignment, etc. Some situations are really bad, but fortunately constitute only a very small percentage of wall coverage."[237]

One series of photographs is on the assembly of the panels, in keeping with the different methods of the two contractors, and the differing quality of the imprints left on the concrete. The series of V lines, defined as "indentations," with perfect edges, is considered "not typical," as is indicated on the photograph: "better than average."[238] The result has probably been obtained with the plastic coating used by the Inter-Dominion Construction Company. The photograph of the inclined panel, to suggest that it is about to be removed, against the backdrop of the concrete wall with V indentations with perfect edges, demonstrates that it is the system of the Inter-Dominion Construction Company that produces the best results.[239] The detail of three V joints on a perfectly smooth surface gets the following comment: "better

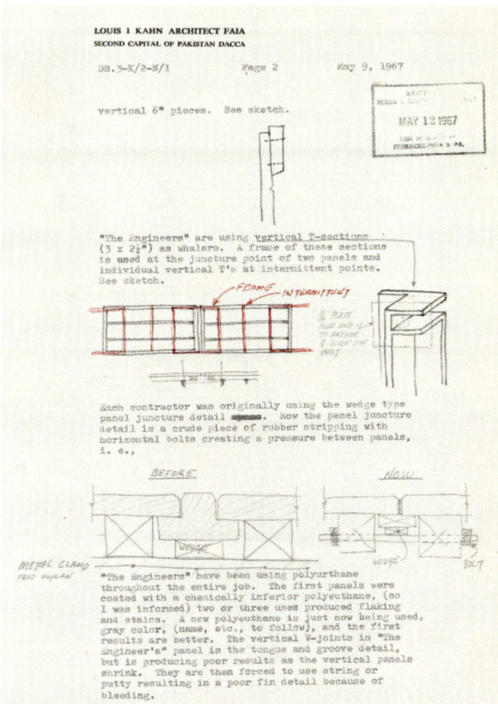

LOUIS I KAHN ARCHITECT FAIA
SECOND CAPITAL OF PAKISTAN DACCA

DS.3-K/2-N/1 Page 2 May 9, 1967

MAY 12 1967

vertical 6" pieces. See sketch.

"The Engineers" are using vertical T-sections (3 x 2½") as whalers. A frame of these sections is used at the juncture point of two panels and individual vertical T's at intermittent points. See sketch.

FRAME
INTERMITTENT

Each contractor was originally using the wedge type panel juncture detail. Now the panel juncture detail is a crude piece of rubber stripping with horizontal bolts creating a pressure between panels, i. e.,

BEFORE
NOW
WEDGE
METAL CLAMP
WEDGE
BOLT

"The Engineers" have been using polyurthane throughout the entire job. The first panels were coated with a chemically inferior polyeuthane, (so I was informed) two or three uses produced flaking and stains. A new polyeuthane is just now being used, gray color, (name, etc., to follow), and the first results are better. The vertical V-joints in "The Engineer's" panel is the tongue and groove detail, but is producing poor results as the vertical panels shrink. They are then forced to use string or putty resulting in a poor fin detail because of bleeding.

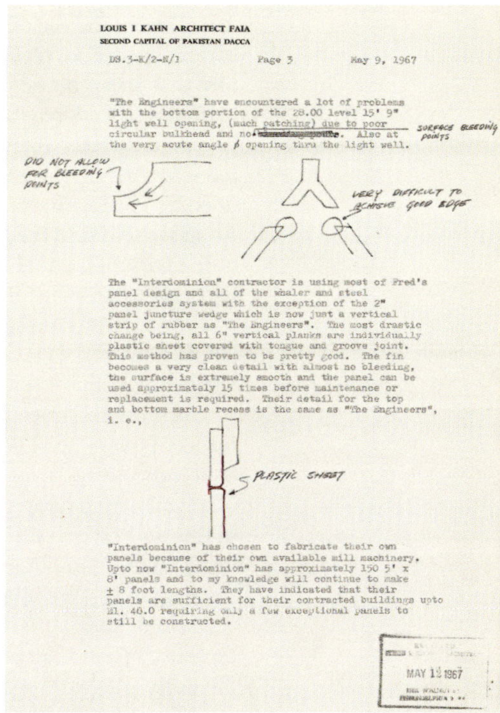

LOUIS I KAHN ARCHITECT FAIA
SECOND CAPITAL OF PAKISTAN DACCA

DS.3-K/2-N/1 Page 3 May 9, 1967

"The Engineers" have encountered a lot of problems with the bottom portion of the 28.00 level 15' 9" light well opening, (much patching) due to poor circular bulkhead and no [struck out]. Also at the very acute angle ø opening thru the light well.

SURFACE BLEEDING POINTS
DID NOT ALLOW FOR BLEEDING POINTS
VERY DIFFICULT TO ACHIEVE GOOD EDGE

The "Interdominion" contractor is using most of Fred's panel design and all of the whaler and steel accessories system with the exception of the 2" panel juncture wedge which is now just a vertical strip of rubber as "The Engineers". The most drastic change being, all 6" vertical planks are individually plastic sheet covered with tongue and groove joint. This method has proven to be pretty good. The fin becomes a very clean detail with almost no bleeding, the surface is extremely smooth and the panel can be used approximately 15 times before maintenance or replacement is required. Their detail for the top and bottom marble recess is the same as "The Engineers", i. e.,

PLASTIC SHEET

"Interdominion" has chosen to fabricate their own panels because of their own available mill machinery. Upto now "Interdominion" has approximately 150 5' x 8' panels and to my knowledge will continue to make ± 8 foot lengths. They have indicated that their panels are sufficient for their contracted buildings upto El. 46.0 requiring only a few exceptional panels to still be constructed.

MAY 12 1967

[234] *Ibid.* On the possible meanings connected with the theories of Gottfried Semper of the strips of marble see, William Curtis, Jr., *Cosmos and State, The National Assembly in Dhaka*, in, "A.V. Monografias de Arquitectura," monographic issue, *Louis I. Kahn with contributions by William Curtis, Martin Filler, Rafael Moneo, Juan Navarro, Vincent Scully*, 1993, No. 44, (pp. 101-103), p. 102.

[235] See the photograph, LIK, 030.IV.A.650.2.63.

[236] Louis I. Kahn, *Remarks*, 1965, now in, Latour, *Louis I. Kahn: writings, lectures, interviews* cit., (pp. 191-207), p. 200.

[237] Gustav Langford, letter to Henry N. Wilcots, 27 April 1967, LIK, PAC – Correspondence – RoyGus January 1967 thru May 1967.

[238] See the photograph, 1 May 1967, LIK, 030.IV.A.650.4.32.

[239] See the photograph, 1 May 1967, LIK, 030.IV.A.650.4.54.

than average surface."[240] Instead, criticism is reserved for the solution documented by the photograph of the panel with wider joints between the boards, partially filled with putty – "typical contraction of 6" plank. Putty solution is bad."[241] The imprints left on the concrete by this panel, used by The Engineers, are shown in the photograph of the V joints with chipped edges – "worst" is the remark.[242] The double adjacent V lines of the joint between the panels, as in the Salk Institute, have marks in one picture – "panel joint 50% are like this"[243] – while another photograph of the same detail documents the improper alignment of the V joints – "this is joint between panels."[244] The note on the picture of the edges of two walls ruined by bleeding reassures about the percentage of such defects on the worksite: "worst condition – only this one."[245] Another photograph of the edge of a particularly imperfect wall documents what is considered a special, "unusually bad" case.[246] A wall from the series of those photographed to document the surface finish is indicated as "close to average."[247] The wall documented by another picture that accentuates, thanks to the light, the perfect execution of the V imprints, is considered "another consistent result."[248]

Several photographs have been taken to show Kahn the various structures applied on the back of the panels. One picture shows the system developed by The Engineers to hold two consecutive panels together – "panel joint" – a system different from that of the experimental panels.[249] On the photograph of the back of a panel used by the Inter-Dominion Construction Company, Langford makes the following note: "techniques have degenerated to this."[250]

The photograph of two wall sectors with the circular opening and the laborers at work is taken to show the effects on the surface before and after hand polishing.[251]

An entire series of pictures forms a sort of catalogue of the various forms of patching. The mark on the lower sector of one wall is not considered indicative – "average wall better than this (slightly)."[252] A similar shot of a wall with a larger mark is deemed "*much* worse than typical."[253] Other marks and defects documented by another photograph get a similar comment: "worst condition."[254] "Very typical" is written on the shot of a wall with several marks,[255] while "not typical – worst" indicates another wall with accentuated marks and flaws[256]. "A typical better wall – or 2 leaks per wall" is the note on another picture.[257] Other flaws come from inept vibration of the pour – "over vibration still a problem," one note reads.[258]

As indicated in Langford's letter, certain pictures show the various defects of alignment of the horizontal lines produced in the assembly of the

[240] See the photograph, 1 May 1967, LIK, 030.IV.A.650.4.70.
[241] See the photograph, 1 May 1967, LIK, 030.IV.A.650.4.40.
[242] See the photograph, 1 May 1967, LIK, 030.IV.A.650.4.39.
[243] See the photograph, 1 May 1967, LIK, 030.IV.A.650.4.62.
[244] See the photograph, 1 May 1967, LIK, 030.IV.A.650.4.82.
[245] See the photograph, 1 May 1967, LIK, 030.IV.A.650.4.58.
[246] See the photograph, 1 May 1967, LIK, 030.IV.A.650.4.76. The quality of the concrete, according to the observations of Gustav Langford, is different - "average" is noted on a photograph (*ibid.*).
[247] See the photograph, 1 May 1967, LIK, 030.IV.A.650.4.89.
[248] See the photograph, 1 May 1967, LIK, 030.IV.A.650.4.69.
[249] See the photograph, 1 May 1967, LIK, 030.IV.A.650.4.84.
[250] See the photograph, 1 May 1967, LIK, 030.IV.A.650.4.20.
[251] See the photograph, 1 May 1967, LIK, 030.IV.A.650.4.56.
[252] See the photograph, 1 May 1967, LIK, 030.IV.A.650.4.63.
[253] See the photograph, 1 May 1967, LIK, 030.IV.A.650.4.72.
[254] See the photograph, 1 May 1967, LIK, 030.IV.A.650.4.78.
[255] See the photograph, 1 May 1967, LIK, 030.IV.A.650.4.66.
[256] See the photograph, 1 May 1967, LIK, 030.IV.A.650.4.68.
[257] See the photograph, 1 May 1967, LIK, 030.IV.A.650.4.80.
[258] See the photograph, 1 May 1967, LIK, 030.IV.A.650.4.76.

254-258. National Assembly Building, Dhaka, 1962-83.

254. Photograph of the formwork panels built by the Inter-Dominion Construction Company (LIK, 030.IV.A.650.4.83).

255. Photograph of the formwork panels by the Inter-Dominion Construction Company, with the note: "bulkhead w[ith] oil cloth covering" (LIK, 030.IV.A.650.4.85).

256. Photograph of the concrete wall, with the note: "better than average surface" (LIK, 030.IV.A.650.4.73).

257. Photograph of the concrete wall, with the note: "worst" (LIK, 030.IV.A.650.4.39).

258. Photograph of the formwork panel by The Engineers, with the note: "typical contraction of 6'' plank – putty solution is bad" (LIK, 030.IV.A.650.4.40).

259-263. National Assembly Building, Dhaka, 1962-83.

259. Photograph of the concrete wall, with the note: "panel joint, 50% are like this" (LIK, 030.IV.A.650.4.62).

260. Photograph of the concrete wall, with the note: "this is joint between panels" (LIK, 030.IV.A.650.4.82).

261. Photograph of the concrete wall, with the note: "unusually bad" (LIK, 030.IV.A.650.4.75).

262. Photograph of the concrete wall, with the note: "average" (LIK, 030.IV.A.650.4.74).

263. Photograph of the concrete wall, with the note: "another consistent result" (LIK, 030.IV.A.650.4.69).

panels between contiguous wall sectors – "top & bottom alignment consistently poor," says the note on the print.[259]

Following the photographic documentation of the defects produced up until April, Philadelphia Kahn's office sends a request to Langford to clarify the technical details of the formwork used by the two contractors, the wood used for the panels, the type of coating, the additive used in the compound, the usage modes of the vibrators and the type of scaffolding.[260] It is clear that Kahn's staff in Philadelphia want to take stock of the formwork on the Dhaka worksite, with possible corrective measures in mind.

The answers are contained in the letter with drawings Gustav Langford sends to his brother and to Wisdom on 9 May, and they are accompanied by another series of photographs to document the details discussed in the letter.[261] From the information provided on the methods of the contractors to make the formwork and the concrete, it appears that: 95% of the wood for "all form panels" is champa; the contractors can make use of the sawmill machinery of the Public Works Department; Pozzolith is no longer used because the engineers of the Public Works Department believe it causes

[259] See the photograph, 1 May 1967, LIK, 030.IV.A.650.4.65. Other photographs, also dated 1 May 1967, document: the large vertical groove in the manner of Frederick Langford and the wall with the quarter circle of the large opening - in this case indicated as "typical" - (LIK, 030.IV.A.650.4.37); the reinforcement of a wall with a circular opening (LIK, 030.IV.A.650.4.29); a similar wall, but already poured (LIK, 030.IV.A.650.4.33); the case of the construction of a staircase (LIK, 030.IV.A.650.4.64); or the lower sector of the circular wall of the WW (LIK, 030.IV.A.650.4.32). Gustav Langford also takes some shots from the top of the walls of the National Assembly Building under construction (LIK, 030.IV.A.650.4.42; LIK, 030.IV.A.650.4.46; LIK, 030.IV.A.650.4.53), and sends many photographs of constructions in brick.

[260] The request is sent by "cable request, PT 370" as written in, Gustav Langford, letter to David P. Wisdom and Frederick Langford, 9 May 1967, LIK, PAC – Correspondence – RoyGus January 1967 thru May 1967.

[261] Gustav Langford, letter to David P. Wisdom and Frederick Langford, 9 May 1967, LIK, PAC – Correspondence – RoyGus January 1967 thru May 1967.

"air bubbles" and "trapped water"; for the scaffolding, neither contractor has as yet supplied drawings or models, and while the Inter-Dominion Construction Company believes it is possible to continue the construction of the upper walls with bamboo scaffolding, The Engineers underlines the need to deploy a "sophisticated scaffolding system"; the 23 vibrators of the Public Works Department have all proven to be unreliable; The Engineers now has 4 electric vibrators and one gas-powered unit, while the Inter-Dominion Construction Company uses one electric vibrator and two running on gas, all considered reliable.

Langford goes into the technical details of the changes to the experimental panels made by each contractor according to their own necessities, in order to simplify the operations.

Of about 300 panels possessed by The Engineers, 50 have a length of just 4 feet (half that of the experimental panel), introduced, as Langford explains, to make the panels easier to handle. The structure on the back has thus been modified to hold the panel segments together. Though the company has not altered the imprint pattern generated by the experimental panel, the configuration of the pieces and their assembly are no longer those of the original, so much so that at times the V imprints in the concrete are the result of simple sculpted grooves, no longer corresponding to the joint between boards and posts. It is clear that the contractor has interpreted the criticisms expressed regarding the panel formulated by the Inter-Dominion Construction Company lacking the horizontal V imprints, and has reduced the number of connections of the pieces and simplified the cutting of the horizontal wood parts without altering the final effect of the imprints of the experimental panels. But at this point there is a separation between the constructive process of the panel and the pattern of imprints that undermines the foundation of Kahn's theoretical principles for architectural concrete.

On the contemporary worksite of exposed concrete, with the panels of The Engineers we see a reproduction, for the same practical reasons, of what happened in ancient times when the joints of regular blocks of the *opus isodomum* became lines sculpted on masonry of irregular blocks (though the real joints, in any case, remained perceptible). The substantial difference between the two walls lies in the process of fabrication through imprints that makes the separation invisible in the exposed concrete. Thus the panel of The Engineers, at this point, tends to become a *sculpture moulée* that depicts a pattern of lines reflecting the original assembly of the panel pieces.

264-267. National Assembly Building, Dhaka, 1962-83.

264. Photograph of the formwork system by The Engineers, with the note: "panel joint" (LIK, 030.IV.A.650.4.84).

265-266. Site construction photographs (LIK, 030.IV.A.650.4.29; 56).

267. Photograph of the concrete wall, with the note: "no typical – worst" (LIK, 030.IV.A.650.4.29.68).

268-271. National Assembly Building, Dhaka, 1962-83.

268. Photograph of the concrete wall, with the note: "worst condition" (LIK, 030.IV.A.650.4.78).

269-270. Photographs of the walls construction, with a notes: "over vibration still a problem" and "top & bottom alignment consistently poor" (LIK, 030.IV.A.650.4.76; 65).

271. Construction site (LIK, 030.IV.A.650.4.32).

There is little doubt that The Engineers are approaching the goal of the perfect fake that can be perceived only by those who know the panel or can guess, from the perfect execution of the edge of the V joint, that the joint is fake because it no longer has the role of permitting the bleeding of the compound, which would usually produce a jagged edge.

Langford does not explain the changes made by The Engineers in the critical terms with which he had warned Kahn and his staff about the modifications introduced by the Inter-Dominion Construction Company with the first panel with boards coated in plastic sheeting. It is probable that he leans towards accepting the result of the pattern of lines as an accurate interpretation of the design of the experimental panel.

"This change in length," he writes, "was decided by them after encountering too many problems in manipulation. [...]. They still need a few exceptional panels to be constructed. The Engineers are using most of the same panel design as Fred's, with the exception of the horizontal marble recess detail, i.e., the top and bottom piece is not tongue and grooved to receive the vertical pieces, it is only screwed on to the vertical 6" pieces."[262]

Also with the aim of simplifying the construction processes, The Engineers and the Inter-Dominion Construction Company change the joint between the panels, eliminating the special piece used by Langford and based on the system of the Salk Institute – the "filler strip." Now the joint happens by simply abutting the panels, sealed with strips of rubber nailed to their edges. To tighten the seal between two panels, bolts with nuts are applied on the back of the structure. The result between the two panels is a single V imprint, but perceptibly wider than the others, due to the thickness of the two rubber strips, in spite of the tightening of the bolts. In this case the clear visibility of a fundamental phase of the construction process of the formwork vanishes (the assembly of the panels next to each other), and Kahn's system is again altered in its theoretical underpinnings.

Langford reports on another particular way of assembling the panel parts, which shortly thereafter causes other substantial modifications made by The Engineers to Kahn's system. The assembly of the boards with the tongue and groove method does not produce perfect joints, due to shrinkage of the wood, generating defects in the pour ("is producing poor results as the vertical panels shrink"). The Engineers attempt to correct the problem in an empirical way: "They are then forced to use string or putty resulting in a poor fin detail because of bleeding."[263] Other difficulties in the placing of the concrete encounter by The Engineers have to do with the circular openings and the acute angles of the walls, "to pour circular bulkheads" and due to the absence of "surface bleeding points."[264]

When he describes the modifications introduced by the Inter-Dominion Construction Company to his colleagues in Philadelphia, Langford again praises the results in the quality of the concrete surfaces thanks to the coating with sheets of plastic. By this time Inter-Dominion is using a different panel from the one proposed in November 1966. The new panel has a

[262] *Ibid.*
[263] *Ibid.*
[264] Gustav Langford also explains that The Engineers has initially used a polyurethane of poor quality that has produced "flaking and stains" on the panels. Only recently has the firm used a new type of polyurethane, with better results; it has also changed the structure on the back of the panel: "The Engineers are using vertical T-sections (3 × 2 1/2") as walers. A frame of these sections is used at the juncture point of two panels and individual vertical T's at intermittent points" (*ibid.*).

272-276. National Assembly Building, Dhaka, 1962-83.

272. Photograph of the formwork by The Engineers, with "two types of panel design" (LIK, 030.IV.A.650.2.69).

273. Photograph of the formwork panel, built by The Engineers, with the note: "over" (LIK, 030.IV.A.650.2.67).

274. Photograph of the formwork panel, built by The Engineers, with the note: "one panel; seasoned wood?" (LIK, 030.IV.A.650.2.65).

275. Photograph of the formwork, built by the Inter-Dominion Construction Company, with the note: "strip of rubber. How is this for precision?" (LIK, 030.IV.A.650.2.71).

276. Photograph of the formwork, built by the Inter-Dominion Construction Company, with the remark: "note guide wires for anchoring" (LIK, 030.IV.A.650.2.43).

horizontal post broken down into two parts, to have the horizontal V imprint parallel to the recess for the marble insert, as with the experimental panel. "This method has proven to be pretty good. The fin becomes a very clean detail with almost no bleeding, the surface is extremely smooth and the panel can be used approximately 15 times before maintenance or replacement is required."[265]

"Some photos to follow tomorrow," is the conclusion of Langford's letter.[266] On the photographs, he makes pithy comments and adds the initials of the contractor concerned (E for The Engineers and ID for Inter-Dominion Construction Company). The pictures are taken to show the technical details of the panels. One of the crucial questions has to do with the V joint, as to whether it is a joint between boards or a sculpted indentation. The most telling photograph about the modifications of the V joint is the one in which Langford shows a comparison of the profiles of two panels used by The Engineers: that with the pieces still cut and assembled in the manner of the experimental panel, and that in which the separation has happened between the form of the original panel and the practical systems for the making of the same imprints – in this case, the horizontal V groove sculpted on the board is clearly visible.[267]

To avoid assembly of the boards according to the system of Frederick Langford and to eliminate the defects that have arisen during the pour with the tongue and groove connections, the workers of The Engineers construct the panels with boards larger than those initially envisioned, large boards on which they sculpt the vertical V-shaped lines.[268] Gustav Langford simply notes the cracking produced in certain boards: "seasoned wood?,"[269] without discussing the meaning of the fake joints. But it is clear that this way of making the V joints marks the extreme point of the transformation of Frederick Langford's panel from something that expresses the constructive logic sought by Kahn to a *sculpture moulée*.

Just as he photographed the panels of The Engineers in their two versions, so Gustav Langford also documents the type of panel used by the Inter-Dominion Construction Company, with the plastic coating.[270] For the first time since he saw and praised this coating in November 1966, Langford

[265] Inter-Dominion Construction Company constructs its own panels, as it has the suitable machinery; until now about 150 panels measuring 5 x 8 feet have been made. According to Gustav Langford's information, the firm will continue to produce panels with the same measurements. With these panels the works can reach the level of 48.00, with the need to make only "a few exceptional panels" (*ibid.*).

[266] *Ibid.*

[267] See the photograph with the note "2 types of panel design," LIK, 030.IV.A.650.2.69. The detail of the sculpted horizontal groove is also visible in the photograph that documents the joint between the excessively wide boards (LIK, 030.IV.A.650.2.67).

[268] See the photograph, LIK, 030.IV.A.650.2.65.

[269] *Ibid.* In the photographs of 1 May 1967 a number of notes are provided on the results of the pouring. See for example: the photograph, LIK, 030.IV.A.650.4.41, with the note: "techniques have degenerated to this"; the photograph, LIK, 030.IV.A.650.4.63, with the note: "average wall better than this (slightly)"; the photograph, LIK, 030.IV.A.650.4.62, that shows the pour with two poorly made close imprints (on the photograph there is the note: "panel joint 50% are like this"); the photograph, LIK, 030.IV.A.650.4.73, with the note: "better than average surface"; the photographs, LIK, 030.IV.A.650.4.74, 75, with the notes, "average" and "unusually bad"; the photograph, LIK, 030.IV.A.650.4.76, with the note: "over vibration still a problem"; the photograph, LIK, 030.IV.A.650.4.82, that documents the skewed imprints: "this is joint between panels"; the photograph, LIK, 030.IV.A.650.4.85, with the note: "bulkhead with oil cloth covering"; and the photographs, LIK, 030.IV.A.650.4.68, 69, 70, 80.

[270] See the photograph, LIK, 030.IV.A.650.2.60.

PANEL DESIGN

OVER

SEASONED WOOD?
ONE PIECE

STRIP OF RUBBER
HOW IS THIS FOR PRECISION?

NOTE GUIDE WIRES FOR ANCHORING

points out that when the sheets are ripped patches are applied that leave visible signs on the concrete.[271]

The photographs also serve Gustav Langford to obtain judgments on certain details in the making of the panels. One of the fundamental changes is that of the joint between panels using rubber strips, documented by a photograph of the back of the panel, on which Langford writes: "How is this for precision?," indicating the detail with an arrow: "strip of rubber 1/14"."[272]

The photographs of April, those of May and the information supplied by Langford outline a situation in the fabrication of the concrete of the National Assembly Building that must have seemed alarming to Kahn. A few days after having received the last photographs, Kahn decides to send a member of his staff to the worksite.

Recommendations of Wilcots and guidelines of Palmbaum for the concrete

At the start of June 1967, Wilcots reaches Dhaka, for a stay of about two weeks, to check on the progress of the exposed concrete work of the National Assembly Building and to set up the previously almost non-existent coordination among the various entities responsible for the construction: Kahn's office in Dhaka, the two contractors and the engineers of the Public Works Department. His impression regarding the progress of the work leaves no doubt about the supervision that does not effectively have any control over the results: "There prevails a quietness about the site like an uncontrolled growth ... The hum of excitement and expectation is missing."[273]

Wilcots proposes organizing coordination of the information transmitted from Kahn's Dhaka office and the execution of the works by the contractors, under the control of supervisors of the Public Works Department who have been suitably trained and have enough authority to allow for "rejecting any and all improper work or deviation from the intention of the drawings."[274] He also wants Kahn's office and Gholam Mohammad to collaborate with the contractors on the development of the details of the formwork, and he expects the contractors to share their respective experiences in the construction of the panels, in order to "eliminate such problems as bleeding, alignment, erection and other causes arising from poor formwork." His goal is to create a unified formwork system, starting with a uniform treatment of the panel surfaces, for which he recommends the solution applied by the Inter-Dominion Construction Company. "[...] we have found that portions

277-278. National Assembly Building, Dhaka, 1962-83.

277. Photograph of the corner in the formwork system by The Engineers (LIK, 030.IV.A.650.2.44).

278. Photograph of the formwork system by Inter-Dominion Construction Company, with the covering in "plastic sheet" and the symbol (+) to indicate a "patch" (LIK, 030.IV.A.650.2.66).

of the work formed by plastic sheet covered form panels have proven to produce a more favorable finish to the concrete, in color, texture and detail."[275]

The portrait of the owners of the two companies provided by Wilcots helps to explain certain differences in the work procedures of Inter-Dominion Construction Company and The Engineers.

"Basically the problem with the N[ational] A[ssembly] concreting is that there are two contractors (they refuse to take tea together). One is a hustler, the other a refined engineer. The first has developed the plastic sheet covered form panels with satisfactory results and the latter realizes his work is not acceptable, acknowledges the results his colleague is obtaining, but refuses to strive for better concreting. He is completely convinced that plastering of concrete shall be the finish. Now, there is the feeling he is losing his shirt and before long will disappear from the project."[276]

The willingness of the Inter-Dominion Construction Company to comply with and interpret the experiments done by Kahn is confirmed not only by the impressions of Wilcots, but also by the position taken by the firm's owner when the engineers of the Public Works Department instruct him, though not in writing, to overlook Kahn's indications regarding certain floor slabs – Kahn wants continuous floor slabs, as opposed to the Pakistani practice of beam structures. "[...] it will not be wise to take such a risk in such an important National work against the direction of Architect. This work should not be treated in the same way as other works," is the owner's opinion.[277]

For the construction of the floor slabs it is precisely Inter-Dominion Construction Company that receives the commission to conduct trials, immediately after the departure of Wilcots, with three types of formwork to evaluate the quality of the design on the surface of the concrete: "exposed wood – 6" plank"; "plastic sheet" and "metal sheet." The results of the metal shutters meet with Langford's approval because the main parameter of judgment is the degree of similarity to the wall surfaces: "The metal sheet has produced the smoothest finish and the same color as the walls. The joints between the sheets are scotch taped and almost invisible."[278] A few days later, Kahn too approves the "sample slab finish" of the metal formwork.[279] It is worth noting that the metal of the formwork is salvaged material, in keeping with an attitude typical of any type of work in Pakistan. "[...] the 'metal sheet' being used is beer can tin,

[271] This is documented by a photograph with the handwritten note: "plastic sheet + patch" (LIK, 030.IV.A.650.2.65).

[272] The photograph shows a panel by the Inter-Dominion Construction Company (LIK, 030.IV.A.650.2.71). In the structure on the back, for the "panel juncture," due to the length of the bolts with nuts fragments of the concrete spreaders equipped with holes are used (see the photograph, 1 May 1967, LIK, 030.IV.A.650.2.69). To make perfect corners no longer marked by the defects caused by bleeding, The Engineers develops a special piece, photographed and defined by Gustav Langford as a "90° corner panel" (see the photographs, LIK, 030.IV.A.650.2.61 and 44). For the other pictures taken during this shoot see: LIK, 030.IV.A.650.2.43, 45, 58, 62, 70, 71, 72.

[273] Henry N. Wilcots, letter to David P. Wisdom, 8 June 1967, LIK, PAC – Correspondence – to/from Gus June 1967 thru December 1967.

[274] Henry N. Wilcots, letter to M. G. Siddiqui, 8 June 1967, LIK, PAC – Correspondence – to/from Gus June 1967 thru December 1967.

[275] *Ibid.*

[276] Henry N. Wilcots, letter to David P. Wisdom, 8 June 1967, LIK, PAC – Correspondence – to/from Gus June 1967 thru December 1967.

[277] Director of Inter Dominion Traders, Ltd., letter to Executive Engineer, 10 June 1967, LIK, PAC – Correspondence – to/from Gus June 1967 thru December 1967.

[278] Gustav Langford, letter to Henry N. Wilcots and Dave P. Wisdom, 27 June 1967, LIK, PAC – Correspondence – to/from Gus June 1967 thru December 1967.

[279] Donald Barbaree, letter to Henry N. Wilcots, 28 July 1967, LIK, PAC – Correspondence – to/from Gus June 1967 thru December 1967.

and I have my doubts as to its durability and resistance to denting," Kahn's staffers write.[280] So the Inter-Dominion Construction Company has managed to locally procure two materials, plastic sheets and recycled sheet metal, that when skillfully used have made it possible to reach the goal of smoothness desired by Kahn through other methods, confirming the potential of intelligent Pakistani *bricolage*. But this approach, inevitably filled with unexpected risks, is not the one Kahn intends to apply in the subsequent worksite phases.

Responsibility for the reorganization of the work in keeping with the indications of Wilcots is assigned to Palmbaum, who returns to Dhaka on 13 July 1967, accompanied by Kahn. The goal of their visit is to draw up a detailed report on the construction processes deployed by the two contractors and their results, and to indicate changes to improve the quality of the exposed concrete.

The document "National Assembly Building Construction Report," written by Palmbaum, contains the guidelines for resumption of work with methods, techniques and materials chosen to introduce a kind of mechanization of all the phases of construction. The fact that Kahn is there with Palmbaum indicates that the contents of the report were also discussed directly with him. Given the underdevelopment of Pakistani worksites in the area of machinery and equipment, in certain cases Palmbaum provides specialized literature or advertising brochures on different industrial products on the market, like "climbing cranes," "lifting hoists" on wheels to raise formwork, pumps for raising and pouring of concrete, special metal devices for formwork, from bolts to spreaders to walers.

By now it is clear that Kahn and his team are trying to regain control over the worksite, and to introduce all the products of American industrial companies, in order to obtain perfect architectural concrete, without having to rely on experimentation with local materials, or various types of do-it-yourself improvisation. Thus the report marks a turning point with respect to the research begun by Frederick Langford with the original panel, of that system based on integration of local materials and practices that in some ways characterizes the approach taken by the Inter-Dominion Construction Company. Not by chance, in the official correspondence of August 1967 we see a return to the description of the Dhaka worksite as a foundation experience for a new way of using concrete in Dhaka. But precisely due to the widespread mechanization of the various worksite phases, many of the proposals contained in the report are destined to remain on paper.

The overview of the worksite outlined by Palmbaum makes it possible to specify some methods of formwork deployment practiced during 1967. The Engineers and Inter-Dominion Construction Company have a number of panels that does not permit their assembly in continuous rows, to make long wall sectors, then raising them to continue the pour on the next level up, as urged by Palmbaum. So the panels have to be moved and reassembled horizontally, with a "loss of position and sequence" and a degree of wear on the panels themselves. "Build the forms in wider units and consequent-

[280] *Ibid.*
[281] Harry M. Palmbaum, "National Assembly Building Construction Report," n.d., [July 1967], LIK, PAC Various Notes Dhaka Jan 66.
[282] *Ibid.*
[283] "A record of remarks and suggestions made by Mr. Harry Palmbaum, structural engineering consultant to the architect, during his field visit of July 1967," LIK, PAKCAP – Schwartz, Benard & Assoc. Correspondence.
[284] See "Comments by Mr. Harry Palmbaum accompany Williams Form Catalog," 10 August 1967, LIK, PAC Various Notes Dhaka Jan 66. The company was indicated

ly have fewer units and less fitting and jointing. This will save many hours of set-up time," is Palmbaum's advice.

The list of defects in the report includes: irregular profiles of the vertical recesses; vertical construction joints in an "unacceptable place without required marble insert recess"; lack of curvature of the floor slab formwork and beams prior to the pour ("1:40 camber"), because "forms shored with bamboo settle unpredictably"; end of openings and walls out of plumb due to imperfect anchoring of formwork of the bulkheads, which shift under pour pressure; misalignment of horizontal imprints; and defective execution of the wooden boxes inserted in the formwork to create the gaps to contain the beams of the floor slabs.[281]

During the inspection by Kahn and Palmbaum one of the questions discussed is probably the construction of the formwork panel in the different solutions proposed by the two contractors, given the substantial modifications to Kahn's logic introduced in the assembly of the pieces. It is significant that the report contains an opinion completely different from those of Gustav Langford and Wilcots regarding the efficacy of the plastic sheeting used by the Inter-Dominion Construction Company, if we consider the fact that precisely the application of those sheets has led to less certainty in the connections of the experimental panels. Palmbaum points to the need to return to the use of polyurethane to spread on a panel built like that of Frederick Langford, due to the difficulties involved in mending possible tears in the plastic sheets (such as those documented by Gustav Langford). "Forms wrapped with plastic film," Palmbaum writes, "have produced very bad surface markings where the plastic has been torn or otherwise damaged. Rejection of damaged plastic film would be more costly and take more time than polyurethane coating."[282] In the original version of the report, among the defects caused by the plastic coating we also find the size of the joints – "also very large V-grooves."[283] This version also contains explicit support for the experimental panel as opposed to all the subsequent variations: "The original architectural panels made in the Spring of 1966 were polyurethaned and had an acceptable V-groove of 1/8" maximum and this therefore sets an acceptable standard to be followed and improved upon and is *not* to be disregarded."

Starting with the report, Palmbaum then supplies other recommendations on the formwork accessories produced by the Williams Form Engineering Corporation, including: a device known as the "pigtail" for the correct positioning of formwork in correspondence to the slabs; the steel wedge that if used with the she-bolts, in place of the "threaded rods and nuts," "gives exact wall dimensions"; special horizontal walers and steel scaffolding with guides for precise panel alignment.[284] Also for the purpose of mechanizing the worksite, Palmbaum pressures Siddiqui for the purchase of a reconditioned crane from Heede International, Inc., with a radius of 125 feet, for which he studies the positioning and use for the worksite of the National Assembly Building in keeping with the choice made by the two contractors of using pours of "three cubic yards" to perform with "smaller booms."[285]

in the original version of Palmbaum's report.
[285] Harry M. Palmbaum, letter to M. G. Siddiqui, 15 September 1967, LIK, PAKCAP – Schwartz, Benard & Assoc. Correspondence.

Notes on concrete

On the trip back to Philadelphia, Kahn is accompanied not only by Palmbaum, but also by Gustav Langford, who is replaced at the worksite by Donald Barbaree. Langford stays in Philadelphia for two weeks to concentrate, with other staffers of Kahn's office, on all the as yet unresolved questions regarding the formwork and concrete of the National Assembly Building.

Precisely to have time to study the measures to be taken, on 4 August Kahn asks the engineers of the Public Works Department to suspend all pouring of the walls of the National Assembly Building, advising them that he wants to examine all the options for improving the concrete. "We are all anxious," is the reply of Abdul Wazid, Superintending Engineer of the Public Works Department, "to produce a high quality concrete the like of which has not yet been produced in this country, subject to prevailing conditions and physical limitations of those concerned with the work."[286]

Actually, the engineers of the Public Works Department worried that the announced "modified methods of forms work" that should "ensure economical, correct and expeditious pouring of wall concrete," indicated by Kahn, will cause extra costs and delays that would not be accepted by the Pakistani government authorities.[287] Kahn's office sends orders to Barbaree to allow work to proceed only on the floor slabs, in the belief that the samples seen by Kahn on the worksite for the pouring of the floor slabs will guarantee acceptable results.[288] But even the pouring of the floor slabs turns out to be defective, due to the use on the part of the Inter-Dominion Construction Company of "thin gauge metal sheet." "The finished ceiling surface is uneven and undulates due to the inadequate thickness of the 'sheet metal,'" Barbaree reports, proposing use of thicker metal and recommending that the contractor be ordered to remake the already prepared panels.[289]

Kahn's office prepares several documents summarizing the principles outlined following the inspection by Kahn and Palmbaum for the fabrication of the formwork and the concrete, written to communicate new directives to be given to the contractors for the continuation of the job. The "Notes on concrete" contain the recommendations with which to recommence the making of Kahn's concrete (the document probably dates back to September 1967).

"Notes on concrete" is a true manual organized point by point, written to precisely convey all the techniques to be applied in the various phases, the result of the many years of experience, at this point, of Kahn and his staff in the making of exposed concrete. All the necessary operations are listed to avoid all types of defects that may be produced by the formwork, the compound and the pour.

[286] Abdul Wazid, Superintending Engineer, letter to Louis I. Kahn, 22 August 1967, LIK, PAKCAP – Correspondence – PAK PWD 1966 – 1967. For an account of the construction processes of Kahn and Langford in Dhaka see *Construction and Kahn's Capital. Haroon Rashid interviews Abdul Wazid [...]*, in, "Mimar," March 1991, pp. 38-39.

[287] Abdul Wazid, letter to Louis I. Kahn, 22 August 1967, cit.

[288] Gustav Langford, letter to Donald Barbaree, 11 August 1967, LIK, PAC – Correspondence – to/from Gus June 1967 thru December 1967.

[289] Donald Barbaree, letter to Henry [N. Wilcots] and Gustav Langford, 17 August 1967, LIK, PAC – Correspondence – to/from Gus June 1967 thru December 1967. Among the things requested from Philadelphia we also find "concrete books."

[290] "Dump concrete into the face of concrete already in place. Deposit concrete close to its final position. Do not move it horizontally in forms. Layers: 18" to 24" deep. Segregation appears as excess water or laitance at the surface. Do not continue placing concrete when segregation is apparent. Correct before proceeding. Use a drop chute (elephant trunk). Do not place concrete on hot reinforcement. When concreting walls, start at the ends and work toward the center. [...]. Always use vibrators. Insert vibrator head deep enough into previous concrete layers (several inches). This layer should still be plastic. Insert vibrator vertically. Do not hold vibrator head against reinforcing. Pull vibrator heads *slowly* out of concrete. Leave vibrator in for about 15 seconds. At this time mortar starts to appear on the surface and large coarse

Following the flaws caused by shrinkage of the boards, it is recommended to keep panels wet at all times during the 12 hours prior to the pour, to keep the wood from shrinking, which leads to widening of the V joints. To reuse the panels the "working surfaces" must be carefully cleaned, treated with formwork oil and dried to the point where they are "only slightly greasy to the touch." To avoid marks on the concrete, it is recommended to "apply limewater or whitewash to new forms that have excessive tannin in the surface." The use of damaged or scrap wood is forbidden for the construction of the panels; any panel roughness is to be puttied; and care is urged to avoid marring of panels with vibrators or other tools.

Other recommendations have to do with the concrete mix, such as the addition of Dorex to the compound, as an "air-entraining agent," and in the case of bleeding the pouring of the upper sectors with less water. The sections on "Concrete methods," "Vibration" and "Cleaning walls" of "Notes on concrete" offered detailed explanations of all the techniques and procedures required to obtain the concrete finish desired by Kahn.[290]

The "Notes on concrete" are accompanied by the "Additional specifications [for] concrete" which contain a number of technical tenets regarding only the construction of the floor slabs, ranging from advise on the curvature of the formwork of the floor slabs and beams, to compensate for their flexing under the weight of the pour, to the shoring for the formwork and the reinforced concrete structure immediately after stripping, or "reshoring."[291] These "Additional specifications" become necessary precisely due to the defects during the construction of the first floor slabs, in August, as reported by Palmbaum, some of them related to the lack of curvature of the formwork.

On the worksite, starting in August 1967

The main operations of construction of the National Assembly Building in the last days of August 1967 are those of the floor slab at the level of 38 feet. The pouring of the slab does not happen without a series of flaws caused by the material of the formwork. The contractor uses thicker sheet metal than in the past, but it is still not rigid enough to prevent the warping of panels under the weight of the pour, leading to undulations in the ceiling. Therefore the contractor looks for even thicker sheet metal.[292] The consequences on the concrete of the ceiling caused by changes in the thickness of the sheet metal of the formwork are summed up by Barbaree using an expression of Kahn's: the various pours will make "the growth of the building" visible, even in the floor slab.[293] This expression, placed in quotation marks by Barbaree, refers to the meaning assigned by Kahn to the horizontal grooves with the marble

aggregate particles start to sink into the mass. Do not over vibrate. [...]. When the wall is stained, or not of uniform color, do the following: Mix 1 part Portland cement to 1-1/2 parts fine sand with sufficient water to produce a grout like thick paint. Use as much white cement as necessary to give the color desired. Wet the surface sufficiently to prevent absorption of the water from the grout. Apply the grout with brushes, completely filling all bubbles and holes. *Immediately* float the surface with a cork float, scouring the wall vigorously. While the surface is plastic, finish the surface with sponge rubber float removing all excess grout. Take care *not* to pull the grout from any holes. If this happens, wait a little while longer. Allow surface to dry throughly. Rub with dry burlap to completely remove any dried grout. No visible film of grout should remain after this rubbing. The entire cleaning operation for any area must be completed the day it is started. No grout is to be left on overnight. If any dark spots or streaks remain, do the following: Wipe lightly with a fine abrasive hone. Use *no* water. Be very careful not to rub deeply as this will change the texture of the concrete" (Donald Barbaree, letter to Henry [N. Wilcots] and Gustav Langford, 17 August 1967, LIK, PAC – Correspondence – to/from Gus June 1967 thru December 1967).

291 "Additional specifications concrete," LIK, PAKCAP – Schwartz, Benard & Assoc. Correspondence.

292 Donald Barbaree, letter to Henry N. Wilcots, 24 August 1967, LIK, PAC – Correspondence – to/from Gus June 1967 thru December 1967.

293 *Ibid.*

inserts, as visible signs of the pour phases. The fact that even all the attempts of the contractors to find materials suitable to make Kahn's architectural concrete are incorporated in the concept of the "growth of the building" does not mean that Kahn himself is willing to view those defects according to Barbaree's outlook, because at this point the technical attempts of The Engineers and the Inter-Dominion Construction Company are no longer producing fossils like those hidden in the underground spaces of the Salk Institute, but are producing traces visible everywhere of the conflict in progress on the worksite of the National Assembly Building between the will to achieve perfection in every single detail and the local practices based on empirical solutions deployed case by case with local materials and unskilled labor.

As soon as he returns to Dhaka in September, Gustav Langford finds other flaws in the construction of the concrete walls: the surface of the panels in contact with the pour has not been planed according to the indications of Kahn's office, thus producing V imprints that are larger than they should be; the thicknesses of the strips that produce the grooves are not correct, nor are the alignments between the horizontal imprints of the various wall segments at the level of 43 feet (to avoid repetition of this latter defect, the top of the wall is evened to place the panels at the same height).

Langford also observes the results of the pouring of the floor slab done under the supervision of Barbaree. The flaws on the surface are caused by the thin sheet metal, which bends under the weight of the concrete; but the use of thicker metal, in his view, has led to "very good results, joints taped and straight – surface even and consistent in alignment, texture and color."[294]

The question of the measures to be taken to achieve "better concrete results" is discussed by Langford with Siddiqui.

"He is so pessimistic," Langford reports, regarding Siddiqui, "he asked me if I really expected something substantial to have been done – I said, 'yes, we have faith in your wanting better concrete and we believed your promise that the form panels would be modified to help achieve more accurate and precise construction.' I have badgered him for days to allow me to experiment and remake 10 panels of each contractor. If successful, he will try and make both contractors change accordingly, i.e. same panel construction and all polyurethane. I'm working on Haque (plaster sheet contractor) first. Haque will change to polyurethane if experiment is successful."[295]

Langford's goal, probably discussed with the staff of Kahn's office during his time in Philadelphia, is to prepare new experimental panels that can eliminate the defects caused by the panels used by the two contractors. Langford wants to make the following changes to 10 panels: planing of the

[294] Gustav Langford, letter to David P. Wisdom and Henry N. Wilcots, 19 September 1967, LIK, PAC – Correspondence – to/from Gus June 1967 thru December 1967.

[295] *Ibid.*

[296] *Ibid.* In the correspondence of September between Philadelphia and Dhaka various booklets and manuals on concrete, formwork and scaffolding are sent (see "During the Month of September 1967," LIK, PAC Transmittals – from Dhaka office Nov. 1 '65 thru 1/19/68).

[297] The list of the formal and technical characteristics of the experimental walls includes, significantly, those defective details observed by Gustav Langford, Palmbaum and Kahn in the works done after Frederick Langford left the worksite: "top and bottom edge of marble recess is straight, properly aligned and smooth," "plumb corners (within 1/8" tolerance)," "constant overall color texture," "straight and plumb wall surface," "minimum bleeding at panel juncture," "plane of wall (within 1/8" tolerance)," "minimum of honeycombing - skillfully patched," "minimum bleeding at end wall conditions" ("During the Month of September 1967," cit.).

surface; use of the tongue and groove system to join boards; revision of the structure on the back of the panel; insertion of three screws, not two ("all screw holes to be wood plugged – no pudding (putty)"); adjustment of the depth of the recesses for the marble; and, finally, coating of the panel face with three layers of polyurethane. One new factor is the proposal to also paint the back of the panels with an "aluminium metallic paint" whose reflective properties would facilitate improvement of the concrete pour: "The paint is highly reflective, radiant energy will be reflected and not absorbed, therefore less moisture loss and shrinkage from the back."[296]

Defects and remedies in the fabrication of the concrete according to Gustav Langford

In the weeks following his return from Philadelphia, Gustav Langford takes stock of the formwork system for the concrete of the National Assembly Building, starting with the experimental walls. The document takes the form of a letter sent to Siddiqui on 9 October 1967, but it belongs to the genre of the reports on concrete prepared by Frederick Langford for the construction of the Salk Institute and for Dhaka.

The short summary of the history of the Dhaka worksite, starting with the arrival of Frederick Langford, sets the stage for the present changes introduced by The Engineers and the Inter-Dominion Construction Company, and for the redefining of the responsibilities of the engineers of the Public Works Department. Listing the results achieved with the experimental panels and the experimental walls of 1966 is the first step towards questioning the strategies used by the two contractors, and supported by the government engineers, aimed at making the fabrication of the panels quicker, at the cost of losing the perfection of the details, thus compromising the results of the concrete. It is clear that Gustav agrees with the critical indications outlined by Palmbaum in the "National Assembly Building Construction Report." Confirmation of the importance of the experimental walls as models is one of the purposes of Langford's document, because those sample walls demonstrate the qualities of the concrete towards which the two contractors must again strive, as already indicated by Palmbaum.[297]

"The concrete work executed by Fred Langford in 1966," we read in Gustav's letter, "was done on an experimental level with the intention of reaching a form panel design and erection system that would lead you to quality concrete work in the National Assembly Building. [...] the major intention of this work was not to set a maximum criterion for architectural concrete; on the contrary, it was to indicate a quality of concrete workmanship that was possible to achieve and from which better concrete could evolve. Another intention was to help establish a form panel construction that would give the panel a high reuse factor, a factor which must be considered if the National Assembly construction is to proceed at a reasonable pace. [...] the experimental panels were not intended as mandatory samples, but the modifications which were made by the contractors were not made to improve the concrete, but were made to facilitate construction of panels. Consequently two techniques have evolved, neither of which have produced accuracy of workmanship comparable to the 1966 experiment, not to mention the lack of uniform finish on the same building."

The central part of the document contains a detailed list of the defects found in the works in progress built by the two contractors. The greatest obstacle to obtaining the quality achieved by the experimental walls is the lack of technical details to permit standardization of the formwork system, along with irregular deployment of the panels, the difficulties in their systematic use caused by defective construction, and the lack of a design for the scaffolding.[298]

The list of shortcomings is followed by a series of recommendations to improve the formwork system based on the principle already indicated by Palmbaum of assembling panels along an entire wall sector to prevent their horizontal movement (the sequence of panels is simply raised, again thanks to an appropriate scaffolding system), to obtain a regular composition of the holes of the spreaders, the alignments between the recesses for the marble strips, more stable and well-sealed joints between the panels, standardization of the formwork accessories, improved positioning of the bulkheads and reduction of processes of maintenance and manipulation of the panels. The pour phases also have to be programmed and improved thanks to regular scaffolding.

"*Elimination of Indiscriminate Positioning* – vertical panel movement avoids many of the previously mentioned poor construction and time-consuming situations – such as additional bolt holes. *Same Vertical Bolt Layout* – this precludes the time-consuming practice of redrilling top and bottom marble recess member; a practice which is undeniably one of the main causes for the present panel misalignment. Another disadvantage of this redrilling practice is that the center dimension of bolt holes never coincides with the bolt holes of the previous pour. Bolts are then unduly bent making vertical panel alignment and tightness between form panel and previous pour practically impossible. Poor alignment and insufficient tightness due to this particular bolting practice is the major cause for bulkhead slippage and cannot be remedied within the present techniques of erection. A general lack of tightness causes excessive bleeding. The practice of drilling additional bolt holes weakens considerably the top and bottom marble recess members, thus decreasing panel reuse factor. *Consistent Panel Junction Relationship* – is maintained by vertical movement. Consistent panel relationships lead to the making of a more standardized detail at the panel juncture and thereby eliminating or controlling unsightly bleeding. A consistent relationship between panels means that they are exposed to the same stresses during repeated vertical

[298] "In the present systems of panel construction and erection, panels are being used in indiscriminate, unprogrammed positions," Langford explains. "This positioning requires that a new series of top and bottom bolt holes be made, a time consuming practice which weakens the form panel considerably and ultimately requires replacement of top and bottom marble recess members. Unfortunately these members are never replaced with the same dimensional accuracy as when originally assembled. Panels, when constantly used in unprogrammed positions, will never achieve a water-tight connection; consequently a maximum of bleeding occurs at most panel junctions. Panels are also put under different stresses and strains in every panel erection sequence, this tends to rack each panel. Accuracy and alignment of construction are impossibile under these practices and again, the contractors are faced with the time consuming jobs of lowering and transporting panels to the shops for tightening, realignment and checking of dimensions. Neither contractor has developed a system of parts and accessories which are standardized in size and shape, the lack of which slows construction by requiring new pieces and parts to be temporarily made for an immediate use, and then discarded uneconomically. Most of the horizontal stud members which help to make the frame on the back of the panels are of different sizes. This requires that wooden shims of all sizes be used in an attempt to create equal inward pressure against the panel. When equal inward pressure against the panel is not maintained poor results of the following nature occur: wall bulges, door, window and end of wall bulkhead slippage, vertical misalignment at panel juncture and bleeding of concrete down onto the lower walls. Neither contractor is using the eleven foot vertical walers which are bolted at the lower bolt position of the previous pour. The major function of these walers was to secure one side of the wall panels in their final, level, and plumb position. The opposite side of the wall panels then would be bolted through and tightened into their required plumb position. Presently guide wires are being used to secure panels, but, they do not sufficiently hold the panels in their required plumb position: thus, there is loss of vertical accurancy. Another disadvantage of the wire system is that it pulls the panels from the top and racks them. Splits and openings occur at the panel carpentry joints. The eleven foot waler which projects down over the previous 5' pour is structurally sufficient to secure the panels in their level, plumb and straight position, thus making guide wires unnecessary. Guide wires will be impossibile to use as the building goes up. The eleven foot waler takes advantage of the previously made bolt hole and by virtue of its stability lends itself to the making of a scaffolding

erection sequences. *Standardization of Parts and Accessories* – can be readily achieved by consistent panel erection techniques and common denominators in general construction conditions. Conversely standardization of parts helps to create positive and consistent erection techniques; time saving factors. Standardization of correctly dimensioned accessories helps to minimize human error during all phases of panel erection. *Vertical Panel Erection* allows the contractors to use the 11' walers resulting in a speedier erection system. *Panel Re-use Factor is Increased* – racking of panels is greatly minimized in solely vertical movement. Accuracy of construction is more easily achieved by virtue of dimensional accuracy maintained in the panel. Quality of form panel condition in all respects contributes to the panel re-use factor. *Minimum Form Panel Handling* – a timesaving factor which minimizes weakening of form panel joinery. General abuse is always the result of excess form panel handling. Vertical form panel movement decreases general abuse and wear. *More Accurate Vertical and Horizontal Construction* – Precision in the making of construction joints becomes more possible as the same panel performs the same vertical duty throughout. The horizontal marble recess must be made by an accurate integral detail in the panel construction which remains constant with each subsequent vertical pour. The vertical marble recess is also accurately made by a similar integral detail. The panel responsible for the vertical recess is always in the same position during its vertical movement. Presently the system is to tack a marble recess member on to an arbitrarily chosen panel which invariably results in misalignment and damage to the panel surface. *Racking and General Panel Deterioration are Minimized* – Panels moved in vertical positions only are not exposed to alternating stresses as they are presently in their unprogrammed positions. The panel construction remains tight and its life-span is greatly increased. *Constant Relationship Between Form Panel Bulkheads* – Form panels in vertical movement only are exposed to the same positioning sequence and therefore lend themselves to additional bolt use. These additional bolts will prevent bulkhead slippage and misalignment which at present is our constant problem in all door, window jamb, and wall conditions."[299]

system. Neither contractor has indicated a scheme of scaffolding, without which a reasonable pace of construction cannot be maintained. At present each contractor is attemping to reduce excess bleeding at panel junctures by securing panels together with 2 horizontal bolts. The bolt holes are most often misaligned which pulls the panels out of horizontal alignment and therefore does not reduce bleeding. Each contractor with his present panel design and erection technique is constantly producing an inaccurate type of construction. Improvement of concrete and construction pace are virtually impossibile. The present panel design and their methods of construction require that a large percentage of panels be always in the shop for *repairs such as: tightening, and replacement of members* and resurfacing. At present each contractor is constructing slab and beam shuttering for elev. 38.00. Most shuttering is being made with the intention of being used on each subsequent floor. The evidence of vertical inaccuracies to date will preclude any positive re-use factor which the contractors should be working towards. The floor and beam shuttering will require constant alterations and repairs to cope with the present type of inaccurate vertical construction. It is conceivable that a tolerance greater than ½" for vertical alignment will render a majority of floor shutterings to be altered for each subsequent pour. Both contractors have assembled to date approximately 20.00 square feet of wall panels. Both their techniques of panel construction and panel finish allow the panels a small re-use factor of about 6, after which, dismantling and refinishing of the panels is required" (*ibid.*).

[299] After the pouring of the walls at level 38 feet the results are not satisfactory: "The majority of the walls [...] are extremely inaccurate" in the vertical and horizontal recesses for the marble inserts, the openings, the corners, the cavities for the beams, the joints between the panels, in the "plane of wall," due to the "panel misalignment," and for an excess of gravel clusters. All these defects lead to difficulties in the installation of the casements or the marble inlays, and promote deterioration of the concrete surfaces. "The irregular plane of the wall," Langford writes regarding this latter aspect, "due to panel misalignment, is not only unsightly on the interior, but also the cause, since it is so jagged and deeply textured, that will encourage fungus growth. A smooth planed wall which prevents fungus growth is a very important design factor. The honeycombing and rock pockets occurring under present pouring techniques have been so poorly patched that the result is a scar much worse than the original poor condition" (member of Kahn's staff, letter to M. G. Siddiqui, 9 October 1967, LIK, PAKCAP – Correspondence – PAK PWD 1966 – 1967).

Langford's document leaves no doubt about the fact that to guarantee the quality of Kahn's concrete it is necessary for the engineers of the Public Works Department to ensure control over all the phases of concrete fabrication, from the construction of the panels to the pour to the stripping, and that the formwork should all be produced in the same workshop of the Public Works Department, and no longer left up to the individual contractors. "The panels now in use need not be deemed a waste if new, quality panels are constructed. Though they cannot be refurbished for visible pours, the lumber of these panels can be re-assembled for making panels used on hidden pours and shuttering of floor slabs. All of these lumbers can be economically reused in this manner, thus there will be no loss to the contractor or to the PWD."

The document is sent by Langford to Kahn's office in Philadelphia, accompanied by copies of drawings of the panels necessary for the new system of vertically raised formwork.[300]

Siddiqui's proposal for the plywood formwork

Gustav Langford's reasoning on ways to improve formwork making never challenges the basic criteria of the experimental panels outlined by Frederick Langford, including their formation by assembling wooden boards. But certain empirical solutions already attempted by The Engineers have proven to be effective to eliminate certain assembly defects, thanks to the use of large wooden panels on which to engrave the motifs favored by Kahn, based on the joints between the boards in the experimental panels. The criticisms of the formwork of The Engineers and the Inter-Dominion Construction Company expressed in the summer of 1967 by Palmbaum and Langford pave the way for the idea of reconsidering the use of plywood to make the panels and to improve the concrete of the National Assembly Building, or even of hypothesizing a drastic change to the system of construction of the buildings of the capitol, passing from exposed reinforced concrete walls to a skeleton in reinforced concrete with masonry filler and final cladding.

One of the problems posed by the panels built on the basis of the experimental panels is the movement of the boards assembled with the "tongue and groove" method, leaving diversified imprints on the concrete. The communication of Gustav Langford to the effect that the Public Works Department is looking for plywood with the proper characteristics to make formwork, therefore a plywood fabricated "with waterproof glue," is of the utmost importance, because it indicates an orientation of the Pakistani engineers towards the simpler solution for the fabrication of the panels, the one indicated by Kahn himself from the early phases of the project, which would solve all the problems arising from the assembly of the boards. But the plywood solution disrupts what Kahn, at this point, sees as the fundamental sign of the concrete of Dhaka, the sequence of the V imprints – "a faint V joint which

[300] Gustav Langford has constructed a "sample panel" with attached drawing (Gustav Langford, letter to Henry N. Wilcots and David P. Wisdom, 17 October 1967, LIK, PAC – Correspondence – to/from Gus June 1967 thru December 1967).

[301] *Ibid.* See also Gustav Langford, letter to Henry N. Wilcots, 20 October 1967, LIK, PAC – Correspondence – to/from Gus June 1967 thru December 1967.

[302] "During the month of October 1967," LIK, PAC Transmittals – from Dhaka office Nov. 1 '65 thru 1/19/68.

[303] "Minutes of the meeting held in Addl. Chief Engineer's room on the 23rd October, 1967 to discuss about the formwork for National Assembly Building at Ayub Nagar," document attached to, "Ayub Nagar Capital Committee Meeting to be held on 22nd Nov '67. Working Paper," LIK, PAKCAP – Correspondence – PAK PWD 1966 – 1967.

[304] *Ibid.*

Lou has indicated as not objectionable," according to Langford. Improvement of the panels made with boards is possible, again in Langford's view, by means of the coating with three layers of polyurethane on the surface in contact with the pour, and the painting of the aluminium on the back.

"My question at this point is – if plywood is made available, its use will definitely give a much different impression i.e. enormous block-like result, so, shall I press for plywood purchase or shall I start preliminary construction on the enclosed panel design? I know nothing will be done in any one direction for at least 10 days, so please send me your reaction to enclosed panel design versus use of plywood."[301]

The plywood hypothesis continues to have the support of the Pakistani engineers, to the point where on 23 October 1967 Langford writes to the Philadelphia studio: "PWD considering plywood use with progress print marble recess detail [;] need LIK comment immediately."[302]

Convinced, like Langford, of the "unsatisfactory quality and also progress of the work" of the concrete of the National Assembly Building, following the document of 9 October, Siddiqui organizes a meeting with representatives of the two contractors, and the presence of Gholam Mohammad, to examine the proposals of Kahn and his staff "for improvement in the manufacture of forms and consequent improvement in the work."[303] The state of maintenance of the formwork is summed up in one phrase: "About 10,000 Sft of forms have been manufactured by each of the contractors, most of which have already been damaged."[304]

Siddiqui continues to propose plywood as an alternative, "if good plywood is available," to be able to rapidly make the panels and complete the works in 1970 (the original deadline foreseen was June 1969), though the option of "seasoned teak champi panel and chapalish or garjan stiffener with normal grooves"[305] remains valid.

In the first days of November, Kahn takes a position on the panels of the formwork for the National Assembly Building, rejecting the use of plywood. "Lou's latest cable ref. N[ational] A[ssembly] form panels came as quite a surprise," Langford remarks. "I thought that the PWD's desire and attitude to the purchase of plywood would have made everybody happy. I must admit plywood with waterproof glue is at the moment still unobtainable."[306] Following Kahn's decision, the engineers of the Public Works Department and the contractors agree that the formwork system has to be made in such a way as to permit only raising of the panels, in keeping with what is laid out in the document of 9 October. To raise the formwork and improve the concrete pouring operations, the representatives of the contractors ask the Public Works Department to import the proper cranes, not available in Dhaka, again as indicated in the document of 9 October.[307] Langford prepares a list of the technical characteristics of

[305] *Ibid.* Certain technical details discussed at the meeting have to do with the solutions of the corners of the forms – "11-0" long angle/Tee iron should be used for keeping the form in proper position." Haque says he is willing to purchase the wood for board panels that can be cured in the oven of the Public Works Department; Rahman foresees difficulties in being able to make the 35,000 formwork on time, and requests the help of the Public Works Department for their fabrication.

[306] Gustav Langford, letter to Henry N. Wilcots, 10 November 1967, LIK, PAC – Correspondence – to/from Gus June 1967 thru December 1967.

[307] "Minutes of the meeting held in Addl. Chief Engineer's room on the 23rd October, 1967 to discuss about the formwork for National Assembly Building at Ayub Nagar," cit.

the panels, to proceed with their construction and to deploy the new formwork with only vertical movement.[308]

All the various problems caused by the fabrication of the concrete also prompt Siddiqui to reconsider, as an alternative to the complex construction methods introduced by Kahn, the system in use in Dhaka, based on a reinforced concrete skeleton, brick filler and plaster facing. His proposal of building Kahn's design for the hospital using this local approach is a direct result of the difficulties encountered in the construction of the National Assembly Building.[309]

The discussion on the making of the formwork also regards, starting at the end of 1967, the methods of manufacture on the part of The Engineers and the Inter-Dominion Construction Company of the panels, in plywood, for the construction of the roofing shell in the assembly hall.[310]

Work chronicles

The Philadelphia office sends a letter in November 1967 indicating modifications to the size of certain beams, also with an eye on the positioning of the recesses for the marble inserts in relation to the ceilings.[311] The letter also contains a warning about the methods to be used when work resumes, to obtain perfect concrete: "everyone should be aware of the concreting problems and methods for their correction before proceeding with construction. The key word and operation is 'alignment.' The value of time and materials consumed in forming to achieve true line and dimension, securing well-strutted and tied joints which are flush and tight, is of the utmost necessity. For every formwork flaw will be reflected in the finished structure as we all have witnessed in the work."[312]

In the early months of 1968, in spite of the initiatives and urgings of Gustav Langford on the metal formwork for the floor slabs and beams,[313] and the various concrete manuals brought from the United States,[314] the works of The Engineers and the Inter-Dominion Construction Company do

279-280. National Assembly Building, Dhaka, 1962-83. Construction site photographs (LIK, 030.IV.A.650.5.26; 34).

not stand out for significant improvements in the quality of the concrete. The reports of Kahn's staff clearly tell the story.[315] Langford communicates the results of recent work:

"The concrete from El. 38.0 to 48.0 has improved only slightly. I believe there is somewhat less bleeding at the panel junctures, wall bulges have been 99% eliminated, beam pockets are still being blocked out, but the PWD is taking greater care in retaining their correct alignment. The bleeding at the beam

[308] "All new panels to be constructed as per my last progress print. (Construction drawings to be sent back for your comment). Same panel construction for both contractors. Panels to be 5' x 4' or 5' x 6' (I don't know yet). All wood to be correctly seasoned teak chambal. All wood members to be immersed in kerosene for 24 hours. All panels to have polyurethane on concrete surface. All panels to have aluminum paint on backs. Vertical movement only for panels. Old panels phased out by elevation 48. PWD mill to be used for construction for Dr. Rahman's panels i.e. AC, SW, and PE. Mr. Haque will construct his own under same supervision. This I don't like, am trying to get all supervision under one roof" (Gustav Langford, letter to Henry N. Wilcots, 10 November 1967, LIK, PAC – Correspondence – to/from Gus June 1967 thru December 1967). The decision to remake the panels is approved by Kahn (Gustav Langford, letter to Henry N. Wilcots, 15 November 1967, LIK, PAC – Correspondence – to/from Gus June 1967 thru December 1967).

[309] See Gustav Langford, letter to Henry N. Wilcots, 3 November 1967, LIK, PAC – Correspondence – to/from Gus June 1967 thru December 1967.

[310] "Ayub Nagar Capital Committee Meeting to be held on 22nd Nov '67. Working Paper," LIK, PAKCAP – Correspondence – PAK PWD 1966 – 1967.

[311] Member of Kahn's staff, letter to Gholam Mohammed, 24 November 1967, LIK, Second Capital – Pakistan Correspondence – Miscellaneous. "This allows for a full five foot form panel with the bottom of form which sets the beam soffit to be located at the top line of this marble recess," the letter states.

[312] *Ibid.*

[313] Gustav Langford, letter to Henry N. Wilcots, 19 January 1968, LIK, PAC – Correspondence To/From Gus January 1968 thru Dec 31, 1968.

[314] See Gustav Langford, letter to Henry N. Wilcots, 25 January 1968, LIK, PAC – Correspondence To/From Gus January 1968 thru Dec 31, 1968.

[315] Gustav Langford, letter to Henry N. Wilcots, 8 January 1968, LIK, PAC – Correspondence To/From Gus January 1968 thru Dec 31, 1968. "A brief history of N.A. form panels follows," is a synthesis of the measures indicated as necessary starting with the document of 9 October 1967.

281-283. National Assembly Building, Dhaka, 1962-83.
Construction site photographs
(LIK, 030.IV.A.650.5.76; 73; 27).

soffit edges is slightly reduced and these beam block outs have been installed with slightly better accuracy. The marble recess alignment has been improved more than slightly. The recess still is not straight enough to accurately receive the marble. I repeat, improved but not enough. I wish I could say we have achieved good concrete but slightly better is the only phrase that can be used right now. The PWD I believe has a better attitude towards wanting to achieve better results but their supervision is poor and inexperienced. Each contractor is operating at a very slow pace and is concentrating on more accurate erection but necessarily improving or changing their techniques."[316]

The decision to eliminate the plastic film runs up against the problem of procuring polyurethane. Bringing the synthetic resin needed to produce polyurethane to Pakistan will take from 6 to 8 months (the advice is to enquire as to whether the Toch Brothers in Paterson, New Jersey, can ship polyurethane to Pakistan, Bangkok or Southeast Asia).[317] In any case, 50 gallons of polyurethane are expected to arrive in 2-3 weeks (and 100 more gallons in about two months). "The sample form panels that I have built," Gustav writes, "have worked out very well in terms of construction but have not yet been tried as polyurethane was not available. However the new panels have now been coated with polyurethane (a small amount which the PWD had on hand), and will be tested in a couple of days. I believe if we have good results we can then go into new form panel production. They are in the process of kiln drying lumber which will be used in the new panels, and these new panels I hope and will try to use from El. 48.0 up."[318]

[316] Gustav Langford, letter to Henry N. Wilcots, 9 February 1968, LIK, PAC – Correspondence To/From Gus January 1968 thru Dec 31, 1968.
[317] Gustav Langford, letter to Henry N. Wilcots, 8 January 1968, cit. "Dr. Rahman's construction," Langford writes, "has been stopped because of unacceptable results and Mr. Haque is progressing slowly to finalize his buildings up to 48 including slab. El. 48.00 is the magic number at which new form panels are to be introduced and we all want polyurethane on them, not plastic table cloth."
[318] Gustav Langford, letter to Henry N. Wilcots, 9 February 1968, LIK, PAC – Correspondence To/From Gus January 1968 thru Dec 31, 1968. This document is followed by the shipment of worksite photographs.

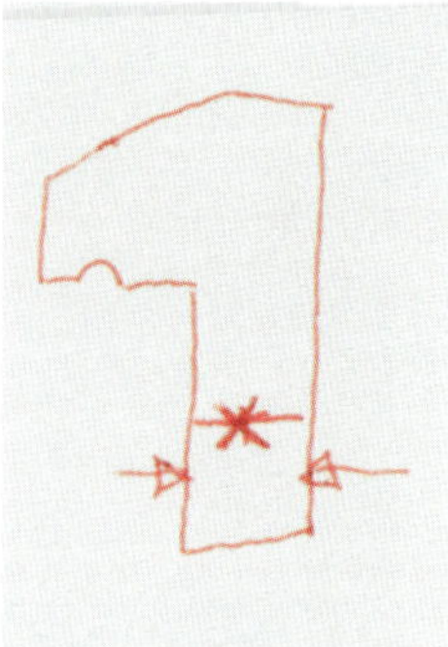

284-286. National Assembly Building, Dhaka, 1962-83.

284. Sketch for the marble strips (G. Langford, letter to H. N. Wilcots, 12 January 1968, LIK, PAC – Correspondence To/From Gus January 1968 thru Dec 31, 1968).

285. "Developed Section Showing Circulatory Walls, Elevations of Light Courts' Walls," 9 July 1965 (LIK, 030.I.C.650.38).

286. "Elevations of Circulatory and Ambulatory Walls," 16 January 1968 (LIK, 030.I.C.650.38).

"From the enclosed photographs, you can see that the quality of concrete is no better and in many cases is worse than when Lou was here last," writes Neil Thompson, the member of Kahn's staff who has joined Gustav Langford in Dhaka since the start of 1968.[319] To seal the joints between the formwork panels "polyfoam" has been purchased, and the results of the use of this product are shown in the photographs. The various defects observed in the works confirm the inability still demonstrated by the engineers of the Public Works Department to control the operations: "This caliber of concrete is due entirely to poor supervision and lack of interest by PWD to pursue an immediate answer."[320] The photographs dating back to March document the state of the work, corresponding to Thompson's descriptions.

The large indentations imprinted in the concrete walls and the marble strips for their filling continue to cause a range of problems. The design of the horizontal cornices with the dripstone causes the marble to chip on the flat part. Gustav proposes increasing the thickness of the band, which therefore requires a deeper recess: "This will also require a revised detail on the new form panels."[321] The length of the pieces indicated by Kahn also raises issues.[322] Gustav's recommendation at this point is to have pieces of reduced length, not only because they break less easily, but also because they can "be used on curved walls" (in his view, the ideal length is from 20 to 30 inches).[323]

The question of the horizontal cornices is again discussed when Kahn recommends corrections for the positioning of the marble according to the spaces: "LIK agreed to five foot lifts for concrete construction of PE, but suggested variations of spacing exterior horizontal marble cornices to interior spaces. (This articulation could be developed to reflect interior spaces of all periphery buildings)."[324]

Kahn expresses his own reservations regarding the quality of the concrete being produced: "LIK felt the N[ational] A[ssembly] concrete has improved, but expressed concern for the irresponsible attitude of everyone concerned with the construction. N[ational] A[ssembly] construction problems still

[319] Neil Thompson, letter to Henry N. Wilcots, 6 March 1968, LIK, PAC – Correspondence To/From Gus January 1968 thru Dec 31, 1968.
[320] *Ibid.*
[321] Gustav Langford, letter to Henry N. Wilcots, 12 January 1968, LIK, PAC – Correspondence To/From Gus January 1968 thru Dec 31, 1968. On the cornices, the marble and its working see also: Roy Vollmer, Jr., letter to [Kafiluddin] Ahmad, 22 March 1966, LIK, PAKCAP – Correspondence – PAK PWD 1966 – 1967; Gholam Mohammad, letter to Louis I. Kahn, 4 August 1966, LIK, PAC – Correspondence – RoyGus July 1966 thru December 1966; and M. G. Siddiqui, letter to Louis I. Kahn, 17 August 1967, LIK, PAKCAP – Correspondence – PAK PWD 1966 – 1967.
[322] Gustav Langford, letter to Henry N. Wilcots, 12 January 1968, cit.
[323] *Ibid.* The position of the vertical bands is indicated in the drawing "Wall elevations. National Assembly" made on 18 August 1966, with revisions in 1968; the revision of 16 January 1968 has to do with the "located vertical marble recess" (LIK, 030.I.C.650.038).
[324] Donald Barbaree, "Notes of LIK Dhaka Visit 29 July - 3 August 1968," LIK, PAC – Correspondence To/From Gus January 1968 thru Dec 31, 1968. 15 October 1969 is the date of the drawing of the "Section, Assembly Chamber, National Assembly," with modifications until 1971. Notes appear on this drawing regarding the vertical marble inserts. In the Assembly Chamber the vertical strips are indicated as having the same function as their horizontal counterparts: "vertical marble @ const. pour-typ." (LIK, 030.I.C.650.038). The same drawing also includes the hypothesis of using a horizontal cornice to "supply air distribution into chamber." For the question of the marble strips see also Neil Thompson, letter to Henry N. Wilcots, 15 June 1968, LIK, PAC – Correspondence To/From Gus January 1968 thru Dec 31, 1968. Attached to this letter there is also the "Drawing of marble layout for NA office sectors as per LIK sketch (sample layouts to be ready for LIK next visit in NE sector)."

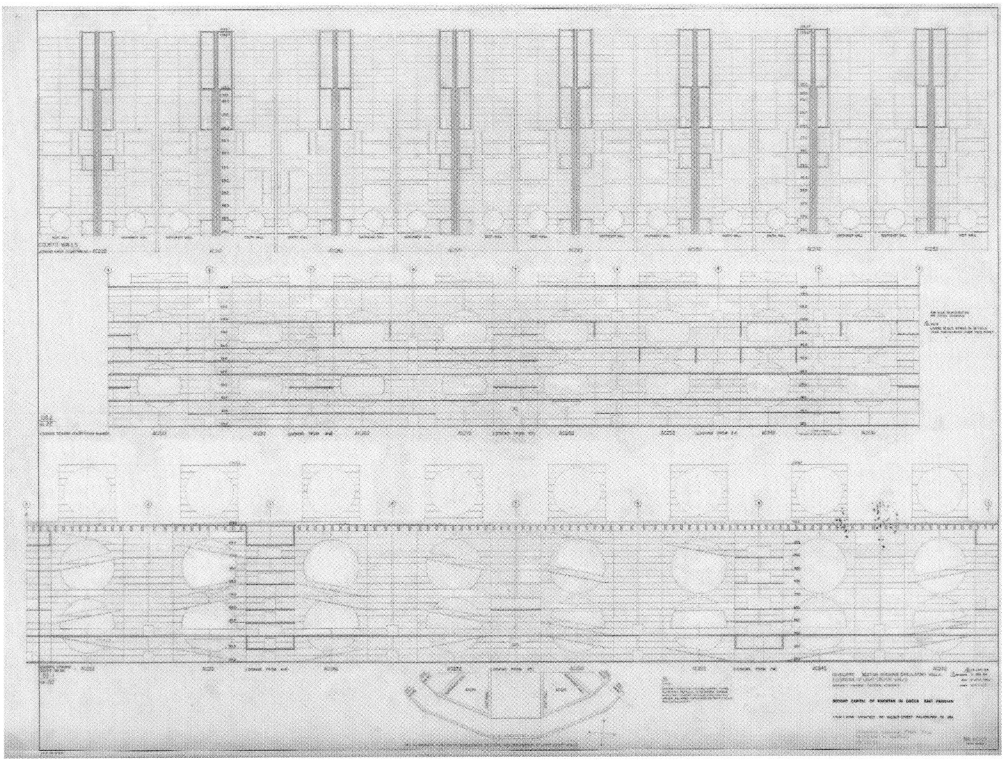

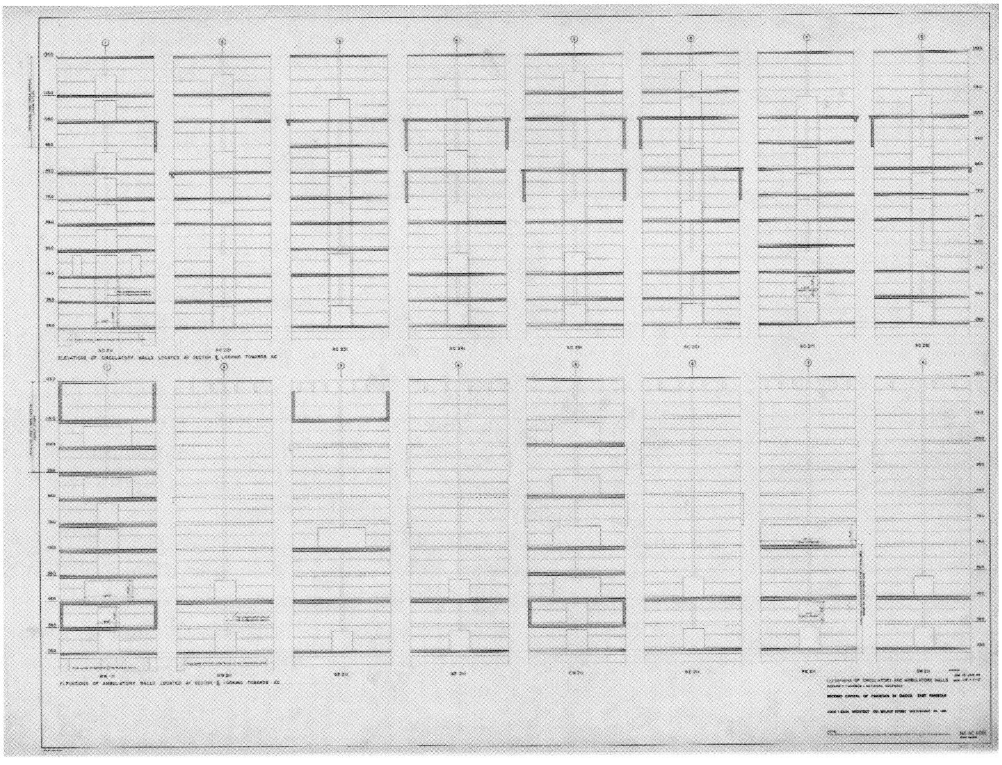

287-288. National Assembly Building, Dhaka, 1962-83. Construction site photographs (LIK, 030.IV.A.650.6.16; LIK, 030.V.E.21.7).

existing: patching, misalignment of forms, failure of forms, failure of shoring, bleeding, and the general apathy of supervision. A letter from LIK to M.G. Siddiqui could be constructive in attempting to resolve these problems."[325]

The photographs received by Kahn's office in Philadelphia at the start of May 1969 clearly document the fact that the construction is proceeding based on the usual local, empirical criteria, with the same method of transport of the concrete by hand, the usual bamboo scaffolding and the panels stripped one by one, then moved horizontally.[326] For the transport of the materials to the different levels, continuous rectilinear ramps are made, again with bamboo poles, through the large openings intentionally left in the walls, so that the workers can easily ascend with the load of concrete on their heads.[327] This is the date in which the cranes appear on the worksite.[328] The fact that Kahn's staff continue to be astonished by the Pakistani worksite and its equipment is demonstrated by certain photographs, like the one that shows the dense weave of the bamboo poles of the scaffolding.

A document from the same period as the photographs sums up some of the most evident defects in the concrete of the National Assembly Build-

ing: "WW formwork has still some defects pointed out before for rectification. Sheets are not laid geometrically. Leakage gaps still exist. In most places tin sheets are warped. That will result in a wavy surface of finished conc.[rete]."[329] The recesses for the marble strips also continue to be made with flaws, in spite of a certain improvement. Their irregular shape prevents correct sealing with the marble cornices. Again for this reason, the mode of attachment of the cornices at level EL 28 is deemed "unsatisfactory." "It is not possible to properly fix the marble unless the recesses are properly prepared for receiving marble."[330] Precisely to be able to correct by hand the shape of the recesses poorly made using the formwork, special tools are requested, which must be imported, like a "high-speed concrete drill, saw and grinder."

The worksite supervised by David Wisdom & Associates and Wazid

The definitive designs of the two main roofs of the National Assembly Building, that of the mosque and that of the assembly hall, are developed during the course of 1970. For the mosque, Kahn proposes a pyramid roof, with the pitches striped by marble cornices.[331] The fact that Kahn talks about the effect of this roof in the viewing of the monument against the backdrop of the landscape would lead us to believe that he is thinking about cladding the whole surface of the pitches, not just the simple cornices shown in the model.[332] But the pyramid form is criticized by persons in Pakistan, who say that the faithful would experience it as having an ecclesiastic character – the "stamp of church." So Kahn is asked "to consider the possibility of providing a flat roof or a dome." "The architect was of the view that the conical dome [sic][333] provided by him was the best aesthetic solution of the problem. He

[325] Donald Barbaree, 13 August 1968, "Notes of LIK Dhaka Visit 29 July - 3 August 1968," LIK, PAC – Correspondence To/From Gus January 1968 thru Dec 31, 1968.
[326] See the photographs taken in March 1969, LIK, 030.IV.A.650.6.1, 2, 3, 4, 5, 11, 12, 13, 14, 15, 31, 32, 33, 34, 35.
[327] See the photograph, LIK, 030.IV.A.650.6.16.
[328] See the photograph, LIK, 030.IV.A.650.6.23.
[329] "Programme and progress of work being done in the local office of Prof. Louis I. Kahn," LIK, PAK PWD Correspondence 1969.
[330] *Ibid.*
[331] The surface of the four pitches exposed to the weather is generically described as "covered by marble" ("Minutes of the meeting held on 19th january 1970," LIK, PAK PWD Correspondence 1970).
[332] *Ibid.*
[333] In the same document, the roof is described as a "pyramidal roof" (*ibid.*).

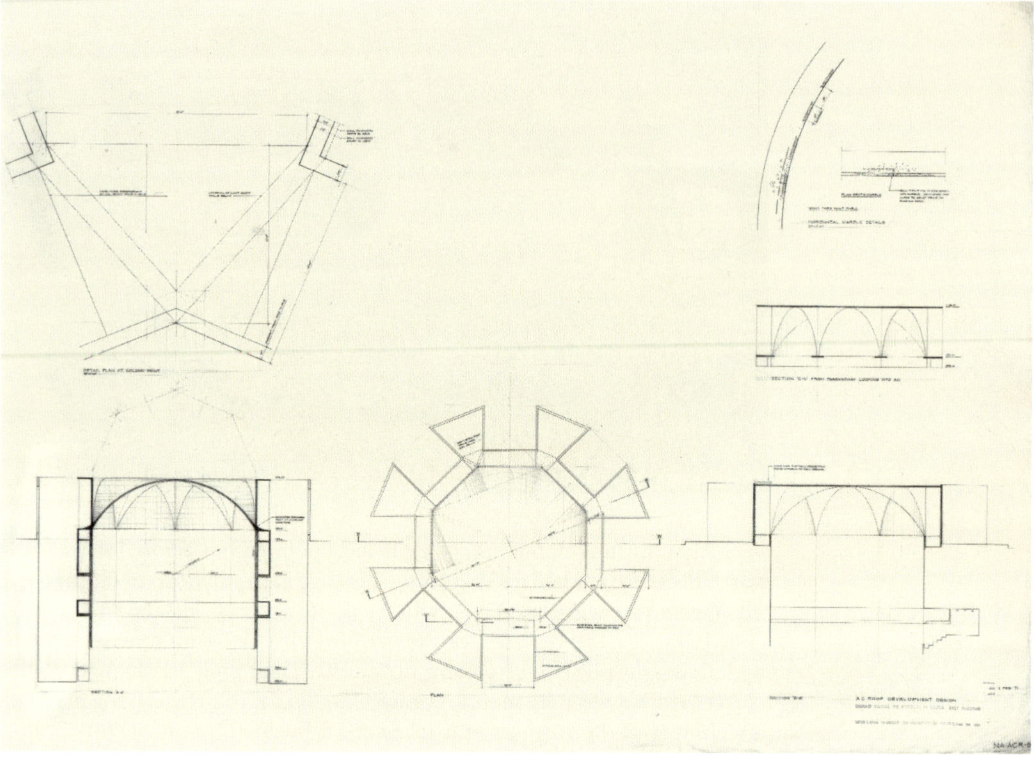

explained that after being covered by marble it will present a very pleasant appearance from the South Plaza. The Addl. Chief Engineer [Siddiqui] maintained that the sentiments of those who will offer prayers inside the mosque were far more important than any aesthetic architectural consideration or the impression created on the viewers."[334]

The definitive design of the roof of the mosque has four inclined pendentives cut by semicircular openings for light, vestiges of the original pyramid roof. These pendentives contribute to support the new flat roof. To make these pendentives it is decided to build a special roof model to "enable the contractor to understand formwork requirements."[335] The number of horizontal marble strips is doubled with respect to that of the walls, in order to distinguish, with the denser sequence, the function of the pendentives in the structure of the roof.

For the roof of the assembly hall, after the hypothesis of a light structure with steel cables, metal panels and concrete cladding, based on the one

[334] *Ibid.* The design of the pyramid occupies Kahn for nearly all of 1970. The decision is made only in September: "After much debate and the receipt of yours and Wazid's cable/letters Lou has decided it best for all to not construct the pyramidal roof of the PE. We sought to find something historical that would reaffirm our thoughts and Lou also began a letter on the development of the PE. In summary, the pyramidal form was an architectural offering and was not fulfilling the spiritual needs of the space" (Henry N. Wilcots, letter to Donald Barbaree, Richard F. Garfield and Das, 15 September 1970, LIK, 030.II.A.10.16).

[335] "Notes," 16 August 1970, LIK, PAK PWD Correspondence 1970. On the new roof, called a "dome," see Abdul Wazid, letter to Richard F. Garfield, 4 September 1970, LIK, PAK PWD Correspondence 1970.

[336] For the description see member of Kahn's staff, letter to Peter Rhinehart, of American BBR of Chicago, 2 July 1970, LIK, Suppliers PAC.

[337] Richard F. Garfield, letter to Harry M. Palmbaum, 9 September 1970, LIK, PAC – Correspondence To/From Dhaka Office – 1969 thru 10/6/71. For a negative opinion on the possibility of "shutter[ing] in quadrants" for the "dome" see also Richard F. Garfield, letter to

289-290. National Assembly Building, Dhaka, 1962-83.

289. "A C Roof Development Design," 1 February 1971 (LIK, 030.I.C.650.59).

290. Design of the formwork system with the construction sequence, 5 April 1975 (H. N. Wilcots, letter to A. Wazid, 10 April 1975, LIK, Dhaka PWD Correspondence National Assembly 1976).

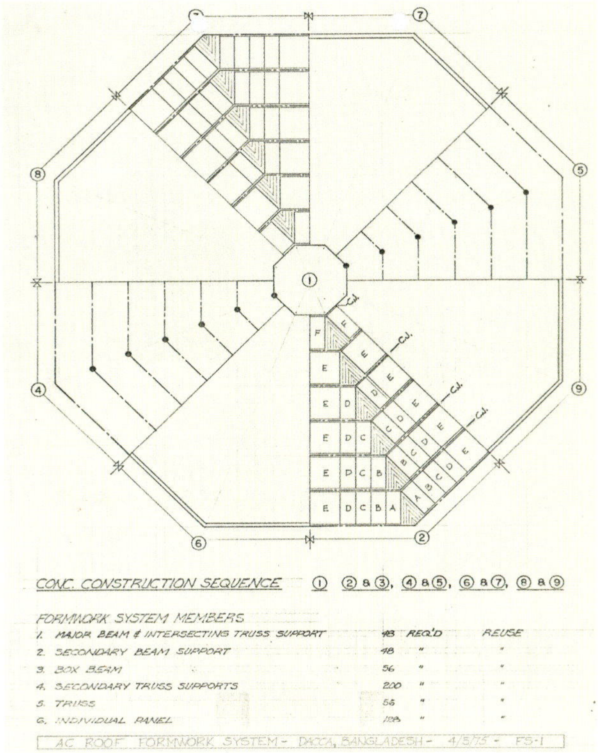

originally envisioned by Komendant and Keast & Hood,[336] the design goes back to the solution of the reinforced concrete segments, interpreting geometric and structural figures typical of the 1950s and 1960s. The technical questions have to do with the construction of the scaffolding and the formwork, and the methods of pouring the concrete.

Kahn, his staff in Philadelphia and Dhaka, Palmbaum and the engineers of the Public Works Department discuss the way of building formwork for the segments, and whether or not it is possible "to shutter in sections"[337] (in this period Richard F. Garfield is in charge of Kahn's office in Dhaka).

To fasten the marble cornices inside the concrete segments the sheets are to be cut with inclined sides, placed prior to the pour, with a metal fastener between one piece and the next to anchor them to the reinforcement bars.[338] In the meantime, the placing of the cornices in the concrete walls begins. According to Barbaree, again referencing Kahn, that marble "tells the story of the construction."[339]

Abdul Wazid, 9 September 1970, LIK, PAK PWD Correspondence 1970. Palmbaum's view on the method of constructing the vault in sectors is reported by Wilcots: "Harry says shuttering in quadrants is definitely out. He suggests either eighths or even sixteenths and in the removal to strike every other panel. But we fear this will leave an impression in the works of the panel which is supported by the shoring. Also, without question the shoring is to be many and full remaining in place until the concrete has reached its proper strength for shoring removal" (Henry N. Wilcots, letter to Richard F. Garfield, 18 September 1970, LIK, PAC – Correspondence To/From Dhaka Office – 1969 thru 10/6/71). For the progress of the worksite in the month of November see Richard F. Garfield, letter to Henry N. Wilcots, 24 November 1970, LIK, PAC – Correspondence To/From Dhaka Office – 1969 thru 10/6/71.

[338] See "AC Roof Development Design 1 feb 1971," LIK, 030.I.C.650.059.

[339] Donald Barbaree, letter to Henry N. Wilcots, 8 December 1970, LIK, PAC – Correspondence To/From Dhaka Office – 1969 thru 10/6/71.

291-294. National Assembly Building, Dhaka, 1962-83.

291. Steel formwork for the dome (F. Fusaro, *Il Parlamento e la nuova capitale a Dacca di Louis I. Kahn 1962-1974*, Roma, Officina, 1985, p. 109).

292. View of the dome (photograph by the author, 2010).

At the start of 1971 the formwork is roughly repaired, because it is no longer possible to make new panels, since the pouring of the walls is nearing its conclusion. Garfield sends a "sketch showing 'state of pour' – formwork is pretty rough from sitting around and from use." The crane at WW Block is out of order, "so that pours there are suffering from extensive down time and handling between mixer and formwork." "But a little 'Joy Bangla' goes along way and workmen respond to the 'new politics' with a healthy attitude toward the work." At the same time, experiments are conducted for the patching of the concrete of the SW Block. Garfield describes the interventions and strategies applied to correct visible defects after stripping:

"The control of color has improved immensely, but whether that can be maintained on a large scale is certainly questionable. Daily, I've been with Gholam Mohammad, who is willing to try things out and is relatively communicative. Slowly, we've moved from bagging, through bagging and stoning (rubbing) (both PWD suggestions) to filling. Filling not entirely successful yet, but we've moved from trowels to largish sculptor's tools and hope to get down to dentist's tools soon. In spite of this advance, I'm afraid a system of patching relatively grossly and then flushing with large quantities of water to remove fines from good surfaces will be favored as it is quicker. Also, since Mr. Wazid wants all holes filled, even those evenly distributed small ones which I associate with relatively good pours, the scope of corrective work envisioned is galactic. All this is by way of asking your advice concerning other approaches especially with reference to details such as shades of wet versus set concrete, flushing, wiping techniques."[340]

The work on the semicircular south wall of the WW is "terrible"; in February the crane has yet to be repaired, and the pours proceed slowly.[341]

[340] Richard F. Garfield, letter to Henry N. Wilcots, 30 January 1971, LIK, PAC – Correspondence To/From Dhaka Office – 1969 thru 10/6/71.

[341] "Northern semicircle has all forms, steel, electric in place for slab at El. 128, but waiting for crane repair for pouring" (Richard F. Garfield, "Notes on progress of concrete pouring," 10 February 1971, LIK, PAC – Correspondence To/From Dhaka Office – 1969 thru 10/6/71).

[342] Henry N. Wilcots, letter to Abdul Wazid, 28 January 1975, LIK, Dhaka PWD Correspondence National Assembly 1976. In April Wilcots requests, from Canadian Metal and Federated, Ltd., based in Toronto, an estimate for the materials needed for the finishing of the surface of the roof of the assembly hall (Henry N. Wilcots, letter to The Canadian Metal and Federated, Ltd., 7 April 1975, LIK, Dhaka National Assembly. Roof Specifications). See also Abdul Wazid, letter to David Wisdom & Associates, 28 February 1975, LIK, Dhaka PWD Correspondence National Assembly 1974-75.

[343] Abdul Wazid, letter to David P. Wisdom and Associates, 20 March 1975, LIK, Dhaka PWD Correspondence National Assembly 1974-75. See also Abdul Wazid, letter to Henry N. Wilcots, 24 March 1975, LIK, Dhaka PWD Correspondence National Assembly 1974-75, and Henry N. Wilcots, letter to Abdul Wazid, 24 March 1975, LIK, Dhaka PWD Correspondence National Assembly 1976.

[344] The correspondence shows that the formwork is being assembled in Dhaka, though the details for the same formwork are still under study in Philadelphia. On this issue, Wisdom and Wilcots urge Wazid to make "one

293. View of the arcade (photograph by the author, 2010).

294. View (photograph by the author, 2010).

In March 1971 work is interrupted and Kahn's office in Dhaka is closed, following the outbreak of the war that will lead to the independence of East Pakistan and the creation of Bangladesh. Only in 1972 is the hypothesis of resuming work for the construction of the capital examined, which in the meantime has been renamed Sher-e-Bangla Nagar. But construction does not resume until 1974. After Kahn's death, on 17 March, the work is supervised by David Wisdom & Associates, a firm that also includes two other members of Kahn's staff, Wilcots and Reyhan Tansal Larimer.

At the start of 1975 the technical details are studied for the construction of the vault segments of the National Assembly Building, especially "the formwork necessary for installing the interior marble at the time of casting the concrete."[342] In the spring the metal scaffolding with the platform is assembled, and wooden formwork is prepared which should have the surface in contact with the pour covered with "steel sheet as was done for the other roofs."[343] No perfect coordination exists between the works done on the site, supervised by Wazid, and the studies done in Philadelphia by Wilcots.[344]

The methods of attachment of the marble strips remain to be defined, and Langford is asked for his opinion.[345] The Philadelphia office develops a system which is never built: the cornices are installed after the pour, in the proper recesses, and hitched to devices placed prior to the pour.[346] The cornices break up the webs into stripes parallel to the horizontal lines of the walls, but closer together. Between one web and the next the strips are interrupted to leave the edges exposed on the vaults. There is no trace, in the pattern development, of that pursuit of a constructive logic that instead accompanied the design of the marble strips in the walls. Only once the motif has been decided on do we find clues that point to the search for a logic,

segment of the formwork" on the ground, in order to "discover and solve all problems of cutting, erecting and securing the work," to have the sample inspected by Gustav Langford, who would go to Dhaka for that purpose, and then to use it as a prototype for the construction of the other segments (David P. Wisdom, Henry N. Wilcots, letter to Abdul Wazid, 4 April 1975, LIK, Dhaka PWD Correspondence National Assembly 1976). Then the design of the formwork system with the assembly phases is sent to Dhaka (Henry N. Wilcots, letter to Abdul Wazid, 10 April 1975, LIK, Dhaka PWD Correspondence National Assembly 1976). This is followed by a series of recommendations, including the pattern of assembly of the sheet metal coating: "jointing of metal to metal and securing in place must respect the ordained position of the marble" (Henry N. Wilcots, letter to Abdul Wazid, 22 April 1975, LIK, Dhaka PWD Correspondence National Assembly 1974-75). "The segments," Wazid replies, "were prepared on ground after laying out the parabolic curve in full scale and are being put in position. It is expected that the formwork will be completed by the end of this month and the steel sheet surface covering on the wooden forms will be laid by the first week of May" (Abdul Wazid, letter to Henry N. Wilcots, 23 April 1975, LIK, Dhaka PWD Correspondence National Assembly 1974-75).

345 Abdul Wazid, letter to Henry N. Wilcots, 23 April 1975, cit.

346 *Cf.* Henry N. Wilcots, letter to Abdul Wazid, 20 May 1975, LIK, Dhaka PWD Correspondence National Assembly 1976. Attached to the letter is a "Time-saver Products catalog (Concrete Forms Corp.)."

295-298. National Assembly Building, Dhaka, 1962-83.

295-296. Views (photographs by the author, 2010).

297-298. Details of the concrete walls with the marble strips (photographs by the author, 2010).

also for those strips, in the construction process of the concrete. One fundamental consideration in this sense is made by Wazid in June 1975, when he announces, "the entire roof has been covered with formwork."[347] The concrete is to be spread in horizontal layers, starting with the feet of the webs and terminating at the summit of the vault, using the marble strips as a means of covering the pour joints. "In these sequences," Wazid explain, "there shall not be any construction joint perpendicular to the vault except a small portion near the groin & there will not be any construction joint in the groin beam except near the crown. The construction joints shall be along the marble inserts and will be covered by the marble. Wherever the construction joint comes, the bulkhead will be placed below the steel in the exact position to give a defined line."

The definitive design for the metal coating of the formwork is done to take advantage of the presence of the marble cornices to create a design of the surface with selected imprints – only the horizontal marble cornices and the edges of the webs remain visible. To obtain the design of the surface of the webs, the metal sheets measuring 6 × 3 feet are applied on the boards of the formwork in horizontal bands, and are nailed where the marble strips will be placed, allowing the latter to conceal the imprints of the nails – "the joints of the sheets will be covered by the marble along with the vault."[348] Instead, the joint between one sheet and the next along the horizontal bands is made imperceptible thanks to its sealing with "plastic tapes." A type of design of vaults with a hierarchy between horizontal and vertical joints was common in the stereotomy of the 1600s and 1700s for the construction of vaults in France.

"As regards removal of formwork," Wazid explains, "it has been considered to remove the parabolic truss and then the sheets with planking in

[347] Abdul Wazid, letter to Henry N. Wilcots, 3 June 1975, LIK, Dhaka PWD Correspondence National Assembly 1976.
[348] *Ibid.*

299-302. National Assembly Building, Dhaka, 1962-83.
Details of the concrete walls with the marble strips (photographs by the author, 2010).

sections from the top in the reversed sequence of pouring concrete. The groin and the vault will be temporarily supported with props where the forms have been removed and then finally these props will be released simultaneously."[349]

The vault is stripped in February 1976. "The marbles are being cleaned and finished. The surfaces of the roof have come out fairly good with slight unevenness here and there, due to expansion and contraction of the tin sheets," Mohammad writes.[350]

When the masonry work has almost been completed, in 1978 the state authorities remark that "the surface finishing of off-the-form concrete works in the National Assembly Building is very poor," saying that it is impossible "to keep the interior faces of the building as it is." Following their requests "for improvement of the interior of the building," former members of Kahn's staff send Dhaka a series of booklets on "different tools to be used for finishing the building."[351]

The attempt of Kahn and his staff to erect a new kind of concrete for Pakistan, controlled in the development of each of its parts to be left exposed to view, remains a failure if the standard for the quality of the work is the same as that applied for the other works of architecture in exposed concrete built by Kahn in the United States. If it is compared to the *malfaçons* of the Unité d'Habitation in Marseille, the visible defects in workmanship in the National Assembly Building even take on tragic overtones for Kahn's vision of an architectural concrete marked by perfect execution of every detail of joining between boards and panels, and by full control of every line between one pour and the next. The work demonstrates that not even the heroic efforts of Kahn's faithful disciples and expert technicians can make it possible to achieve a coherent materialization of the philosophy of the formwork and pouring system invented by Kahn, without the support of skilled and experienced craftsmen and suitable mechanical equipment, precisely because that philosophy is rooted in the technical and constructive context of American architectural concrete, which already requires education itself to appreciate and put into practice the details designed by Kahn, if the aim is to fully achieve the goal of perfection, as proven by the experience at La Jolla. The sophisticated devices of Kahn's formwork and pouring system could not help but generate, given the technical conditions of construction sites in Dhaka, results such as those that are still visible today in the National Assembly Building, where the V joints continuously vary in their relief, to the point of vanishing in certain sectors, and where the marble bands that were supposed to attenuate the perception of defects in the concrete become, in turn, as in the semicircular volumes, lines composed of small inserts of irregular size, clumsily set into poorly made recesses. Precisely the concrete of the National Assembly Building demonstrates the extent to which the elimination of cladding pursued by Kahn requires painstaking control of the worksite unless there is a willingness to venture, as Kahn always refused to do, into the poetics of the *malfaçon* and of the "as found" championed by the theorists of international Brutalism and their imitators.

[349] *Ibid.*
[350] Gholam Mohammad, letter to Henry N. Wilcots, 14 February 1976, LIK, Dhaka PWD Correspondence National Assembly 1976.
[351] "Minutes of the meeting with the Secretary, Ministry of Public Works and Urban Development," 2 June 1978, LIK, Dhaka PWD Correspondence National Assembly 1978.

303-308. National Assembly Building, Dhaka, 1962-83.

303-305. Details of the concrete walls with the marble strips (photographs by the author, 2010).

306. View of the interior (photograph by Arnout Fonck).

307-308. Views of the interior (photographs by the author, 2010).

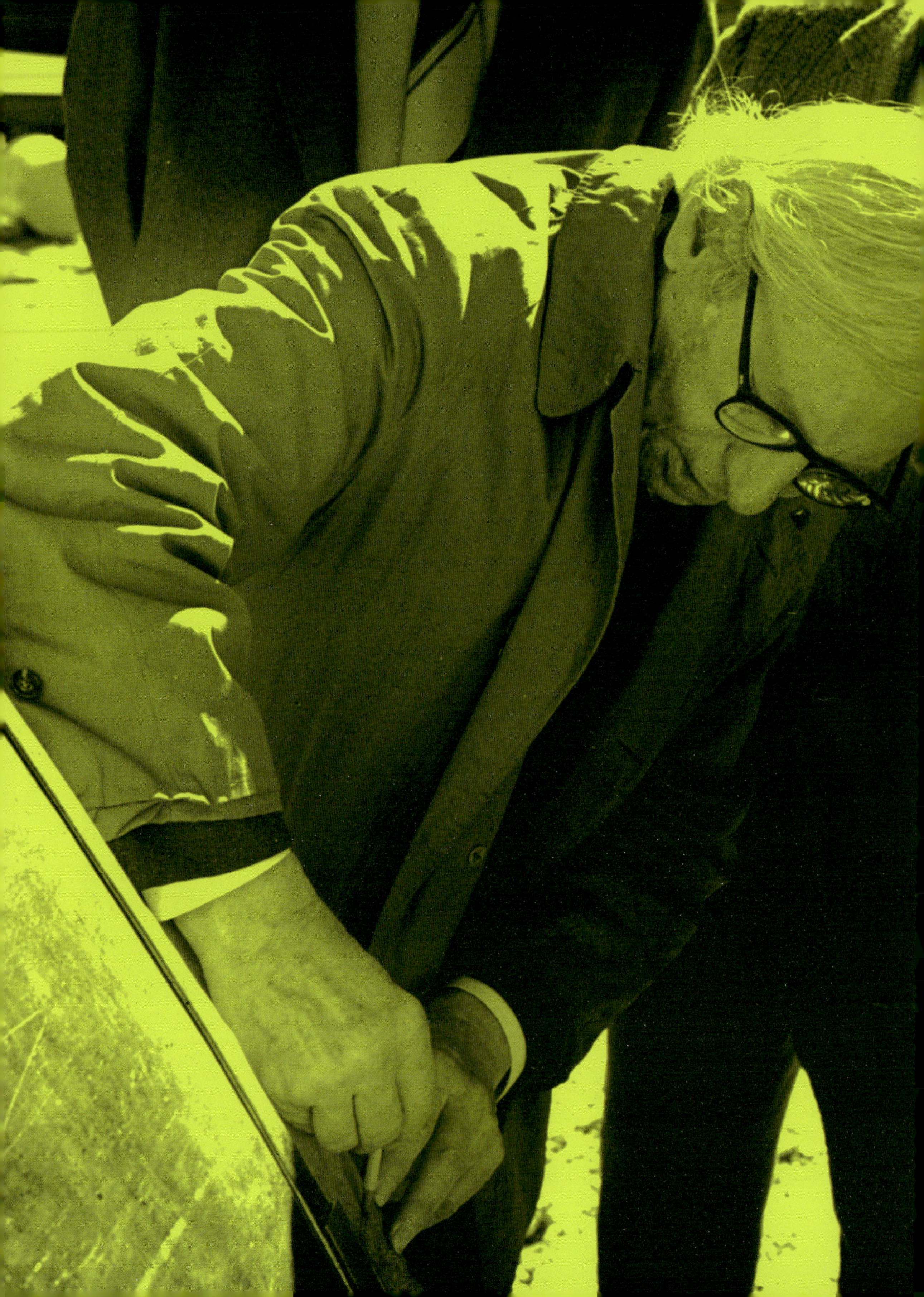

chapter four
walls and vaults of the kimbell art museum

Skeleton, "cavity curtain walls" and the cycloid curve

When all the concrete works of the Salk Institute have just been completed and the construction of the National Assembly Building is entering its crucial phase, in October 1966 the director of the Kimbell Art Foundation, Richard F. Brown, commissions Kahn to design a museum to be built in a park in Fort Worth. The local architecture and engineering firm of Preston M. Geren is hired to handle the permits, the preparation of drawings for the contractors, and worksite supervision.

The museum is destined to become another crucial work of Kahn's concrete, a variation on the research on the true nature of the material conducted with the Salk Institute and increasingly oriented towards erasure of any imprint that is not justified by a process of construction controlled in every phase. In this perspective, the practical expedients introduced by The Engineers and the Inter-Dominion Construction Company, precisely in the fall of 1966, in the system organized by Frederick Langford for the worksite in Dhaka stand out as deviations from the logic pursued by Kahn in the design of the imprints on the concrete. But the character of Kahn's particular vision of architectural concrete also leads the American professionals and builders involved in the construction at Fort Worth into attempts to translate his patterns into pragmatic technical solutions, not unlike the ones applied in Dhaka, again bordering on merely decorative patterns.

Kahn is looking for a reinforced concrete structure configured to control the light and space of the gallery at the main level of the museum, that of the park (the lower level, accessed from Arch Adams Street – now Van Cliburn Way – is set aside for storage, workshops, offices, services and an auditorium).

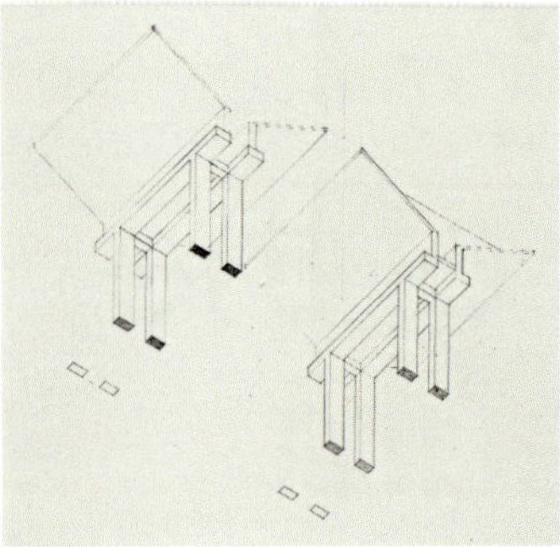

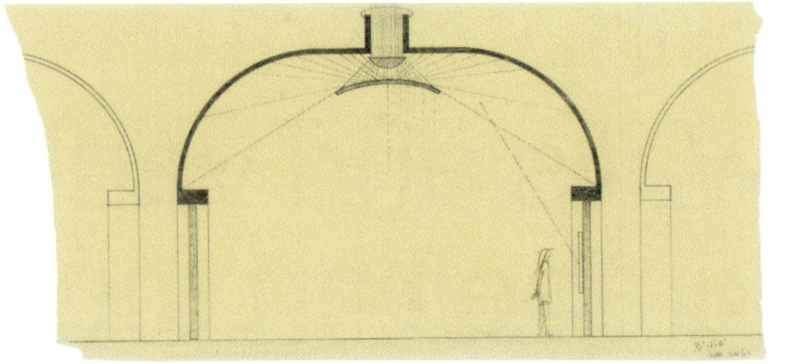

309-311. Kimbell Art Museum, Forth Worth, Texas, 1966-72.

309. Isometric view of the early structure solution (LIK, 030.I.C.730.3).

310. Section of the vault, fall 1967 (LIK, 030.I.C.730.3).

While in the Yale Art Gallery the cavities of the tetrahedral floor system were designed to contain the artificial lighting system, in the Kimbell Art Museum, from the first project submitted in March 1967, Kahn envisions structural units to install in parallel rows, as in a warehouse, each with a three-pitch roof cut for zenithal lighting controlled by suspended diffuser screens (more light is provided by special patios created in the series of structures, and originally designed with colors). The design of the roof profile becomes the fundamental theme for the lighting quality, and it takes on the configuration of the vault, explored by Kahn in the studies conducted in the summer and the fall.[1]

The search for the exact profile of the vault leads Kahn to employ an extremely depressed elliptical curve, then transformed into a cycloid curve, starting from the examples contained in the book by Frederick Angerer, *Surface Structures in Buildings*, shown to Kahn in October by Marshall D. Meyers.[2] Komendant's involvement in the design of the "surface structure" of the Kimbell Art Museum, starting in December,[3] is further confirmation of the evolution of Kahn's vault into a slender reinforced concrete membrane of the shell-type structures that are spreading in the work of engineers at the time. It is Komendant who provides, thanks to his knowledge of vaulted and thin structures, an initial estimate of the cost of what he calls "shell construction," a system he recommends for its potential as opposed to the conventional "slab and beam" approaches – in his view, Kahn's vaults would

[1] "I spoke with Lou about the walls and vaults before he left for Pakistan," a member of Kahn's staff writes in July. "He made a sketch [...]. In it the walls are thought of as cast concrete structural elements providing for solid wall infill of appropriate material [...]. The vaults would spring from behind the edge of the top of the wall – thus the light on the vaults is a kind of ambient light belonging really only to the ceiling, therefore to the space as a whole and not to the painting which would be carefully artificially illuminated. The whole ceiling is becoming more integral" (David C. S. Polk, letter to Richard F. Brown, 13 July 1967, LIK, 030.II.A.37.18). For the sketch see *The Louis I. Kahn archive: personal drawings* cit., vol. V, p. 213, No. 730.39. For the Museum construction see *Louis I. Kahn. The Construction of the Kimbell Art Museum*, Milano, Skira, 1997.

[2] The book contains a series of examples, from Germany in particular, starting with the experiments conducted by Franz Dischinger. *Cf.* Fred Angerer, *Bauen mit tragenden Flachen*, Munich, G.D.W. Callwey, 1960, Eng. tran., *Surface Structures in Buildings. Structure and Form*, New York, Reinhold Publishing Corporation, 1961. On the structure of the Kimbell Art Museum see Guy Nordenson, *The Lineage of Structure and the Kimbell Art Museum*, in, "Lotus International," 1998, No. 98, pp. 28-48.

[3] In December 1967, Kahn involves Komendant "to prepare a typical structural section of the proposed concrete shell indicating reinforcing, prestressing and an outline specification" (Marshall D. Meyers, letter to Preston M. Geren, 11 December 1967, LIK, 030.II.A.37.5).

[4] Marshall D. Meyers, letter to Preston M. Geren, 21 December 1967, LIK, 030.II.A.37.5.

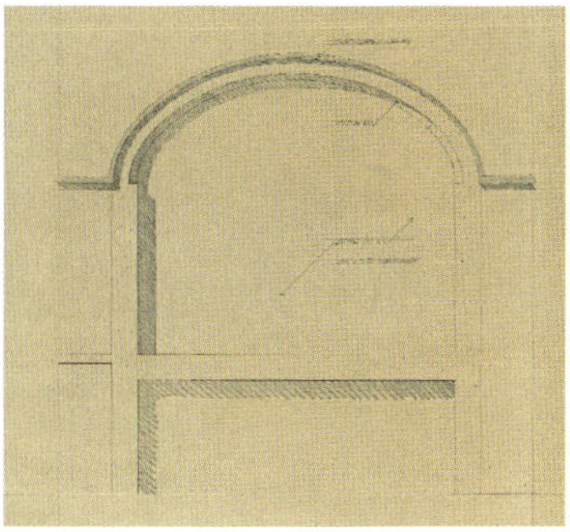

311. "Typical End Wall Elevation," 7 December 1967 (LIK, 030.I.C.730.006).

lead to savings on construction, thanks to their "long span system" and "repeated use of forms."[4] After the discovery of the cycloid curve and the preliminary assessment by Komendant, all that remains for Kahn and his staff, including Komendant himself, is to develop the formwork system with centering, the mix of the concrete to be left visible, and the type of metal reinforcement for spans that range, at first, around 116 feet. The luminosity discussed by Kahn regarding this phase of the design depends precisely on the color of the light reflected off the exposed concrete surfaces: "I felt that the light in the rooms structured in concrete will have the luminosity of silver."[5]

The other element indicated by Kahn in July as fundamental for the Kimbell Art Museum, namely the wall, is similarly destined, like the vaults, to go through extensive variations regarding its technical nature and its chromatic shading, always evaluated in the context of the exposed reinforced concrete. From the earliest studies the element indicated in the correspondence as the wall is actually a system formed by a skeleton of reinforced concrete, connected to the cycloid vaults, and by infill sectors equipped with an interspace that justifies the particular terminology used in certain documents: "cavity curtain walls."[6] In the preliminary phases of the project the support of this paneling is specified as "cast concrete structural elements." In the versions of the design up until the summer of 1968, the system of the exposed concrete skeleton and the paneling with slabs of stone is extended to the two levels of the museum (the plan of the lower level is crossed by a dense array of pillars that support the floor slab of the exhibition gallery).

The design of the stone slabs, of increasing size in the various projects, reflects the desire to express the role of closure of the walls, while at the same time creating a suitable backdrop for the works of art on display. The opening made between the vault and the wall, known as the "arc de lumière naturelle,"[7] also underlines the function of non-load-bearing closure of the panels. The fact that in the first phases of the project Kahn is oriented towards a system of skeleton, infill and vaults made expressive by the exposed concrete and the slabs of stone is confirmed by Brown's advice regarding the possibility of extending the arched opening of the vaults along the pillars, to obtain a more evident "true structure."[8]

A decisive degree of monumentality for the work would be provided by the fine workmanship of the surfaces and the finishing materials.

[5] Louis I. Kahn, *Space and Inspiratations*, 14 November 1967, now in, Latour, *Louis I. Kahn: writings, lectures, interviews* cit., (pp. 224-230), p. 228.

[6] [Marshall D. Meyers], "Tentative schedule of finishes. Kimbell Art Museum, Forh Worth, Texas," 21 December 1967, LIK, 030.II.A.37.5. The concrete walls mentioned in the document ("Exterior walls to be architectural finish concrete") refer to the garden, as shown by the handwritten addition of the word "garden" between "exterior" and "walls" (see, [Marshall D. Meyers], "Tentative schedule of finishes. Kimbell Art Museum, Forh Worth, Texas," 21 December 1967, with handwritten corrections of 3 December 1968, LIK, 030.II.A.95.5).

[7] Richard F. Brown, letter to Louis I. Kahn, 22 July 1968, LIK, 030.II.A.37.18.

[8] It is significant that Brown criticizes the absence of the beam connecting the pillars. "[...] such a device might even better express the true structure of the building by enframing and emphasizing the diaphragm nature of these end walls, and make even more appreciable the bold statement of these walls as mass, color and texture – qualities to relish in their own right as aesthetic objects, as well as means to enclose a space. The need for stability could be very frankly expressed by horizontal transom bars between each side of the wall and the adjacent pier" (Richard F. Brown, letter to Louis I. Kahn, 22 July 1968, LIK, 030.II.A.37.18). Brown even advises Kahn to extend the opening of the arch of light to the floor, to underscore the role of closure of the walls. See also [Richard F. Brown], "For Kimbell Art Museum, Attachment to R. Brown's letter to Marshall Meyers, Jan. 22, 1969," LIK, 030.II.A.37.18.

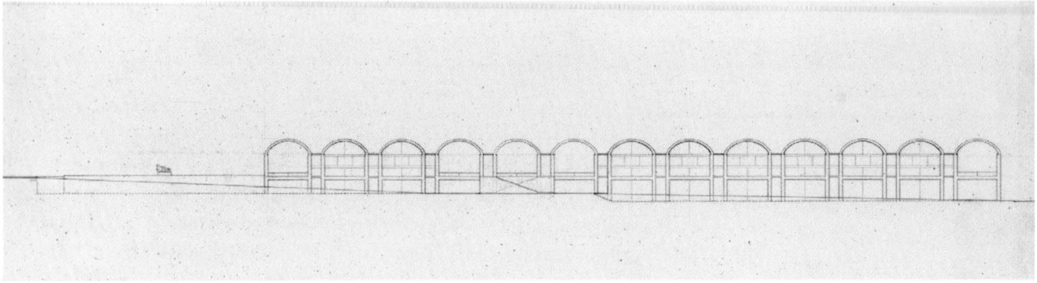

312-314. Kimbell Art Museum, Forth Worth, Texas, 1966-72.

312. Section, 2 January 1968 (LIK, 030.I.C.730.6).

313-314. Plans, 25 September 1968 (LIK, 030.I.C.730.009).

From the first discussions on costs, Kahn insists on the "quality of finishes,"[9] including the "exposed concrete" of the "shell with flat slab" ("no rubbing, no wash, no filling") and of the columns ("very good formwork, no grout, no wash, no rubbing or after treatment of any kind"), the "special fixed bronze frames" with Miesian overtones for the openings, with "polished plate glass in very large slabs," the "large pieces" of local stone for the walls ("Texas Limestone"), and the slabs of marble for the floors of the gallery, cut in large pieces ("perhaps extra thick to permit some extra large sizes not otherwise possible").[10]

The definitive version of the project begins to emerge in September 1968, when Kahn organizes the plan of the museum in three sectors, where the central portion features an entrance court (the entire functioning of the spaces is oriented differently than in the previous versions). It is in this design that perimeter walls in reinforced concrete, to be left exposed, are introduced on the lower level. The ambiguous meaning of these walls determines the different design of the two sides: the side facing inward is marked by the protrusions that continue in the pillars on the level above; while the side facing outward features an assembly of rectangular panels of the formwork leaving no glimpse of structural supports.

Certain exceptions are introduced in the structure of the floor slab of the gallery, as in the "prestressed concrete slabs" supported, at the center, by a pillar to eliminate the forest of pillars of the lower level and make their layout more functional, as per indications supplied by Brown.[11]

During October and November the measurements of the structure of the cycloid shell are modified by Kahn and Komendant (passing from a span of 24 × 116 feet to one of 20 × 150 feet[12]), while the "plexiglas mirrored reflection" and the linear device of artificial lighting "for illuminating the underside of the cycloid shell at night" are also designed.[13]

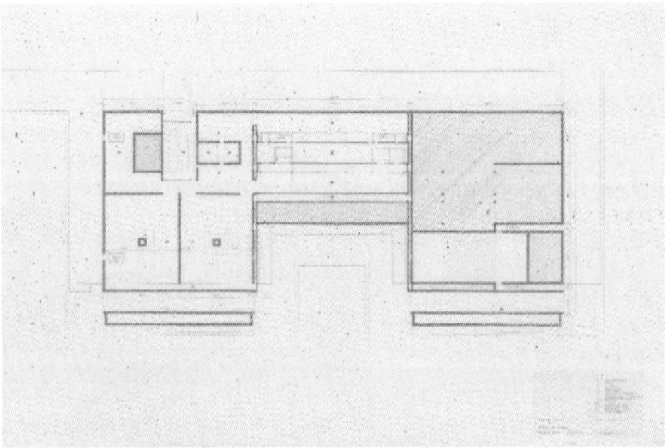

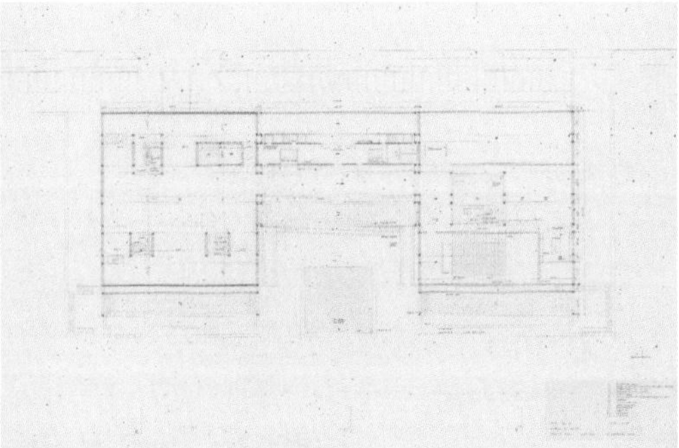

The concrete according to the specifications

Once the constituent parts of the project have taken on a nearly definitive configuration, towards the end of 1968 Kahn's office begins to prepare the documents needed for their construction. The contractor Thomas S. Byrne, Inc., of Fort Worth, with A. Thomas Seymour in charge, is assigned the job, while Texas Industries, Inc. supplies the concrete.[14] The technical rules for the fabrication and use of the formwork and concrete continue to form the main chapter of the specifications for Kahn's works of architecture. Other specifications are prepared during the course of 1969 by Komendant.

It may seem surprising that the architect who more than any other in the United States is contributing to the rise of a new way of thinking about architectural concrete feels it is appropriate, in the preparation of the specifications for the concrete of the Kimbell Art Museum, to copy the part on "Architectural Concrete" contained in the specifications drawn up for the construction of the International House of the University of Pennsylvania, at 3701 Chestnut Street in Philadelphia, a work by John A. Bower and Frederick M. Fradley (designed in 1968, the student residence in exposed poured-in-place concrete was opened in 1970). The fact that Kahn's office makes use of the specifications of the International House instead of the ones prepared for the Salk Institute can probably be explained by the systematic organization of the text of Bower & Fradley. The documents on the concrete of the National Assembly Building, still under construction when Kahn designs the Kimbell Art Museum, contain descriptions of techniques that are too particular with respect to American concrete construction methods, so no part of that material can be used. Nevertheless, a common root can be seen among the various specifications for architectural concrete, given the fact that the document of Bower & Fradley contains phrases on the quality of exposed concrete that, in term, seem to have been borrowed from the specifications written by Kahn and his staff for the Salk Institute.

Bower & Fradley have written a paragraph that is useful to understand the difficulties, in the United States, regarding acceptance of *béton brut* (this paragraph is eliminated in Kahn's specifications).

"A great deal of thought has been devoted to methods of achieving superior concrete work at competitive costs. In general this is to be accomplished by more reuse of superior, more sophisticated formwork than is ordinarily used on a concrete building. During the bidding period there will be a conference to explain to bidders the formwork system and economies to be gained thereby and to review the concrete work in the demonstration construction being built in the field. It is considered essential that bidders

[9] Marshall D. Meyers, letter to Preston M. Geren, 21 December 1967, LIK, 030.II.A.37.5.
[10] [Marshall D. Meyers], "Tentative schedule of finishes. Kimbell Art Museum, Forh Worth, Texas," cit.
[11] "Outline Specification for preliminary working drawings for Kimbel Art Museum," 30 December 1968, LIK, 030.II.A.37.21. In the version of the project of November 1968 the secondary entrance for visitors reaching the museum by car also appears.
[12] Marshall D. Meyers, letter to August E. Komendant, 27 November 1968, LIK, 030.II.A.37.24.
[13] Marshall D. Meyers, letter to Richard Kelly, 27 November 1968, LIK, 030.II.A.37.25.
[14] Thomas S. Byrne, Inc., Engineers & General Contractors, was founded in 1923 by Thomas Sneed Byrne and had constructed, in Fort Worth, from 1958 to 1961, the Amon Carter Museum of Western Art designed by Philip Johnson, a building completely clad in pale stone with surfaces marked by small cavities or fossil imprints. Johnson's museum is at 3501 Camp Bowie Boulevard, near the site chosen for the construction of the Kimbell Art Museum.

familiarize themselves with the techniques necessary to achieve the quality of concrete work desired prior to submitting their bids."

In the passage taken from the specifications of Bower & Fradley, which includes various considerations on the "quality of the concrete," its "uniform color" and the "appearance of form contact surfaces exposed to view," on "Exposed Architectural Concrete" (Kahn's staff copies these expressions, including the upper case letters), Kahn and his staff add the technical characteristics that generate the formal quality of his architectural concrete, starting with the Salk Institute. These additions include the "V-type" profile and the treatment of the surfaces of the plywood formwork with the product used in the Salk Institute – "Burke Form Coating (for Raw Wood only)" of Burke Concrete Accessories, Inc.[15] The emphasis Kahn puts on the tools for making the concrete is demonstrated by the fact that the title of the section of the specifications of Bower & Fradley, "Architectural Concrete," is first changed to "Formwork for Concrete" and then to "Formwork and Concrete Placement." In the more general text of the specifications for all the works of the museum, the "Outline Specification," we find indications on another characteristic of Kahn's architectural concrete: "Form ties are to be recessed cone type snap or screw ties regularly spaced, left partially filled."

The decision to make no modifications to the concrete surface after stripping leads Kahn to consider "acceptable" the "fins broken in stripping."[16] In the note written by Seymour in the margin of the stipulation that no intervention should be allowed on the concrete after stripping, the typical adjective of the concrete of Le Corbusier is used: "will look rough – ok?"[17] In one of the versions of the "Outline Specification" we also find a statement of principle regarding the nature of the concrete of the museum: "Concrete is to be and look rugged and direct and economical."[18]

Research on the color of concrete: the question of the pozzolan of the Salk Institute

During 1969 the technical characteristics of the concrete of the Kimbell Art Museum are defined, from its degree of workability to its color. One of the basic requirements of the compound is a result of the methods of its placing for the construction of the vaults. For this purpose, Texas Industries, Inc., recommends "hardrock mix design for the post-tensioned arches," with smaller aggregate than usual to facilitate pumping.[19] According to the company, this type of compound is also ideal for the finish of the intrados of the vaults desired by Kahn: "This should provide better pumping and a better-looking exposed concrete, since the concrete is not going to be rubbed or 'dressed up' after the forms are removed."

The other question being examined starting in the spring has to do with the color of the materials. The exposed concrete, the Travertine stone of the panels, the surface of the roofing of the vaults, and even the pavements around the museum, are decisive factors for the finishing of the work.

[15] "Specifications for formwork and concrete placement, Museum for Kimbell Art Foundation, Ft. Worth, Texas," LIK, 030.II.A.95.5.
[16] "Outline Specification for preliminary working drawings for Kimbel Art Museum," 30 December 1968, cit.
[17] Thomas S. Byrne, Inc., "Notes and questions on outline spec.," 1 January 1969, *ibid.*
[18] "Outline Specification for preliminary working drawings for Kimbel Art Museum," 30 December 1968, cit. The phrase is used verbatim in, "School of Fine Arts, Fort Wayne Fine Arts Foundation. [...] Outline specification," 19 January 1968, LIK, 030.II.A.17.6. On a sheet of handwritten notes with sketches, we read: "Nature of column tie was explored (tie between close spaced column on 8'-0" centers). Columns could be cantilevered from foundations. However better to have tie at top of columns so they can move together. Tie need be only 6" to 8" deep from top of column" ("MDM [Marshall D. Meyers] to Komendant," 11 February 1969, LIK, 030.I.C.730.014).
[19] Robert E. May, of Texas Industries, Inc., letter to A. Thomas Seymour, 21 February 1969, LIK, 030.II.A.95.4.

In the various documents of the preliminary design phases the chromatic nature of the concrete is never specified. The first clue as to the desire to obtain a particular color is found in an interview with Kahn, done during a visit to Fort Worth. The journalist from the local newspaper "Fort Worth Star-Telegram" publishes, on 3 May 1969, the words of Kahn regarding the color of the concrete of the Kimbell Art Museum: "the concrete will be treated chemically to give it a rose tone."[20] During the same visit to Fort Worth, Kahn talks with Seymour about the same issue of the color of the concrete ("there was some conversation pertaining to a rose colored concrete"[21]). The indication of the rose color to be obtained by means of pigments demonstrates that Kahn, after the experiences in La Jolla with a compound containing pozzolan, is willing to experiment with coloring effects in the compound achieved with chemical treatments.

The statement made to the "Fort Worth Star-Telegram" takes the representatives of Texas Industries, Inc., by surprise, and they turn to Kahn directly for more information: "From the specifications that we have, there is no mention of colored concrete that we have been able to discover. It was Mr. Seymour's understanding that your intentions were to use a type of pozzolith or other admixure to give the rose tone. I would appreciate your sending me the specifications covering this or your thoughts on what type of admixure you had in mind [...]."[22]

But Kahn does not return to the subject of the rose color of the concrete of the Kimbell Art Museum, in spite of other requests from Texas Industries, Inc.[23] Instead, he receives an invitation from Architectural Concrete Consultants, Inc., of Houston, Texas, to take part in a convention in Houston, from 2 to 4 June, on "ideas in architectural concrete." The particular feature of the samples of the many products presented at the convention lies in the coloring of the concrete through control of the aggregate, not the addition of pigments. The explanation of the characteristics of the samples, sent to Kahn by the president of Architectural Concrete Consultants, Inc., contains a principle, "painting with stone," that cannot help but reinforce Kahn's belief in the validity of the experiments done for the Salk Institute on the concrete mix.[24]

Though the specifications of the Kimbell Art Museum indicate no particular shade for the concrete, after the interview with the newspaper in Fort Worth Kahn is back in contact with the suppliers and the companies that made the concrete of the Salk Institute, whose color remains the goal to achieve for the concrete of the new museum, in his view. It is clear that he now wants to proceed in keeping with the principle of "painting" with the components of the concrete. The fact that he has declared his intention of creating a rose tone suggests that he is thinking of making concrete similar to that of the Salk Institute.

Many documents of 1969 on the choice of the concrete for the Kimbell Art Museum offer glimpses of the success achieved in the American technique

[20] Doug Clarke, *Kimbell Museum. Contract Granted*, in, "Fort Worth Star-Telegram," 3 May 1969, Saturday.

[21] Thomas S. Justiss, Jr., of Texas Industries, Inc., letter to Louis I. Kahn, 7 May 1969, LIK, 030.II.A.37.21.

[22] *Ibid.*

[23] On 23 May Justiss again contacts Kahn to get a response about the "rose colored concrete," given Kahn's silence on this aspect – neither do the representatives of Byrne, Inc., have any further information on the subject (Thomas S. Justiss, Jr., letter to Louis I. Kahn, 23 May 1969, LIK, 030.II.A.37.21).

[24] "We display architectural concrete panels," the president of Architectural Concrete Consultants, Inc., explains, "which show the *principles* we use in 'painting with stone' to achieve a multitude of effects and colors without the use of pigment. These are not typical exposed aggregate panels to be seen at a precast concrete suppliers' yard. Many samples represent the effects to be achieved on prominent buildings already under construction" (James M. Shilstone, president of Architectural Concrete Consultants, Inc., letter to Louis I. Kahn, 26 May 1969, LIK, 030.II.A.37.21).

of concrete as produced at La Jolla. From these documents we learn that the particular composition developed by Kahn, Komendant and the various companies has become a true product, also used on other worksites by the Riverside Cement Company, though only in California; a sort of artistic concrete with its own color, and therefore comparable to the International Klein Blue (IKB) invented by the painter Yves Klein. Kahn's mixture even becomes known, for the clients of the Riverside Cement Company, as the Salk Mix, and it is also used – under the aegis of the same company – for the construction of the Hoffman Hall by Pei at the University of Southern California. But when Kahn gets back in touch with the Riverside Cement Company, by now the pozzolan of the Airox Company, which was used in the concrete of the Salk Institute, is no longer being produced, due to American legislation to prevent air pollution. These regulations have imposed modifications to manufacturing processes, changing the factor that is Kahn's primary concern, namely the color of the product. "[…] Airox's Pozzolan is now made by a different process resulting in a gray or neutral color," a member of Kahn's staff warns.[25] Instead, a pozzolan similar to the original Airox type is now made by the Riverside Cement Company and marketed under the name Riverside Pozzolan. The characteristics of Riverside Pozzolan are described by Walter H. Price, Vice-President of the American Cement Corporation of Los Angeles, Riverside Division, in a document prepared in September 1968 that reaches Kahn in July 1969, when he is working on the composition of the concrete for Fort Worth.

After reminding the reader that pozzolan of volcanic origin was used by the ancient Romans to make hydraulic cements, and that pozzolan can also be artificially produced, Price lists the "advantages of using pozzolan in concrete": "1. Pozzolans decrease leaching by combining with lime liberated during the hydration of Portland cement. 2. Portland cement-pozzolan mixtures will produce lower heat of hydration and develop higher ultimate strength than straight Portland cement. Pozzolan is used in massive structures where temperature control is important to reduce cracking. 3. Concrete is made more impermeable by the addition of pozzolans. 4. Pozzolans reduce disruptive expansion and popouts caused by reaction between certain aggregates and alkalies in the cement. 5. Pozzolans improve the resistance of concrete to sulfate attack. 6. Pozzolans improve the workability of the fresh concrete and reduce bleeding and segregation and facilitate placing operations. 7. Concrete is made slightly more acid-resistant by use of pozzolans."[26]

Produced by firing "raw material" at 1400-1600 Fahrenheit, Riverside Pozzolan is described as the "best pozzolans produced in this country," capable of making a concrete that after seven days takes on strength of 1600-2200 pounds per square inch (psi), and after 28 days "above 100% as compared to the straight Portland cement mixes." It should not be overlooked, where Kahn's interest in the chromatic quality of concrete is concerned, that Price underlines the effects of pozzolan added to the compound: "Also, pozzolan having the color of Riverside Pozzolan has been used to obtain a beige color of concrete surfaces for architectural purposes."[27]

The beige color of the Riverside Pozzolan is not that of the original pozzolan of the Airox Company, nor is it the rose color Kahn has indicated for the concrete of the Kimbell Art Museum. Nevertheless, Kahn puts the Riverside Cement Company and Texas Industries, Inc., into contact, to conduct trials on concrete with the addition of Riverside Pozzolan, for the construction of the Kimbell Art Museum. In July, Kenneth R. Merrigan, manager of the Riverside Cement Company (or the Riverside Division of the

American Cement Corporation) has five bags of Riverside Pozzolan sent to Arlington, Texas, to the attention of Guy Newman, technical director of Texas Industries, Inc., for study of the composition of the concrete for the Kimbell Art Museum. Merrigan guarantees that the Riverside Pozzolan "is almost identical to the Airox Pozzolan that was used on the Salk Institute." On this occasion he also sends copies to Kahn's office of the documents on Riverside Pozzolan, including the one written by Price, with a certificate issued by the United States Department of the Interior attached, regarding tests performed on a sample of Riverside Pozzolan.[28]

"The primary purpose for its use on the Salk job," Merrigan explains to Newman, "was to meet the color requirements as specified by the architect, and this was accomplished by blending 12% by weight pozzolan with our Type II cement. In our mix design was did not consider the pozzolan as contributing to the concrete's compressive strength – it was merely an addition. However, our tests have shown that it does increase compressive strengths significantly after about 4 days. I am sure trial mixes using your materials will bear this out. As I mentioned, adding the pozzolan at the ready-mix plant has worked satisfactorily on several jobs subsequent to the Salk job."[29]

Kahn's office also gets information on an additive for the concrete to obtain specific hues, Colorfull, a registered trademark of the Owl Manufacturing Company, based in Arcadia, California. The fundamental ingredient of Colorfull is a pozzolanic material.[30] On one page of the advertising booklet of Colorfull sent to Kahn, we find phrases on the question of the color of the concrete obtained with pozzolan at the Salk Institute: "Pozzolan color now changed due to different process"; "Pozzolan added about 4/yd. US $6.00 for color in about 1962."[31]

While Kahn's office, thanks to Merrigan and Newman, explores the possibilities of artificial pozzolan as an ingredient for the coloring of the concrete of the Kimbell Art Museum, and while Kahn also obtains information about Colorfull, Geren's office, for the definition of the color of the concrete of the same building, starts testing, with the cooperation of the technicians of Architectural Concrete Consultants, who in Fort Worth are working on the construction of the City Hall designed by Edward Durell Stone in 1967 (Geren's office has similar responsibilities on that project to those it has been assigned for the Kimbell Art Museum). Thanks to the Architectural Concrete Consultants, four concrete samples are quickly prepared with different compositions, to show Kahn "the color of locally available cements and aggregates."[32] These samples become the points of reference for later

[25] Marshall D. Meyers, letter to Thad E. Harden, Jr., 11 July 1968, LIK, 030.II.A.37.5.

[26] Walter H. Price, "Riverside Pozzolan," 30 September 1968, LIK, 030.II.A.37.21.

[27] *Ibid.*

[28] See Graydon E. Burnett, Chief Research Scientist of the United States Department of the Interior, Bureau of Reclamation, in Denver, Colorado, letter to Walter H. Price, stamped received 12 February 1969: "We have completed tests on a sample of Riverside pozzolan"; the tests cover "physical properties" and "chemical analysis" (LIK, 030.II.A.37.25).

[29] Kenneth R. Merrigan, of the Riverside Division, American Cement Corporation, letter to Guy Newman, of Texas Industries, Inc., 23 July 1969, LIK, 030.II.A.37.21.

[30] "Colorfull," protected by a patent in 1962 by Elmer O. Rodeffer – "Material and method for coloring Portland cement concretes" – is "a mixture of lime-proof, non-fading pure synthetic mineral-oxide pigments, a pozzolonie-type, predispersing carrier, and a water reducing admixture"; it comes in different shades of "reds, tans, golds, yellows, and green" ("Colorfull" advertising booklet, March 1969, LIK, 030.II.A.95.6).

[31] *Ibid.*

[32] Thad E. Harden, Jr., letter to David P. Wisdom, 7 July 1969, LIK, 030.II.A.37.5.

choices of the concrete for the museum. The first two samples have Bridgeport Limestone and Bridgeport Limestone Sand as aggregate, supplied by Texas Industries, Inc., and use the cement of Texas Industries, Inc., (sample No. 1) and the cement of Gifford-Hill (sample No. 2). The third sample is supposed to be made with the cement of the Trinity Portland Cement Company, one of the most important manufacturers in Dallas. Since that type is not available, the sample is made with the cement of Texas Industries Inc., mixed with Bridgeport Limestone and Aledo Sand. The fourth sample is made with Bridgeport Limestone, the cement of Texas Industries, Inc., and a mixture of Bridgeport Stone Sand and Aledo Sand.

All the surfaces in contact with the mold are sanded, because this is the usual treatment of exposed concrete done by the technicians of Architectural Concrete Consultants. For this reason, Thad Harden of Geren's office warns Kahn that only "the edge and back" can "give you an understanding of what the as-cast color and condition would be."[33] But Kahn cannot adequately judge the precision of the color of the four samples because, as Meyers explains, the sanding of the smooth surface and the roughness of the back alter the perception of the exact tone, and the edges are not large enough to get a clear idea of the color. The idea that Kahn wants to obtain, as in the case of the Travertine as well, the same chromatic qualities as in the Salk Institute, is borne out by the fact that Meyers makes a pithy account of the research conducted at La Jolla, concluding with the hope that Geren's office too can launch similar experimentation, given the approaching start of construction of the Kimbell Art Museum. The basic question to obtain the color of the Salk Institute remains that of procuring the original pozzolan produced by the Airox Company. It is worth noting that Meyers does not mention the new Riverside Pozzolan, whose beige color does not correspond to that of the original Airox pozzolan – the one that possesses, as he emphasizes, the "old color."

"The local concrete available in the San Diego area for the Salk Institute," Meyers writes, "was too green in color and Pozzolan was added to the mix to give it a warmer tone and at the same time aid the strength of the concrete. Various other materials and admixtures were investigated at that time such as synthetic mineral-oxide pigments, but no one could get the confidence that these materials were truly 'non-fading' as the manufacturers implied. Pozzolan was also cheaper than the pigments. [...]. Mr. Merrigan said, however, that they (Riverside) can supply the correct Pozzolan and also have been supplying a concrete called the 'Salk Mix' for local jobs [...]. Since they can supply Pozzolan of the old color it may be advisable to have Tom Seymour or Texas Industries get a hold of some for a test panel or panels. [...]. It also might be a good idea to get Texas Industries or Architectur-

[33] *Ibid.*
[34] Marshall D. Meyers, letter to Thad E. Harden, Jr., 11 July 1968, LIK, 030.II.A.37.5.
[35] See J. S. Radney, letter to Don W. Baumgarner, 15 August 1969, LIK, 030.II.A.37.34, and Don W. Baumgarner, letter to August E. Komendant, 18 August 1969, LIK, 030.II.A.37.34.
[36] Don W. Baumgarner, letter to August E. Komendant, 18 August 1969, cit., with handwritten notes indicating Kahn's choices.
[37] August E. Komendant, letter to Don W. Baumgarner, 26 August 1969, LIK, 030.II.A.37.24.
[38] Marshall D. Meyers, letter to Thad E. Harden, Jr., 21 August 1969, LIK, 030.II.A.37.6.
[39] A. Thomas Seymour, letter to Louis I. Kahn, Preston M. Geren, August E. Komendant, 22 August 1969, LIK, 030.II.A.37.34.
[40] [Marshall D. Meyers], "Tentative schedule of finishes. Kimbell Art Museum, Forh Worth, Texas," 21 December 1967, with handwritten corrections made on 3 December 1968, cit.

al Concrete Consultants to see if Pozzolan can be obtained locally that would provide the color Mr. Kahn got at Salk. […]. We have received correspondence from a Mr. Thomas S. Justiss, Jr., of Texas Industries, requesting information about this and I would appreciate your telling him what we are doing about it. It also would be good to have whoever pursues this get information from Merrigan about the complex mix: type of cement, aggregate, etc."[34]

The technicians of Texas Industries, Inc., prepare different types of compounds for the concrete of the Kimbell Art Museum containing Pozzolith and also pozzolan, prepared with the cement of Texas Industries, Inc., or of Gifford-Hill, to submit to the scrutiny of Komendant and Kahn.[35] Kahn selects Mix I, made with the cement of Texas Industries, Inc., Paolo Pinto Limestone, Limestone Concrete Sand and Pozzolith; instead, he ranks in seventh (last) place the compound containing 9% pozzolan (the only one with this additive).[36] Komendant receives the seven samples towards 26 August: "In the following weeks I would like to perform several tests on these samples and will carry out also some microscopic investigation."[37] In the meantime, Meyers tells Harden the exact composition of the concrete of the Salk Institute.[38] Though Kahn has excluded the compound prepared with pozzolan, Texas Industries, Inc., makes samples of "'pozzolan' tinted concrete" and sends them to Komendant and Geren on 21 August – they are charged with passing the samples on to Kahn.[39]

Color and working of the Travertine

The same care that goes into the selection of the ingredients to obtain a specific color of the concrete is also seen in the choice of the cladding material for the walls of the museum. When several materials indicated in 1967 are changed (such as stainless steel in place of bronze), for the walls Kahn begins to consider the hypothesis of Roman Travertine, which is the material used in the finished work.[40] Only for the wall above the long beam of the entrance to the museum from the parking area is the Travertine replaced by concrete, for reasons of insulation (at first the concrete is envisioned in prefabricated panels[41]).

Also in the case of the Travertine, the model remains that of the Salk Institute. Walker & Zanger, which supplied the material for La Jolla, is again contacted by Kahn for samples and an estimate for the Travertine of the museum.[42] The handwritten note "Near Rome!" placed on a letter of Walker & Zanger refers to the Travertine desired by Kahn for its chromatic hue, as is borne out by what Meyers writes: "Mr. Kahn looked at the samples and noted that the Travertine color was 'rosier' or more 'peach colored' than the Travertine used on the Salk Institute. The Travertine used on the Salk Institute was 'yellower.' Mr. Kahn said that there were several

[41] See Marshall D. Meyer, letter to A. Thomas Seymour, 25 March 1969, LIK, 030.II.A.37.34. The text of the specifications is modified to include the option of the two materials for the cladding of the upper level, Travertine stone "or exposed concrete" ("Outline Specifications for Kimbell Art Museum, Kimbel Art Foundation, Forth Worth, Texas," 7 April 1969, LIK, 030.II.A.95.5). See also Marshall D. Meyers, letter to Thad E. Harden, Jr., 20 May 1969, LIK, 030.II.A.37.5. During the course of 1969 there is discussion on the support of the Travertine slabs, using webs of reinforced concrete, as suggested by Seymour, or "precast separations and braces," as urged by Wisdom (David P. Wisdon, letter to Thad E. Harden, Jr., 27 May 1969, LIK, 030.II.A.37.5).

[42] John Iberti, of Walker & Zanger, letter to Louis I. Kahn, 3 April 1968, LIK, 030.II.A.37.14.

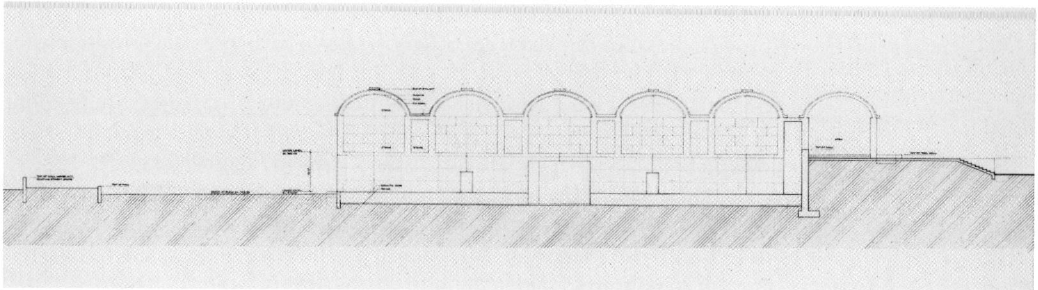

315-316. Kimbell Art Museum, Forth Worth, Texas, 1966-72.

315. "North Elevation," 29 August 1969 (LIK, 030.I.C.730.11).

quarries for Italian Travertine but he thought that what was used on Salk was from a quarry near Rome. [...]. The finish on these samples should also conform to the finish used on Salk."[43]

The correspondence between Meyers and Walker & Zanger focuses on finding the exact color of the Travertine used in the Salk Institute. Walker & Zanger guarantees that the Travertine of the samples sent to Kahn comes from the same quarry as that of the Salk Institute, and that the difference of color noticed by Kahn can be attributed to a different surface treatment, given the fact that the Travertine of the Salk Institute was "rough sawn," while that requested for the Kimbell Art Museum has a "honed finish."[44] "Dave! I think Lou wants 'rough sawn' Travertine here," Meyers writes.[45] In the meantime MacAllister, who is still in La Jolla to supervise the completion of the job, is asked to send a sample of the Travertine of the Salk Institute for comparison.[46] But it is Kahn himself who gets ahold of the sample: "Enclosed is one-half of a piece of Travertine Mr. Kahn brought back from the Salk Institute on his last visit. This is the kind of Travertine he wants for the Kimbell Art Museum. He said that the lighter side of the sample was bleached in the sun."[47] Just as the concrete with pozzolan is called the Salk Mix, so the sample sent by Kahn is referred to as "Light Roman Travertine-Salk" by Walker & Zanger.[48]

To choose the Travertine, Wilcots goes to Italy in July 1969, to visit the suppliers of Walker & Zanger in Carrara and in the vicinity of Rome.[49] The company sends Kahn's office two samples of Travertine corresponding to the stone used at the Salk Institute, indicating the reason for the visible differences: "It is actually difficult to place the sample of the Travertine from the Salk job against the samples that I have sent to you face to face, and to appraise them as being *exactly* equal. However, if you would break the Salk sample and break my sample and put the two exposed broken edges together, you will see that these Travertines are exactly the same. The Salk sample is weathered on one side and stained on the other and the true color can only be seen in a cross section of the piece."[50]

Kahn compares the two samples to that of the Travertine used in La Jolla and decides to change the type of workmanship, to make the Travertine of the museum identical to that of the Salk Institute. "He has selected the 'Light Roman Travertine-Salk' with the sand-sawed finish as our Salk sample, rather than the diamond-sawed finish as your two samples," Wilcots advises.[51] But this is still not the definitive choice.

Kahn's ideas on the chromatic tuning between the materials, starting with the concrete, are reported by Seymour in every nuance, including the value of the grain of the Travertine. Once again, the fundamental role of concrete in Kahn's color schemes is confirmed.

"The Architect, Mr. Louis I. Kahn, prefers a slightly darker type of Travertine from the immediate Rome area which is slightly grayer so as to blend nicely with the cement gray color of the exposed concrete which would be adjacent to the Travertine. At the same time, Mr. Kahn wants a minimum of contrast between the light and dark veining of the marble itself. In other words, he would prefer the Travertine and concrete to blend together in tone which would mean the Travertine should be of a grayer tone rather than a yellow-orange tone and should have a minimum of whitish strata in same so as to minimize the contrast in the veining itself."[52]

The color of the Travertine for the panels influences that of the compound chosen by the landscaper charged with developing the design of the park, George E. Patton. In order to find a chromatic match with the stone slabs of the museum, the "stabilized gravel" he has chosen for the pavement under the portico and around the building has a large percentage of "yellow or buff color" (the binder is "bituminous concrete"[53]).

For the facing of the vaults, Kahn envisions the application of a material that will be in tune in terms of color with the others, and can be easily

[43] Marshall D. Meyer, letter to John Iberti, 4 April 1969, LIK, 030.II.A.37.14.
[44] John Iberti, letter to Marshall D. Meyers, 7 April 1969, LIK, 030.II.A.37.14.
[45] Marshall D. Meyer, handwritten notes in, John Iberti, letter to Marshall D. Meyers, 7 April 1969, cit.
[46] See John Iberti, letter to Marshall D. Meyers, 13 May 1969, LIK, 030.II.A.37.14.
[47] Marshall D. Meyers, letter to John Iberti, 21 May 1969, LIK, 030.II.A.37.14. For the granite, Walker & Zanger sends a sample of "Italian Granite" considered "*too* noisy for us," requesting a "light or medium gray (cold-gray)" with a "more subdued pattern." On the granite see John Iberti, letter to Marshall D. Meyers, 9 June 1969, LIK, 030.II.A.37.14.
[48] John Iberti, letter to Marshall D. Meyers, 27 May 1969, LIK, 030.II.A.37.14.
[49] For the choice of the stones, Wisdom goes in person first to Carrara, in June, and then to Rome "to visit Travertine quarry" in July, always with representatives of Walker & Zanger (see John Iberti, letter to Henry N. Wilcots, 17 June 1969, LIK, 030.II.A.37.14, and Henry N. Wilcots, letter to Walker & Zanger, 19 August 1969, LIK, 030.II.A.37.14).
[50] John Iberti, letter to Marshall D. Meyers, 26 September 1969, LIK, 030.II.A.37.14.
[51] Henry N. Wilcots, letter to John Iberti, 1 October 1969, LIK, 030.II.A.37.14.
[52] Thomas A. Seymour, letter to The Domestic Marble Company, 20 May 1970, LIK, 030.II.A.37.34. The whole paragraph is accompanied by a question posed to Kahn to get confirmation on the chromatic relationships between the materials.
[53] George E. Patton, letter to Marshall D. Meyers, 23 April 1969, LIK, 030.II.A.37.34. On the "yellow colored aggregate" and the composition of the product see, "Stabilized gravel paving," n.d., received by Kahn's office on 23 April 1969, LIK, 030.II.A.37.34. The granite for the parapets and other details is chosen by Kahn, comparing the material – "Sierra White" – with the Travertine samples of the Salk Institute. When the Travertine turns out to be perceptibly different from that of the Salk Institute, a sample is requested to be sent to the worksite so Kahn can reconsider his choice (Marshall D. Meyers, letter to Frank Sherwood, 10 June 1970, LIK, 030.II.A.37.6).

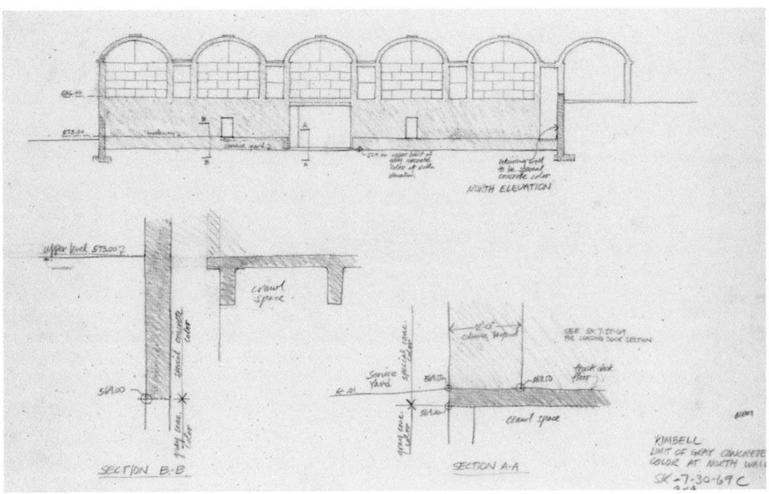

316. "Limit of Gray Concrete Color at North wall," 30 July 1969 (LIK, 030.I.C.730.15).

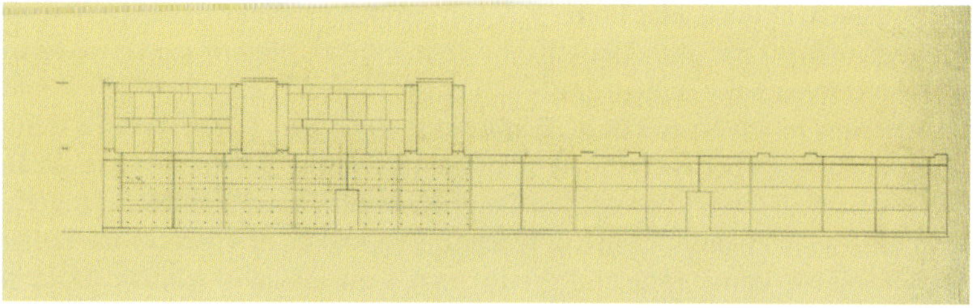

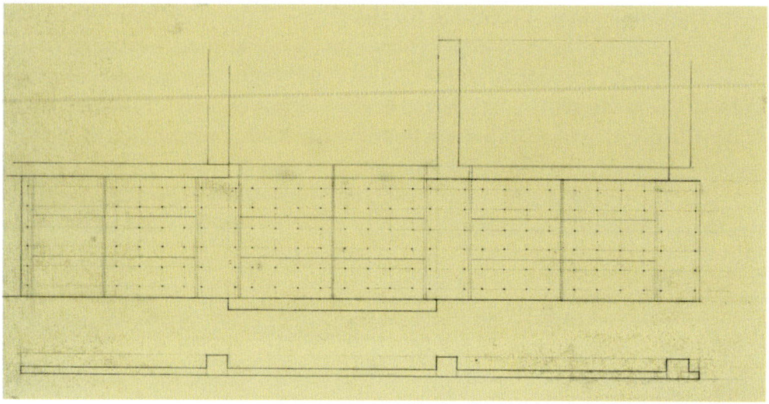

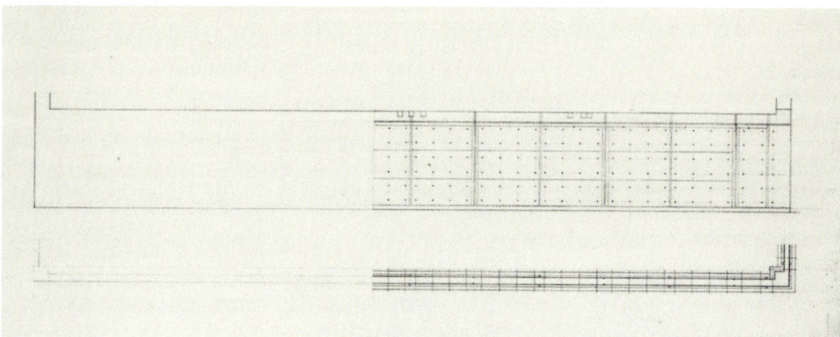

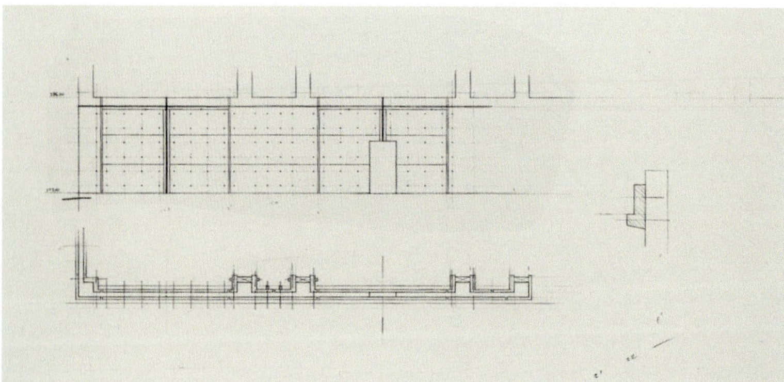

317-320. Kimbell Art Museum, Forth Worth, Texas, 1966-72. Formwork studies (LIK, 030.I.C.730.15; 18).

adapted to the curvature: lead. Kahn has already used lead as a seal for the tie holes of the Salk Institute. Harden is critical of this choice for the facing of the vaults, because lead is expensive and the conditions of its durability "in direct contact with concrete" are not known. "Our experience," he adds, "has been that a chemical reaction sets up between the lime in the concrete and the lead such that the lead deteriorates."[54]

The formwork for the walls

When the excavations for the foundation begin, on the 1° July 1969, evaluation is underway of the technical aspects and details of the formwork for the walls. Meyers supplies information to Harden on the materials needed for the treatment of the panels to obtain a perfect smooth concrete surface – the "plastic coated plywood" of the Burke Concrete Accessories Company. Once again, the model of reference is the Salk Institute.[55]

When the foundations are poured Kahn's office has not yet provided the drawings for the retaining walls the contractor is supposed to start building in the last week of August. Geren's office takes the initiative of preparing the missing drawings.

"With the prints of the job that are to be mailed today," Geren writes to Kahn, "you will find that we have drawn wall elevations and have drawn the rustication joints and form joints to the best of our understanding. We asked you many weeks ago for the pattern of the joints that you wanted, but since we have not received any direction from you, we have done the best we can. We do not now expect to receive correction to what we have done."[56]

After the initiative of Geren's office on the formwork and after having received the specifications of Komendant on the "reinforced & prestressed concrete,"[57] Kahn decides to send Frederick Langford to Fort Worth, on 3 September, "to discuss with Seymour the formwork construction which leads to the valid imprints on the concrete."[58] During his visit, Langford also shows "certain photographs of similar jobs" (most probably of the Salk Institute).[59] Again in this case, as in Dhaka, Langford offers his services as

[54] Thad E. Harden, Jr., letter to Louis I. Kahn, 20 October 1969, LIK, 030.II.A.37.6.

[55] "Generally there are two types of plastic coated plywood: one has a lamination of plastic film, the other is a film which is painted on the plywood. The laminated film does not produce as good a surface, causes red blotches, is prone to damage and delamination and is not repairable. The painted film is a polyurethane film which can be repaired and gives a better surface. On the Salk job we used a polyurethane form coating manufactured by the Burke Concrete Accessories Company of Burlingame, California. The material was applied with rollers in a two-coat process, the second coat applied within 24 hours of the first. It is important to watch out for bubbles and the plywood must be dried on racks in a horizontal position to prevent running" (Marshall D. Meyers, letter to Thad E. Harden, Jr., 5 August 1969, LIK, 030.II.A.37.6). Meyers also proposes products similar to polyurethane, like Aquabar Durofilm, a "moisture-cure ployurethane" produced by Construction Specialities, Inc., of Philadelphia.

[56] Preston M. Geren's office, letter to Louis I. Kahn, 22 August 1969, LIK, 030.II.A.37.6. Geren's office provides Byrne, Inc., with indications on the methods of placement of the panels – the "steel pan formwork" of the entire "Lower Level concrete floor" (Preston M. Geren, letter to Thomas S. Byrne, Inc., 22 August 1969, LIK, 030.II.A.37.6). On 22 August, 95% of the excavation has been completed and 31 foundation piles have been driven. "We will not frame any 'rustication' joints into this wall pour at this time" (Thomas A. Seymour, letter to Louis I. Kahn, Preston M. Geren, August E. Komendant, 22 August 1969, LIK, 030.II.A.37.34).

[57] With the letter of 24 July, Komendant outlines his responsibilities and those of Geren, and the last decisions are made on the dimensioning of the columns and walls of the lower level to make it possible for construction to begin (August E. Komendant, letter to Preston M. Geren, 24 July 1969, LIK, 030.II.A.37.24).

[58] Louis I. Kahn, telegram to Thad E. Harden, Jr., 28 August [1969], LIK, 030.II.A.37.6. Identified as a "form consultant," Frederick Langford will interact with Seymour on "polyurethane and methods" ([Thomas A. Seymour], "Memorandum," 29 August 1969, LIK, 030.II.A.37.34).

[59] Thomas A. Seymour, letter to Marshall D. Meyers, 3 September 1969, LIK, 030.II.A.37.34.

321. Kimbell Art Museum, Forth Worth, Texas, 1966-72. "South Elevation of Sample Walls," 9 October 1969 (LIK, 030.I.C.730.14).

a consultant for the formwork, to be hired with a separate contract.[60] The drawing of the pattern of the joints and every technical detail of the formwork is now required, and Seymour asks the offices of Geren and Kahn for the necessary documents, though he also emphasizes that the quality now required for the concrete was not covered in the initial estimates, when the walls were supposed to be clad on all levels.[61] During the summer, until the early days of September, Khan and his staff (Langford, in particular) are engaged in the study of the patterns of the assembly of the large plywood panels, arranged horizontally with the V joints developed in the Salk Institute, both in the simple version obtained by abutting panels and the double version resulting from the use of the key strip. Other studies document the introduction in the basic pattern of a large band with a slender relief strip, the so-called "gang-strip with shrinkage control molding."

Sample walls: in pursuit of the perfect line

The concept of growth through a process of experimentation that happens on the worksite, leaving visible traces on the finished work, is supposed to also determine the character of the construction of the Kimbell Art Museum, like that of the Salk Institute or the National Assembly Building.

The samples to test the design of the formwork and the mix design of the concrete are planned, at first, for the walls of the lower level of the museum. For this purpose, Kahn has the drawing FA-2 modified "to include the forming of the sample pours at the North wall."[62] The drawing also contains the details of the "vertical joints, both chamfered ('V' Type) and flush butt ('FB' Type)."[63] The indications of all the details for the making of the panels demonstrate the degree of perfection achieved in the construction of the formwork, thanks to the experience of the Salk Institute. What Meyers writes in September 1969 on the methods for making the edges of the panels that appear in the design can be seen, at this point, as part of an ideal technical treatise on Kahn's architectural concrete:

"Where chamfers occur between sheets, the plywood is to have an edge thickness of 1/8" before sanding (minimum of 3/32" after sanding). The chamfer is to be 45° with 1/4" on each side as shown. This dimension for the chamfer was arrived at after much experimentation on the Salk Institute job and satisfied both aesthetic and construction requirements. Best results were obtained by chamfering edges of each sheet individually with a Stanley router before panel assembly. Plywood below Elevation 573.00 may be standard 3/4" exterior plywood where joints are 'FB' (Flush Butt). Where joints are chamfered, plywood must have 1/8" edge thickness."[64]

[60] *Ibid.* Geren dies in the month of September and his place is taken by Preston M. Geren, Jr. From this moment on, questions are raised concerning the expense reimbursements and consulting fees of different specialists (*cf.* Patricia Cummings Loud, *The art museums of Louis I. Kahn*, Durham, Duke University Press, 1989, p. 153).

[61] "What we want from Geren on the subject of special form pattern will be a complete layout of all intended joints on all exterior surfaces where they may occur as well as a typical section showing the physical dimension and size of each joint. We will also need a specification from you or Mr. Langford stating the type of plywood, type of polyurethane material, type of each kind of joint filler, etc. that you think you are going to require on this project" (Thomas A.Seymour, letter to Marshall D. Meyers, 3 September 1969, LIK, 030.II.A.37.34). "On August 27th we asked for you to furnish us the pattern for the joints in the plywood forming on this building. As you know, we have drawn the elevations with a certain pattern and are prepared to make changes, but we need your guidance" (Preston M. Geren, Jr., letter to Marshall D. Meyers, 17 September 1969, LIK, 030.II.A.37.6).

[62] Marshall D. Meyers, letter to Thad E. Harden, Jr., 26 September 1969, LIK, 030.II.A.37.6.

[63] *Ibid.*

[64] *Ibid.* Meyers recommends "Burke Form Coating (for Raw Wood Only)" to apply on the panels according to the instructions contained in the Burke Concrete Accessories Handbook No. 500 ("Follow all instructions on labels of cans and as written in Burke's catalogue Page 105").

[65] One source indicates that the composition of Mix I is as follows: Cement TXI type I;

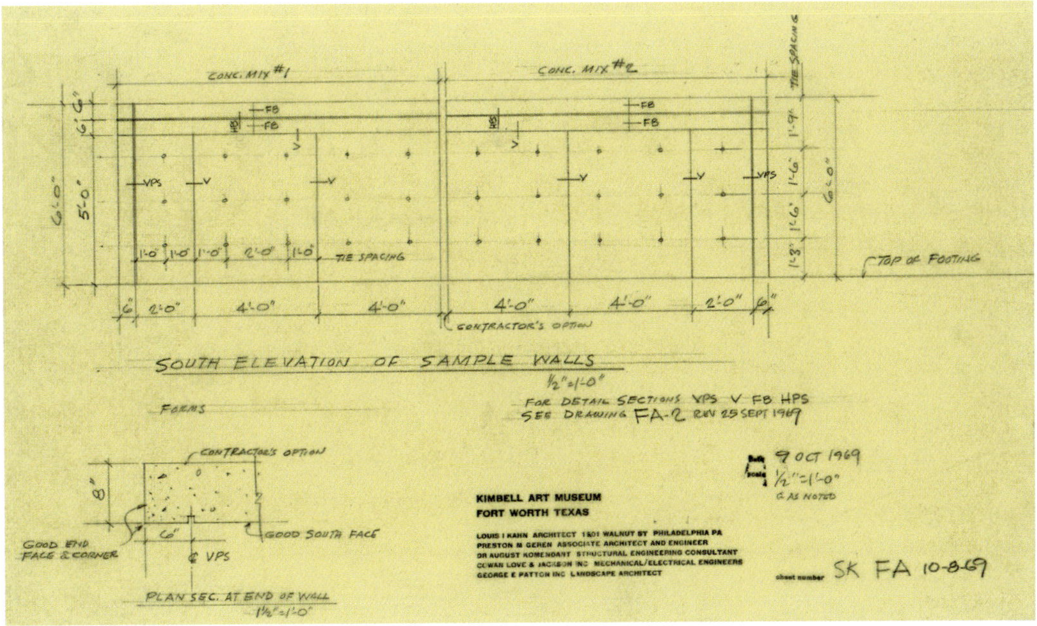

The samples requested by Kahn also have the purpose of testing the effects of the concrete compound on the formal quality of the lines and surfaces. The trials are to be done using the same formwork design for two "sample pours," both prepared with the concrete of Texas Industries, Inc., using the mixes prepared in August Mix I and Mix II.[65] Kahn's office asks Walker & Zanger, again at the time of the two sample pours, for at least three slabs of Travertine measuring 3 × 6 feet, "so this color may be seen with the concrete color."[66] For the composition of the concrete, Kahn envisions using Pozzalin, a volcanic ash normally used to increase the strength and workability of the compound. Seymour asks for specification of the quantity of the Pozzalin, to make a new overall cost estimate for the system of formwork and concrete.[67]

Brown's intervention prevents trials from being conducted on wall sectors that are to remain in place.[68] Therefore it is decided to make two special sample walls, isolated at the edge of the worksite, along Camp Bowie Boulevard.[69] The drawing prepared by Kahn for the making of these samples shows two separate wall sectors to be built with vertically assembled

Paolo Pinto Limestone; Limestone Concrete sand; Pozzolith; and that of Mix II is: Hydrate Lime; Cement TXI type I; Paolo Pinto Limestone; Limestone Concrete sand; Pozzolith (Don W. Baumgarner, letter to August E. Komendant, 18 August 1969, LIK, 030.II.A.37.34).

[66] Marshall D. Meyers, letter to Thad E. Harden, Jr., 26 September 1969, LIK, 030.II.A.37.6.

[67] Thomas A. Seymour, letter to Marshall D. Meyers, 3 September 1969, LIK, 030.II.A.37.34. The preparation of the two sample pours is described by Seymour: "In about two weeks we will be pouring the remainder of the lower portion of the north wall that is not on 'hold.' This pour will be divided into two sides (east side and west side of loading dock) of approximately 10 cu. yds. each. One side to be made up of your Mix I and the other side about the same quantity of your Mix II" (Thomas A. Seymour, letter to Texas Industries, Inc., 7 October 1969, LIK, 030.II.A.37.34).

[68] Thomas A. Seymour, letter to Texas Industries, Inc., 9 October 1969, LIK, 030.II.A.37.34. The letter contains instructions that modify those of the letter of 7 October: the lower north wall will no longer be made with the two mixes, but the owner requests, for 17 or 20 October, "sample wall pours at the upper level out of the way of construction." "The walls will involve possibly 4 to 5 cu. yds. of each type of both Mix I and Mix II," the letter indicates.

[69] See the worksite photograph, LIK, 030.IV.A.730.1, and Thomas A. Seymour, "Memorandum," 10 October 1969, LIK, 030.II.A.37.34. A handwritten note beside the "Memorandum" indicates that on 23 October the sample wall "will be poured & inspected by Kahn."

panels (two of 4 × 5 feet, and one narrower panel of 2 × 5 feet), where the spreaders are placed horizontally, at a level of 2 feet. For the back of the two wall sectors, the choice of the type of formwork is left up to Byrne, Inc.[70] The formwork are made with plywood panels ("3/4" regular plyform backing"[71]). The fact that the test is divided into two sectors leads to the names "Wall No. 1" and "Wall No. 2," each of which has one type of concrete: the TX1-No. 2 Mix for the left sector, or "Wall No. 1"; the TX1-No. 1 Mix for the right sector, or "Wall No. 2."[72] The pump used by Byrne, Inc., for the concrete is one of the best on the market, the Thomsen Model 640, which however "has pumped all other mixes of concrete except the Type I mix using the crushed limestone sand."[73] In the sample wall the ties holes are "filled with plastic clay and lightly greased."[74]

To judge the samples an "appropriate design team" is created, including Kahn, Langford,[75] Brown[76] and representatives of Byrne, Inc., and of Geren's office. The team examines the two wall sectors on 23 October. On this occasion, Kahn chooses the TX1-No. 1 Mix for the quality of the surface obtained with that type of compound. "In wall No. 2 flat surfaces in general were very satisfactory," Langford writes.[77] The team also discusses certain improvements needed in the pour method, also considering the difficulties encountered by the pump with the TX1-No. 1 Mix. For this reason, it is decided to make another sample, "Wall No. 3," also with modification of certain parts of the formwork system. "We will pour a third wall sample of same design as first sample," Seymour reports, "and try to perfect forming technique as discussed at job; [and] we will pump the 'Type I' mix to see if there is any problem involved in pumping same."[78]

322. Kimbell Art Museum, Forth Worth, Texas, 1966-72. View of the sample wall (LIK, 030.IV.A.730.1.18).

For the definitive choice of the Travertine, Walker & Zanger sends four slabs of the stone to the worksite, measuring 3 × 6 feet each, worked in different ways, to be observed together with the concrete sample walls. Two slabs have been specially prepared for the trials ("AR Diamond Sawn - Current"); the others are from the shipment sent to the Salk Institute ("AR Sand Sawn - Salk"), and they are "somewhat aged and would represent the samples after exposure to the weather for a certain period."[79] On the same day as the inspection of the sample walls, 23 October, Kahn also chooses the stone, selecting the "diamond cut travertine sample."[80]

Following the observations of Kahn and Langford on the sample walls, expressed on 23 October, photographs are taken by Brown to fully document their defects.[81] Though "Wall No. 3" is still being built, when he receives the photographs Kahn decides to make another sample wall, the fourth of the series – namely "Wall No. 4" – which takes the necessary corrections into account to obtain the perfect finish of every detail of his architectural concrete, according to the indications supplied on 23 October, which Langford now has to specify using the photographs sent by Brown. "Lou has suggested," Langford reports, "that I make written comments referring to photographs of the existing walls so that a new wall could be poured incorporating the necessary corrections to the formwork built by Thos. S. Byrne, Inc."[82]

In the definition of the technical details, Kahn does not only indicate the Salk Institute as the model for the composition of the concrete and the quality of the Travertine. Aware of the difficulties that can arise on the worksite during the deployment of his system of architectural concrete, he "also suggest[s] that a tour of the Salk job in La Jolla should be organized as soon as possible," with Langford as guide, to show Seymour, Brown, Thad Harden and the Job Superintendent of Byrne, Inc., Virgil Earp, "problem areas, solutions and of course, to get a general feel for the concrete work" (the trip is set for 17-18 November).

The new wall sample requested by Kahn is to be built following the indications supplied by Langford in the letter of 6 November, another fundamental document on the fabrication of Kahn's exposed concrete.[83] "Most of the following comments were made by Lou or myself at the site on 23 October 1969," Langford points out in the letter. Langford discusses all the details of the formwork system and the placing of the concrete, from vibration to the product to be used to coat the panels, starting with the defects observed in Wall No. 1 and Wall No. 2 on 23 October.

[70] See the drawing dated 9 October 1969, LIK, 030.I.C.730.014.
[71] Thomas A. Seymour, "Memorandum," 5 December 1969, LIK, 030.II.A.37.34.
[72] Frederick Langford, letter to Marshall D. Meyers, 6 November 1969, LIK, 030.II.A.37.6.
[73] Thomas A. Seymour, "Memorandum," 5 December 1969, cit.
[74] Marshall D. Meyers, letter to Preston M. Geren, Jr. and Frank Sherwood, 11 May 1970, LIK, 030.II.A.37.6.
[75] *Cf.* Frederick Langford, letter to Marshall D. Meyers, 6 November 1969, LIK, 030.II.A.37.6.
[76] *Cf.* Thomas A. Seymour, "Memorandum," 10 October 1969, cit.
[77] Frederick Langford, letter to Marshall D. Meyers, 6 November 1969, LIK, 030.II.A.37.6. For Wall No. 2 see the photographs of the wall, LIK, 030.IV.A.730.1.
[78] Thomas A. Seymour, "Memorandum," 23 October 1969, LIK, 030.II.A.37.34.
[79] John Iberti, letter to Thomas A. Seymour, 15 October 1969, LIK, 030.II.A.37.14.
[80] Thomas A. Seymour, "Memorandum," 23 October 1969, cit.
[81] Frederick Langford, letter to Marshall D. Meyers, 6 November 1969, cit. See also, Marshall D. Meyers letter to Thad E. Harden, Jr., 6 November 1969, LIK, 030.II.A.37.6, and Thad E. Harden, Jr., letter to Thomas A. Seymour, 10 November 1969, LIK, 030.II.A.37.6.
[82] Frederick Langford, letter to Marshall D. Meyers, 6 November 1969, cit.
[83] *Ibid.*

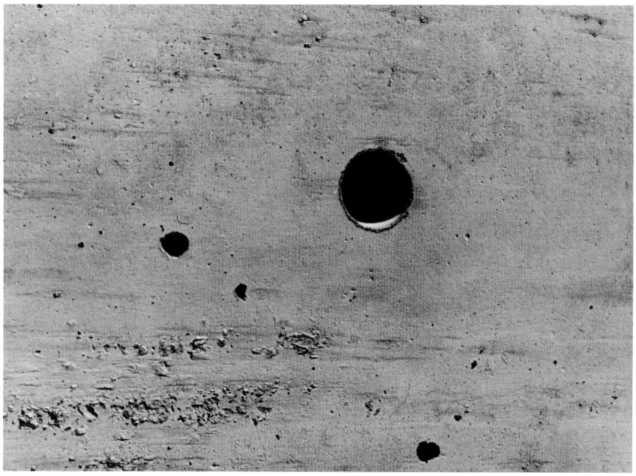

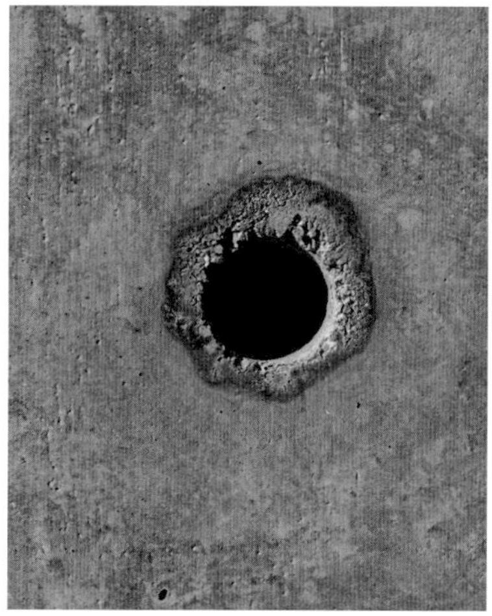

323-327. Kimbell Art Museum, Forth Worth, Texas, 1966-72. Details of the sample wall (LIK, 030.IV.A.730.1.24; 21; 23; 20; 19).

In Langford's notes, the intensity of vibration appears to be a very important parameter for the quality of the finish: "A need for improved vibration technique was evidenced in four places." The vibrators have produced burns on the formwork, which have then marked the concrete surface "in several places (not shown on photos)." Therefore Langford recommends the use of "rubber tipped vibrators." Insufficient vibration of Wall No. 1 is the cause of a "rock pocket at the top" of the wall.

Many of Langford's considerations have to do with the protruding V lines on the concrete of the museum, which in the two trials have a nibbled appearance, even for long segments. The criticisms range from the lack of compliance with the "proper dimension" indicated in detail number 8 of the drawing FA-2, to the imperfect making of the "lower horizontal V-joint" in Wall No. 2, attributed to too much vibration – the line is jagged due to lack of adherence of the concrete to the groove in the formwork.

"Many V-joints were broken," Langford writes. "A certain percentage of this can be expected, however a little more sanding between coats of resin will help to avoid this defect. Horizontal V-joints that occur at the top of the wall should be kept clear of splatter accumulation that dries ahead of the main pour. Elephant trunks should be used if free fall exceeds."[84]

The construction of the low base and the two wall samples demonstrates the need to avoid the phenomenon of bleeding of the pour, by means of "a neoprene tape" to insert between the formwork of the two parts.[85] Bleeding has also caused the imperfections in the corners of the two samples. "The corners of the wall ends," Langford observes, "show excessive bleeding. [...]. Proper placement and tightening of bolts thru bulkhead will eliminate this condition."[86]

A photograph of the detail of a spreader is particularly vivid proof of the surface defects produced in the process of the pouring of the compound.[87] This is the photograph addressed by Langford's comment: "Photo 3 shows a well formed tie hole. Voids produced by air or water bubbles are acceptable, however, approximate diameter or depth should not exceed the size of a penny. The scar at the upper left of the photo appears to be splatter accumulation. This can be minimized with elephant trunks."[88]

The type of ties of the spreaders can influence the quality of the details of the concrete, as Langford points out with reference to the photograph of another hole in the sample wall: "Photo 4 shows a tie hole that has bled excessively. This is a result of not tightening the hairpin wedge enough. 5000 lb. ties should be used so that the cone can slightly dent the plywood surface. The surface marks to the left of the hole were probably caused by bubbles in the resin application. Small amounts of this are acceptable."

Other defects on the concrete surface have been caused by the resin coating of the panels, not suitably treated with sanding to eliminate the flaws produced by air bubbles. The fact that the defects generated by the resin were not seen in the construction of the Salk Institute leads Langford to hypothesize that in recent years Burke Company has altered the composition of the product – "this should be investigated," is his advice. In the two sample walls, in any case, the resin applied to the panels makes the wood grain invisible, eliminating the

[84] *Ibid.* See the photograph, LIK, 030.IV.A.730.1.
[85] See the photograph of the detail, LIK, 030.IV.A.730.1.
[86] For the photographs Langford lists in the letter see, LIK, 030.IV.A.730.1.
[87] See the photograph, LIK, 030.IV.A.730.1.
[88] Frederick Langford, letter to Marshall D. Meyers, 6 November 1969, cit. See the photograph, LIK, 030.IV.A.730.1.

328-329. Kimbell Art Museum, Forth Worth, Texas, 1966-72.

328. View of the sample walls and vault (LIK, 030.IV.A.730.1.1).

329. "Transversal Section Through Cycloid," 28 October 1969 (LIK, 030.I.C.730.13).

concept of the imprint of the essence of one material on another, of wood on concrete, as Kahn has successfully done in the Salk Institute. Precisely the quality of the resin produced by Burke Company and the defects of its application to the panels observed in the two sample walls convince Kahn and Langford to change the project, during the course of construction, with fundamental consequences on the final appearance of the concrete.

Harden urges Seymour, for the "new test wall," to follow Langford's indications to the letter "in the construction of the forms and the pouring of the concrete."[89]

The third sample wall, divided like the other one into two sectors, but known as Wall No. 3, is made towards 21 November. To get more fluidity in Mix I, first Byrne, Inc., mixes Concrete Sand with Limestone Sand, increasing the quantity of cement and adding a lubricant - Masterbuilders' MBVR Air Entraining Agent – but without good results. Then the firm decides to avoid the Mix I chosen by Kahn on 23 October, and to use Mix IV, to make it possible to use the Thomsen Model 640. Komendant makes no objection to the use of Mix IV from a structural viewpoint, and this is the mix design that is used to pour the "third sample."[90]

The other fundamental modification introduced in Wall No. 3 has to do with the type of plywood panels of the formwork, which are now 3/8 inch – the so-called 3/8 inch "Marine Plywood overlays."[91] For this reason, the new wall sector is also indicated as the 3/8 inch "skin system."[92] Byrne, Inc., turns to these panels due to the problems encountered in procuring the 3/4 inch panels with the necessary characteristics to make Kahn's concrete, as Seymour explains: "In regard to type of form material we have diligently searched both national and local markets including the American Plywood Association and have been unable to find a 3/4" fir or pine plywood of any kind, thick face or regular face with a solid core that is untreated and in 10 to 12 foot lengths that is in any current stock anywhere. From what we have learned to date it will require a special order to obtain lengths beyond 8 foot and will take somewhere from 3 to 6 weeks to receive same according to the best available information after ordering."[93]

Wall No. 3 produces good results, according to Seymour: "The forms are stripped on this third sample and, in our opinion, produced satisfactory results both as to appearance of color of concrete as well as quality of form techniques."

After the test, Byrne, Inc., recommends the 3/8 inch plywood to Kahn because it is available in lengths of 8, 10 and 12 feet without special orders, it reduces water leakage (on the back, each joint can be reinforced with 3/4 inch plywood backing), it is more economical and easier to work. Then Byrne, Inc., requests, yet again, the drawings and specifications for the ex-

[89] Thad E. Harden, Jr., letter to Thomas A. Seymour, 10 November 1969, LIK, 030.II.A.37.6.
[90] Thomas A. Seymour, "Memorandum," 5 December 1969, cit.
[91] *Ibid.*
[92] Frederick Langford, letter to Marshall D. Meyers, 6 November 1969, LIK, 030.II.A.37.6. "The wall that is now being built," Langford writes.
[93] Thomas A. Seymour, "Memorandum," 5 December 1969, cit.
[94] *Ibid.* For the corrections to the "Memorandum" see Thomas A. Seymour, "Memorandum," 10 December 1969, LIK, 030.II.A.37.34. Seymour urgently requests the "wall pattern work" (Thomas A. Seymour, "Memorandum," 20 November 1969, LIK, 030.II.A.37.34).
[95] It is significant that "columns and walls have been provided at strategic positions [...] to reduce the slab spans" (Marshall D. Meyers, letter to Thad E. Harden, Jr., 20 May 1969, LIK, 030.II.A.37.5).

posed concrete surfaces, in order to calculate the number of panels that will be needed. Seymour proposes making the fourth sample, the one requested by Kahn, with the 3/8 inch plywood of the Marine Plywood type (in the letter he writes 3/4 inch, by mistake), promising that he will order the 3/4 inch plywood if Kahn does not approve the sample. Seymour himself recommends that prior to judgment, the wall should remain exposed to weathering for about one week. The fourth sample wall is poured in December. The compound is made with "concrete sand and coarse aggregate from Bridgeport, Texas quarry similar to Texas Industries' Mix IV except the coarse aggregate is from Bridgeport in lieu of Pablo Pinto [...]."[94] Kahn's visit to Fort Worth to evaluate the results is slated for 19 December.

The slab structure: exposed or with suspended ceiling

While the structure of the gallery roof of the Kimbell Art Museum has been developed in the initial design phases, that of the floor slab is still uncertain when construction begins. During the course of 1969 a certain empirical criterion emerges for the layout of the support points on the lower level, to bear the load of the gallery floor slab.[95]

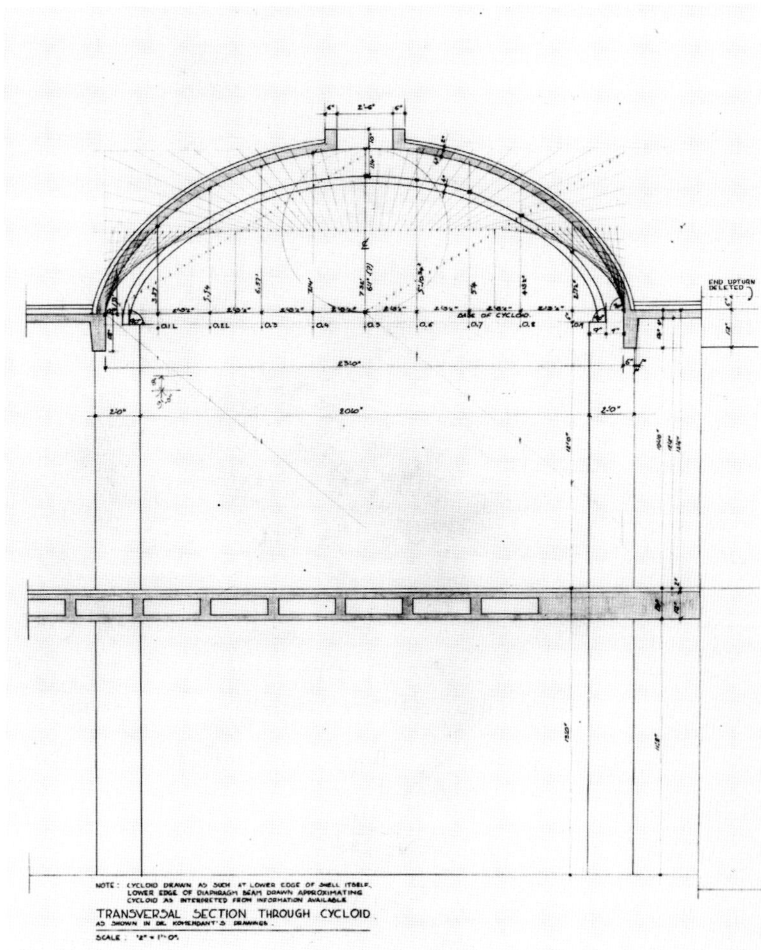

For the design of the floor slab structure Komendant considers a "post-tensioned slab."[96] The waffle slab with coffers applied by Kahn for the Dormitory at Bryn Mawr also returns among the proposals for the Kimbell Art Museum,[97] though Kahn is leaning towards a "solid slab."[98] The evolution of the floor slab structure takes the direction of a breakdown into different sectors when Komendant proposes three "alternate floor systems": waffles with visible structure; waffles with "metal lath" for a ceiling to be made with "cement plaster enclosure or prefabricated infills 30" × 30" in size"; and a "hollow core slab" containing insulation. The three systems do not constitute alternatives, but solutions to be applied in the various particular cases.[99] Still seeking the character of each part, Komendant calls for post-tensioned "long span beams" and other structural features.[100]

Komendant prepares his own "Specifications for reinforced & prestressed concrete Museum for Kimbell Art Foundation, Fort Worth, Texas" which are received by Kahn on 27 August. It is in this document that we find the first summary of the various forms for the parts of the museum, including those for the waffles: "The molds for columns, beams, slabs and cycloids shall be of plastic covered plywood or plywood painted with form armor [...]. The forms for waffle type floors shall be fiberglass [...]."[101] The indication of fiberglass for the coffers of the slab suggests the desire to obtain the smooth surface required for Kahn's exposed concrete. It is worth noting that during this same period Kreier obtains the patent to make thin panels of the "waffle slab" type, prefabricated with molds in fiberglass.[102]

The question of the slab structure in relation to the visibility of the concrete has not yet been resolved. Geren's studio wants to conceal the concrete surfaces as fully as possible, in order to simplify the construction. Though they are aware that Kahn would prefer to leave the concrete exposed, it is emblematic that with respect to the three "alternate floor systems" of Komendant, Geren's studio recommends the waffle "with a uniform plaster ceiling over the entire surface." "Also," they add, "it would mean that the concrete used for floor framing need not be colored with Pozzolan [sic]."[103] Kahn's team discusses with Seymour the solution to apply to a waffle ceiling. "[...] from conversations with Mr. Meyers in Kahn's office, it appears that the underside of the upper floor will be closed off with some type of plaster," reports Seymour, who nevertheless makes a point of recalling Kahn's dissenting view: "It was argued successfully by Mr. Kahn that all lower floor offices would not have any ceiling."[104]

[96] See Marshall D. Meyer, letter to Thomas A. Seymour, 25 March 1969, LIK, 030.II.A.37.34.

[97] On 7 April 1969 the new "Outline Specifications for Kimbell Art Museum, Kimbell Art Foundation, Fort Worth, Texas," are prepared, LIK, 030.II.A.95.5.

[98] See Preston M. Geren, letter to August E. Komendant, 19 July 1969, LIK, 030.II.A.37.24.

[99] "It is understood that each of these systems can be used to overbridge a particular area of the floor," Komendant explains (August E. Komendant, letter to Louis I. Kahn and Preston M. Geren, 15 August 1969, LIK, 030.II.A.37.24).

[100] *Ibid.*

[101] August E. Komendant, "Specifications for reinforced & prestressed concrete Museum for Kimbell Art Foundation, Fort Worth, Texas," n.d., [August 1969], LIK, 030.II.A.37.24.

[102] The patent is No. 3,520,967 of 21 July 1968, titled *Method for making thin concrete panels*, and based on the use of a "flexible fiberglass." A newspaper announces that Kreier "is preparing to introduce the floor slab process in Brazil, where operations are planned in Rio de Janeiro and Sao Paulo" (*Dr. George J. Kreier Jr.*, in, "Daytona Beach Morning Journal," vol. XLV, 25 July 1970, No. 177, p. 2).

[103] Preston M. Geren, letter to August E. Komendant, 29 August 1969, LIK, 030.II.A.37.6.

[104] [Thomas A. Seymour], "Memorandum," 29 August 1969, LIK, 030.II.A.37.34.

330-331. Kimbell Art Museum, Forth Worth, Texas, 1966-72.

330. Construction site photograph, 15 November 1969 (LIK, 030.IV.A.730.1.2).

331. Construction site photograph (LIK, 030.IV.A.730.1.6).

On the question of the waffle structure, exposed or concealed by a suspended ceiling, Seymour insists on his intention of discussing it personally with Kahn. From the positions taken by Geren's office, Seymour and Kahn it is clear that what is at stake is not just the visibility of the reinforced concrete structure, and therefore its perfect fabrication and color, but also the very concept of the arrangement of the space of the offices, which in the case of the exposed waffle would no longer be "free," since the ribbing would dictate the layout of the offices.

"We recommended and later 'begged,'" Seymour writes, "that all the air conditioning ducts and conduits for electrical work be done in the normal manner and that the ceilings for the lower area be furred down so that ducts

332-333. Kimbell Art Museum, Forth Worth, Texas, 1966-72.

332. Intrados of the concrete floor slab (Kimbell Art Museum Archive, Forth Worth).

333. Construction site photograph, 30 October 1969 (Kimbell Art Museum Archive, Forth Worth).

and conduits, etc. could pass through the space and feed down into offices and areas receiving them. Everyone knows we do not have anything in our budget to furr any ceilings. Now it is being contemplated to spend money not in the budget to 'close off the cells' of the unsightly waffle slabs where partitions do not exactly fall on a waffle pan legs. The job is being delayed right now because of the great amount of time it takes to plan all of the openings accurately in the lower floor slab. Why not change direction at this point and decide to furr the ceilings, and run the air conditioning ducts, conduits, etc. overhead feeding down into the offices in the normal manner? This will eliminate the expense of feeding the ducts up through the floor and the savings in the less expensive type of construction would probably save a good portion of the furred ceiling expense. Since it is being proposed to do it anyway but not have a furred ceiling this makes a lot more sense to us."[105]

For the construction of the lower floor slab metal formwork are used, shaped like a U and assembled reversed, with which to create a reinforced concrete structure with long rectangular openings.[106] For the construction of the floor slab of the gallery, in the parts designed as a "combination" of "waffle and two-way flat slab" the formwork consist of about 2400 pieces of 8 × 30 × 30 inches expanded polystyrene, produced by United Floation Industries of Irving, Texas.[107]

When the Southern part of the floor slab is about to be poured, Byrne, Inc., proposes using the No. 4 Mix of concrete, to pour by machine. The de-

[105] *Ibid.*
[106] See the photograph of the construction of the floor slab, LIK, 030.IV.A.730.1.
[107] See Thomas A. Seymour III, *Contractor Challenged by Kimbell Art Museum Design*, in, Latour, *Louis I. Kahn, l'uomo, il maestro* cit., (pp. 321-29), p. 323.
[108] Thad E. Harden, Jr., letter to Thomas A. Seymour, 2 December 1969, LIK, 030.II.A.37.6.
[109] "A non-staining type accessory such as galvanized, stainless steel or copper footed is usually not noticeable after form stripping" (Thomas A. Seymour, letter to Preston M. Geren, Jr., 8 December 1969, LIK, 030.II.A.37.34). Kahn provides indications for the positioning of each expansion joint "in the center of the 3' space" in the floor slab (Thad E. Harden, Jr., letter to Louis I. Kahn, 14 November 1969, LIK, 030.II.A.37.6, and Preston M. Geren, Jr., letter to Louis I. Kahn, 25 November 1969, LIK, 030.II.A.37.6).
[110] Thomas A. Seymour, "Memorandum," 7 November 1969, LIK, 030.II.A.37.34. Kahn makes decisions on the measurement of each exposed concrete element based on a given design. It is significant that he decides to position to exit door at the center of a "block wall"

cision causes delays because Texas Industries, Inc., has to find the required aggregate for the preparation of that compound. Harden promises that this change in the mix design will not cause a different color of the concrete: "the difference in the color of the concrete would not be distinguishable [...]."[108]

Seymour expresses concerns regarding the possible use of plastic accessories, like the reinforcement ties produced by Dayton Sure-Grip Shore Company Plastic Protected Accessories, because they would be visible on the inner surfaces of the main floor slab and the cycloid vaults, where the concrete is to be left exposed.[109]

The "Fred Langford Formwork Systems"

The technical details of the design of the exposed concrete were not covered in the original specifications, because Kahn had initially planned to also have the lower level of the museum clad with stone slabs. But after the making of the first sample walls Kahn changes the organization of the project and the work on the exposed concrete. Seymour begins to insistently request "elevations and details for all patterned, exposed concrete both exterior and interior," which Kahn had promised to provide when he visited the worksite on 23 October 1969.[110]

While Wall No. 3 is being completed, and against the backdrop of the exchange of technical information on the sample walls between Kahn's office, Geren's office and Byrne, Inc., detailed formulation is given to Langford's proposal, already put forward on 3 September, to operate as a professional consultant "assisting and advising" Kahn and Geren in the "design and execution of the formwork of all formed concrete to be placed in the Kimbell Art Museum."[111] This confirms the fact that after the experience of the Salk Institute and the National Assembly Building, in Kahn's office a true specialization has emerged, that of the design and construction of formwork, supervised for the most part by Frederick Langford. It is clear that Kahn intends to control a part of the working design usually assigned to the contractors, but at the same time to avoid assuming responsibility, on the part of his office, for the costs involved in this type of work. This is why, after the proposals to the Public Works Department of Pakistan, Langford creates his own office at the service of Kahn for the construction of the museum in Fort Worth, Frederick Langford & Associates, Construction Systems Research & Development.[112]

The construction of the sample walls represents the pretext for the drawing up of a contract, prepared by Langford and signed by Kahn on 24 November, which stipulates the responsibilities of Langford. The text of the contract lists Langford's functions:

"To confer with Louis I. Kahn in Philadelphia to prepare formwork drawings and formwork specifications which interpret the structural

and with the "lintel" symmetrical. "However, in this case he would make the lintel the full width of the corridor to eliminate the small pieces of block at each end of the lintel" (Marshall D. Meyers, letter to Thad E. Harden Jr., 10 November 1969, LIK, 030.II.A.37.6).
[111] Frederick Langford, letter to Louis I. Kahn and Preston M. Geren, Jr., 24 November 1969, LIK, 030.II.A.37.2.
[112] See the letterhead, [Frederick Langford], letter to Marshall D. Meyers, 6 November 1969, LIK, 030.II.A.37.6). The stationery shows that the office of Frederick Langford & Associates, Construction Systems Research & Development is in Philadelphia, in the Western Savings Fund Building; but on the document in question the address is crossed out, and replaced in handwriting with: 320 Walnut Street (Kahn's office is at 1501 Walnut Street).

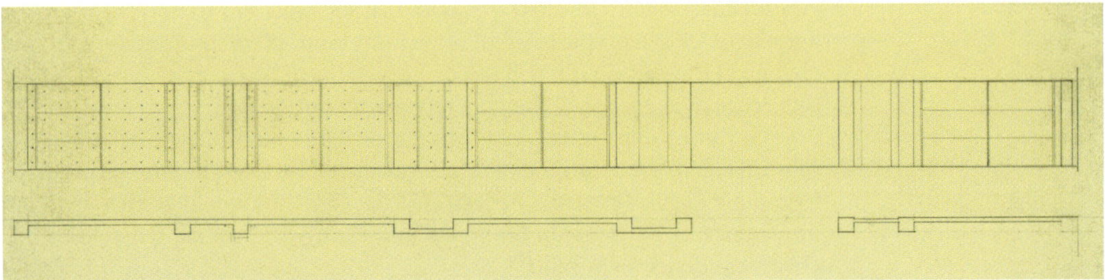

334-336. Kimbell Art Museum, Forth Worth, Texas, 1966-72.

334. Formwork study (LIK, 030.I.C.730.14).

335-336. Construction site photographs, 15 May 1970 and 1 April 1970 (Kimbell Art Museum Archive, Forth Worth).

demands and the construction process as an Architectural expression. These drawings shall be of such detail as to adequately illustrate from layout and details necessary to obtain the desired excellence of architectural concrete. To consult with the Architect and Contractor to advise on the construction of sample walls in order to familiarize the Contractor with the formwork details and the desired results. To evaluate the results of these sample pours and advise the Architect and Contractor of any revisions or modifications to the drawings and specifications arising from discussion of these sample walls and to incorporate, wherever possible, those methods and recommendations of the Contractor which give acceptable results. To evaluate the Contractor's formwork shop drawings in order to assist in the proper implementation of the detail drawings and specifications."[113]

Kahn sends the contract to Geren's office, now directed by Preston M. Geren, Jr., taking his father's place after the latter's death in September. On

this occasion, Kahn emphasizes the experience gained by Langford at La Jolla and Dhaka, as well as the need for his presence in Fort Worth to obtain an "excellent Architectural Concrete." In Kahn's letter we find the reasons behind the proposal of hiring Langford as a consultant for the design of the Kimbell Art Museum:

"If we relied on the usual arrangements between Architect and Builder we would not get the high quality of concrete we desire. However, we have nothing in our Contract with the Owner to take care of this specialty and this could be an oversight on my part. Because this work is very rewarding aesthetically it would be fair to our Commission if the Owner were to recognize this specialty as outside the terms of our Contract or as an Addendum to our

[113] Frederick Langford, letter to Louis I. Kahn and Preston M. Geren, Jr., 24 November 1969, LIK, 030.II.A.730.2.

337. Kimbell Art Museum, Forth Worth, Texas, 1966-72. Construction site photograph (Kimbell Art Museum Archive, Forth Worth).

Contract and we should consult with him on this. At present, we must absorb this fee together in the interest of the building."[114]

The first reaction of Geren to the contract sent by Kahn is to suggest the candidacy of the technicians that collaborated with his office on the construction of the Fort Worth City Hall, in which architectural concrete, as in the case of the Kimbell Art Museum, had not been planned at the start: "In that case we used 'Architectural Concrete Consultants Inc.' of Houston, Texas, a very capable firm of considerable more experience than Mr. Langford."[115]

Submitted officially on 24 November by Kahn, the contract is never signed[116] and Langford's role as a consultant for the design of the formwork ends up creating conflicts regarding technical responsibilities between Geren's office, Kahn's office, Komendant and Byrne, Inc.

Up to the end of 1969, Kahn does not supply the formwork drawings, also awaiting the definition of the relationship between Langford and the Kimbell Art Foundation. Halfway through December, Langford is on the payroll of the Kimbell Art Foundation to work on the formwork and the concrete, though not in keeping with the terms outlined in the draft of the

[114] Louis I. Kahn, letter to Preston M. Geren, Jr., 26 November 1969, LIK, 030.II.A.730.2.

[115] Preston M. Geren, Jr., letter to Louis I. Kahn, 1 December 1969, LIK, 030.II.A.37.6.

[116] Frederick Langford, letter to Bowen King, business manager of the Kimbell Art Foundation, 20 May 1970, LIK, 030.II.A.37.34.

[117] Frederick Langford is hired by Brown for the formwork and concrete. A letter from Meyers to Brown confirms (Marshall D. Meyers, letter to Richard F. Brown, 16 December 1969, LIK, 030.II.A.37.16).

[118] Thomas A. Seymour, "Memorandum," 19 December 1969, LIK, 030.II.A.37.34. For the drawings of the "exposed concrete walls at the Lower Level South Wing" see

contract.[117] This constitutes official recognition of the particular character of Kahn's architectural concrete within the American construction process. On 19 December, when Seymour sends yet another request for drawing of "form patterning,"[118] Langford has prepared a series of new drawings for the concrete patterns (with more complex designs than those of the studies he produced towards September[119]). All the drawings bear the indication "Fred Langford Formwork Systems."

Langford's system is based on that of the Salk Institute, and it makes use of the V joints, the spreaders arranged in a regular pattern, and the vertical elements that imprint bands and grooves on the concrete that reflect the assembly of the formwork: the narrow "key strips" with the edges cut to produce two adjacent V lines, and the wider strips, known as "gang-strips," again with cut edges to form V lines on the concrete, but with a different configuration, depending on whether they are used to shape inward facing walls (in this case the surface is flat) or outward facing walls (where a narrow strip is applied to imprint a groove, and the "gang-strip" is made "with s.c. [shrinkage control] molding"). The set of drawings called "details" sums up the basic parts of the "formwork systems."

The question of the ambiguous nature of the exposed concrete wall of the lower level of the museum is the focus of the project documented by Langford's drawings. The arrangement of the panels, the "key strips," "gang-strips" and spreaders is designed to render visible, along the external wall face, the division of the wall into sectors of closure and sectors with protruding pillars, towards the museum interior. Each wall sector, on which the pairs of pillars that support the vaults rest, is bordered by two "gang-strips with s.c. [shrinkage control] molding" and made with only three long panels assembled horizontally, while the positioning of the spreaders offers a glimpse of the two pillars in continuity with those of the upper level. Every closing wall sector is made with two rows of three short panels mounted horizontally and joined at the center by means of a key strip.

Langford's official entry in the Kimbell Art Museum project as a consultant for formwork and concrete is sanctioned by his participation in the meeting on 22 December 1969 at the office of Brown, together with Geren, Kahn, Seymour and Bowen King, business manager of the Kimbell Art Foundation, to show the drawings and "to discuss materials, methods and formwork documents."[120] On this occasion, the fourth sample wall is also examined. Langford is assigned the task of preparing the formwork drawings, with the exception of the technical details,[121] but not without difficulty, due to the reactions of Geren and Seymour, who are also critical of the technical quality of his drawings, though for different reasons.

Geren continues to express reservations about Langford, suggesting the technicians of Architectural Concrete Consultants, as he writes to King on the day following the meeting: "I had discussed Mr. Langford's services with you and Dr. Brown in Dr. Brown's office and advised that I would have to rely on Mr. Kahn's recommendation as I am not familiar with

Dewayne N. Manning, of the Geren office, letter to Louis I. Kahn, 4 December 1969, LIK, 030.II.A.37.6, and also Thomas A. Seymour, "Memorandum," 9 December 1969, LIK, 030.II.A.37.34.

[119] Marshall D. Meyers, letter to Richard F. Brown, 16 December 1969, LIK, 030.II.A.37.16. A document by Langford is attached to the letter, indicating the work already done (Frederick Langford, 16 December 1969, LIK, 030.II.A.37.16).

[120] Frederick Langford, letter to Richard F. Brown, 9 January 1970, LIK, 030.II.A.37.16. See also the handwritten note in, "Specifications for formwork and concrete placement," LIK, 030.II.A.37.11.

[121] *Cf.* Frederick Langford, letter to Bowen King, 20 May 1970, LIK, 030.II.A.37.34.

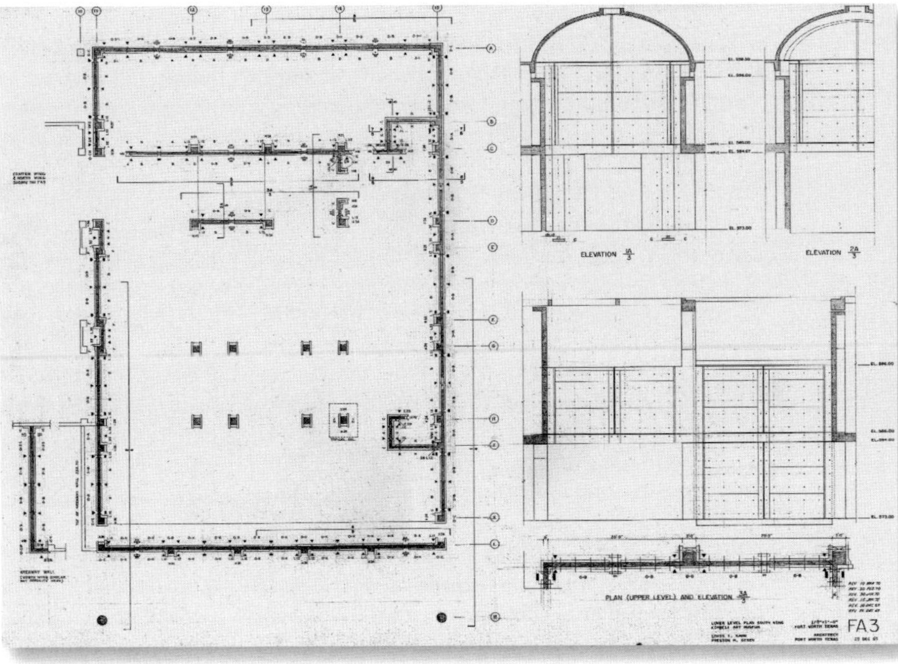

338-340. Kimbell Art Museum, Forth Worth, Texas, 1966-72.

338. "Lower Level Plan South Wing," 19 December 1969 (owned by F. Langford).

339. "South Wing," 19 December 1969 (owned by F. Langford)

340. "Details," 19 December 1969 (owned by F. Langford).

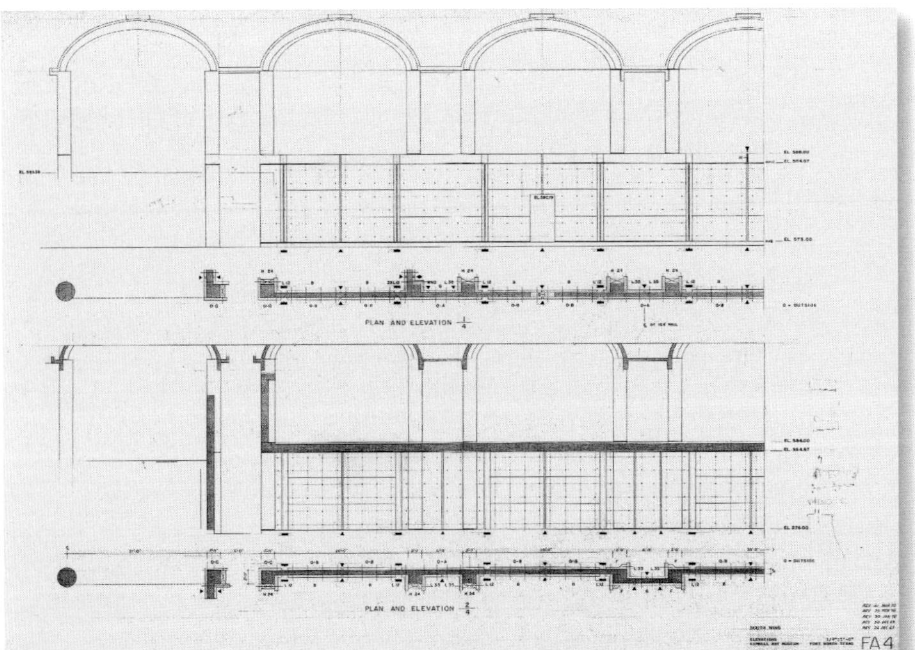

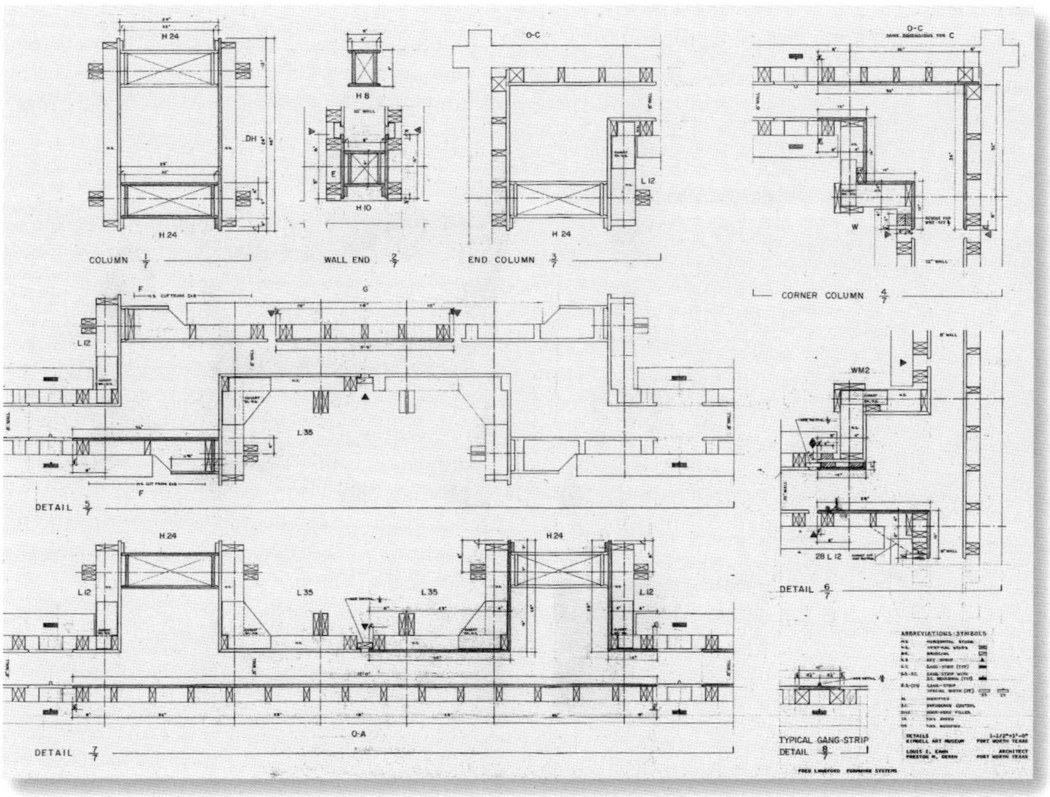

Mr. Langford's qualifications. There is a very good firm in Houston with considerable experience, however, Mr. Langford has been working with Mr. Kahn for some time now and it would seem to me that it would be logical to proceed with his service."[122]

The fact that Geren wants to remove control of the design and construction of the formwork from Kahn's loyal staff member is demonstrated by the recommendation, sent to King, to not have Langford make other trips from Philadelphia to Fort Worth, thus eliminating the expenses involved. Where Langford's drawings are concerned, he has other objections, because he feels that drawings like the one for the "South Lower Section" presented on 22 December are useless. One of his statements further clarifies the critical stance of the Fort Worth office, seeking to undermine the very foundations of Kahn's vision of architectural concrete as the result of a process of construction of the formwork panels: "The intricate details showing form construction on the drawings presented in the Kimbell Foundation Office on Monday, December 22, should not be drawn for the remainder of the work to be accomplished by Mr. Langford. It is necessary only that he show the design of the pattern on the concrete, not the construction of forms."[123] So again in Fort Worth, we can observe the same misunderstandings regarding the logic of Kahn's formwork that arose on the Dhaka worksite, for similar practical reasons.

[122] Preston M. Geren, Jr., letter to Bowen King, 23 December 1969, LIK, 030.II.A.37.6.

[123] *Ibid.*

For his part, Seymour believes that Langford's drawings do not contain sufficient indications to proceed with the construction of the concrete, also because they have not been reviewed either by Geren's office or by Komendant, and are thus lacking indications on the position of the expansion and contraction joints.[124]

Formwork details according to Seymour

At the start of 1970, in spite of the consulting work of Langford, who on 13 January proposes a new contract, again never signed,[125] Byrne, Inc., continues to play an active part in the development of all the formwork details.

After the testing done with the various sample walls and the observations of Byrne, Inc., on the plywood panels and the availability on the market of their different thicknesses, modifications are made in January to the text from 1968, "Specifications for formwork and concrete placement."[126] The changes also have to do with the formwork.

In the sections of the specifications all indications on possible suppliers of plywood and patching products, and for the sealing of the formwork panels, are eliminated.[127] An addition reveals the details of the protruding V lines, calling for careful preparation of the plywood, to be cut along the edges, putting the inside of the panel in contact with the pour. The explanation added to the specifications precisely to achieve perfect lines has to do with the treatment of the angled surfaces created by the beveled cut along the edges of the panel. By now Kahn's system demands unusual care in the making of the details of every single component. "Where chamfer 'V' joint

[124] "Yesterday," Seymour writes to Geren, "we were handed by Mr. Kahn and Fred Langford incomplete preliminary details showing a portion of the south wall and column forming which supports the upper floors. We do not consider these complete plans nor do we accept them as being the plans which we are supposed to have issued to us to build the building by, in as much as neither your office nor Dr. Komendant had chance to review same. [...] we need completed pattern joint layout clearly shown on elevations with each type of joint detail. Mr. Kahn's drawings indicated full-size pattern of the joint, however some of them had not been marked yet and others had not been approved as to shape" (Thomas A. Seymour, letter to Preston M. Geren, Jr., 23 December 1969, LIK, 030.II.A.37.34). During December 1969, Langford revises certain drawings (FA-3, FA-4, FA-5, FA-6, FA-7, FA-8), takes part in discussions with Kahn, Meyers and Komendant, and visits the worksite, on 22 December, "to view sample walls with Louis I. Kahn, Thomas Seymour and others of Thomas S. Byrne Company to discuss materials, methods and formwork documents" (Frederick Langford, letter to Richard F. Brown, 9 January 1970, LIK, 030.II.A.37.16). The original proposal of 14,000 dollars for his work done in December is modified (Louis I. Kahn, letter to Richard F. Brown, 9 January 1970, LIK, 030.II.A.37.16), and the payment amounts to only 2470.50 dollars (Frederick Langford, letter to Richard F. Brown, 9 January 1970, LIK, 030.II.A.37.16). Another contract is proposed by Langford to Brown for "design and execution of the formwork of all formed concrete to be placed in the Kimbell Art Museum" (Frederick Langford, letter to Richard F. Brown, 13 January 1970, LIK, 030.II.A.37.16). Following the criticisms for the lack of the opinion of the engineer on the formwork design, Langford agrees to remake the drawings on the basis not only of his discussions with Kahn, but also of those with Komendant (*ibid.*).

[125] Frederick Langford, letter to Bowen King, 20 May 1970, LIK, 030.II.A.37.34.

[126] "Specifications for formwork and concrete placement," LIK, 030.II.A.37.16. For the date see Thomas A. Seymour, letter to Preston M. Geren, Jr., 19 January 1970, LIK, 030.II.A.37.34. For the modifications made to the document in April see Preston M. Geren, Jr., "Addendum No. 1 to the specifications for the Museum for Kimbell Art Foundation," 28 April 1970, LIK, 030.II.A.37.11.

[127] "Specifications for formwork and concrete placement" cit.

341-344. Kimbell Art Museum, Forth Worth, Texas, 1966-72.

341. Formwork sketches, with the study of the insertion of the lead disks to seal the holes (F. Sherwood, letter to Thompson, 6 May 1970, LIK, 030.II.A.37.6, verso).

342. Construction site photograph, 1 October 1970 (LIK, 030.IV.A.730.1.5).

343. Construction site photograph of the auditorium, 1 November 1970 (Kimbell Art Museum Archive, Forth Worth).

344. Detail of the concrete wall (photograph by the author, 2010).

345-349. Kimbell Art Museum, Forth Worth, Texas, 1966-72. Details of the concrete walls (photographs by the author, 2010).

exposes non-solid plywood core, voids shall be patched with wood dough or wood putty as approved by Architect," Kahn's staff writes.[128]

In spite of the indications of Byrne, Inc., the version of the specifications with the first corrections still confirms the 3/4 inch thickness of the panels. But in the handwritten modifications made by Seymour, the thickness of the plywood is correct to 3/8 inch – the 3/8 inch marine plywood, as approved during the meeting on 22 December 1969.[129]

In the light of the recommendations that have emerged from the inspection of the various sample walls regarding the methods of pouring the concrete into the formwork, we can grasp the message of the note added in the margin

[128] *Ibid.*
[129] "Specifications for formwork and concrete placement," with handwritten modifications made by Seymour, received by Geren's office on 9 January 1970 and by Kahn's office on 21 January 1970, LIK, 030.II.A.37.34.

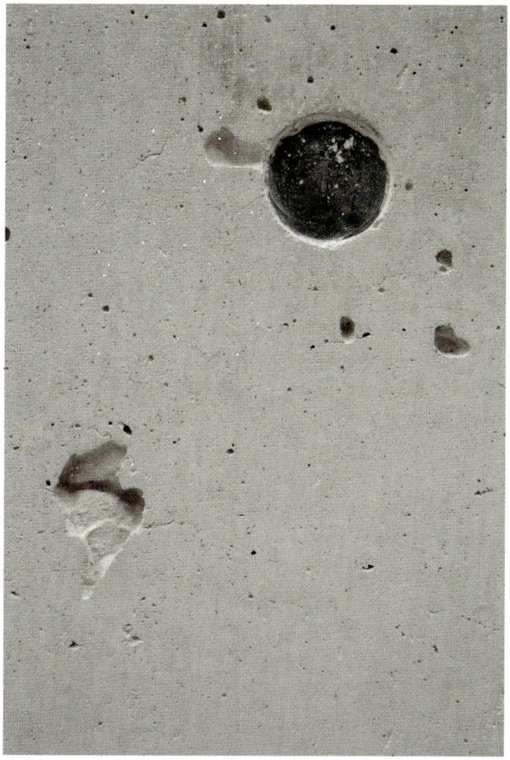

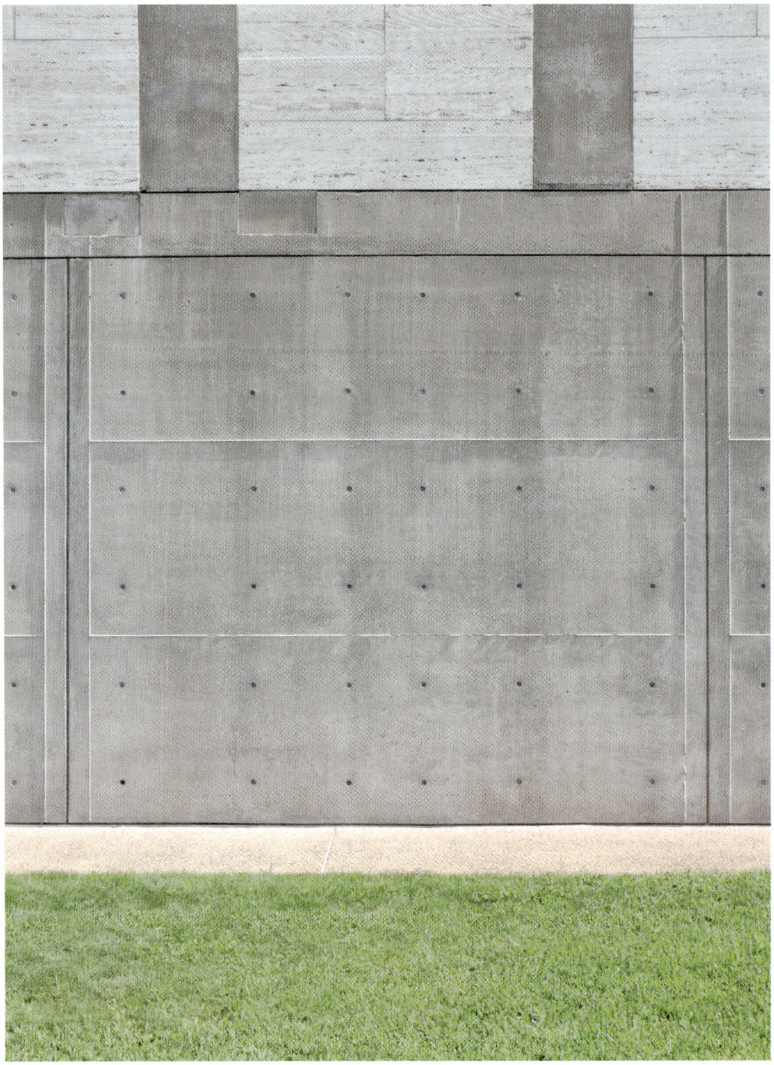

350-352. Kimbell Art Museum, Forth Worth, Texas, 1966-72. Details of the concrete walls (photographs by the author, 2010).

by the paragraph on concrete placement: "pumping or crane." Traces of the discussions on the responsibilities of Langford in the technical design of the formwork are found in the margin of the section on "Formwork design drawings," where Seymour indicates that his company will construct the formwork based on the "approved sample," but using its own system, "method to be ours," a reflection of the criticisms already expressed regarding "Fred Langford Formwork Systems." All the gray plastic devices used to position the reinforcements of the floor slab will be visible, Seymour warns: "We will use the gray plastic coating accessories but everybody must understand that we fully expect the pattern of this to be noticeable when viewed from the underside [...]."[130]

Once again, in the design process for the formwork a separation happens between the construction of the system and its effects on the concrete. As in Dhaka, in Fort Worth the contractor intends to build the formwork system in such a way as to achieve the formal results desired by Kahn, but without necessarily respecting his logic of assembly, thereby undermining the theoretical basis of his architectural concrete.

"We are not going to go by the very detailed systems indicated for forming this concrete," Seymour declares. "We have made our point very clear both to Mr. Kahn, Mr. Geren and Mr. Lan[g]ford. We merely want to know the joint patterns desired and once we have this completed information, we will use in our best judgment the proper method to give the desired effect as indicated on the approved wall sample that Mr. Kahn reviewed on December 22nd, 1969. We are not going to submit formwork shop drawings and we are not going to use the overall 3/4 inch plywood forming system. We will use the 3/8 inch marine plywood liner as approved by Mr. Kahn on December 22, 1969. We will not use Burke form coating but will use 'form-cote' manufactured by Preco Chemical Corporation."[131]

130 Thomas A. Seymour, letter to Preston M. Geren, Jr., 19 January 1970, LIK, 030.II.A.37.34.

131 *Ibid.*

Two details of the formwork are the subject of discussion between Kahn's office and Byrne, Inc. While Kahn and his staff favor the product by Burke, Inc., for the protective coating of the panels in contact with the pour, in spite of the uneven results achieved using that product in the first two sample walls, Seymour proposes Form-Cote made by the Preco Chemical Corporation of New York.[132] To inform Kahn of his proposal, he sends the office in Philadelphia an advertising booklet for Form-Cote.[133] Unlike "plastic coated wood forms," Form-Cote is applied with a brush or roller to create a "thick protective film" that stands up to the chemical agents in the water, and to abrasion, when applied to the formwork panels. "The smoothness of concrete cast against FORM-COTE panels is equivalent to the fineness of finish obtained when using plastic panels as forms."[134] It is Seymour who makes the correction to the text of the specifications regarding formwork coating – Form-Cote in place of Burke Company.[135]

The detail of the plastic cones for the formwork continues to be a crucial ingredient of Kahn's concrete, due to the pattern of holes to be sealed with lead. The quality of the plastic cones produced by Burke Company, is questioned by Seymour. He adds a question mark to the specifications regarding the use of that cone, as indicated in the text, or the use of cones produced by Superior Concrete, Inc., with the Richmond Screw Anchor Company, Inc., which he thinks are stronger, though he has never tried them in practice.[136] On this question, he sends Kahn a sample of the cone of Richmond and Superior Concrete, for comparison with the cone made by Burke Company.[137]

While Seymour thinks that the different form of the cone of Richmond and Superior Concrete should have no consequences on the imprint in the concrete, since the hole is to be sealed with lead, Kahn's staff raises technical questions regarding the "more acute angle" of that cone as opposed to that of Burke Company, because the angle would make it hard "to hold the lead filler plug in place."[138] This detail too confirms the level of sophistication reached by Kahn's system. The suggestion of Meyers, were the cone proposed by Seymour to be used, is to develop a modification in order to create an imprint more suitable for sealing. "It may be necessary," he points out, "to roughen or deform the surface of the conical hole to allow the lead to grip it. A detail of the lead plug and the tool to install it will be provided by Fred Langford."[139]

Starting with the modifications to the specifications discussed in January, a new document is drawn up, now with the title "Architectural concrete formwork and placement," which incorporates many of the notes added by Seymour, from the Form-Cote to the cones of the Richmond Screw Anchor Company.[140] These new guidelines also indicate that the holes left by the cones must be "plugged with cast lead cones driven into the holes to leave a 1/4" deep recess." "The lead cones may be cast at the job site," the document says.

[132] Thomas A. Seymour, letter to Louis I. Kahn, 15 January 1970, LIK, 030.II.A.37.34.
[133] "Form-Cote. Preco Chemical Corporation," handwritten date 9 January 1970, LIK, 030.II.A.37.34.
[134] *Ibid.*
[135] "Specifications for formwork and concrete placement," with handwritten modifications made by Seymour, received by Geren's office on 9 January 1970 and Kahn's office on 21 January 1970, LIK, 030.II.A.37.34.
[136] *Ibid.*
[137] "We have noticed," he writes, regarding the defects of the Burke Cone, "that these cones tend to have slight cracks along the face and in some cases crack all the way through. We have even found where part of the back of the cone disintegrated slightly and we are concerned about whether or not this cone is actually strong enough to hold up even though they will be used only once" (Thomas A. Seymour, letter to Louis I. Kahn, 15 January 1970, cit.).

The difficulties encountered by Kahn to gain acceptance for the formwork design have not yet been fully resolved on 1 February, when following the drawings by Thompson for the formwork of the Ground Floor Level, Kahn has to write to Geren to explain the cultural reasons for the imprints to be made on the concrete as traces of the construction process.

"Fred Langford's reasoning in arriving at the markings," Kahn explains, "is based on conveying to the eye the record of events of forming and pouring. He believes as I do that every building should have a logical system in keeping with the material, aesthetic, and operational conditions. Understand that there is no disagreement in the use of 3/8" surfacing plywood. The sample certainly proves Virgil's good interest and skill. This, however, is only one aspect of the conditions. Such concerns as columns, corners, pilasters, cold joints, etc., which Fred makes suggestions about, are the problem areas not yet demonstrated. The layout shown in unexposed spaces is intended to offer the builder panel interchange because of preordained tie-hole points which makes for plug-in readiness."[141]

But Geren's office continues not to willingly accept either the technical principles of Kahn's concrete, or the fact that Langford provides the guidelines for the construction of the formwork. In the reply to Kahn's letter, Geren – like Seymour – continues to insist on the possibility of a distinction between the design of the imprints and the construction of the formwork:

"Mr. Kahn has written concerning his intense desire to have the concrete cast in forms assembled of panels similar to those shown on drawings prepared by Fred Langford. Mr. Kahn feels that 'every building should have a logical system in keeping with the material, aesthetic, and operational conditions.' Although no one would know whether a section of wall was poured with one or more panels making up the total form once they are removed. It seems that Mr. Kahn needs this to justify the pattern."[142]

Against the backdrop of the disagreements concerning the value to assign to the process of construction of the formwork in the design of the concrete imprints, Earp develops drawings to document the "wall form methods" which Byrne, Inc., intends to use, starting from the remarks of 22 December with Kahn, Komendant, Brown and Geren.[143]

All the materials required for the formwork have already been purchased by Byrne, Inc., which begins to erect the panels for the concrete walls using Langford's design in mid-February, starting with the "south interior walls," as requested by Thompson, and not the walls with exposed faces along the museum. The fact that Seymour does not intend to comply with the technical guidelines provided by Langford is confirmed by what he writes to Geren's office: "We do not recommend Langford's system and would not be responsible for the results produced by using it; thus, as far as we are concerned we have no use for Langford's details sheets other

[138] Marshall D. Meyers, letter to Thomas A. Seymour, 27 January 1979 [sic: 1970], LIK, 030.II.A.37.34.
[139] *Ibid.*
[140] "Architectural concrete formwork and placement," 9 February 1970, cit.
[141] Louis I. Kahn, letter to Preston M. Geren, Jr., 1 February 1970, LIK, 030.II.A.37.16. For the works of Langford see Frederick Langford, letter to Richard F. Brown, 1 February 1970, LIK, 030.II.A.37.16.
[142] Preston M. Geren, Jr., letter to Thomas S. Byrne, Inc., 5 February 1970, LIK, 030.II.A.37.6. Geren insists on the need to keep the imprint patterns desired by Kahn and the construction system of the formwork separate, requesting from Byrne, Inc., an estimate for "panels similar" to those of the drawings of Langford, but with a "method" better suited to the contractor.
[143] Thomas A. Seymour, letter to Preston M. Geren, Jr., 11 February 1970, LIK, 030.II.A.37.34.

than occasional reference, since your office is providing us with necessary information shown on *your* working drawings."[144]

The meeting on 12 February in Brown's office in Fort Worth, with Kahn, representatives of the office of Geren and of Byrne, Inc., on hand, is held to decide on the formwork system, and to validate the position of Geren and Seymour. Not by chance, it is established that the contractor "shall construct forms of his own design so long as they produce the concrete design shown on the working drawings prepared by the Joint Architectural offices and the form drawings prepared by Mr. Langford."[145] After the meeting, another revision is made to the "Specifications for formwork and concrete placement," in order to have a single text that incorporates the points on the formwork contained in the specifications of Komendant and the indications of Langford.[146]

"About the concrete": Kahn's position

The construction of the concrete walls starts in March 1970, according to the criteria indicated by Byrne, Inc., but the first flaws immediately become perceptible, as a result of the conflicts and misunderstandings between the various protagonists of the worksite.

At the start of April, Kahn decides to replace the snap ties with screw ties, to avoid having imperfect holes in the concrete walls. This is one of the first clear consequences of the underestimation of the degree of perfection required in the assembly of the formwork for his architectural concrete, as opposed to the usual techniques of contractors. According to the views of Kahn's team, the defects caused by the snap ties can be blamed on the lack of expertise of Byrne, Inc.

"Neil Thompson has reported," Meyers writes, "that many of the snap-tie cones are not tight against the inside face of the form and that it is necessary to break away the thin concrete wall which covers the cone in order to extract it. This makes for a ragged edge and an unsightly hole. It also appears that where screw-ties are used, this problem does not occur. In the past we have not had this problem with snap-ties. The contractor's method of layering the formwork may be one of the causes of this problem."[147]

The administrators of the Kimbell Art Foundation believe that the defects in the first walls built by Byrne, Inc., can be attributed to the fact that Seymour has not understood the difficulties involved in the fabrication processes demanded by Kahn's concrete.[148] Those defects reflect, to some extent, the atmosphere of hostility that accompanied Langford's involvement in the design of the formwork, and the belief on the part of Byrne, Inc., and Geren's office that it is possible to build the concrete without the help of Kahn's team.

The documents offer no clue as to Kahn's personal position on the question of control of the concrete of the Kimbell Art Museum, until the last days of April, when he writes a letter to Geren containing a "stab" on

[144] *Ibid.*
[145] Preston M. Geren, Jr., letter to Richard F. Brown, 16 February 1970, LIK, 030.II.A.37.6. At the meeting of 12 February Geren also raises the doubt that the "concrete previously approved" is not "good"; for this reason, Kahn and Brown visit the worksite to "review the sample."
[146] Preston M. Geren, Jr., letter to August E. Komendant, 24 February 1970, LIK, 030.II.A.37.6. For the indications provided by Komendant with a letter dated 27 February 1970, including the elimination of the section "Depositing concrete" because it does not correspond to the pumping process, see the handwritten notes in, "Architectural concrete formwork and placement," 9 February 1970, LIK, 030.II.A.37.24. Komendant proposes connecting the pairs

the way the contractor makes the concrete without following his indications.[149] Kahn's position taken in the letter, which is not conserved in his archives, can be grasped from the observations of Meyers, who after having read the letter sends his own comments to Kahn, before Kahn writes a new letter to Geren "about the concrete."[150] This new letter is also needed due to the lack of satisfaction expressed by the administration of the Kimbell Art Foundation regarding the first works done in concrete – "they are looking *to you* for guidance," Meyers tells Kahn.

"I think the weight you give to the problem of the architect-contractor relationship is self-defeating," Meyers writes to Kahn. "You are criticizing the builder's attitude and your only solution is to say that he should respect the architect's image of the building as a kind of norm of practice. This merely restates the problem without giving the parties concerned a solution or a 'way out' – and gets another dig in at the provincial ways of Texas. Your use of the word 'consensus' may also give a false impression of majority decision which is not what you mean at all. A better word, and to my mind the crux of the problem, is the word 'cooperation.' By this, I mean the understanding of your image and the desire and 'agreement' to give the effort to attain it. In this light you and Dave should call Komendant and get a firsthand report on what he accomplished in Ft. Worth. We understand that he tutored the builder on the proper way to vibrate the concrete, depth of vibration etc., to eliminate the blotchiness in the concrete. He also insisted that the contractor fill the cones with the plaster material as *was originally instructed for the sample wall.* Whether this will correct the tie-cone problem remains to be seen. But what was evident here was the fact that the contractor *was not living up to his part of the bargain* and was also not being supervised to do so. We also understand that Komendant likes the corners (it is his detail). After talking to Komendant you and Dave should call Dr. Brown and Bo King to determine how they will support your efforts to get the parties to cooperate and what you propose to do. This should guide you in your letter to Geren: 1. To restate that the construction in general is very good. 2. To say that greater perfection and improvement are still necessary to ensure the total effect of concrete Travertine and the other materials. 3. To say that this perfection has been attained by others in the past and is easily within reach of the skills of our contractor as evidenced by his work to date. 4. To propose that you are ready to send Fred Langford to the site immediately to work with all concerned to achieve this improvement, which is no more than bring the job up to the level demonstrated and approved on *the contractor's sample wall.* You have no other course open to you. It is the only course you can rely on and which wastes no time. The Contractor can make counterproposals – but he has exhausted confidence in his judgment. The concrete work to date is mute testimony to his inability to achieve the achievable. If he refuses any help, he brands himself uncooperative."

of columns supporting the vaults with a beam (Preston M. Geren, Jr., letter to August E. Komendant, 17 March 1970, LIK, 030.II.A.37.24).

[147] Marshall D. Meyers, letter to Preston M. Geren, Jr., 30 March 1970, LIK, 030.II.A.37.6. It is decided, in Kahn's presence, to use poured in place concrete to build the supports on which the Travertine slabs are to be applied when facing outward, and to use "light-weight concrete blocks" as the supports for the same slabs when they face inward ("Kimbell Art Museum. […]. Summary of conclusions reached at meetings during week of April 13, 1970," LIK, 030.II.A.37.18).

[148] Marshall D. Meyers, letter to Louis I. Kahn, 29 April 1970, LIK, 030.II.A.37.6.

[149] *Ibid.*

[150] *Ibid.*

Recommendations of Komendant for the making of the concrete

Komendant is also involved in the decisions on the formwork system and the quality of the concrete. His name is mentioned in the correspondence and he is also called into play by Meyers. The worksite inspection made by Komendant towards the end of April 1970 should be seen against the backdrop of the criticisms of the works of Byrne, Inc. On that occasion, Komendant provides instructions to the contractor which are summed up in points by Meyers, in a document that effectively helps to improve the quality of the concrete. "The essential thing," Meyers writes, "is that the builder must understand that we expect the standard of workmanship produced on the sample wall to be produced on the building itself."[151]

Komendant notices certain shortcomings in the methods of vibrating the pour, which create undesired chromatic effects on the concrete surface: "Vibrating too close to the top of the concrete placement pulls up all impurities and water resulting in a darker color which contrasts with the next placement." As a remedy, he recommends the use of "plate vibrators," application of vibration at the center of the wall sector, and preparation of a more fluid compound. To prevent the formation of layers of water in contact with the formwork, favored by vibration, he recommends taping on the formwork one hour after pouring; in any case, he believes that "outside surface form vibrators" should not be used.

One of the crucial aspects for the proper building of Kahn's concrete, the precision in the placement of the tie holes, is discussed by Komendant with Kahn. While Kahn believes that the screw ties give the best results, in Komendant's view both types of ties used – the initial snap ties, then replaced with screw ties based on Kahn's decision – produce details of insufficient quality, because whereas the cones were treated in the sample wall, they are being deployed for the walls of the museum without any such measures. To improve this detail, Komendant agrees with the contractor's initiative of using "rubber washers," also to prevent loss of water with resulting "discoloration." Nevertheless, he thinks the seals should be thicker than those used to date, and that the cones should be "covered with grease"; he also suggests conducting a trial on an "unexposed wall face." Kahn, on the other hand, has doubts about the efficacy of the washers due to their movement around the cones, which would lead to infiltration of the pour, requiring intervention after stripping. Komendant himself, in any case, thinks it would be better to use the snap ties, following the procedure applied in the sample walls – "cones filled with clay, lightly greased and secured tightly."

"In summary, then Lou went along with Dr. Komendant's discouraging of the use of screw ties, instructed that the tie cones be greased and filled as on the sample walls; objected to the gasket and instructed that its use be stopped except for trial use in unexposed shaft space."

Another defect on the concrete surface observed by Komendant during the inspection are the marks produced by the oil placed on the panels to

[151] Marshall D. Meyers, letter to Preston M. Geren, Jr., and Frank Sherwood, 11 May 1970, LIK, 030.II.A.37.6.
[152] *Ibid.*
[153] *Ibid.*
[154] Marshall D. Meyers, letter to Frank Sherwood and Preston M. Geren, Jr., 14 May 1970, LIK, 030.II.A.37.6.
[155] *Ibid.*
[156] Bowen King, letter to Frederick Langford, 15 May 1970, LIK, 030.II.A.37.16. Geren communicates that he is still waiting for the "patterns of the cycloid forms" (Preston M. Geren, Jr., letter to Louis I. Kahn, 20 May 1970, LIK, 030.II.A.37.6).
[157] "I have convictions," Langford writes, "that a thoughtful approach to the problems and requirements of Architectural Concrete buildings can result in three

facilitate stripping. Besides indicating how to eliminate the flaws from the concrete by using "soap and water," he proposes his own method to avoid the use of the oil: "'Water repellent' or even plain water on the forms is enough to break the bond. Even without water the forms can be easily removed by waiting 5 or 10 minutes after tie removal before forms are stripped. The forms will then loosen themselves."

Still in order to obtain a perfectly smooth concrete surface, another expedient suggested by Komendant is that of using artificial curing, by means of heat, of the "form coating" spread on the panels. The solution is based on the methods used to make the molds of precast concrete.

From the information gathered by Thompson on 4 May, it appears that Byrne, Inc., is not following the instructions supplied by Komendant and Kahn. "The contractor was still proceeding with gasket idea on exposed surfaces and he was not greasing or filling the tie-cones." Another concern has to do with the color of the concrete, which according to Thompson is "very 'green.'"[152]

The pursuit of perfection in the design of the holes is not limited to the making of the imprints of the cones on the concrete. The other question, also under discussion in May, has to do with the detail of the sealing of the holes with lead disks, as in the Salk Institute. Not having heeded the advice of Meyers and Langford on the need to modify the cones to get holes suited for the insertion of the seals, the contractor encounters the first difficulties in attaching the lead disks, because they are not thick enough with respect to the depth of the holes. The contractor thus proposes applying putty in the hole prior to inserting the disk,[153] meeting with the critical reaction of Kahn's staff. To avoid puttying, due to the fear that it would separate over time, and also for concerns of correspondence of the technical details with coherent principles in every phase of the construction process, Langford draws the "shape and size of the lead plugs that fill the tie-cone holes."[154]

For the insertion of the lead disks, the solution of the special bar used at the Salk Institute is considered, but there is also the hypothesis of creating another "round bar" shaped in such a way as "to incise a cut in walls of the hole that the lead could be forced into when the plug is installed."[155]

Formwork for the "site work concrete walls"

Langford continues to produce working drawings for the formwork. Halfway through May 1970 he is asked to make drawings for the cooling tower, the intrados of the cycloid vaults and the retaining walls.[156] But Langford does not fail to emphasize the difficulties involved in his participation, ranging from late payments for his work to the impossibility of directly controlling the implementation of the designs.[157] His complaints conclude with a request for better definition of his role, an official meeting takes place between him, Geren, Kahn and Seymour, and a personal meeting with Seymour precisely "to discuss his intentions and methods for forming and pouring the cycloid vaults."[158]

major benefits. (1) Excellent quality throughout (2) Considerable cost savings over conventional forming methods and (3) The final articulation of the surface becomes an honest and meaningful record of the construction process. I have proven these three points on previous projects, but I have not been allowed to do so in Texas. And from what I hear, the Museum continues to be constructed without realizing even *one* of these three benefits" (Frederick Langford, letter to Bowen King, 20 May 1970, LIK, 030.II.A.37.34).

[158] Frederick Langford, letter to Bowen King, 20 May 1970, LIK, 030.II.A.37.34.

353-355. Kimbell Art Museum, Forth Worth, Texas, 1966-72.

353. Formwork study for the garden wall, 26 January 1971 (LIK, 030.I.C.730.2).

The most urgent question related to the formwork design has to do with the drawings for the construction of the retaining walls of the garden, and for the cycloid vaults.

Kahn's instructions for the retaining walls point to his desire to develop a true language of exposed concrete, equipped with a range of variants capable of expressing the character of the different structures through the varied arrangement and form of the same parts. For the invention of differentiated rhythms of certain imprints that determine the character of the walls, Kahn exploits the part of the formwork destined not to leave direct marks on the exposed concrete: the structure that supports the panels. To distinguish the retaining walls from those of the museum, he wants a pattern in which the panels are joined without the use of the typical V joint, and he indicates that the arrangement of the spreaders should be more open.[159]

At the start of 1971, Kahn explains his idea of the concrete surface for the walls in the garden, emphasizing the need to "distinguish the site walls from the building walls by making them more 'plain'" thanks to the elimination of the V joints, to be replaced by "butt joints" and the use of "tie holes" that are not "recessed."[160]

To make the holes of the site walls as desired by Kahn, Byrne, Inc., experiments with an "unconventional form tie washer" that makes it possible to cut the "tie rod" without scratching the surface of the concrete.[161] But the effect sought of absence of relief is not achieved with that type of washer, which in Kahn's opinion "leaves a recess that will be too 'loud' and too similar to the tie holes on the building."[162] To avoid pronounced cavities, Kahn and Meyers recommend the use of a "flat washer" which nevertheless requires particular care in the stripping phase, when the tie is cut, "because it may move and spall the concrete around the washer." "After grinding, the washer would be removed and any burrs or sharp edges on the exposed tie rod would be eased off." Meyers advises that Earp "should experiment

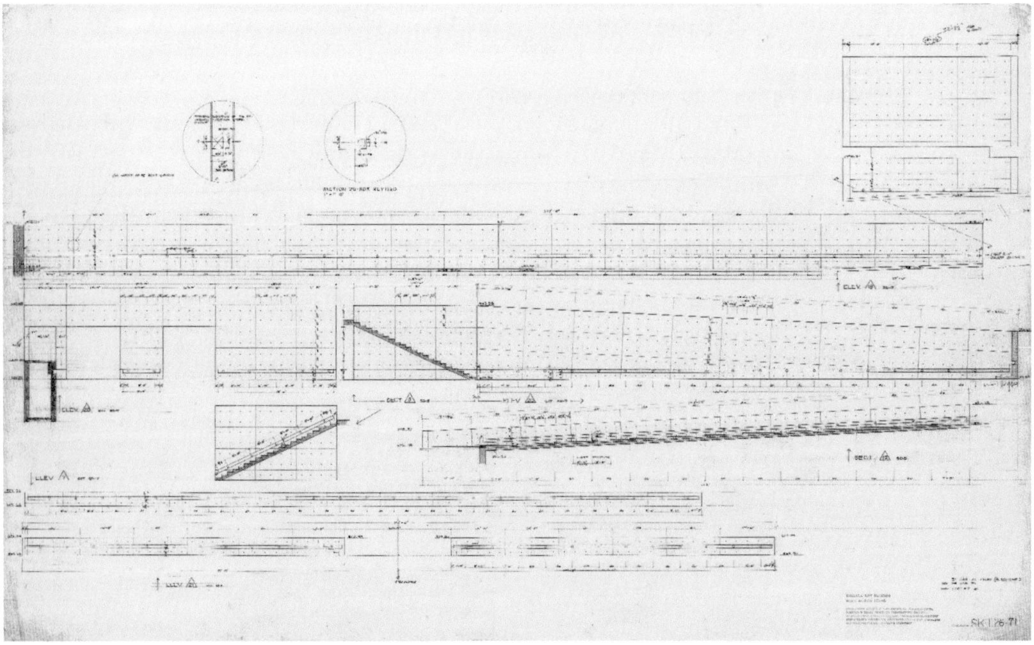

354-355. View and detail of the garden wall (photographs by the author, 2010).

with this to get the best result."[163] This is the solution eventually used by Byrne, Inc.

Kahn prepares the drawing of the "pattern for site work walls," identified as SK 1-26-71, which Sherwood sends to Byrne, Inc., on 28 January.[164] The documented version of that drawing (revised on 31 January) shows the spreaders arranged in a grid of 23 × 24 inches, apart from certain exceptions, while the panels measure 7 feet-8 inches × 4 feet-0 inches.[165] Certain indications on the drawing reflect Kahn's desire to obtain walls with a "plain" character: "All joints to be butt joints," is clearly specified. The openings for lighting are positioned based on the constructive logic of the formwork ("typical location of tie rod through lighting recess," is the title of a detail shown in the drawing).

For the "site work concrete walls" it is decided to use the "coated 3/8" exterior plywood lining" and 1-1/4 inch "indented washers on stainless steel ties for recessed break-back."[166] The exposed aggregate is a compound formed by Cactus Green Marble Chips, produced by the Bilbrough Marble Company of Burnet, Texas, and Gray Portland Cement Matrix.

The site walls are built during the spring and summer of 1971. The concrete surfaces have the "plain" look desired by Kahn, thanks to the lack of

[159] "As for the Retaining Walls," Meyers explains, "Lou wants the tie-hole pattern to be different from that on the building. He proposes that the Retaining Walls be made from the plywood used on the building ceilings and that the ties be placed farther apart (no less than twice as far vertically and horizontally). This will necessitate heavier walers and studs. The joints between the plywood sheets will be taped, butt joints" (Marshall D. Meyers, letter to Preston M. Geren, Jr., 22 May 1970, LIK, 030.II.A.37.6). The greater distance of the spreaders in the retaining walls, "approximately twice the distance as on the present building," leads to a "stronger form construction" (Frank Sherwood, letter to Thomas A. Seymour, 8 June 1970, LIK, 030.II.A.37.6). The series of drawings of 29 June 1970, "Retaining Walls elevations," confirms the indications of Kahn.

[160] Marshall D. Meyers, letter to Frank Sherwood, 15 January 1971, LIK, 030.II.A.37.7.

[161] Frank Sherwood, letter to Louis I. Kahn, 4 January 1971, LIK, 030.II.A.37.7. Sherwood reminds Kahn that it was decided to use the method of cutting the tie rods "as close to the surface of the concrete as possible," rather than "snapping them."

[162] *Ibid.*

[163] *Ibid.*

[164] Frank Sherwood, letter to Thomas S. Byrne, Inc., 28 January 1971, LIK, 030.II.A.37.7.

[165] See the drawing dated 26 January and revised on 31 January 1971, LIK, 030.I.C.730.002. Also see the drawings, SK-2-2-71, 2 February 1971, LIK, 030.I.C.730.002; SK-2-24-71, 24 February 1971, LIK, 030.I.C.730.002; and SK-1-8-71, 8 January 1971, LIK, 030.I.C.730.002.

[166] Preston M. Geren, Jr., "Bulletin No. 40," 12 March 1971, LIK, 030.II.A.37.26.

356-357. Kimbell Art Museum, Forth Worth, Texas, 1966-72. View and detail of the porch (photographs by the author, 2010).

V-joints and deep holes. In this perspective, one fundamental factor is the use on the part of Byrne, Inc., of slender, slightly rounded steel washers that produce a delicate cavity, with the cut tie still visible at the center.

Another form of surface, again to be obtained with the formwork imprints, distinguishes the concrete walls designed for the Conservator's Court, at least according to the drawings discussed in May 1970 and marked by the absence of all marks left by panels and spreaders.[167] A drawing dated 3 November clears up the enigma of the concrete walls of the Conservator's Court, showing them bearing the same motifs as the other walls – V-joints and the pattern of the tie holes.[168]

Certain errors are made by Byrne, Inc., in the arrangement of the spreaders on the concrete surfaces of the Kimbell Art Museum. Kahn would have liked the columns to be free of such signs, confirming the fact that in his vision both the V-joints and the holes are imprints characteristic of the wall, not of the framework of pillars and beams. But the surfaces of the columns of the two porticoes, poured in continuity with the exposed concrete walls in the court on the lower level, are marked by tie holes. No modification is made to those four columns, though it is emphasized that the "tie holes in portico columns L-1, L-4, L-11 and L-15 are incorrect" and Kahn's intervention is announced: "There should be no tie holes in any columns. Mr. Kahn will present solution to this problem."[169] The grooves that show the pour of the floor slab, at the top of the walls, are also erroneously reproduced by Byrne, Inc., on the portico columns, though this detail has not been indicated by Kahn.

"At exterior concrete pours," Meyers writes, "the museum floor slab is expressed by two concave pour joints: one at the bottom of the floor slab

and one at the top of it. Because the base of the 2' × 2' concrete portico columns is lower than the main floor slab, the contractor introduced these same two indentations ringing the base of the column even though, at this location, they no longer signified an event in the concrete pour sequence. No one from the Philadelphia office was at the site at that moment, nor had the contractor asked how to form the base of these longer columns. He remembered those two indentations in the concrete at the floor slab elevation, so he put two little indentations in the columns at the same point. Of course Kahn left it that way and did not insist on deleting this decision made by another sensibility."[170]

Concrete shells of cycloidal shape

When the waffle slab is introduced for the Kimbell Art Museum, the specifications of April 1969 indicate the type of structure of the vaults: "The roof structure consists of a series of longspan *post-tensioned* concrete shells (of cycloidal shape)."[171] Detail studies for the system of construction of the cycloid vaults are conducted only in the spring of 1970. Alongside Byrne, Inc., Komendant is involved in the creation of the centering system, and Kahn's office is in contact with him to decide on the pattern of the formwork (besides Langford, Thompson is also involved).[172] Komendant's presence at the worksite becomes increasingly important, in view of the start of construction of the post-tensioned structures, and starting with the inspection, in early August, of the "first pouring of concrete of upper level slab."[173]

For the design of the concrete surfaces of the cycloidal shells visible from the gallery, at first, during June, Kahn follows the same tenets applied in the exposed concrete walls of the museum.[174] Each shell is to be made in

[167] "You will also note," Sherwood writes to Seymour, "that around the Conservator's Court area there are two concrete walls filling the arch of the cycloid with no form marks or ties on them. This will require the use of screw-tie forms in the wall below the spring line to attach, probably, steel channel or I-beam walers running vertically with horizontal studding to support the forms in order that this section of the wall may be poured without form ties. I believe there are two such walls" (Frank Sherwood, letter to Thomas A. Seymour, 8 June 1970, LIK, 030.II.A.37.6).

[168] See the drawing R-FW-1, 3 November 1970, LIK, 030.I.C.730.017.

[169] Marshall D. Meyers, "Kimbell Art Museum. Items from Inspection Trips of May and June 1972," 13 July 1972, LIK, 030.II.A.37.29.

[170] Marshall D. Meyers, *Interview with Alessandra Latour, [...], March 19, 1983*, in, Latour, ed., *Louis I. Kahn, l'uomo, il maestro* cit., (pp. 71-88), p. 81.

[171] "Outline Specifications for Kimbell Art Museum, Kimbell Art Foundation, Fort Worth, Texas," 7 April 1969, LIK, 030.II.A.95.5. On the calculation of the thicknesses of the structure of the cycloid and the columns, also to accommodate the "Air Supply and Return Ducts," see Preston M. Geren, Jr., letter to Marshall D. Meyers, 22 May 1969, LIK, 030.II.A.37.5. On the "return air" that "will be taken at the ends of cycloids in general," see David P. Wisdom, letter to Thad E. Harden, Jr., 27 May 1969, LIK, 030.II.A.37.5. See also, for various solutions, Thad E. Harden, Jr., letter to Louis I. Kahn, 3 June 1969, LIK, 030.II.A.37.5.

[172] Marshall D. Meyers, letter to Preston M. Geren, Jr., 22 May 1970, LIK, 030.II.A.37.6.

[173] Frank Sherwood, letter to Richard F. Brown, 5 August 1970, LIK, 030.II.A.37.7. In the meantime, Geren's office makes drastic decisions on the preparation of the detail drawings, cutting Kahn off from any form of control to avoid costs and delays (see, among other documents, Preston M. Geren, Jr., letter to Louis I. Kahn, 20 July 1970, LIK, 030.II.A.37.7). "Preston Geren has refused to continue to work if Louis Kahn is given the opportunity to check the shop drawings. [...]. Dr. Brown apologetically stated the only way Louis Kahn can have the opportunity of reviewing shop drawings is for him to review them in Fort Worth" ("Notes taken during meeting of 7 August 1970 in Fort Worth," LIK, 030.II.A.37.15).

358-361. Kimbell Art Museum, Forth Worth, Texas, 1966-72.

358. "Formwork," 17 June 1970 (LIK, 030.I.C.730.18).

359-361. Study for the concrete vault centering (LIK, 030.I.C.730.18).

three sectors with four types of centering (a fifth special piece is envisioned when the shell has to terminate in an end wall). Each sector, in turn, is broken down into three parts: two large lateral centerings and a narrower one at the position of the cut in the shell for overhead lighting. The lateral centerings are formed by narrow panels – nine of them, plus one special panel in the two end sectors of the vault, and twelve in the middle sector. For the construction of a shell, it is sufficient to prepare two sectors of a complete centering (with the shorter one it is possible to make the two ends of the shell by inverting the arrangement of the centering). In the assembly of the centering sectors Kahn calls for insertion of a strip that imprints lines on the concrete to express the formwork system, as in the case of the walls. In the drawing dated 17 June 1970, showing the "form parts" of the system, this special piece is called the "closure strip."[175]

Kahn and his staff study the structure of the centering and the methods of moving and reassembling the various sectors. Several drawings document the research on the various structures for centering broken down into three sectors, with a joint between the sectors to be made visible using the typical T strip which would leave longitudinal imprints – "V-joint" is indicated on the sheet of sketches.[176] Mounted on wheels, the various sectors can be moved by hand by the workers, from one span to the next, through the large openings.[177]

For the construction of the cycloids, in October, alongside the sample walls, a full-size model is prepared of a shell segment with half of the end wall (the other end is completely open). The model is placed on the ground, on two special wide concrete bases that imitate the beams in order to also test the making of the horizontal groove separating the shell and the beam. For the large rectangular panels that shape the inner surface of the shell, marine plywood with three layers is used, to make it easier to bend in keeping with the curvature of the cycloid. The panels are nailed onto traditional wooden centering. Then the tubes for post-tensioning are laid, along with the network of steel reinforcement. To separate the metal bars from the panels, devices of steel wire are inserted, known as "chairs," with four posts (the feet of the posts remain visible on the intrados of the concrete). This detail includes the earlier suggestions of Seymour on the method of separating the bars from the intrados. Then pairs of overlaid and bent boards are inserted, running parallel above the panels and reinforcements, remaining detached thanks to steel brackets, while longitudinal boards nailed onto the bent ones stiffen the system. The bent boards remain in place to contain the insulation panels and the covering in lead sheet. The longitudinal boards allow the workers to move with ease on the curved structure during the pouring operations, and are removed one by one depending on the progress of the work.

The pouring of the concrete for the model happens on 26 October. To prevent concrete bleeding at the base of the shell, two rows of rectangular panels are assembled between the bent boards, nailed in place thanks to two square plates of plywood attached at their extremities (but this operation is

[174] The pattern of the formwork of the vaults is designed in June 1970 (see the drawing, "Formwork. 2-SK-FA6-17-70," LIK, 030.I.C.730.017).

[175] See the drawing, R-FW-1, 3 November 1970, LIK, 030.I.C.730.017. For the elevation, see the development drawing, n.d., LIK, 030.I.C.730.018.

[176] See the development drawing, n.d., LIK, 030.I.C.730.017.

[177] *Ibid.*

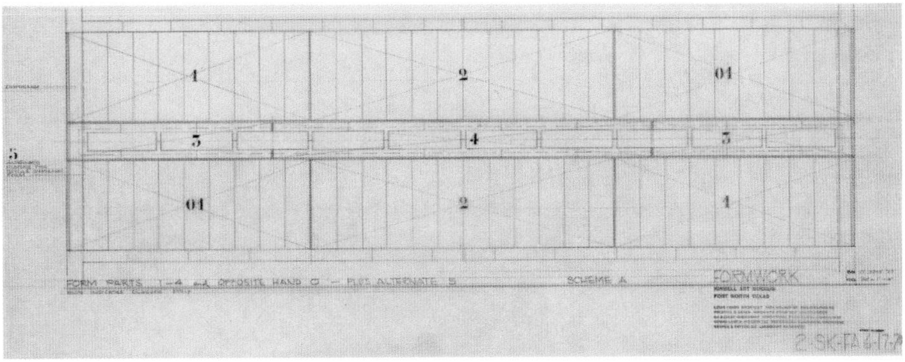
1
2
01
3
4
3
01
2
1
SCHEME A
FORMWORK

3
1
01

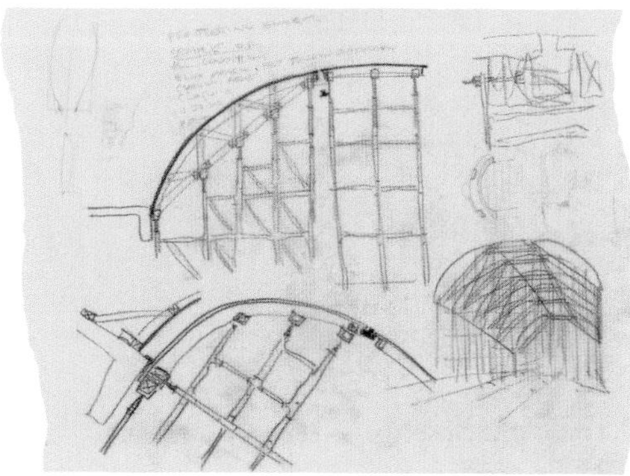

362-365. Kimbell Art Museum, Forth Worth, Texas, 1966-72. Construction of the sample vault, 26 October 1970 (Kimbell Art Museum Archive, Forth Worth).

improvised and not systematic). The concrete is poured first in the upper part of the shell with a feed hopper raised by the crane. The workers make the concrete slide down with shovels, making sure it slips into the various compartments formed by the structure of the boards, and they pack it by using the vibrator, to make it penetrate inside the network of steel reinforcements and post-tensioning tubes.[178] But after this first operation, trials are conducted with the hopper lowered further down, at the height of the sides of the shell, as a result of the difficulties involved in positioning the concrete in the interspaces of a curved structure.[179] Once the concrete has been uniformly distributed, the workers smooth the extrados with a metal trowel.

[178] See the worksite photograph conserved in the Kimbell Art Museum Collection, Fort Worth.
[179] See the photograph of the construction of the model of the shell conserved in the Kimbell Art Museum Collection, Fort Worth.

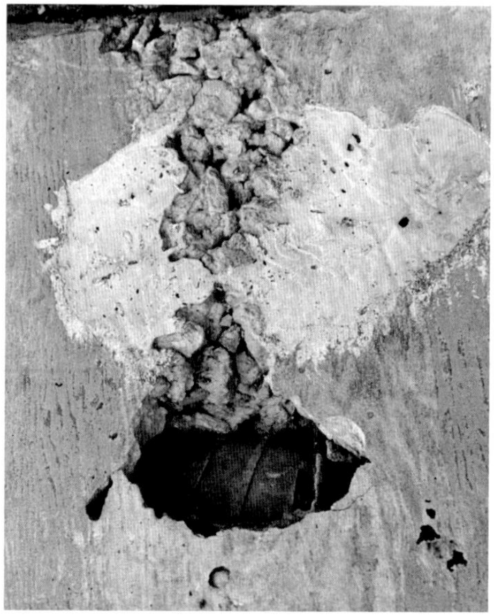

366-372. Kimbell Art Museum, Forth Worth, Texas, 1966-72.

366-367. View and detail of the sample concrete vault, 1 November 1970 (Kimbell Art Museum Archive, Forth Worth).

368. Kahn making drawing on the sample vault, 15 November 1970 (Kimbell Art Museum Archive, Forth Worth).

Kahn first examines the "cycloid Mock-Up" on 15 November, when all the concrete work is complete, and then towards the beginning of December, when the roof layer has also been applied.[180] Because the concrete surface is disfigured by many gravel pockets, it is decided to modify the formwork system to improve the tamping of the pour. Thus the solution of the rectangular panels with the plywood plates attached to the curved boards is extended to the whole shell, forming the interspace for the pour along with the large plywood panels (a similar solution, but with thinner strips, is used to shape the edge beams of the shell[181]). To direct the pour, which can no longer be made as a model, a rigid canvas funnel is applied to the hopper, to be placed at the mouth of the interspace. During the pour a worker pushes the concrete into the opening using a specially constructed wooden shovel.[182]

At the start of November the carpenters are making the various parts for the centering of the shells and laying them out on the spans on the worksite. For the construction of one entire shell segment wooden centering is used with curved panels. The centering is subdivided into two parts, connected by a metal structure at the end of a fixed central bridge, for easy removal, movement and reuse. These centerings rest at their level on scaffolding poles[183] (the device on wheels imagined by Kahn's staff is not made). For the construction of one shell four sectors are needed, as well as the two half centerings at the two ends,[184] instead of the three sectors as specified in June. First the columns with the walls and the long beams are constructed.

[180] "Memorandum," 18 November 1970, LIK, 030.II.A.37.34. See also Thomas A. Seymour, letter to Preston M. Geren, Jr., 18 November 1970, LIK, 030.II.A.37.34.

[181] See the worksite photograph, M. D. Meyers Archive, AAUP, Philadelphia, 136.54.12. For the use of the layer of panels to contain the pour, recommended in shells "from the forms on slopes running steeper than about 45 degrees," see Hurd, *Formwork for concrete* cit., p. 12-10.

[182] See the photograph of the construction of the shell conserved in the Kimbell Art Museum Collection, Fort Worth.

[183] See the worksite photograph of the construction of the vaults conserved in the Kimbell Art Museum Collection, Fort Worth.

[184] See the aerial photograph of 24 March 1971, LIK, 030.IV.A.730, and that of 27 April 1971, LIK, 030.IV.A.730.

369-372. Construction site photographs, 4 November 1970, 10 November 1970, 26 January 1971, November 1970 (Kimbell Art Museum Archive, Forth Worth).

373-374. Kimbell Art Museum, Forth Worth, Texas, 1966-72.

373. Construction site photograph, 18 January 1971 (LIK, 030.IV.A.730.1.1).

374. Construction site photograph, 26 January 1971 (Kimbell Art Museum Archive, Forth Worth).

Then the centering formwork is raised and placed, and the operations tested with the model are carried out, also with the layer of external formwork. After the setting of the concrete, the bars inside the sheaths are tensioned and then stripping proceeds, moving the two hinged sectors of the centering. When the upper part of the formwork has been stripped, the insulation panels are installed. The first shells are poured on 26 January 1971.

"An unusual feature of the Byrne-developed roof forms," Seymour later recalls, "was the mobility of the 19' long curved form sections, which could be 'jacked down' to roll under completed previously poured adjacent cycloids or lifted out by crane to be placed in another location. The forms were partially consumed in the building by re-using two layers of 3/8" marine plywood form sheathing as permanent decking over curved ribs of 2" × 4" lumber which kept the forms from spreading during pouring operations and at the same time gauged the concrete shell thickness and remained permanently attached to the curved cycloid shell surface as both for the 1-1/2" fiberglass roof insulation and nailers for the sheathing."[185]

To seal the hollows made in the arches at the ends of the cycloid shells, where the ends of the post-tensioning cables are located, special concrete pieces are made.[186] Given the fact that the lead sheeting of the roof is laid over the shell in such a way as to leave the load-bearing structure of the arches visible at the two extremities, the concrete of those arches is perfect smooth, and then made as impermeable as possible to rainwater, by means of an electric planer.[187]

After the making of the foundation walls, the pillars, the arches and the shells, and after the laying of the roof covering,[188] Kahn has only to control the installation of the Travertine slab cladding, to apply on the special concrete partitions erected to close the perimeter gaps between the load-bearing members (on the side towards the gallery a concrete block wall is erected first).[189] The photographs that document the Kimbell Art Museum prior to the application of the Travertine show the different treatment of the walls to be left exposed or to be covered.[190] Kahn demands an arrangement of the Travertine slabs with "plumb" joints: "the cement can have some irregularities, it is one of its characteristics, but the space between the Travertine and the cement must not exceed 6 mm."[191]

The entire Kimbell Art Museum is woven with chromatic tunings between materials, and by the junction lines of the lead slabs, the Travertine slabs and the imprints left by the making of the concrete. Other lines, still designed to link the parts into a coherent whole, are those of the horizontal grooves that cross the walls at the basement level to the level of the slab of the gallery, and those that run continuous to the imposts of the shells, also continuing onto the pillars, to mark the presence of the beams. Other grooves correspond to true joints, like that of the neoprene that runs between the cycloid shells and the pillars. All those lines demonstrate what Kahn had indicated at the start of the 1950s: "buildings are not monoliths."

[185] Thomas Seymour III, *Contractor Challenged by Kimbell Art Museum Design* cit., p. 323. "The forms for cycloids," Komendant would explain, "are designed in sections to simplify stripping and handling for reuse. The skin of the forms consists of three layers of plywood so that they can be bent more easily to match the cycloid curvature. The skin is armor-coated to avoid damage to the plywood during stripping and handling. For outside forms, where the curvature is rather sharp, slip forms can be successfully used. Since narrow-spaced welded wire fabric is used, there is no difficulty in keeping the concrete from sliding down. As the dimensions of the 16 cycloids are the same, only two sets of molds are actually required for pouring the shells. The two cycloids at the center will be formed at the same time, and after these shells are post-tensioned, the forms can be stripped and reassembled for construction of the adjacent shells" (Komendant, *Contemporary Concrete Structures* cit., p. 507).

[186] See the drawing, "S/S Drip between Sectors," 7 November 1970, R-/S/S-1, LIK, 030.I.C.730.029.

[187] See the photograph of a worker on the job, conserved in the Kimbell Art Museum

Collection, Fort Worth.
[188] The lead is supplied by the Murdock Lead Company, of Dallas, and installed by the McClung Roofing Company of Fort Worth (Thomas Seymour III, *Contractor Challenged by Kimbell Art Museum Design* cit., p. 323).
[189] See the worksite photograph showing the installation of the Travertine slabs, conserved in the Kimbell Art Museum Collection, Fort Worth. On the Travertine see: John Iberti, letter to Thomas A. Seymour, 28 October 1969, LIK, 030.II.A.37.34; John Iberti, letter to Marshall D. Meyers, 1 December 1969, LIK, 030.II.A.37.14; Len Enninga, of the Carthage Marble Corporation, letter to Thomas S. Byrne, Inc., 30 March 1970, LIK, 030.II.A.37.34; Frank Sherwood, letter to Thomas S. Byrne, Inc., 14 September 1970, LIK, 030.II.A.37.7.
[190] See the slide, M. D. Meyers Archive, AAUP, Philadelphia, 136.128.52.
[191] Marshall D. Meyers, letter to Frank Sherwood, 12 March 1971, III, 8.1.70-4.71, File Box LIK 37, now in, Patricia Cummings Loud, *The art museums of Louis I. Kahn* cit., p. 159.

375-379. Kimbell Art Museum, Forth Worth, Texas, 1966-72.

375-376. Construction site photographs, 26 January 1971 (Kimbell Art Museum Archive, Forth Worth).

377-378. Construction site photographs (M. D. Meyers Archive, AAUP, 136.54.13; 12; 136.128.52).

380-384. Kimbell Art Museum, Forth Worth, Texas, 1966-72.

380-382. Construction site photographs (M. D. Meyers Archive, AAUP, 136.176.36; 136.142.12; 13).

382-384. Kimbell Art Museum, Forth Worth, Texas, 1966-72.

380-382. Construction site photographs (M. D. Meyers Archive, AAUP, 136.142.12; 136.128.52; 136.142.13).

383. Finishing of the concrete surface of a vault, photograph dated 23 June 1971 (Kimbell Art Museum Archive, Forth Worth).

384. Detail of the vault (photographby the author, 2010).

385-388. Kimbell Art Museum, Forth Worth, Texas, 1966-72.

385. Construction site, photograph take during the Kahn visit (M. D. Meyers Archive, AAUP, 136.128.44).

386. Implementation of the travertine cladding, photograh dated 8 November 1971 (Kimbell Art Museum Archive, Forth Worth).

387-388. Views (photographs by the author, 2010).

389-392. Kimbell Art Museum, Forth Worth, Texas, 1966-72.

389-391. Vault details (photographs by the author, 2010).

392. Detail of the connection between a pillar and a vault of the porch (photograph by the author, 2010).

chapter five
concrete and brick

Mikveh Israel Synagogue, or the cladding of concrete

At the start of the 1960s composite masonry of brick, concrete blocks and reinforced concrete, poured in place or precast, has become one of the possible forms of construction in Kahn's works of architecture. The presence or absence of concrete blocks laid together with clay bricks depends on the structure of the wall. While in the First Unitarian Church and School in Rochester these two materials continue to exist, for Kahn other combinations are possible between brick and concrete, as is also demonstrated by the project in Dhaka. The development of a kind of unity between a reinforced concrete skeleton and a brick wall that frees the brick from its role as filler or cladding takes place precisely at the start of the 1960s, with the design of the Mikveh Israel Synagogue, composed of a Sanctuary with a Hall, the Chapel and the School. Studies for the design of the complex begin in 1961 and are abandoned in 1972.

The constant thinking about the hollow column and the genesis of the light in temples, which according to Kahn coincides with the creation of the sequence of the columns, leads to the invention of a device for indirect, diffused lighting in religious spaces, an extreme evolution of the "screen walls" created in 1960-61 for the Meeting House of the Salk Institute: hollow cylinders, now referred to by Kahn not as hollow columns, but as "light chambers," "rooms of light,"[1] "perforated screen walls"[2] or "window rooms."[3]

The appearance of this device for light, with hollow cylindrical towers crossed by walkways to form an ambulatory, dates back to August 1962, when Kahn develops the second version of the project for the Mikveh Israel Synagogue. The fact that the "rooms of light" are derived from the idea of the hollow column is borne out by certain studies made by Kahn for the placement, also on those hollow structures, of grand reticular roofs. During approximately this same period, Kiyonori Kikutake, Kenzo Tange and Arata Isozaki imagine hefty cylindrical columns for their Metabolist megastructures.

Kahn studies different structural systems for the cylindrical towers and the walls, all based on the use of concrete and brick. In 1962 he envisions a slender reinforced concrete skeleton to create supports that are only visible in certain portions, attempting to obtain lines that seem to frame, rather than to contribute to the stiffening of brick masonry to be left exposed on both sides of the walls.[4]

The use of the skeleton determines the horizontal terminations of the narrow vertical openings. "The exterior and interior would be brick," Kahn explains at the meeting on 4 September 1962. "Precast concrete window frames would be used acting as trim to the brickwork. The brick walls would be reinforced with light concrete elements."[5]

The research on arched openings lies behind the use of a different form of structure. Already at the start of 1963, Kahn proposes wider use of reinforced concrete, as seen in the minutes of a meeting held in January: "Mr Kahn stated the possibilities of using concrete. The entire building could be poured or precast concrete with brick forming cavities where necessary."[6]

In the details of the structural system developed later the concrete is to be poured in place, in formwork made with bent panels to create walls freely cut with arched openings. "[...] formwork of large curved wood panels," is written on a drawing from 1965.[7] The brick masonry changes its function and the entire Mikveh Israel Synagogue takes on a different character due to the different relationship Kahn now establishes between the two materials: the concrete is protected by a layer of brick only when it remains exposed to the elements, while it is visible in other cases and defines the atmosphere, light

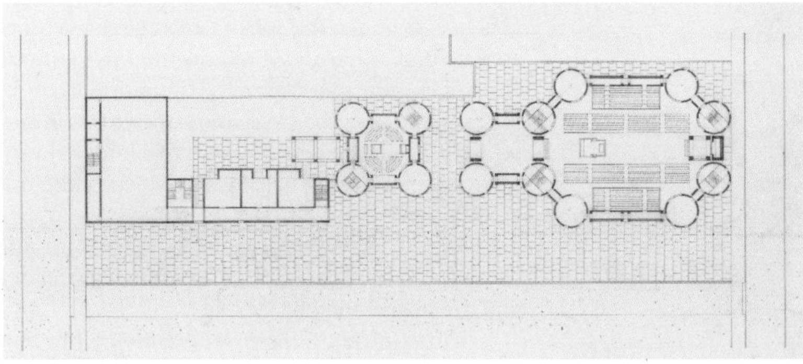

[1] Heinz Ronner, Sharad Jhaveri, Alessandro Vasella, ed., *Louis I. Kahn: Complete Work 1935-74*, Zürich, ETHZ, 1977, p. 221. On the synagogue see Susan G. Solomon, *Louis I. Kahn's Jewish Architecture: Mikveh Israel and the Midcentury American Synagogue*, Waltham, Brandeis University Press, 2009.

[2] Louis I. Kahn, letter to Daniel C. Cohen, 18 December 1968, LIK, 030.II.A.38.22.

[3] Kahn's light chambers have extraordinary affinities with the hollow pylons for indirect light created by Leon Battista Alberti in the church of Sant'Andrea in Mantua.

[4] See the drawing "Wall Section," 22 October 1962, LIK, 030.I.C.615.002, and the elevation, n.d., LIK, 030.I.C.615.001.

[5] Moshe Safdie, of Kahn's office, "Mikveh Israel – Job meeting Memo," 10 September 1962, LIK, 030.II.A.38.13.

[6] Moshe Safdie, "Mikveh Israel – Job meeting Memo," 22 January 1963, LIK, 030.II.A.38.20.

[7] See the drawing 7 June 1965, No. 1, LIK, 030.I.C.615.004.

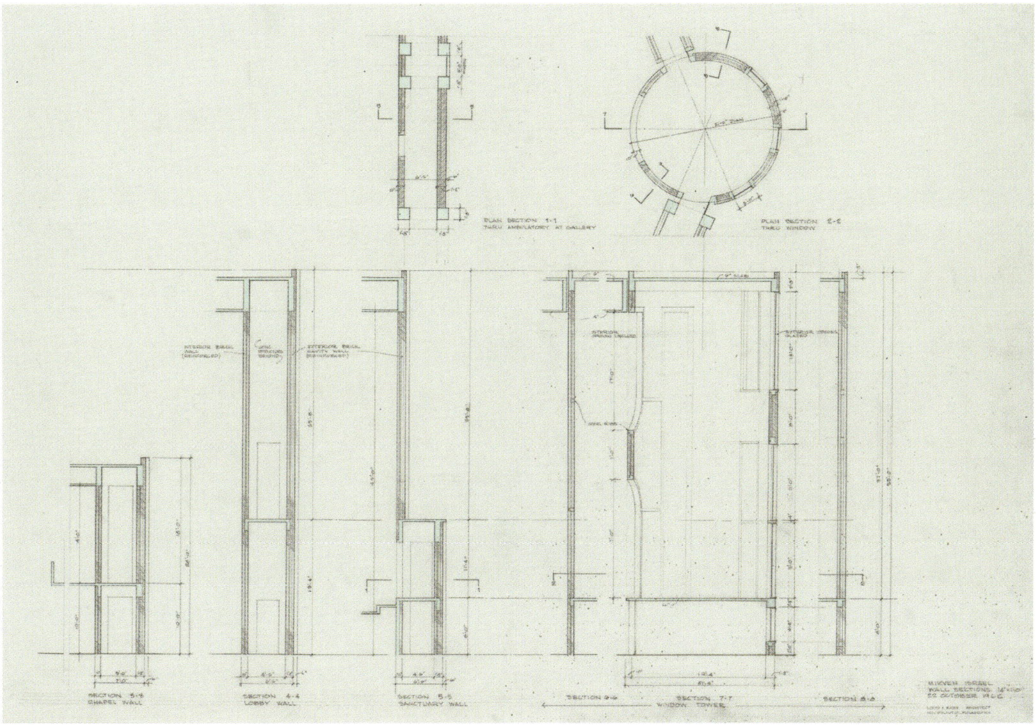

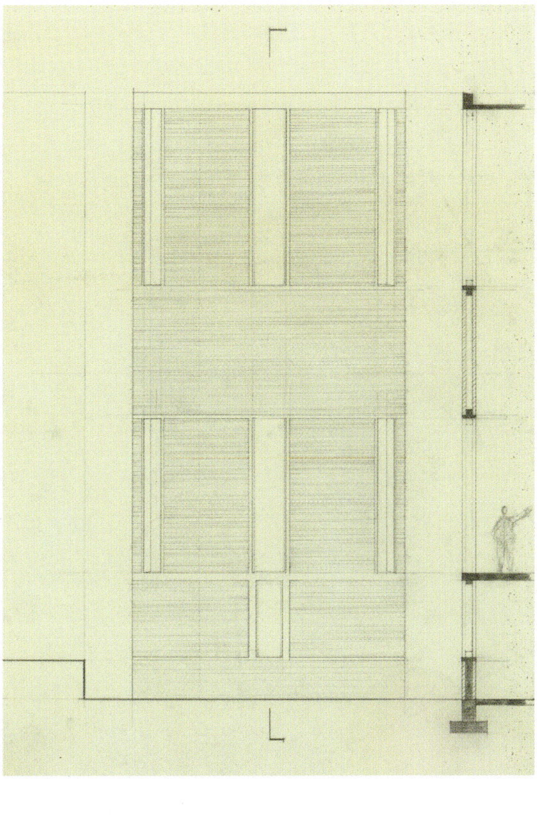

393-395. Mikveh Israel Synagogue, Philadelphia, Pennsylvania, 1961-72.

393. Plan, 31 October 1962 (LIK, 030.I.C.615.2).

394. "Wall Sections," 22 October 1962 (LIK, 030.I.C.615.2).

395. Wall section and elevation (LIK, 030.I.C.615.1).

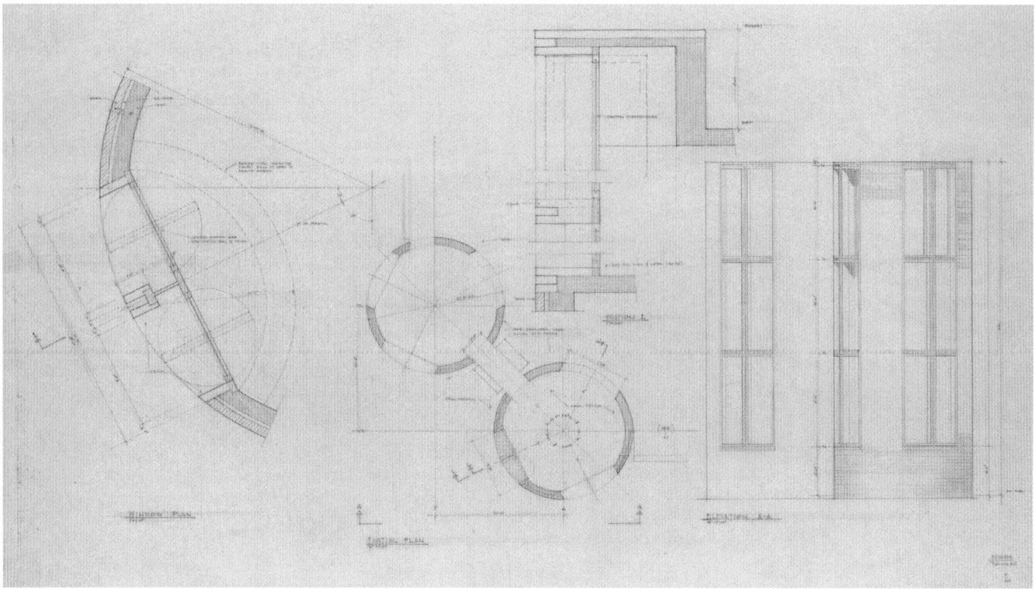

396-400. Mikveh Israel Synagogue, Philadelphia, Pennsylvania, 1961-72.

396. Wall sections, plans and elevations, 7 June 1965 (LIK, 030.I.C.615.4).

397. Wall section, plan and elevation, 7 June 1965 (LIK, 030.I.C.615.4).

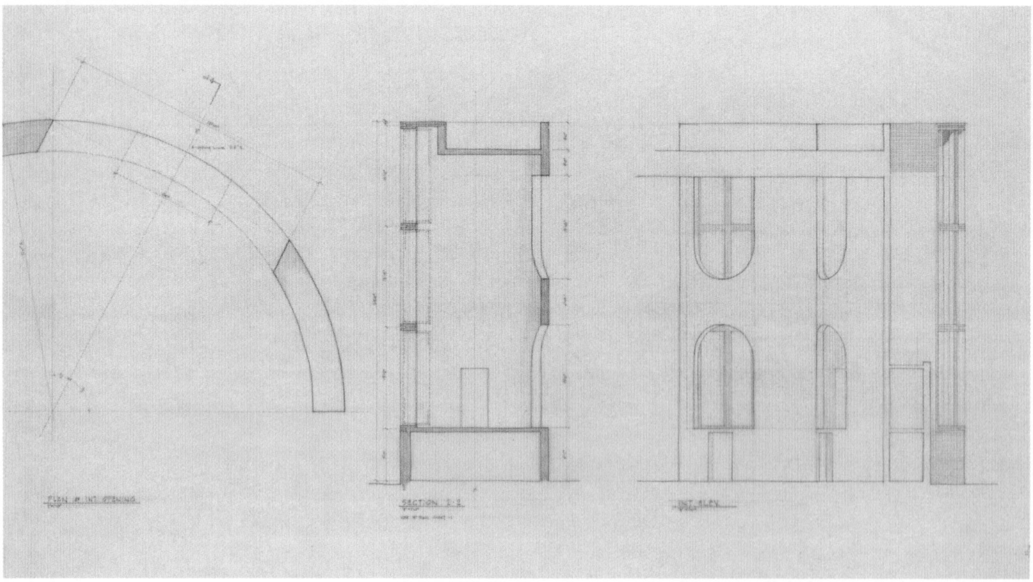

398. Plan (LIK, 030.I.C.615.3).

399. Plan details, 14 July 1965 (LIK, 030.I.C.615.4).

400. Wall plan, section and elevation, 26 July 1965 (LIK, 030.I.C.615.4).

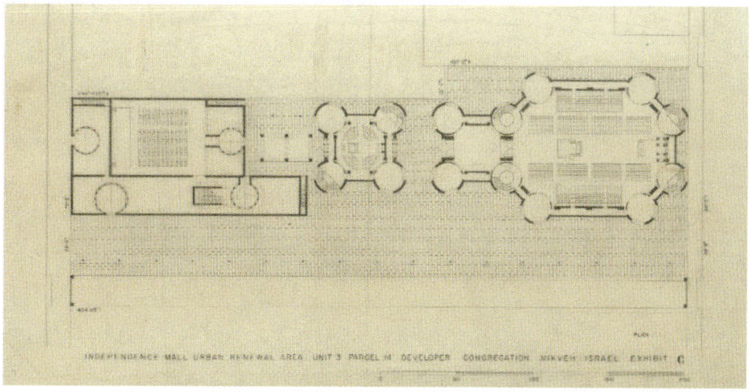

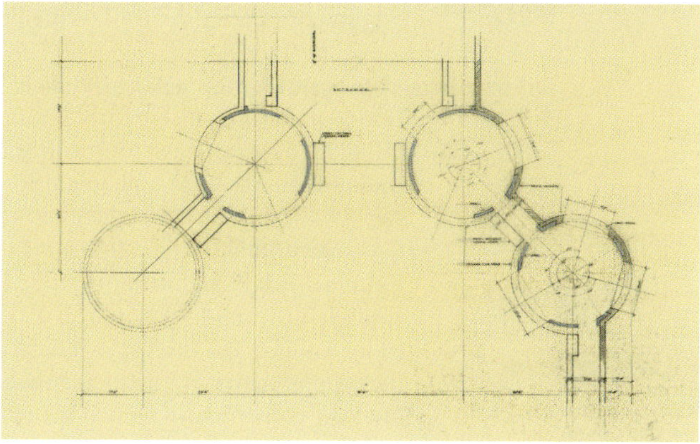

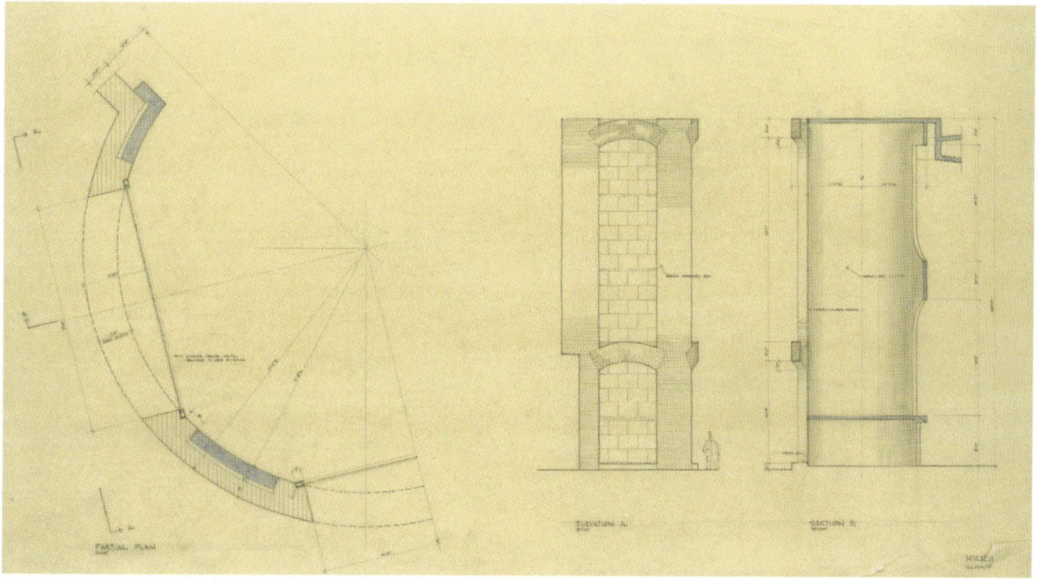

and meaning of the space of the Sanctuary – the hall with the cavities of the cylindrical towers opening onto it. The perspective drawings of the Sanctuary, with the intense shadow on the concrete around the large arched openings, against the backdrop of the glow of the openings in the cylinders,[8] demonstrate that Kahn's main concern is to invent a "room of light" that avoids the unpleasant sensation caused by the phenomenon of reverberation experienced when he was in Luanda.[9] A vertical opening, as in other, earlier works by Kahn, separates the cylinders from the continuous walls to indicate the different roles in the structure: the walls support the roof; the cylinders control the light.[10] The vertical surface of the roof of the Sanctuary is also in exposed concrete. The spindle-shaped section of its structure is invented by starting with the roof of the chapel of Ronchamp and returning to the original design for the roof of the hall in the National Assembly Building.

Antiseismic reinforcements for the Indian Institute of Management in Ahmedabad

Concrete and brick remain the fundamental materials for the construction of the complex of classrooms, library, lodgings, offices and services of the Indian Institute of Management in Ahmedabad, which Kahn begins to design in 1962, with the on-site supervision of Balkrishna Vithaldas Doshi.[11] Construction starts in 1964 and is still in progress in 1974, the year of Kahn's death.[12] All the floor slabs of the complex are made of reinforced concrete, left exposed without the usual plaster and painting, expressly rejected by Kahn.[13] The discussions for the making of the concrete have to do with the type of material for the formwork, in either wood or metal.[14] Kahn defines the construction type of the Indian Institute of Management as a "brick and concrete order": "It is a composite order in which the brick and the concrete are acting together, in not only the floors, but also in flat arches which are restrained by concrete members. It extends the idea of concrete entering the brickwork. I understand that today there is much work on walls and concrete combining to make structures. It is more consciously done, and I think there could be a composite brick and concrete order established as I tried to do here with these restraining members."[15]

401. Mikveh Israel Synagogue, Philadelphia, Pennsylvania, 1961-72. Perspective drawing of the sanctuary (LIK, 030.I.A.615.56).

[8] See the perspective sketches of the interior in, Robert McCarter, *Louis Kahn*, Phaidon, Paris, 2007.

[9] See Florindo Fusaro, *Il parlamento e la nuova capitale a Dacca di Louis I. Kahn: 1962/1974*, Roma, Officina, 1985, p. 81. See also *Louis Kahn*, in, "Perspecta," n.d., [1961], No. 7, pp. 9-28.

[10] Ronner, Jhaveri, Vasella, *op. cit.*, p. 227.

[11] To develop Kahn's project Balkrishna Vithaldas Doshi, the most important collaborator of Le Corbusier for the works in Ahmedabad, coordinates a group of Indian engineers and architects of the National Design Institute of Ahmedabad.

[12] One of the first indications on the nature of the construction is supplied by Kahn to Doshi with the urging to study "the nature of the joint in the geometry of the design – think of concrete and masonry" (Louis I. Kahn, letter to Balkrishna Vithaldas Doshi, 14 July 1963, LIK, 030.II.A.113.3).

[13] See the handwritten notes at the entries on walls and ceilings in, "Typical Tender Schedule, House type 5," n.d., [1964], LIK, 030.II.A.113.37. At the start of 1970 the director of the Indian Institute of Management, Ravi J. Matthai, describes the work with "all the internal surfaces [...] finished with plaster and paint" (Anant Raje, letter to Louis I. Kahn, 24 January 1970, LIK, 030.II.A.113.6).

[14] Chandrasen Kapadia, letter to Balkrishna Vithaldas Doshi, 27 September 1967, LIK, 030.II.A.113.11. "The decision of exposed concrete work here is very vital. Please confirm telegraphically if prof. Kahn would accept wooden shuttering to the main shaft and overhead tank, etc. This would save considerable labour and cost" (*ibid.*). On the formwork for the floor slabs see Chandrasen Kapadia, letter to M. S. Satsangi, 3 October 1967, LIK, 030.II.A.113.11.

[15] Kahn, *Address*, 5 April 1966, now in, Latour, *Louis I. Kahn: writings, lectures, interviews* cit., (pp. 208-220), p. 218.

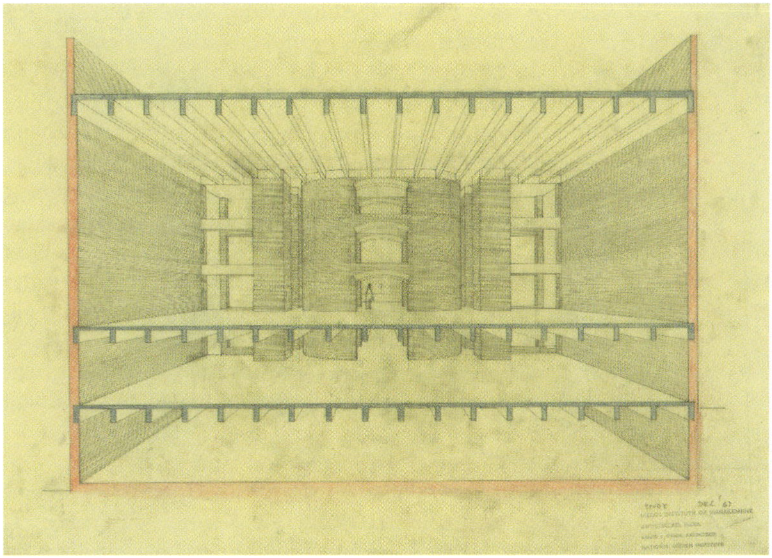

402-404. Indian Institute of Management, Ahmedabad, 1962-74.

402. Perspective section of the Library, December 1967 (LIK, 030.I.C.645.7).

403. The floor intrados of the Library (photograph by the author, 2009).

404. View of the stair of the Library (photograph by the author, 2009).

When the engineer Mahendra Raj, in 1969, following the tremors of an earthquake, proposes reinforcement of the brick masonry with metal bars or with special "reinforced concrete columns,"[16] Kahn opts for the columns in order "to make evident to eye earthquake needs."[17] Given the urgency of intervening in the work in progress with a minimum of substantial modifications, Kahn accepts Raj's suggestion to simply introduce metal bars embedded in concrete inside the brick masonry – scheme E[18] – though he continues to consider the question of the need to make this new skeleton visible, as he had already done during the project for the Bath House. The words Kahn writes are pertinent to this pursuit of truth in the expression of the structure, which he always pursues in the deployment of materials: "I hold to the importance of making manifest the way a thing is made which demands of all elements that they must account for the validity of construction, comprising themselves in the ornaments of its making."[19]

Reinforced concrete masonry: projects for Interama

Kahn's ongoing research on ways of combining clay bricks, concrete blocks and reinforced concrete goes through other variations in the mid-1960s. He decides to abandon the clay brick used in Rochester, Dhaka and Ahmedabad, to create a skin in concrete blocks that is not just a mere cladding. The fact that he returns to the Roman solution of wall claddings with a poured nucleus is not dictated by some passion for the archaeology of construction, but is a logical consequence of his tenacious, obstinate research on the transformation of stone slabs, bricks or blocks into an essential part of the implementation of the concrete, as incorporated formwork that also in some cases has a load-bearing function. Thus Kahn, like Perret before him, proceeds to systematically destroy the concept of cladding as a mask with no role in the making of the structure. Unlike Perret, however, Kahn is even attempting to get beyond the dialectic of skeleton and infill.

The point of arrival, though it remains in the drawing phase, of Kahn's research on a wall made with layers of concrete, is the system of wall surfaces and core, both in concrete, invented for the Inter-American Cultural and Trade Center, or Interama, designed from 1964 to 1969 for Bahia de las Americas in Miami, Florida, as a permanent exposition center, by a group of architects (besides Kahn, the project involves Marcel Breuer, Robert B. Browne, Edward Durell Stone, Irving E. Muskat, Paul Rudolph, José Luis Sert, Frank Watson and Harry M. Weese).

405-407. Project for Interama Community "B", Miami, Florida, 1963-69.

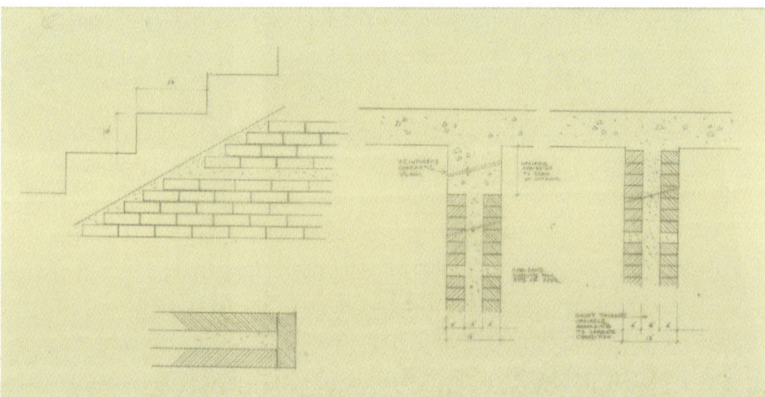

405. Study drawings for a construction system in concrete and brick (LIK, 030.I.C.680.4).

406. "Typical Wall Details," 9 December 1966 (LIK, 030.I.C.680.4).

407. "Exhibition – Auditorium Structural Notes Wall Section," 16 January 1967 (LIK, 030.I.C.680.3).

Certain generic indications on the construction discussed by the group of architects, call for buildings with a "white or light" color, in the material "precast or poured-in-place concrete."[20] The fact that the team includes four of the leading exponents of American architectural concrete – Sert, Rudolph, Breuer and Kahn – explains the decision to make the Interama complex with different forms of concrete.

After the new form of *opus caementicium* studied for the Meeting House in 1961-62, Kahn defines similar technical characteristics for the reinforced

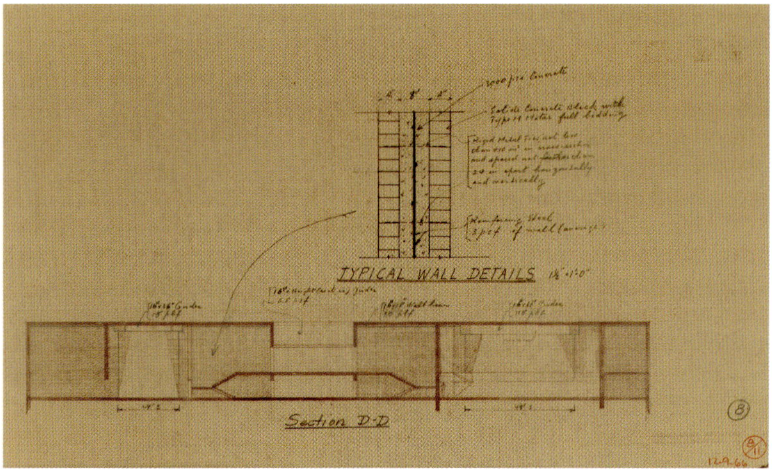

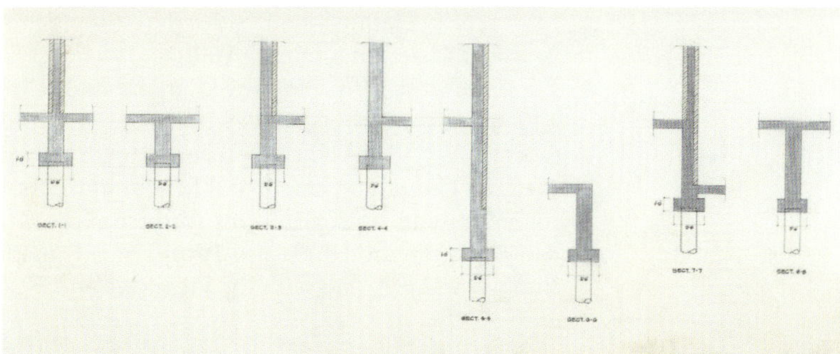

[16] Mahendra Raj, telegram to Louis I. Kahn, 11 July 1969, LIK, 030.II.A.113.8.
[17] "Doshi reference earthquake forces please have Mahendra Raj send immediately idea sketches showing character and intended intervals concrete columns because want to make evident to eye earthquakes needs stop will then judge between use of columns or concealed steel bars in masonry" (Louis I. Kahn, telegram to Vastu-Ship, 15 July 1969, LIK, 030.II.A.113.8).
[18] Raj opts for masonry reinforced with metal bars (Anant Raje, letter to Louis I. Kahn, 1 August 1969, LIK, 030.II.A.113.6). Scheme E of Raj, the one preferred by Kahn, calls for vertical and horizontal reinforcements "embedded in concrete" with the "vertical joint" that becomes a "mullion."
[19] "I like, therefore, Scheme E for its simplicity," Kahn explains. "I trust entirely that the reinforcing mentioned fall in sure position, that they do not conflict with the adjunct concrete of the arches. In regard to Scheme E: does Raj believe that the paneled brickwork between the concrete would control the cracking crack lines in event of an earthquake? If so, this would reinforce the validity of the concrete showing as against interior, unexpressed reinforcing. Altogether, let me say that I am practical enough to see the point of view that the effectiveness of the steel within the wall begins to look to me like straw or cow-hair that reinforces old plaster and I am quite willing to give in to what would keep the job going without undue effort because we are already so far behind. I had to do the same in Dhaka. Maybe we should reserve a more forceful expression for the time when we can incorporate such expression from the beginning" (Louis I. Kahn, letter to Anant Raj, 13 August 1969, LIK, 030.II.A.113.8).
[20] [Edward Durell Stone], "Agenda. Meeting on June 7, 1965," LIK, 030.II.A.113.21.47.

408-409. Library and Dining Hall, Philips Exeter Academy, Exeter, New Hampshire, 1965-72.

408. Section (LIK, 030.I.C.710.1).

409. Section, 1 November 1966 (LIK, 030.I.C.710.1).

concrete walls of his buildings for the Interama, or "reinforced masonry" – the technical detailing is developed with the engineer Paul Weidlinger.[21] By now the wall, still based on the ancient Roman model, becomes a system composed of wall claddings that form the formwork which remains in place after the pouring of the concrete, and is equipped with metal reinforcement. The two wall claddings are in "concrete brick" and the "core" is in reinforced concrete, as indicated in the drawings.[22] Kahn envisions extensions of the core at regular intervals as far as the wall surfaces to create horizontal bands corresponding to the phases of the pour. They are called "pour bands," to be made with special formwork, with the idea of making the core visible, as well as connecting the various layers of the wall.[23] The pattern of the wall would be similar to that of the Tribune Review Building, though the concrete bands would make the different constructive nature of the wall clearly visible.[24]

At a meeting in 1967 the architects involved in the project discuss the quality and finish of the concrete in their respective works: Watson will have "local concrete samples" and "color stucco samples" made; Weese and Stone will use "integral color stucco" to spread on concrete block walls; Sert is considering "stain on his board-finished concrete rather than the paint currently specified"; Rudolph rejects any type of painting or "applied finish," "but will use the lightest of gray cements"; and Kahn "will use the lightest warm gray tone available in cement brick."[25]

Concrete versus brick: the Exeter Library

The dialectic between brick and concrete finds another form of expression in the design and construction of the Phillips Exeter Academy Library in Exeter, New Hampshire. Besides the commission for the library, Kahn is also asked to design a Dining Hall.

The work is first conceived, in the studies from the middle of 1966, as a great mass of eroded clay, where the reinforced concrete is limited to the floor slabs.[26] The walls have a concrete block core, based on those built for the First Unitarian Church and School in Rochester. The choice of a building entirely in brick conforms to the image of the other buildings on the campus, from the 1800s and 1900s, mostly made with brick from the local Eno's Brickyard, laid with the Flemish bond.[27]

[21] For the relations between Kahn and Weidlinger see Mario Salvadori, Paul Weidlinger, "Optimal structure research. A program," New York, typewritten, December 1954, LIK, 030.II.A.66.34.

[22] "Exhibition Building Section A-A," 7 April 1967, LIK, 030.I.C.680.003.

[23] See the drawing, n.d., LIK, 030.I.C.680.004. For the development of the technical detail of the concrete wall by Kahn, in collaboration with Weidlinger, see the drawing "Typical Wall Details" dated 9 December 1966, where the core is brought to 8 inches and the surfaces are connected to each other by "rigid metal ties" (LIK, 030.I.C.680.004). See also, for the various cases of surface and core, the drawing "Structural notes wall section," 16 January 1967, LIK, 030.I.C.680.003.

[24] For some generic indications on the concrete see Louis I. Kahn, "Outline specifications for Construction of Interama Community B, Miami, Florida," 14 May 1966, LIK, 030.II.A.66.21.65, and "Outline specifications Interama Community B," n.d., LIK, 030.II.A.66.21.58. Also for the construction systems in concrete and cement blocks see "Outline specifications for Construction of Interama Community B, Miami, Florida," 20 January 1967, LIK, 030.II.A.66.21.26. See also Frank E. Watson, M.W. Deutschman, H. Samuel Krusé, William J. Lyon of Watson, Deutschman & Krusé Architects and Engineers, "Report on conference held Monday 5 June, 1967, in Philadelphia in the office of Louis I. Kahn," LIK, 030.II.A.66.21.20.

[25] Edward Durell Stone, "June 29, 1967 Meeting. CFA Architects," LIK, 030.II.A.66.21.40.

[26] "Minute 4. Meeting of the buildings and grounds committee, October 3, 1966," LIK, 030.II.A.66.6.1.

[27] For the list of possible colors of the bricks in the various buildings of the campus, all supplied by the Eno Brickyard of Exeter, and mostly laid with Flemish Bond, see, among other sources, "Eno's Harvard Waterstruck Brick," 10 October 1963, LIK, 030.II.A.66.6.2. The document is attached to, Theodore Barry, letter to David P. Wisdom, 4 August 1967, LIK, 030.II.A.66.6.2.

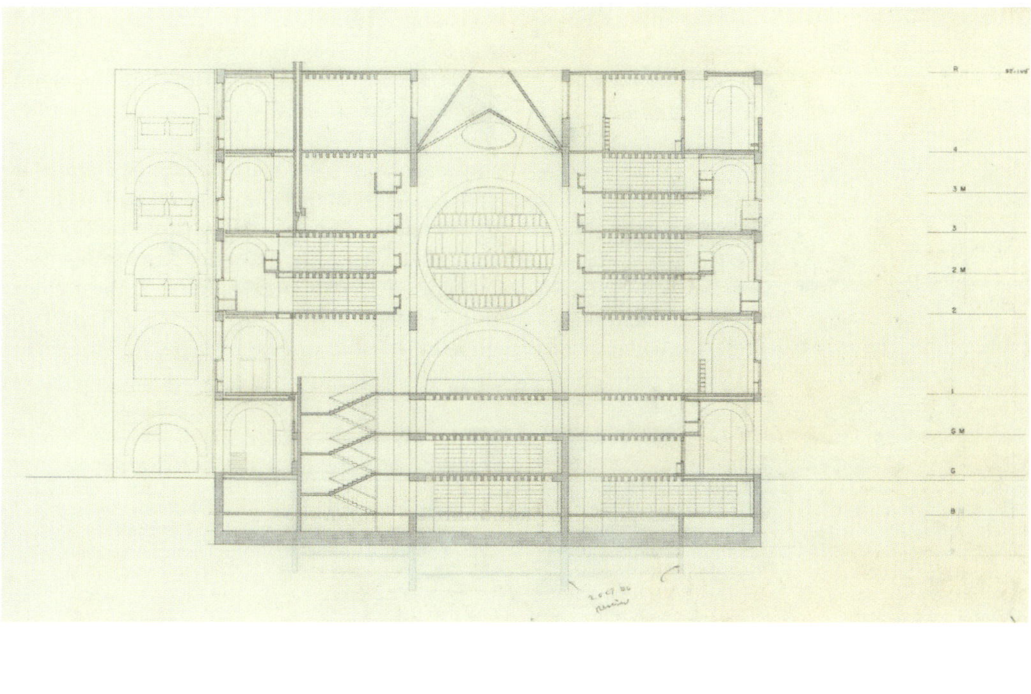

410-411. Library and Dining Hall, Philips Exeter Academy, Exeter, New Hampshire, 1965-72.

410. Section (LIK, 030.I.C.710.2).

411. "Masonry Details," 2 April 1968 (LIK, 030.I.C.710.5).

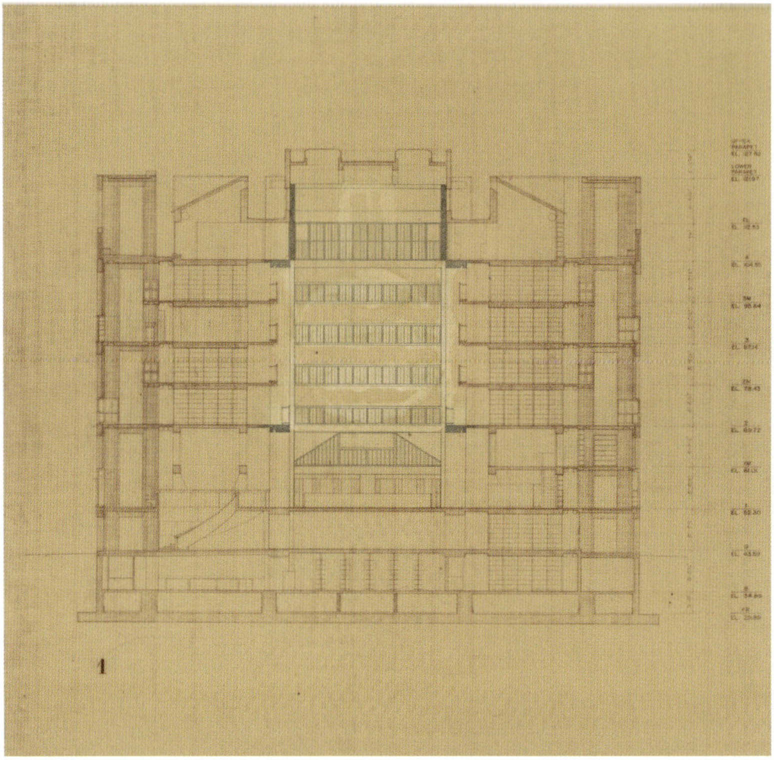

The first decisive change happens in late October-early November, when, in order to reduce the excessive cost of the project, Kahn, MacAllister and the contractor initially hired for the construction, the Macomber Company, develop modifications, including the suggestion of the Macomber Company "to change the building to a straightforward reinforced concrete structure with a masonry veneer exterior."[28] "Change all of the interior masonry to simple structural concrete, including the central core" and other project elements, is the successive recommendation of the Macomber Company, following discussions with Wisdom.[29]

The question of the appearance of exposed concrete prompts immediate critical reactions on the part of members of the committee of Phillips Exeter Academy, who agreed with the choice of brick. "Why can't the bricks be hung on concrete as a veneer?" they ask Kahn.[30] Rodney Armstrong, the librarian who chairs the committee and recommended Kahn as the architect of the new building, asks Kahn to explain – "am I correct in believing that the walls will now be of concrete?"[31] – and to send a letter clarifying this issue to the committee, in which "you state simply what your true feelings are

[28] George B. H. Macomber, letter to Louis I. Kahn, 9 November 1966, LIK, 030.II.A.66.6.1. Among the other modifications, we find that of leaving the concrete of the floor slabs exposed, eliminating the soffits and the floors.

[29] George B. H. Macomber, letter to Louis I. Kahn, 10 November 1966, LIK, 030.II.A.66.6.1. The Macomber Company emphasizes the replacement of the brick walls with reinforced concrete, also in the four main towers and the staircases.

[30] This is the view of Hugh Calkins, reported in, Richard Day, letter to Louis I. Kahn, 14 November 1966, LIK, 030.II.A.6.1.

[31] Rodney Armstrong, letter to Louis I. Kahn, 23 November 1966, LIK, 030.II.A.6.1.

[32] Rodney Armstrong, letter to Louis I. Kahn, 15 December 1966, LIK, 030.II.A.6.1.

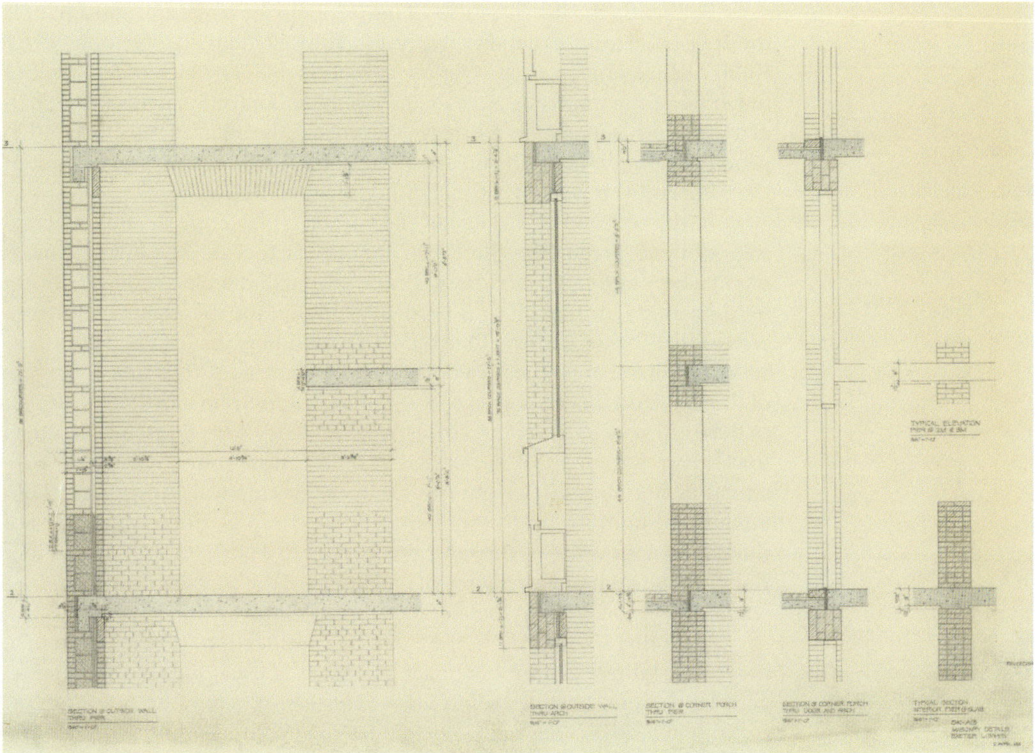

as to the aesthetics of brick for the exterior of the library and further just where you believe brick should appear within the interior."[32]

The solution of the four colossal eyes open to all the levels, which adds strong character to the hall of the library, comes directly from the originally envisioned system of brick construction (Kahn also considers the possibility of their elimination).[33] The evolution of the project, with the intrusion of the concrete, determined by economic factors, still conserves a perimeter of brick forming a protective covering of the core, as in the Mikveh Israel Synagogue, and to gain the thickness of the walls required to insert the study carrels, to achieve a given type of light, and to offer a view of the landscape. At this point the materials create the two fundamental atmospheres of the library, the corners for reading and the zones of shelving, replicating the indication of spaces for different functions organized in Rochester with the same materials. "The space of the periphery harboring the reading rooms towards the light is of brick," Kahn writes. "The interior is of concrete containing a ring of book stack space related to the reading rooms and overlooking a high enclosed central court."[34]

[33] See the section dated 1 November 1966 (LIK,030.I.C.710.001), where the large eyes are set on four arches at the main level of the hall, and the whole structure is still in brick, and the section dated 17 March 1967 (LIK, 030.I.C.710.002), where the structure is now in reinforced concrete – the four large arches below have been replaced by beams in reinforced concrete. For the elimination of the large eyes see the drawing, No. 1, n.d. LIK, 030.I.C.710.002.

[34] Louis I. Kahn, letter to Vappi & Company, Inc., of Cambridge, 23 October 1968, LIK, 030.II.A.6.3.

Following the development of the design – the structural consultant is the Keast & Hood Company, represented by Leidigh – in October 1967 Kahn's office prepares the "Outline specification." The lack of any kind of finishing of the materials after their placement confirms the Brutalist character of Kahn's works, in the sense that materials are used "as found": "Exposed masonry will not be finished, the natural material as it comes out of the forms or as it is laid up constituting the final finish."[35]

In the section on "Concrete" the specifications copy the fundamental lines of those prepared in June 1967 for the factory of the Olivetti Underwood Corporation in Harrisburg, Pennsylvania, but with certain significant modifications, in pursuit of a particular quality for the architectural concrete of the Exeter Library. One telling indication is that of the widest possible measurement for the plywood panels, "largest stock sizes." As in the case of the Olivetti Underwood factory, in this version Kahn entrusts the design of the formwork to the contractor, simply making corrections, and suggests that the arrangement of the panels should follow only the logic of construction, and that the "layout" be "the most economical and structural obtainable without regard to symmetry." If we compare the specifications of the Exeter Library with those of the factory in Harrisburg, it seems that Kahn, in the preliminary phases of the design of the construction details, intends to find the optimal logic of fabrication of the formwork.

The other indications in the "Outline specification" on the concrete of the Exeter Library repeat what have become the usual phrases of Kahn regarding the ban on retouching gravel clusters, patching tie holes or correcting "any fins." The same logic guides the deployment of the other materials. "Do not use any discolored or off-color block in exposed block walls," we read, for example, on the subject of the masonry.

Starting with the text prepared in October, new specifications are drawn up. In the definitive version of 24 November a paragraph is added on the "sequence of erection," to explain the modes of construction of a building subdivided into a "central core" in reinforced concrete and a perimeter band in masonry.[36] In the chapter on the formwork the value of the formwork changes decisively. The assignment of formwork design to the contractor is eliminated. The phrase regarding the possibility of seeking the most logical assembly of the panels, without any symmetry, is also taken out. Now, the formwork design has to be guided by the goal of obtaining large concrete surfaces without joint imprints, as in the case of the Dormitory at Bryn Mawr College. "Formwork shall be designed by the Architect in the formal arrangement, using full-length sheets," is Kahn's crucial indication. In the "formal arrangement" the position of the ties is specified: "screw ties will be lined up." The formwork design even seems to have played a determining role in the definition of the proportions of the whole building, if we consider a phrase whose aim seems above all to underline the importance of this technical detail of the fabrication of concrete: "The building is designed for maximum reuse of forms."

[35] "The Phillips Exeter Academy, Exeter, New Hampshire, Library Building. Outline specification," 10 October 1967, LIK, 030.II.A.6.26. The document states: "The building is a concrete structure with a concrete mat foundation. The outer skin is faced, wood panels, brownstone lintels and insulating glass." The internal partitions are envisioned in concrete blocks or "poured concrete."

[36] "The central core (stack area) is supported by hanger columns from the second floor to the roof trusses. All floors shall remain in the forms until all loads are transferred to the truss, or shall be reshored after stripping. Jacking will be involved in the operation. Build the central concrete structure completely before starting exterior masonry walls. Floor slabs around the perimeter of the central core will be constructed at the time of laying up the masonry walls. Haunches will be provided at the central core to receive the slabs" ("The Phillips Exeter

Rubber gaskets or V joints

During the course of 1968 modifications are developed, also in collaboration with Wood & Tower, Inc., to reduce the construction costs of the Phillips Exeter Academy Library. In the Dining Hall, it is proposed to eliminate "all concrete Retaining Walls," to replace the concrete stairs with steel, and to "change sandstone lintels to brick arches." For the roof, instead of the floor slab poured in place, a precast slab system is considered.[37]

After the "Outline specification" of 1967, the "Construction specifications" are finally issued on 29 November 1968. This is the text in which, on the subject of the quality of the exposed concrete, besides repeating the usual warnings about the absence of chamfers or rock pockets, Kahn explicitly outlines what was to become, after the construction of the Salk Institute, an essential characteristic of his architectural concrete: he states that on the concrete surfaces there should be no "wood grain markings."[38]

The text of the "Construction specifications" contains other useful indications to understand Kahn's idea of exposed concrete not worked after stripping. Following the experience of the Salk Institute, for the Exeter Library Kahn also intends to control the process of the making of the concrete, to reduce defects to a minimum, while in any case leaving what escapes that process visible, as if every pour of concrete in his works were always supposed to document the degree of control achieved in the placement process, without any correction after stripping. This is the particular position taken by Kahn in the sphere of international Brutalism, from which he accepts the rule of "as found" but not the attitude of intentional neglect of constructive details. Kahn is aware that certain defects cannot be eliminated, in spite of all the calculations and technical precautions taken to achieve perfect installation of the concrete:

"The exposed surfaces are expected to show certain planned and certain accidental markings of fins at joints in form work surfaces, some random broken parts of fins inevitably caused in stripping, some positive uniform markings left by system of joints between separate movable form panels and some small voids caused by very small air bubbles trapped against formwork."[39]

While in the Salk Institute, as in the project being built at the same time as the Exeter Library, the Kimbell Art Museum, Kahn resolves the technical detail of the connection between the formwork panels by means of the V-joint, carefully studied to control all the "accidental markings" that can happen along that critical line, now – as is described in the phrase on the "exposed surfaces" of the Exeter Library – he decides to abut the edges of the panels without any cutting. Through this elimination of the V-joint, it is clear that Kahn wants to shift all the control of that connection to the exact making of the structure of the panels, so that the joint between them is practically sealed, preventing infiltration of the pour. It is no coincidence that Kahn also indicates the use of rubber gaskets to prevent leakage, introducing expedients not used for the construction of the concrete of the Salk Institute or the Kimbell

Academy, Exeter, New Hampshire, Library Building. Outline specification," 24 November 1967, LIK, 030.II.A.1.7).

[37] See the documents attached to, Winton F. Scott, Jr., letter to Colin F.N. Irving, 26 February 1968, LIK, 030.II.A.5..27. See also, Wood & Tower, Inc., "Meeting Memorandum," 23 February 1968, LIK, 030.II.A.6.23.

[38] Louis I. Kahn, "Construction specifications. New Library and Dining Hall for the Phillips Exeter Academy, Exterer, New Hampshire," 29 November 1968, LIK, 030.II.A.1.20.

[39] Louis I. Kahn, "Construction specifications. New Library and Dining Hall for the Phillips Exeter Academy, Exterer, New Hampshire," 29 November 1968, cit.

Art Museum. The progress of the worksite, however, demonstrates the technical difficulties involved in the goal of making perfectly sealed joints.

Other indications included in the "Construction specifications" on the quality of the concrete surface confirm the goal of reduction of the joints between form panels, "by the use of large sheets of plywood," as had already emerged in the construction of the dormitories of Bryn Mawr College. In the Exeter Library, too, the holes left by the spreaders are to be plugged not with mortar, but with the lead disks already used for the Salk Institute, to prevent rust from the iron reinforcements from making stains on the concrete. This is why the plug is applied only to the holes in the walls where the concrete remains exposed to "view on exterior work." Kahn allows the holes to be "filled with mortar" only on the outer face of the basement walls, where the concrete does not remain visible.

In the specifications for the Exeter Library, Kahn clarifies what has become a praxis of the formwork design in his works: he asks the contractors to make the drawings of the panels for the concrete not exposed (in this case he simply wants to be able to approve the design), while he maintains control of the design of the formwork for the architectural concrete.[40]

At the time of the "Construction specifications" Kahn has selected, for the formwork panels, a fir plywood, with a 5/8 or 3/4 inch thickness to abut with "tight joints," and in the case of an "exposed concrete surface" he has chosen the plastic-faced plywood called GPX, sold by the Georgia-Pacific Plywood Company. The other instructions on the assembly of the formwork range from coating to the types of ties and cones. With respect to the quality of the concrete, the only indication has to do with the nature of the cement, which must be "light colored."

The novelty of Kahn's construction system, seen in the context of the other buildings of the Phillips Exeter Academy, prompts the representatives of the school, the engineers of Keast & Hood and the technicians of the Macomber Company to take measures to guarantee the precise execution of that "important element of the design" indicated as the "quality of concrete finish."[41] After the decision to appoint a "special inspector" to examine all the components and phases of the making of the concrete, from mix design to pouring to curing,[42] supervision of the Exeter worksite is assigned to Herman G. Protze Materials Technologist, director of a consulting firm based in Newton Highlands, Massachusetts, who has already done similar jobs for the Phillips Exeter Academy.[43] This is the first time Kahn agrees not to insist on the daily presence of his staff on the worksite to supervise the pouring of the concrete.

Effects of the materials, or of the "natural finish"

In the Phillips Exeter Academy Library concrete, brick, granite, slate, Travertine, wood or carpeting are the materials with which Kahn creates refined formal effects, thanks to their particular surfaces and colors. Though it has appeared only in a later phase of the project and remains visible only in the core of the library, the material that sets the ranking of values and determines the type of finish of all the others is concrete. Brick, slate, Travertine, wood and granite are used with surfaces that have different degrees of roughness, all derived from the logic of their fabrication process intentionally interrupted prior to any conventional finishing to enable them to establish a dialogue with the particular nature of Kahn's concrete surfaces that are not worked after stripping. The worksite documents clearly outline this poetics of the deployment of materials, which in any case Kahn has systematically explored since the beginning of the 1950s. We can safely say that specifications like those of the Exeter Library have become true theoretical manifestos of Kahn's architectural concrete.

The brick wrapper gives the work an important monolithic effect, when seen from the park. Kahn wants to accentuate this effect through the use of brick for the paving of the portico and the roof of the library. "Using the same material for walls and floor gives the building a solidness, the appearance of being modeled from a brick-made block," he writes in 1967, also clarifying the meaning of the granite used for the base of the building: "This will be emphasized by the use of the granite paving blocks surrounding the building."[44] The small scale of the granite blocks creates, in his view, "the effect of grass" – the blocks measure 4 × 4 × 6 inches, like those he has seen in the pavements of "European streets." The color of the granite must not be black. "Ordinary gray," he explains, "which blends quietly with the brick yet gives a firmness of edge which brick doesn't have."[45]

The choice of the method of laying the brick is addressed by Kahn in the phases prior to the start of construction. He does not intend to imitate the "Flemish or Dutch bond" because that would require "breaking bricks in half to appear alternately as 'headers'" creating a "false" masonry, given the fact that his brick wall surface is anchored with metal brackets and requires only certain courses with the bricks arranged as "headers."[46]

The value attributed to the granite blocks, connecting the base of the pillars of the portico, is the same one assigned to the slate slabs specified for the various levels of the library between the brick pillars, in the area set aside for reading facilities that wraps the exposed concrete cores. Thanks to those slabs of slate, Kahn intends to correct certain defects in the perception of the structure that could be caused by its breakdown into two different materials, reinforced concrete and brick masonry. Scott, a member of Kahn's staff assigned to the Exeter Library project, significantly writes about the value of the "slate 'carpet edging'" that forms a "border of slate around the carpet where the carpet would otherwise meet the brick piers."[47] What is at stake is the perception of the load-bearing role of pillars and floor slabs, threatened by the carpeting, as well as the technical question of obtaining a "smooth transition in the walking surface" between the carpet and the slate. In the research on how to express the relationship between the base of the pillars and the floor slab, it is significant that Scott suggests, in place of the slate bands, to bring out the concrete of the horizontal structure from the carpet.

[40] The paragraph on the different responsibilities for the design of the formwork reads: "In some areas location of joints in formwork are shown on the drawings. In other areas location of joints is subject to approval" (Louis I. Kahn, "Construction specifications. New Library and Dining Hall for the Phillips Exeter Academy, Exterer, New Hampshire," 29 November 1968, LIK, 030.II.A.1.20).

[41] Colin F.N. Irving, letter to David P. Wisdom, 23 October 1968, LIK, 030.II.A.2.8.

[42] Thomas J. Leidigh, letter to Louis I. Kahn, 29 October 1968, LIK, 030.II.A.2.8.

[43] Winton F. Scott, Jr., letter to Hoffman, 15 January 1969, LIK, 030.II.A.3.10.

[44] Louis I. Kahn, letter to Rodney Armstrong, 11 December 1967, LIK, 030.II.A.6.22. The draft of the letter is dated 7 December 1967 (LIK, 030.II.A.6.22).

[45] *Ibid.*

[46] "In solid masonry walls," he explains "which may or may not be insulated on inside face, the 'headers' are a full brick length and they tie the wall together. Our face wall is tied thru the insulation by non-rusting metal ties. Our structural walls behind are tied together every seventh course with a row of 'headers' or alternate headers and stretchers" (Louis I. Kahn, letter to Rodney Armstrong, 11 December 1967, LIK, 030.II.A.6.22).

[47] Winton F. Scott, Jr., letter to Rodney Armstrong, 22 November 1968, LIK, 030.II.A.6.3.

412-415. Library and Dining Hall, Philips Exeter Academy, Exeter, New Hampshire, 1965-72. Views (photographs by the author, 2010).

This would also make it easier to understand the meaning Kahn has assigned to certain materials that cover the concrete structure.[48]

In the construction of the brick masonry along the facade, Kahn does not hesitate to use the handmade bricks of Eno's Brickyard, with different firings, some even with such imperfect surfaces that they create misshapen outcroppings. In any case, the choice of using the bricks made by Eno's Brickyard[49] implies this type of risk, given the fact that the firm stopped production in 1965, one year prior to Kahn's commission to design the library, and sold all remaining stock in 1966 to the Phillips Exeter Academy, for possible future use in restoration projects.

[48] "If it becomes necessary to accept the alternate for removal of the slate carpet edging there would be two options," Scott explains. "One would be to merely fit the carpet around the brick piers. This would give one the impression that the very heavy loads being carried by the brick piers were being transmitted downward by way of the carpet. This is a rather disturbing relationship which is very real in the case of the freestanding interior brick piers as they stand free in a field of carpet and are not always on the perimeter of this field as in the case of the large 15' long concrete walls. Nor do they receive visual support from the bookstacks as in the case of the 18" square concrete columns. The other option would be to expose and finish sections of the concrete floor around the brick piers. This would serve the same function as the slate edging and result in approximately the same savings relative to slate, although I believe as a finish walking surface it would probably be more controversial and would present certain construction difficulties" (Winton F. Scott, Jr., letter to Rodney Armstrong, 22 November 1968, LIK, 030.II.A.6.3).

[49] "Memorandum," meeting of 18-19 February 1969, LIK, 030.II.A.3.3.

Not only the materials of the Exeter Library are conceived as complementary to the concrete. The cladding technique is also reconsidered by Kahn in terms and with details that offer a glimpse of a precise critical position with respect to the idea of the masking of the structure. Kahn's way of installing slabs of Travertine or slate in the steps and the curved parapets of the two main staircases is emblematic of a particular conception of cladding that sets out to leave large parts of the reinforced concrete structure visible to state the different roles of the materials. It is also fundamental for Kahn, in order to manage to put the different materials into contact with each other, that the quality of the working of the surfaces of the slabs be matte, like the concrete, and not polished to achieve a sheen that his concrete is not finished after stripping.

In the choice of the slate, which he wants in a color identical to that of the slate used for the dormitories of Bryn Mawr College,[50] Kahn looks for slabs made by hand, raising doubts in the potential suppliers, who explain to him that "to cut the slate to your details and maintain the natural cleft finish" would be impossible for all the parts, because the slab would have to be "machine cut" and also "to be finished on a machine."[51] Kahn wants the slabs as they are removed from the quarry, with the surface obtained only by the work of the quarrymen, without any other finishing. The "natural cleft finish" of the slate corresponds to the concrete not worked after stripping.

For the Travertine too his choice, as in the case of the Kimbell Art Museum, is made by comparison with the samples in his possession of the Travertine used in the Salk Institute. Kahn wants a similar degree of working of the slabs, leaving the natural ruggedness of the material visible. As for the bricks and the slate, he forbids polishing, again because the surfaces of his concrete are matte. At first he approves the samples of the Italian Roman Travertine provided by the Freeport Marble & Tile Company.[52] The indications sent to the supplier of the Travertine emphasize that Kahn prefers a "natural sawn" finish, "just as it comes from gang saws."[53] For the Travertine parts that are to cover the curved parapets of the entrance staircase, since that type of finish is not possible, a "sand finish"[54] is proposed (a Travertine sample is prepared with two sides bearing the different worked finishes). When Kahn receives the samples with the honed finish, he repeats that he approves of the type of material, but not the surface finish: "not honed finish - [Kahn] wants sawn finish."[55] For the red stone architraves, the correspondence with the suppliers confirms that Kahn wants "rough-textured finishes": "We specifically do not want a smooth finish [...]."[56]

[50] Kahn will approve the samples proposed by the Bangor Slate Company, consulted for the slate of the dormitories of Bryn Mawr College (Earle W. Bolton III, letter to James B. Harrington Jr., 18 March 1969, LIK, 030.II.A.5.17).

[51] James P. Molyneaux, of James P. Molyneaux Marble, Granite, Slate, Flagstone, of Drexel Hill, Pennsylvania, letter to Louis I. Kahn, 13 December 1968, LIK, 030.II.A.5.27.

[52] See Earle W. Bolton III, letter to James B. Harrington, Jr., 26 May 1969, LIK, 030.II.A.3.10.

[53] H.P. Cummings Construction Company, letter to Earle W. Bolton III, 22 July 1969, LIK, 030.II.A.3.8.

[54] *Ibid.*

[55] James B. Harrington, Jr., letter to Earle W. Bolton III, 5 May 1969, LIK, 030.II.A.3.12.

[56] Earle W. Bolton III, letter to James B. Harrington, Jr., 18 March 1969, LIK, 030.II.A.5.17.

[57] Jules D. Prown, in, Latour, *Louis I. Kahn: writings, lectures, interviews* cit., (pp. 122-139), p. 140. "I have seen him spend an hour trying to explain to a contractor just how he wanted cinderblock joined in the basement, where nobody would ever see it" (*ibid*).

It is this cult of materials with the most natural texture possible that distinguishes the zero degree of the truth of materials sought by Kahn. There is no doubt that in the choices of the materials made for the Exeter Library, which go beyond any Brutalism of fabrication skillfully controlled by Kahn to date, we can even glimpse a psychological trait that may have to do – as has been grasped by those who have listened to Kahn discussing materials – with his face scarred by the accident with the hot coals that happened in his childhood. "If something went wrong in the pouring of the concrete, for example, it would not upset him," reports one witness to his way of operating on the worksite. "He would say, 'That is part of the history of the building.' I suspect that this particular architectural attitude grew out of his own life experience, his face having been quite badly scarred in a boyhood accident."[57]

The concrete worksite: leakage phenomena

After the first contacts with the Macomber Company, though they are decisive for the transformation of the design, the construction of the Exeter Library and the Dining Hall is assigned, instead, to the H.P. Cummings Construction Company, based in Ware, Massachusetts, directed by James B. Harrington, Jr.

Excavation begins in April 1969, when agreements have been made with Kahn regarding the type of formwork for the two types of concrete. For the construction of the basement walls "not exposed to public view" the H.P. Cummings Construction Company proposes using the Uni-Form Concrete Form Panels, documented in a booklet on the "form system" of the Universal Form Clamp Company. "It is also our understanding," the contractor explains, "that all work that is exposed to public view is to be plastic form stock and it is our desire to use 4' × 8' sheets, and we would appreciate your notifying us of any areas where these will not be satisfactory."[58] For this latter type of formwork, it is proposed to coat the panels with a releasing agent, Magic Kote, produced by the Symons Manufacturing Company, to obtain a concrete surface free of possible flaws caused by stripping.[59]

Kahn considers the 4 × 8 feet panels "acceptable," but not for the "visible portions of the central core walls in the Library" and for the "walls around the cooling tower" in the Dining Hall, where he wants panels of 4 × 10 feet "if possible."[60] In the meantime, he has to approve the design of the arrangement of the spreaders "for the exterior wall" of the library, supplied by A.H. Harris & Son, Inc., Concrete Construction Specialties.[61]

To define the type of brick masonry, the structure of the lintel and the shade of color of the mortar, two "Sample Jack Arch Panels," designed in March, are prepared on the worksite.[62]

[58] James B. Harrington, Jr., letter to Earle W. Bolton III, 7 March 1969, LIK, 030.II.A.3.12. A copy of Harrington's letter contains handwritten notes by Thomas J. Leidigh (see James B. Harrington, Jr., letter to Earle W. Bolton III, 7 March 1969, LIK, 030.II.A.3.12).

[59] "It provides easy stripping," we read in the booklet of the Symons Manufacturing Company sent to Kahn, "and reduces or eliminates cleaning forms before recoating for the next pour as it does not react with concrete. It prevents concrete build-up on forms, tools, equipment and machinery" (*Spec. Data. Symons Mfg. Company. Concrete Formwork, form coatings*, November 1968). The booklet is attached to, James B. Harrington, Jr., letter to Earle W. Bolton III, 11 April 1969, LIK, 030.II.A.3.12.

[60] Earle W. Bolton III, letter to James B. Harrington, Jr., 2 April 1969, LIK, 030.II.A.3.10.

[61] James B. Harrington, Jr., letter to Earle W. Bolton III, 17 March 1969, LIK, 030.II.A.5.17.

[62] See James B. Harrington, Jr., letter to Earle W. Bolton III, 18 March 1969, LIK, 030.II.A.3.1, and Earle W. Bolton III, letter to James B. Harrington, Jr., 21 April 1969, LIK, 030.II.A.3-10.

To allow Kahn to choose the color of the concrete, 12 × 12 inches blocks are prepared.[63] Only when construction has already begun does Kahn have the chance to evaluate a "color sample" of concrete "cast against plastic-surfaced plywood."[64] The H.P. Cummings Construction Company sends samples of form ties and a cone of the Symons Manufacturing Company, approved by Kahn,[65] and proposes for the sealing of the tie holes the replacement of the lead disks with a plastic plug produced by the Richmond Screw Anchor Company, Inc. Three samples are sent to Kahn of the Plastic Setback Plug, in different colors, with the comment that this type is "used very successfully on filling coil holes."[66]

To lower construction costs it is decided to use prefabricated concrete planks made by the San-Vel Concrete Corporation, and in the Dining Hall to also use them for a "*finished* ceiling." Because they will have to define the quality of the finish, Kahn's staff emphasize the formal characteristics of the planks, which should have "uniformity of color and texture" and should have no chamfers.[67]

Given the importance for the quality of the works of the "appearance of exposed concrete," recognized both by Leidigh and by Protze, certain control measures are dictated, to apply in all phases of the pouring, from inspection of the coated plywood of the formwork to supervision of the work during "placing, mixing, vibrating and tight form work."[68] The deployment of the formwork and concrete is assigned to the supervision of Protze's office.[69] No members of Kahn's staff are systematically present on the worksite, as happened in the construction of his other works.

The City Concrete Company, Inc., based in Dover, New Hampshire, is hired for the works in concrete, from the preparation of the compound to the assembly of the formwork.[70] But the first parts of the library, at the level of the foundations and the base, have several imperfections:[71] the Universal Form products "do not make tight forms," though the mix design is considered "satisfactory"; the pouring of the compound is done by means of "hoppers and trunks," at intervals of 10-12 feet; vibrators are used, "lifts were 18 to 24 inches and there were no excessive delays in the progress of the placement."[72] Moreover, "some of the exterior forms required additional bracing during casting operations."[73] Also in the Dining Hall, where the formwork of the library is reused, leakage occurs on the walls of the Mechanical Room, and in spite of the fact that the work is accepted, the City Concrete Company is advised that it "is not according to the high stan-

[63] See James B. Harrington, Jr., letter to Earle W. Bolton III, 19 March 1969, LIK, 030.II.A.3.12, and Earle W. Bolton III, letter to James B. Harrington, Jr., 26 March 1969, LIK, 030.II.A.5.17.

[64] Herman G. Protze, letter to Louis I. Kahn, 13 June 1969, LIK, 030.II.A.5.15.

[65] See James B. Harrington, Jr., letter to Earle W. Bolton III, 7 April 1969, LIK, 030.II.A.3.12, and Earle W. Bolton III, letter to James B. Harrington, Jr., 12 April 1969, LIK, 030.II.A.5.17.

[66] James B. Harrington, Jr., letter to Earle W. Bolton III, 15 April 1969, LIK, 030.II.A.5.17.

[67] Earle W. Bolton III, letter to James B. Harrington, Jr., 24 April 1969, LIK, 030.II.A.3.10. Kahn approves the sample of a concrete plank of San-Vel Corporation (H.P. Cummings Construction Company, "Report No. 4," 8 May 1969, LIK, 030.II.A.2.7).

[68] H.P. Cummings Construction Company, "Report No. 4," 8 May 1969, LIK, 030.II.A.2.7.

[69] All the inspection reports for the concrete are prepared by Protze and E. S. Van Buren, R. S. White, E. J. Cameron, Zampine, H. L. Morley, who work in his office.

[70] After having seen the first concrete samples, Kahn requests "further investigation as to a lighter architectural concrete" (H.P. Cummings Construction Company, "Report No. 6," 29 May 1969, LIK, 030.II.A.5.7). Then the contractor has other samples prepared with Penn-Dixie II and Saylors (H.P. Cummings

dard of H.G. Protze."[74] When work is being completed on the base, at the end of July, Protze criticizes the "concrete work." "[...] the overall quality of the work is poor," the consulting firm states, indicating the case as "poor formwork, not in the mix itself, nor in the placement."[75] The main defects come precisely from the connection between the formwork panels, that detail which Kahn, in the works for Exeter, wants done without the V-joint of the Salk Institute, in order to experiment with another system. To prevent the leakage, Protze proposes either improvement of the structure of the formwork (especially making the joint between the panels "tighter"), or the use of a "rubber gasket" to insert between the panels along the joints.[76] Again in this case, the completed operations are approved because they have to do with "non-public areas," with one stipulation: "Future work, however, will be in exposed, public areas and the quality of the concrete work we are now getting would be entirely unacceptable."[77]

The report on the concrete work lists all the defects in the pour of the wall of the staircase in the Dining Hall: "vertical joints in plywood panels leaked"; "boxes for slab recesses leaked"; "vibrator could not be entered into all of the lower 8 ft. of the wall, causing water marks and honeycomb."[78] Kahn is persuaded to permit the City Concrete Company to "patch this wall," though he makes some technical recommendations: "cutting out all defects to sound virgin concrete or 1/2" minimum depth"; "outline of defect to be cut on irregular line rather than straight or geometric lines"; "cut edges normal to the wall surface to prevent feather edges."[79] In any case Kahn reserves the right, after examining the results of the patching, to demand the demolition of the wall.[80] But it is significant that at a meeting held on the worksite, precisely with reference to the wall of the staircase, it is stated that "patching of defective concrete work, although discussed, is not yet authorized."[81]

To avoid the type of defects produced in the wall of the staircase, other recommendations are made (some corresponding to those already advanced by Protze), namely for "tighter forms," especially at the joints, and the use of rubber gaskets, "double lapped corners" and systematic application of vibration.[82] Precisely to prevent the defects of the wall of the staircase in the Dining Hall, Protze's firm recommends that the joints of the formwork presently being used in the library should be tightened well "before concrete placement [is] started," because they look too wide.[83]

During the works done in August, chemical products are used to prevent leakage through the joints of the formwork panels. The report on an

Construction Company, "Report No. 6," 29 May 1969, LIK, 030.II.A.5.7), while Protze writes a technical report on the "concrete mixture"; he prepares, at the start of June, other samples for the architectural concrete (H.P. Cummings Construction Company, "Report No. 7," 5 June 1969, LIK, 030.II.A.2.7).

[71] See R. S. White, 12 June 1969, "Inspection of concrete," LIK, 030.II.A.5.15, and R.S. White, "Inspection of concrete," 20 June 1969, LIK, 030.II.A.5.28.

[72] E. S. Van Buren, "Inspection of concrete," 30 June 1969, LIK, 030.II.A.2.1.

[73] E. S. Van Buren, "Inspection of concrete," 2 July 1969, LIK, 030.II.A.2.1.

[74] E. J. Cameron, "Inspection of concrete," 16 July 1969, LIK, 030.II.A.2.1.

[75] Earle W. Bolton III, letter to James B. Harrington, Jr., 25 July 1969, LIK, 030.II.A.3.10.

[76] *Ibid.*

[77] *Ibid.*

[78] E. S. Van Buren, letter to Hoffman, 31 July 1969, LIK, 030.II.A.2.1.

[79] *Ibid.*

[80] H.P. Cummings Construction Company, "Report No. 14," 31 July 1969, LIK, 030.II.A.5.7.

[81] Earle W. Bolton III, letter to James B. Harrington, Jr., 10 September 1969, LIK, 030.II.A.3.10.

[82] E. S. Van Buren, letter to Hoffman, 31 July 1969, LIK, 030.II.A.2.1.

[83] *Ibid.*

"Inspection of Concrete" shows that for the pour of the first floor slab in Dining Hall, although the "forms were clean, pre-wet and appeared tight," polyethylene has nevertheless been applied on both sides of the formwork to waterproof it and to prevent leakage.[84] The fact that the fabrication of the concrete is supervised by the technicians of Protze explains the use of chemicals to resolve the technical problems arising on the worksite. This is a significant passage in the evolution of Kahn's concrete, since until this episode he has always attempted to solve technical problems by mechanical means, intervening on the assembly of the formwork. It is worth looking ahead to note that after the experience in Exeter, the focus in Kahn's projects returns to the study of the mechanical connections of the formwork panels.

When the work on the first floor slab of the Dining Hall is inspected, in spite of the precautions taken certain defects are found on the surface, attributed to the fact that the "forms were stripped too soon after pouring."[85]

Kahn visits the site on 23 September, accompanied by Leidigh and Bolton. The discussions during the inspection are decisive for the continuation of the work. The defects that continue to appear confirm the validity of the V-joint solution. Kahn is not satisfied with the results of the "concrete pour," whose bad quality is attributed to the "poor formwork."[86] He is probably the source of the suggestion of using "washers between tie and cone, mitering corners of forms, 18" pours and new anchoring method at slab surface as possible methods of stopping leaks and discoloration." The indication of "mitering corners of forms" corresponds to the technical detail of the V-joint. "Architect will not approve concrete for upper floors unless results improve," is the final warning at the end of the worksite visit.

The quality of the "exposed concrete finishes was not totally satisfactory," according to Leidigh, although "the finish of flat surfaces and corners was generally very good – smooth, hard, with minimum bubbles or other markings."[87] The worst finishes are located at the "bases of walls and around Tycones, where leakage of mortar had caused discoloration." Marks have also been produced on already completed walls, from pours done in formwork mounted above them.

Following Kahn's inspection and his criticisms, the technicians of Protze Materials Technologist prepare a report, "Appearance of architectural concrete," on the technical methods for preventing defects, starting with the observation of what has happened thus far on the worksite, summed up as follows: "migration lines caused by leakage at the base of walls and at plastic wall tie cones"; "fins at vertical joints"; "different planes across vertical joints."[88] The remedies proposed do not correspond to those suggested by Kahn on 23 September: "fasten 1" × 4" sole plate to existing concrete floor and set toe of plywood form *against* sole plate"; "use 3M Scotchfoam pressure sensitive tape between plywood form bottom and abutting surface"; "use Richmond washers with plastic wall tie cones"; "tighten all tie bolts to the same torque by means of an indicating torque wrench"; "insert the rubber gasket in the vertical joints and backup joints"; "plastic coated plywood edges are not true nor is the thickness uniform"; place the concrete in pours with a height of 18 inches to reduce pressure against the formwork. "Reduction of head pressure will not subdue leakage per se, only *tight* forms will eliminate this problem," the report states, in any case.

It is worth emphasizing that Protze's office does not consider the hypothesis of the V-joint, but continues to insist on the sealing of the joints between panels, to be done by using rubber gaskets. Nevertheless, even the application of this type of seal for the joints leads to defects on the concrete surface, as recorded during visits to the worksite on 25 and 29 September, when it is seen that the leakage has been corrected only in the case of the form ties, thanks to "torqueing ties with torque wrench to a uniform tightness."[89] During the operation of tightening the ties of the formwork the quantity of rubber used for the seals of the joints and ties has been excessive, emerging on the surface of the panels, which have been only roughly cleaned prior to the pour.

At the start of October, Scott, the member of Kahn's team, observes that "with a few exceptions the quality of the work is quite high," in spite of the ongoing existence of "the problem of leakage and consequent stains at the base of walls, at ties cones, at electrical boxes and at joints [...]."[90] Leidigh agrees, and on the concrete of the library he writes: "pours made most recently have more surface bubbles but less leakage. This is more acceptable to Architect. Tightness of formwork still being adjusted, but improving."[91]

Gustav Langford's glossary of formwork and pouring

At the worksite inspection of the Phillips Exeter Academy Library, on 2 October 1969, after which Scott and Leidigh report their observations, Gustav Langford is also on hand, after recently finishing his period of supervision of the concrete work for the National Assembly Building in Dhaka. In this same period, Frederick Langford has been sent by Kahn to solve the formwork issues on the worksite for the Kimbell Art Museum. The observations of Gustav and the solutions he has proposed are gathered in the form of a short letter-document sent to the H.P. Cummings Construction Company by Scott on 7 October.[92] The attachments to the document include a glossary of terms used to describe the various phenomena encountered on the worksite and in the shop drawings of the formwork, to remedy defects. So after the "Report on concrete work" written by Frederick Langford in 1964 for the work at La Jolla, and after the other document prepared by Frederick in 1966 following the construction of the "experimental walls" in Dhaka, Kahn's staff produces a new document focusing on the question of the formwork.

In the letter sent to the H.P. Cummings Construction Company no mention is made of possible assignment of Gustav Langford to the Exeter project as a concrete expert representing Kahn's office, as happens during the same period in Fort Worth, in the case of Frederick Langford. But it is clear that Gustav's document reflects a position taken by Kahn's office regarding the work being done at the time, and also a declaration of authority and technical expertise on exposed concrete, against the backdrop of the fundamental works for the architectural concrete handled by professional consultants extraneous to Kahn's office. In the autumn of 1969, Kahn finds himself having to cope with two construction projects in which exposed

[84] E. J. Cameron, "Inspection of concrete," 25 August 1969, LIK, 030.II.A.2.1.

[85] Earle W. Bolton III, letter to James B. Harrington, Jr., 10 September 1969, LIK, 030.II.A.3.10.

[86] H.P. Cummings Construction Company, "Report No. 22," 24 September 1969, LIK, 030.II.A.2.7.

[87] Thomas J. Leidigh, letter to Earle W. Bolton III, 1 October 1969, LIK, 030.II.A.2.20.

[88] E. S. Van Buren, "Appearance of architectural concrete. Library Building & New Dining Hall," 30 September 1969, LIK, 030.II.A.2.19.

[89] E. J. Cameron, "Inspection of concrete," 30 September 1969, LIK, 030.II.A.2.19.

[90] Winton F. Scott, Jr., letter to James B. Harrington, Jr., 3 October 1969, LIK, 030.II.A.5.28.

[91] Keast & Hood, "Site visit memorandum," 3 October 1969, LIK, 030.II.A.5.24.

[92] Winton F. Scott, Jr., letter to James B. Harrington, Jr., 7 October 1969, LIK, 030.II.A.2.20.

416-419. Library and Dining Hall, Philips Exeter Academy, Exeter, New Hampshire, 1965-72. Sketches and notes for the formwork system (LIK, 030.II.A.2.20.001; 002; 003; 004).

concrete is of extreme importance, but controlled in Fort Worth by Geren, and in Exeter by Protze.

The methods recommended by Gustav Langford for the construction of the formwork for the works at Exeter cover all the critical points, from the joints to the spreaders, in which leakage is happening of the material during the pouring. Some of the indications are the same as those written by Leidigh on 1 October.

"The majority of the exposed concrete surfaces are very good. Specific areas such as bases of walls, areas around tie cones and panel junctures have resulted in very objectionable leakage and staining. At the wall base the sequence of form construction should be altered to first place a member tight against the existing floor surfaces and then build the forms up from this, rather than erect the forms and then try to seal the joint between the form and the floor.[93] The member could be secured to the floor as shown in SK-1 with power driven fasteners, expansion bolts or cut nails and there should be a gasket between the other member and the floor. Rubber washers at the tie cones together with the use of a torque wrench to obtain uniform tightness of the ties should stop leakage at tie cones. It should be noted that staining around the tie cones to the floor, i.e., those closest to the floor showed a greater tendency to stain than those at the top of the wall. This tendency could be the result of over-vibration in addition to greater fluid pressures. Leakage and staining at panel junctures where tape and even caulking is being used is excessive and objectionable. At the present time adjacent panels are being nailed to a common 4 × 4 with tape between the plywood ends. With this technique no lateral pressure can be exerted on the tape. It is recommended that the detail shown in SK-2 be adopted. This bolted detail will facilitate pressuring the tape between panels and will make the panel system more flexible since damaged panels in a ganged set can be more easily replaced. Leakage and staining was also noted at electrical boxes and at oth-

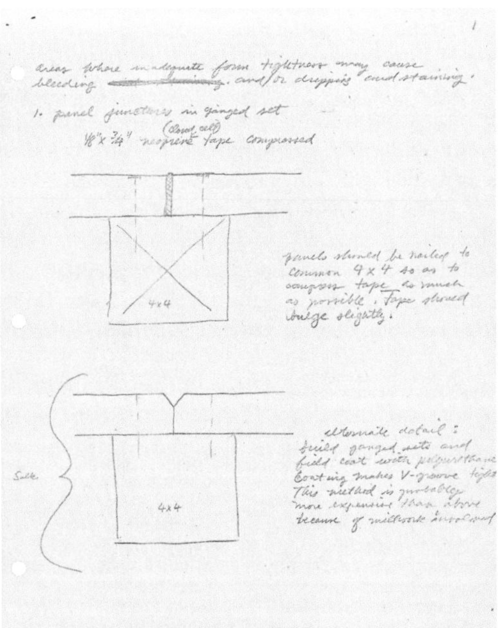

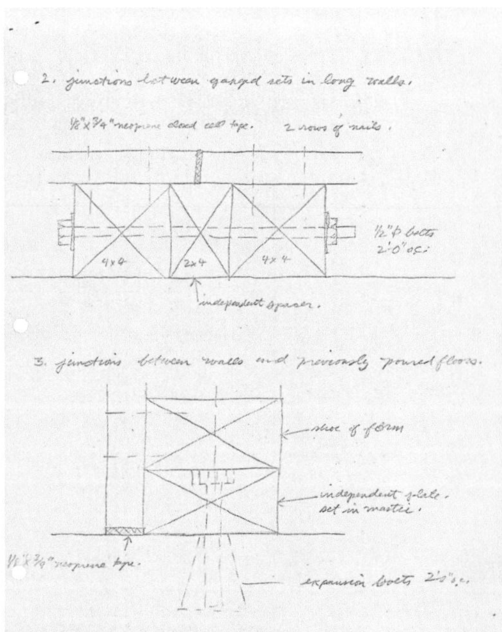

er boxes and sleeves placed in the forms. At these locations it becomes imperative that taping and/or caulking be employed to eliminate staining. Also more pressure should be put on electrical boxes to embed them into the plywood surfaces. This additional pressure together with taping and/or caulking should suffice at these points. Where a slab or beam is being poured over an existing wall or column, drippings from the pour above have stained the surfaces below. These markings when they occur should be removed as soon as possible to minimize their staining effect. It is recommended that, in future pouring, the exposed surfaces below be washed down at the completion of the pour. This technique has been found to be one of the most satisfactory and easiest methods of handling this dripping. On future work, it is also request that plywood edges, where exposed on inside corners or by mitering, should be heavily coated with polyurethane so that they will not soak up water and cause staining. On the curved walls at the ground floor level, plastic coated forms must be used and if a field coating technique is used it must be approved by our office."[94]

On the attached handwritten page titled "Glossary," Langford explains the terms of the system of construction for architectural concrete, which are used in the correspondence and the drawings. The glossary shows there is a substantial difference, regarding the effects on exposed concrete, between "leakage" composed of "water + cement paste" and "bleeding" which is "free water separates." Under the terms "staining" and "dripping" no explanations are supplied. The "pour strips" are the slats of wood to insert between "vertical sections of walls" or between "walls and ceilings." The entry

[93] The entire section, from "the sequence" to "the floor," is written on the basis of the statements of Leidigh (Thomas J. Leidigh, letter to Earle W. Bolton III, 1 October 1969, LIK, 030.II.A.2.20).

[94] Gustav Langford in, Winton F. Scott, Jr., letter to James P. Harrington, Jr., 7 October 1969, LIK, 030.II.A.2.7.

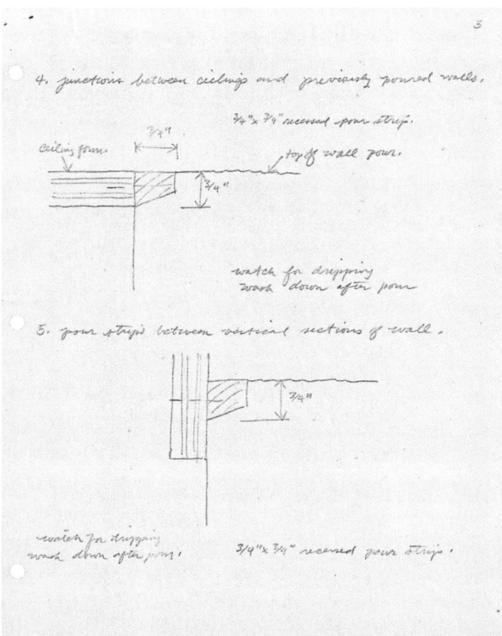

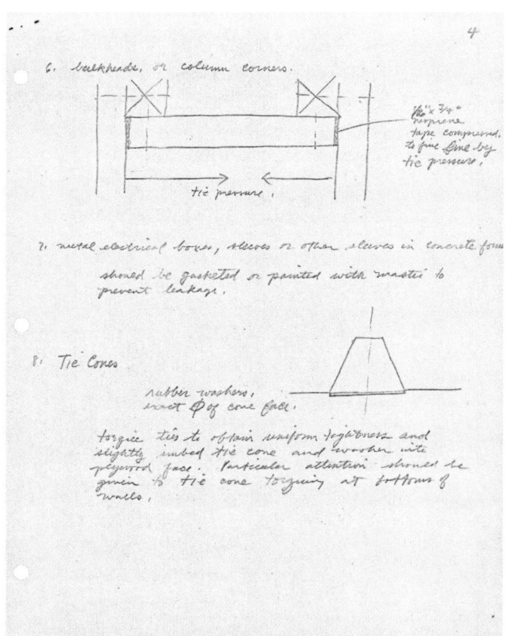

"construction joints" indicates that they are vertical joints in walls. The definition of "ganged set" is a "group of form panels set and removed as a group." The glossary also includes the expressions "panel junctures in a ganged set," "junctures between ganged sets" and "where ganged set is to be placed on concrete floor." The bulkheads are located, the document explains, "at wall terminations." Finally, the glossary also indicates the use of 1/8 × 3/4 inch "compressible neoprene tape."

The other four sheets of Langford's missive contain eight construction details documented by sketches and explanatory notes. All the details have to do with various cases of joining of panels, including the spreaders, where phenomena of bleeding, dripping and staining have been generated.

In the first detail, regarding the ganged set of two panels, Langford supplies two versions, one using neoprene, the other with direct abutting of the panels with the edges cut to obtain a joint called a V-groove, as in the case of the formwork for the Salk Institute – the word "Salk" appears to name this solution. "This method is probably more expensive than above because of millwork involved," is Langford's comment, which partially explains Kahn's abandonment of the V-joint. Three other panel junction solutions (the second, third and sixth) call for the use of neoprene and various types of reinforcements on the back of the panels to guarantee a tight seal (the sixth is for the "bulkheads, or columns corners," with neoprene at the ends, to compress by means of the ties connected to the panels). One solution illustrates the "junctions between ceilings and previously poured walls," to be done with a wooden strip, the "recessed pour strip," placed at the top of the pour at the position of the panel of the formwork of the floor slab. Another solution shows how to avoid marking the surfaces of already placed concrete, in the case of the construction of walls, thanks to the insertion of "pour strips between vertical sections of walls." Once again, the solution is the wooden "recessed pour strip." In the case of "metal electrical boxes, sleeves or other sleeves in concrete forms" (the seventh point), mastic is recommended to seal the joints. The eighth and last case has to do with the "tie cones": the drawing shows rubber washers of the exact diameter as the large head of the cone.

After the various discussions on the possible ways to prevent leakage in the construction of the Library and the Dining Hall, and independent of the indications of Gustav Langford, some improvement is achieved, since during an inspection with Kahn present it is reported that the "last concrete pour in Library where rubber washers and shoe plate with plastic material produced satisfactory results and method will be continued for future pours."[95] The "architect," it is reported, "requested a bit more vibration in an attempt to reduce pin holes."

The design of the formwork for the circular openings

Starting at the end of September 1969, Kahn's office develops the formwork design for the individual parts of the Exeter Library and the Dining Room. In the drawing of the formwork and pour joints of the main staircase, "Formwork patterns core 40,"[96] a note confirms Kahn's aim of designing an order of the wall with few imprints: "Use 10' sheets for interior forms core 40. 8' or less vertical dims. will be accepted only if joint falls behind keyway and is not exposed." In October the design of the "form panel pattern" of the Cooling Tower in the Dining Hall is prepared,[97] and work begins on the drawing of the special formwork for the pouring of the concrete in the four large "eyes" of the library hall. Kahn and his team now have to indicate the

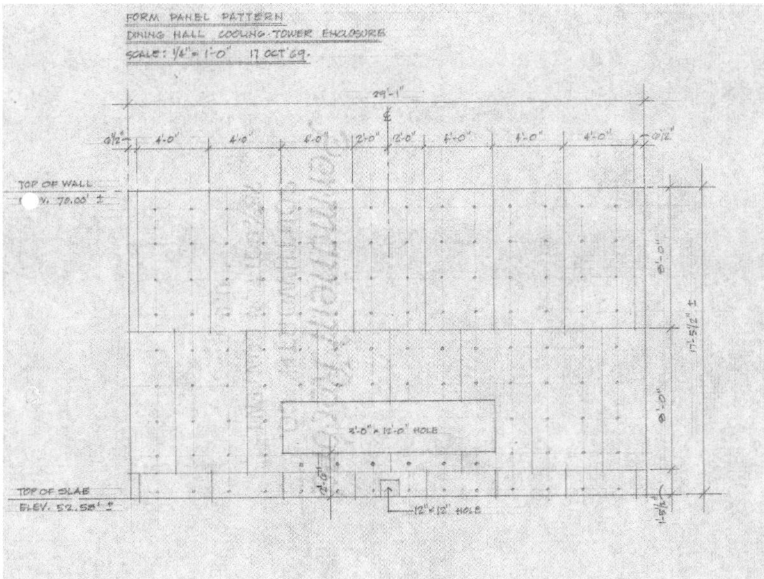

420. Library and Dining Hall, Philips Exeter Academy, Exeter, New Hampshire, 1965-72. "Form Panel Pattern. Dinning Hall Cooling Tower Enclosure," 17 October 1969 (LIK, 030.II.A.3.10.001).

various pour phases for the construction of those eyes, taking the difficulties of shaping all the concrete surfaces only with formwork into account, due to the circular figure. Precisely the detail of the shaping of the surface in the lower section of the circumference of the "circular openings" reveals Kahn's determination to always seek a configuration that expresses each phase of the process of fabrication of the concrete.

The drawing "Formwork - FLS. 1 - 3M" of 29 October documents the elevation of the circular opening.[98] The graphic indications and notes reveal the rigorous system invented by Kahn and his team for the architectural concrete of the central space with the four circular openings. Every line is perfectly calculated to create a weave of imprints of joints between panels and pour lines. The pour joints correspond to the heights of the various floors (corresponding to the lower surface of each floor slab). The seven panels, each 4 feet wide (their height coincides with that of two pour joints), are assembled starting with the central vertical axis, and everything is calculated so that there are four points of contact between the circumference, the pour joint and the imprint of the joint between the panels. "At these points, form joints should precisely intersect pour joints at circular opening. This relationship sets up panel spacing," a note on the drawing explains. In the actual construction, the intersection between pour joint and form joint does not always come at a single point on the circumference, however. The lower beam has a smooth surface; "no form ties in beam pour," is indicated on the drawing.

The question that concerns Kahn in the construction of the four large eyes has to do with the lower part of the circumference, and is explained as

[95] H.P. Cummings Construction Company, "Report No. 24," 8 October 1969, LIK, 030.II.A.2.7.
[96] See the drawing "Formwork patterns core 40," 29 September 1969, LIK, 030.I.C.710.010.
[97] "Form panel pattern, Dining Hall Cooling Tower Enclosure," 17 October 1969, LIK, 030.II.A.3.10.
[98] See the drawing, "Formwork - FLS.1-3m Exeter Library," 29 October 1969, revised on 15 December 1969 and 7 January 1970, LIK, 030.I.C.710.006.

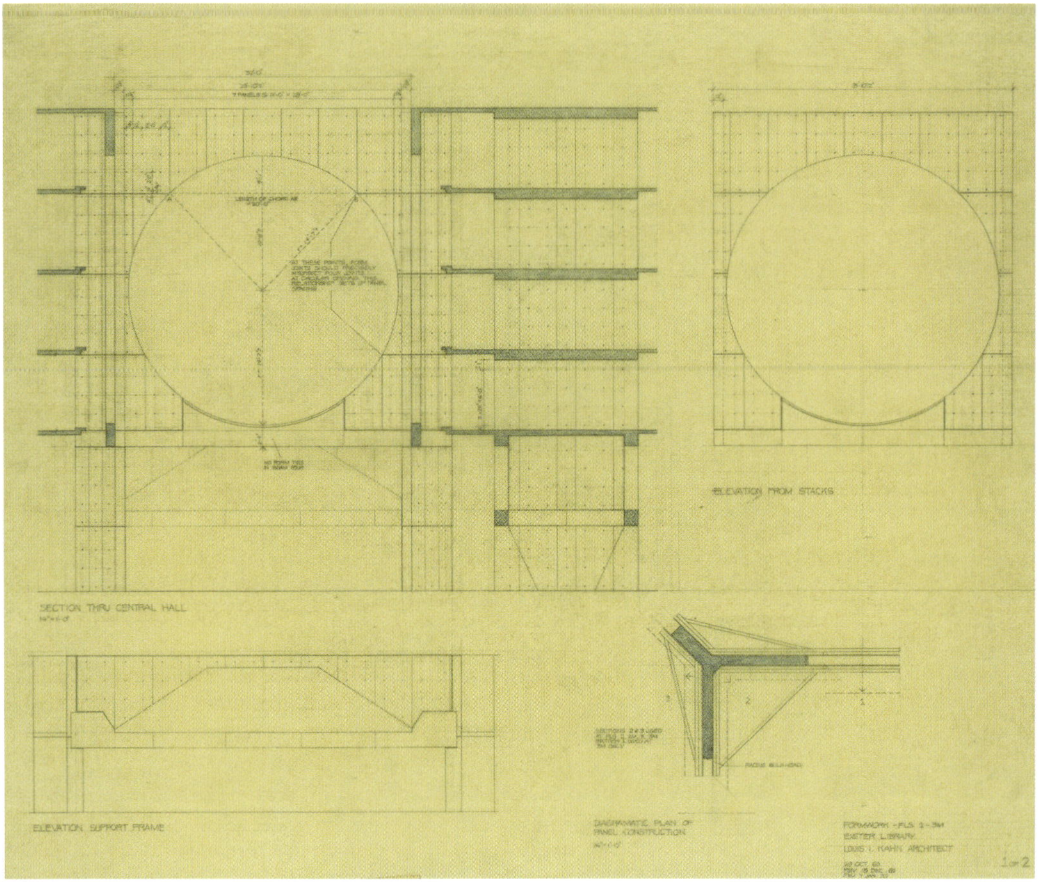

421-422. Library and Dining Hall, Philips Exeter Academy, Exeter, New Hampshire, 1965-72.

421. "Formwork FLS. 1 – 3M," 29 October 1969 (LIK, 030.I.C.710.6).

follows by Scott to the H.P. Cummings Construction Company: "this section of the sloping jamb of the opening must be finished by hand and therefore present a finish different in texture and color from the other concrete surfaces where the color and texture of the finish results from placement against the coated plywood forms."[99] The solution proposed by Kahn to express the particular nature of that sector of the circumference is not implemented, and consists in a change of material, to avoid having two different types of working of the concrete in contact with each other: a "1" thick Travertine facing for the lower half of the jamb of the circular opening." "In this way," Scott comments, "we feel that we can solve a very difficult construction problem and actually improve the total design by making a feature of the situation."

Halfway through December, Kahn develops what becomes the final solution for the construction of the lower section of the circumference. The drawing "Formwork - FLS. 1-3M"[100] covers precisely the detail of the great eye and its system of grooves at the point of passage, along the inner face, between the "formed finish," i.e. the concrete shaped in contact with the formwork, and the "troweled finish," namely the concrete to be smoothed by hand. The point of passage between these two finishes happens along a "vertical pour joint"; this joint does not reach the circumference, but stops short to leave a small "deep slot-opening" that separates the "formed finish" from the "troweled finish." That "deep slot-opening" is another form of the

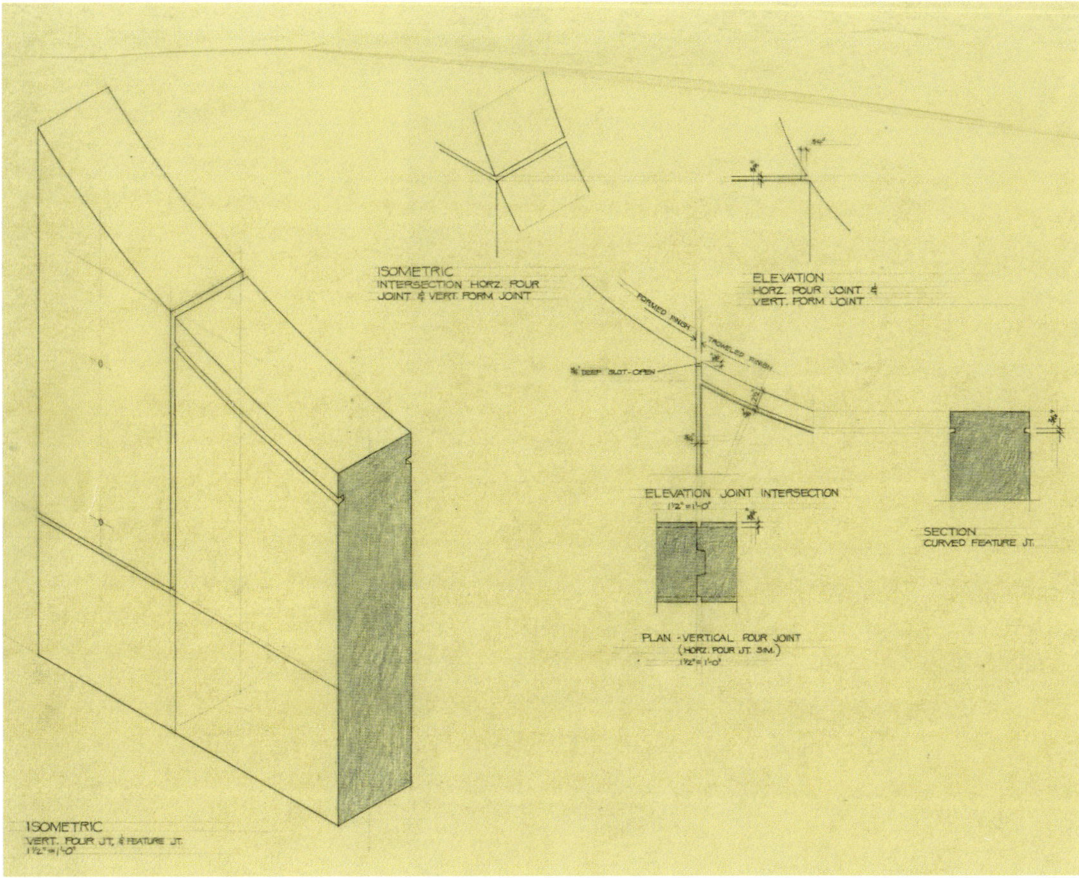

422. "Formwork FLS. 1 – 3M," 15 December 1969 (LIK, 030.I.C.710.6).

original "shadow joint" applied by Kahn in points of passage from one material to another. The drawing also shows cases of encounter between "horizontal pour joints" and "vertical form joints."

The precautions of Kahn and his team, in any case, do not prevent difficulties from arising for the contractor in the making of a figure like a large circle in concrete, which in the case of Kahn's design had first been introduced to be made in brick. The construction of the large eyes also leads to an increase in costs for the material of the formwork and the quantity of the concrete.[101] During the first pours Protze's staff remark on the difficulties of vibrating the concrete in the recesses of the circle.[102]

At the start of December 1969 the Cooling Tower of the Dining Room has been completed, and Leidigh describes its walls as "generally good": "No serious honeycomb. Some difficulty under pockets. Some visual evidence of rather high lifts in pouring and resultant discoloration."[103]

To make Kahn's architectural concrete with uncontaminated surfaces, precautions must be taken on the worksite in the construction phases that

[99] Winton F. Scott, Jr., letter to James B. Harrington, Jr., 30 October 1969, LIK, 030.II.A.3.10.

[100] See the drawing "Formwork - FLS.1-3m Exeter Library," 15 December 1969, LIK, 030.I.C.710.006.

[101] James B. Harrington, Jr., letter to Louis I. Kahn, 5 March 1970, LIK, 030.II.A.3.6.

[102] See Zampine, "Inspection of concrete," 25 November 1969, LIK, 030.II.A.2.19.

[103] Keast & Hood, "Site Visit Memorandum," 10 December 1969, LIK, 030.II.A.5.24.

423-427. Library and Dining Hall, Philips Exeter Academy, Exeter, New Hampshire, 1965-72.

423. Interior view (photograph by the author, 2010).

424. Interior detail (photograph by the author, 2010).

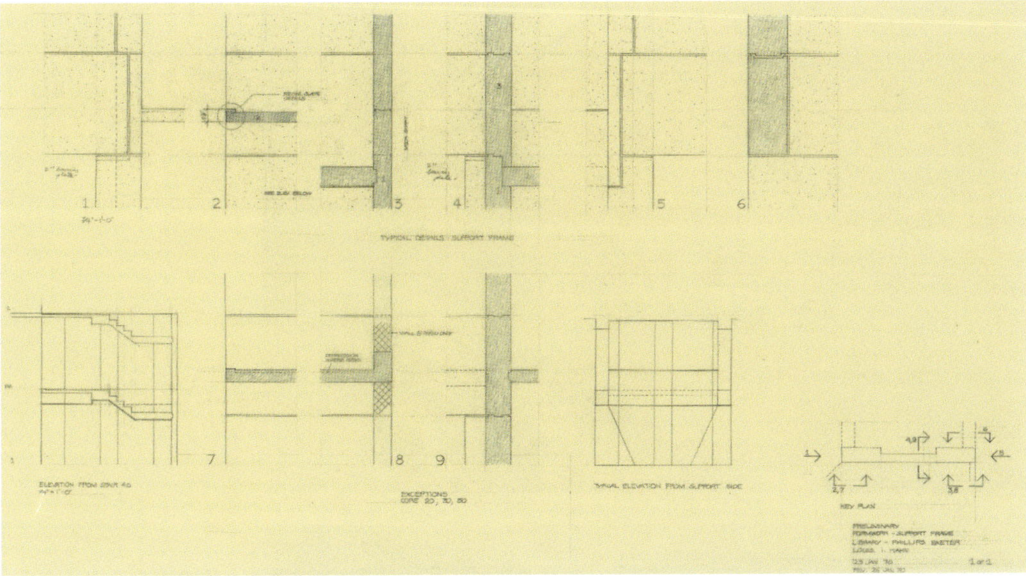

425. "Preliminary Formwork – Support Frame," 23 January 1970 (LIK, 030.I.C.710.6).

426-427. Views of the interior (photographs by the author, 2010).

428-430. Library and Dining Hall, Philips Exeter Academy, Exeter, New Hampshire, 1965-72. Views of the interior (photographs by the author, 2010).

take place during the cold winter in Exeter. In January 1970, Leidigh advises protecting the reinforcements of the concrete to prevent rain and other atmospheric phenomena from causing rust that could stain the already placed parts.[104]

As construction proceeds, Kahn prepares the "Preliminary Formwork - Support Frame,"[105] for the four supports of the beams of the Hall, under the circular openings. The drawing reflects Kahn's conception of a structure in reinforced concrete to install in phases identified by grooves that become tangible signs of the construction process. Thus the whole structure is broken down into parts according to the pour phases that generate a pattern of pour joints. Each part identified by the grooves ends up with the appearance of a colossal monolith, and the whole of the structure appears as an assembly of its parts.

The last significant decisions for the concrete have to do with the material to be used to seal the cone holes of the Cooling Tower in the Dining Room: Kahn chooses the "plastic plugs" with a "dark gray" color, so as to use the "lead plugs that were to be installed there" in "areas selected" by him in the library.[106] For the type of finish of the concrete planks, also in the Dining Hall, namely painted or not,[107] Kahn opts to leave everything "untouched."[108]

After various indications of Bolton or Leidigh regarding the need to intervene on the concrete surfaces to clean away "excess mortar and spots resulting from patching of the forms"[109] and to remove "wood strips" and "pieces of rubber gaskets,"[110] Kahn himself supervises the last refinements: "Mr. Kahn after meeting accompanied General Contractor to site and indicated areas he wished to be touched up and improved."[111]

[104] See Thomas J. Leidigh, letter to Earle W. Bolton III, 14 January 1970, LIK, 030.II.A.3.2, and Earle W. Bolton III, letter to James B. Harrington, Jr., 21 January 1970, LIK, 030.II.A.2.3.

[105] See the drawing "Preliminary Formwork – Support Frame Library – Phillips Exeter," prepared on 23 January 1970 and revised on 26 January 1970, LIK, 030.I.C.710.006.

[106] H.P. Cummings Construction Company, "Report No. 42," 10 June 1970, LIK, 030.II.2.7.

[107] James B. Harrington, Jr., letter to Earle W. Bolton III, 6 April 1970, LIK, 030.II.3.6.

[108] H.P. Cummings Construction Company, "Report No. 40," 20 May 1970, LIK, 030.II.5.7.

[109] Earle W. Bolton III, letter to James B. Harrington, Jr., 22 December 1970, LIK, 030.II.2.3.

[110] Thomas J. Leidigh, "Site Visit Memorandum," 9 March 1971, LIK, 030.II.3.2.

[111] H.P. Cummings Construction Company, "Report No. 75," 10 February 1972, LIK, 030.II.2.7.

The violin in concrete and the case in brick: the Performing Arts Theater

At the start of the 1970s several important works by Kahn, just completed or still under construction, confirm that the complex association of bricks, concrete blocks and exposed concrete is more than a corollary of his particular architectural concrete. But the modes of deployment of the materials and their relationships in the definition of the structure vary from building to building, depending on totally different logics, and therefore do not outline an evolutionary process leading to the discovery of a definite, univocal form of composite masonry structure.

The idea of a powerful and meaningful reinforced concrete structure left exposed and wrapped by masonry in brick or concrete blocks had already taken form in the Unitarian Church and School. But it is only in the Performing Arts Theater, built from 1970 to 1973 at Fort Wayne, Indiana, that concrete, brick and blocks become the materials to create a system of autonomous parts, which each forms a container with certain characteristics. In the passage between the two works it is precisely the Exeter Library that plays a crucial role, due to the clear distinction between a concrete core and a brick enclosure, under construction exactly when Kahn is still working on the design for Fort Wayne. But the reasons that lead Kahn to create two separate containers, each made with its own different material, go well beyond the economic concerns that in 1966 had prompted him to accept the suggestion of the Macomber Company to building the "central core" of the Exeter Library in reinforced concrete.

Designed as part of the Fine Arts Center, which Kahn has been developing since 1961, the Performing Arts Theater takes on its definitive configuration thanks to the contribution of various acoustics experts, who suggest some of the rules for the shaping of the concrete of the Auditorium. In August 1963, Paul S. Veneklasen, the first acoustical consultant involved in the project due to the initiative of Igor Buketoff, orchestra conductor and director of the Philharmonic Hall of the Fine Arts Center, indicates to Kahn a series of principles for the choice of the overall form of the plan and for the design of the walls and ceiling of the Auditorium of the Philharmonic Hall. Some of these principles also become fundamental for the development of the design of the Performing Arts Theater.[112] After certain technical considerations on the propagation of the sound of a violin, on the page of notes from the discussion between Kahn and Veneklasen a jagged line appears, with the following observation: "texturing must be deep if to be effective for round shape."[113]

The fact that Kahn is informed, while he is designing the complex of halls for the Fine Arts Center, about the acoustic characteristics of the Grande Salle in Montreal, which has segmented walls like those recommended by Veneklasen, confirms that questions of reverberation have become the ba-

[112] During the discussion Veneklasen responds to Kahn's questions about the acoustic correctness of the "form" of the cathedral and the church; he indicates the "rectangular shape" as the best from an acoustic standpoint, "for a concert hall," and says that the worst figures include the "dome," the "round" or "semi-circular" structure; on the other hand, "coffers" are acceptable. Veneklasen suggests, to achieve optimal acoustics, the use of walls with a segmented profile: "Texturing of the wall – creating pockets within the masonry to diffuse sound" ("Note from meeting with P. Veneklasen," 19 August 1963, LIK, 030.II.18.9).

[113] "Note from meeting with P. Veneklasen," 19 August 1963, LIK, 030.II.18.9.

[114] Russell Johnson, Bolt Beranek and Newman, Inc., "The Acoustical Design of the Grande Salle, Place des Arts, Montreal, Quebec," 1 November 1963, LIK, 030.II.18.9. Attached is the plan of the hall with the lateral walls segmented like those proposed by Kahn for the Perfoming Arts Theater. The Grande Salle is a work by the architecture office Affleck, Desbarats, Dimakopoulos, Lebensold, and Sise, dating back to 1958.

[115] Kahn quoted in, "Informal

sis for his research on a given configuration of the structure, just as questions of light had guided his choice of the vault for the roof of the Kimbell Art Museum.[114] But the example of the Grande Salle also points to the most widespread solution applied in halls to achieve perfect acoustics, a solution Kahn, instead, is determined to avoid: the cladding of the structure.

Of the entire complex of buildings planned for the Fine Arts Center, the only one to be built is the Performing Arts Theater, a project developed by Kahn from 1966 to 1969. In October 1966, or about one month prior to the fundamental modifications introduced with the concrete "central core" in the design of the Exeter Library, Kahn defines the Performing Arts Theater as a "brick building with a concrete core."[115] It is clear, at least in chronological terms, that the two works in Exeter and Fort Wayne are taking on a similar form with respect to the relationship between their basic materials of brick and concrete.

For the configuration of the ceiling of the Auditorium, Kahn imagines the transformation of a current ceiling type, the "baffled ceiling," into a structural system capable of controlling the acoustics of the hall – "showed how the baffled ceiling construction allows even distribution of sound."[116] He even turns to the archetype of the Greek outdoor theater to be covered with temporary mobile structures, in pursuit of an ideal ceiling, studied at the start of 1967, when not by chance the Performing Arts Theater is defined, in the discussions, as "Greek-type" also due to the layout of the seating for the audience.[117]

Several documents show the disagreements between Kahn and the client regarding the question of the cladding of the concrete, in discussions similar to those that also arose during the design of the Exeter Library. In the case of the theater in Fort Wayne, the challenge is against Kahn's choice of inventing a complex, costly structure in reinforced concrete for the roof of the Auditorium, rather than using a "false ceiling,"[118] which is seen as preferable.

After Veneklasen, Kahn, for the acoustics of the Performing Arts Theater, turns to Cyril M. Harris, one of the most famous American specialists. Harris, after examining the project and discussing it personally with Kahn, prepares a report containing recommendations to obtain perfect acoustics in the hall. He's criticism has to do with the continuity of structure in Kahn's project between the Auditorium and the perimeter spaces: "in a conventional theater the interior wall and ceiling surfaces provide the function of noise isolation. They help to control noise which is transmitted through the exterior structure as well as noise from interior services such as ventilation ducts, etc. For these reasons I recommended an interior skin to provide such isolation – in particular, the use of wood was discussed."[119]

After having received Harris's letter on 7 February, on 12 February Kahn makes four schematic drawings that define the concept of the new

notes: Kahn presentation of theater design for Fine Arts Center," 1 October 1966, LIK, 030.II.17.9.

[116] *Ibid.*

[117] While the first model of the theater was called "Egyptian," perhaps due to the pyramidal effect of the triangular plan with the inclined roof, the version under discussion at the start of 1967 is called the "Greek-type," due to the seating layout and the possibility of open-air performances ("Informal notes: Yalman presentation of new theater model for Fine Arts Center," 17 January 1967, LIK, 030.II.17.9). For the drawing see *The Louis I. Kahn archive: personal drawings* cit., vol. II, p. 273, No. 605.64.

[118] "Questions and/or comments (Theater design, July 11, 1967)," LIK, 030.II.17.5.

[119] Cyril M. Harris, letter to Edward F. Menerth, Jr., 5 January 1968, [sic: February], LIK, 030.II.17.5.

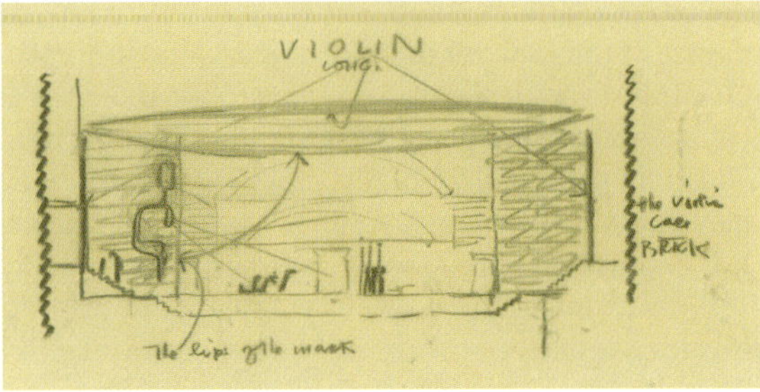

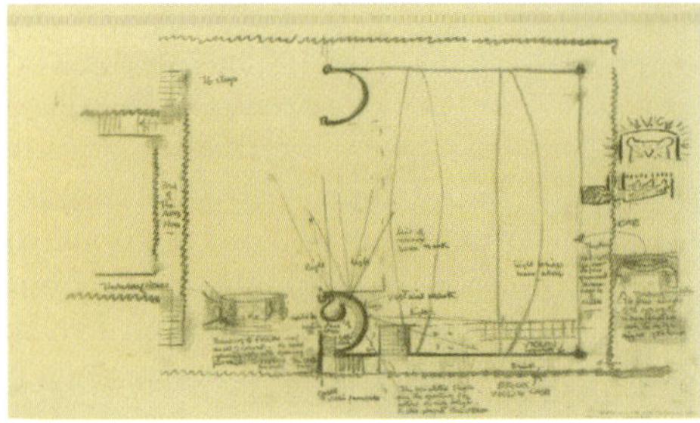

structure of the theater, starting from the suggestion of creating a clear independence between the Auditorium and the wall of the theater, though avoiding the wood cladding recommended by Harris.[120]

Harris's observations help Kahn to define the relationship between the materials of the structure, and thus to assign concrete, clay brick and concrete blocks the task of defining two distinct containers, separated by a gap as an acoustic barrier against the noises of the city. The particular nature of the materials allows Kahn to shape the two containers in different ways: that of the Auditorium is jointed and bent according to curves and broken lines that partially reflect the logic of acoustics and of the beams of spotlights to light the stage, while the one around the Auditorium takes on the configuration of an anonymous box. In this way, Kahn reaches the point of designing, for acoustic reasons, an Auditorium in one material, concrete, which is wrapped and protected by an interspace made possible by walls of clay brick and concrete blocks.

After the recommendations of Harris, Kahn introduces the image of the violin and its case to explain the new configuration of the Performing Arts Theater. The notes written on one of the drawings of February 1968 underline this point of arrival: "The walls of the violin now are all of concrete – the brick appearing only as the opening permits [?]"; "brick violin case" is the indication regarding the walls of the container that protects the hall.[121] "The free shape of the concrete is contrasted with the more rigid of the brick," is another note written by Kahn regarding the potential of the two materials.[122]

The "free shape" permitted by concrete takes on a particular meaning in the context of the images used by Kahn to explain the two containers of the theater. We cannot rule out the idea that the design of the concrete container shifts to the point of defining a roof with a series of waves that take on the profile of the sounding box of a violin, its two f-holes and the scroll.[123] If we accept the existence of this metaphorical character, the studies for the profile of the reinforced concrete structure of the roof of the Auditorium of the Performing Arts Theater seem to go well beyond the logic of acoustics and projection of beams of light often shown in the drawings.

In one schematic drawing from the series of four made in February, Kahn – besides again evoking the metaphor of the violin and its case[124] – continues to emphasize the functional logic behind the configuration of the beams of the Auditorium: "These shapes will be adjusted to be both structural and functional as light *bridges.*"

The technique for the construction of the "free shape" of the roof of the Auditorium is not indicated in the design documents. Nevertheless, one consideration of Kahn, reported by a staffer, in response to the hypothesis of modifying the structure of the roof to lower costs, offers a clear glimpse

431-433. Fine Arts Center, School, and Performing Arts Theatre (now Performing Art Center), Fort Wayne, Indiana, 1959-73. Sketches for the auditorium (LIK, 030.I.A.605.126; 70; 127).

of his intention to make use of the cement gun, as he had also envisioned for the roof of the hall of the National Assembly Building. "[...] we would not change the concept of the independent auditorium structure. Mr. Kahn has stated that he would build the auditorium of concrete block, and a concrete gunite ceiling which would span from the light-bridge beams."[125]

The formwork of the Evans Products Company

The construction of the Performing Arts Theater is assigned to the Hagerman Construction Corporation, based in Fort Wayne, while the concrete is supplied by Erie-Haven, Inc., also of Fort Wayne. Palmbaum is the engineering consultant for the structure, and the architecture office of T. Richard Shoaff, of Fort Wayne, is hired for worksite supervision.

A consideration on the quality of the exposed concrete is contained in the first specifications on the works of the Fine Arts Center, dating back to January 1968, those for the School of Fine Arts. After having indicated that the plywood panels should be "large standard-size sheets as far as possible," that the cone-type snap or screw ties should be "regularly spaced, left partially filled," and that "fins broken in stripping are acceptable," the effect Kahn is seeking for the concrete is stated: "concrete is to be and look rugged and direct and economical."[126] The phrase is repeated a few months later in the "Outline Specification" of the Kimbell Art Museum.

The first documents on the characteristics of the concrete of the Performing Arts Theater, the "Construction specifications" of December 1969, contain the usual generic indications on the formwork, and confirm Kahn's renunciation of the V-joint, already seen in the Exeter project, as well as what is by now his typical pursuit of reduction of the number of the joints.[127] The panels are to be made in plywood and should have the "maximum size," in order to "keep joints to minimum"; they should be assembled with "tight joints

[120] *The Louis I. Kahn archive: personal drawings* cit., vol. II, Nos. 605.70; 605.124; 605.125; 605.126.
[121] See the diagram of the conception of the hall, February 1968, *ibid.*, p. 276, No. 605.70.
[122] *Ibid.*
[123] *Ibid.*, p. 299, No. 605.122; p. 302, No. 605.127; p. 305, No. 605.133.
[124] "The concrete Violin is the structure of the Hall. The exterior of brick is separated from it as is the case of a Violin," Kahn writes in a note on the drawing (*ibid.*, p. 301, No. 610.125).
[125] "Fort Wayne Indiana Theater, Tim Miller called, March 7/69," handwritten, LIK, 030.II.A.18.22. "Diffusion of sound by the side walls, if we can accomplish it, will be done by the configurations imparted by the formwork that holds this concrete of the wall while it 'sets,'" Kahn writes (Louis I. Kahn, letter to Cyril M. Harris, 1 August 1969, LIK, 030.II.10.14).
[126] "School of Fine Arts, Fort Wayne Fine Arts Foudation. [...] Outline specification," 19 January 1968, LIK, 030.II.17.6.
[127] Louis I. Kahn, "Construction specifications. Theater of Perfoming Arts for Fort Wayne Fine Arts Foundation, Fort Wayne, Indiana," 1 December 1969, LIK, 030.II.18.9.

in pattern" and without chamfers, and treated with "nonstaining, alkali-proof lacquer or plastic coating"; "plastic or steel threaded cones" with "coil bolts" are to be used, assembled according to a "pattern" set by Kahn. "Provide neoprene or pvc washer of same diameter as cone face between cone and face of form," is a technical recommendation that also recurs in Kahn's specifications. Another characteristic of Kahn's architectural concrete, also confirmed in the preliminary indications for the Performing Arts Theater, is the lead plug for the holes left by the cones on the surfaces exposed to weathering.

Towards the end of March 1970, Kahn and his staff indicate, with new specifications, the various characteristics of the formwork, from the type of spreaders (cones in plastic or steel) to the product for treatment of the plywood surface (Nox-Crete Form Coating by the Nox-Crete Company), all the way to the "retardant form coating" to use in the "construction joints to ensure good bond" (the recommended product is Sika Rugasol).[128]

After the discussions on the worksite of the library in Exeter on how to avoid loss of material along the joints between the formwork panels, using strips of neoprene, Kahn is oriented towards the same type of joint for the design of the formwork of the Performing Arts Theater, demonstrating that he has definitively put aside the typical V-joint of the Salk Institute, the National Assembly Building and the Kimbell Art Museum.

The specifications contain a table of drawings titled "Typical Concrete Formwork Details" containing notes on the construction of the formwork. The subtitle indicates the purpose of these technical details: "to obtain tight forms, avoiding bleeding and staining of exposed surfaces." The table is prepared starting directly with the freehand drawings of technical details made by Gustav Langford in October 1969 for the Exeter worksite (the terms and titles of the individual drawings are also repeated, but the explanations are expanded). All the joining details are based on the use of a "compressible tape" (without the indication of neoprene, as in Langford's sketches for Exeter). In this area, it is significant that with respect to Langford's drawings, the case of the V-joint still illustrated for the Exeter Library has now been eliminated.

Before tackling the construction of the more complex parts of the Performing Arts Theater, Kahn wants to have a "sample wall panel" of about 10 feet in length and 8 feet in height on the worksite, to evaluate all the aspects of the fabrication of the concrete.

During the course of June 1970 the type of plywood to use for the formwork is modified: "Furnish plastic coated concrete formwork plywood grade B or better for veneers under overlay surfaces and C grade or better for inner plies and overlay on two sides as manufactured by Evans Products Company in lieu of that specified for all exposed concrete finish as shown on Room Finish Schedule."[129]

The product now indicated has technical characteristics that are significant for Kahn's research on the nature of concrete. The Evans Products Company, located in Portland, Oregon, offers a special formwork panel capable of eliminating any imprint of the wood grain on concrete. This is the Evans

[128] "Section 3A. Concrete," extract from the specifications, n.d., [end of March, beginning of April 1970], LIK, 030.II.18.29.

[129] "Change order," 27 June 1970, LIK, 030.II.19.3.

[130] James Schafenacker, letter to Anthony Pellecchia, 15 March 1971, LIK, 030.II.19.1.

[131] Evans Products Company, of Portland, Oregon, "Technical data. Evans Concrete Form 107," n.d., [March 1971], LIK, 030.II.19.3.

Concrete Form 107,[130] which is simply plywood "with a hard smooth phenolic resin impregnated paper overlay," as explained by the firm: "This superior overlay is irreversibly cured in the hot press under heat and pressure to combine with the plywood substrate to form an integral unity of the overlay and the wood." The company guarantees that the panels will absorb less water and can be more effective than those with medium density overlays and even some made with fiberglass.[131] The fact that the characteristics of the Evans Concrete Form 107 emphasized in the advertising brochures include "superior masking quality of wood grain of plywood" and the creation of "a consistent smooth non-glassy concrete surface with no 'pinking'" explains Kahn's interest in this type of panel for formwork, in pursuit of his degree zero.

Concrete procedure and worksite

During the preliminary phases of development of the specifications for the Performing Arts Theater, in 1970, Kahn's office outlines the possible relationships with the contractor for the design of the formwork, in what is called the "concrete procedure": "Go over concrete procedure, we must have shop drawings on formwork patterns indicating location of construction joints. We want to make sure that the contractor is in agreement and aware of determined construction joints. If the contractor isn't we would like to be notified and determine a solution rather than be placed in a position of having the steel and plywood forms in place and finding out that they don't agree with our drawings and not wanting to tell the contractor to rebuild the form (so results in the building are marked with the mistake for life)."[132]

To indicate to the Hagerman Construction Corporation the quality of the concrete to be achieved for the theater, Kahn and his staff continue to make reference to the Salk Institute, to which they add the examples of the Bryn Mawr Dormitories and the Exeter Library, now under construction. Anthony Pellecchia, a member of Kahn's staff, invites Schafenacker to visit these buildings, as examples of "tight plywood forms resulting in a good surface."[133] Then he has photographs sent of the Salk Institute: "This is the quality of concrete that we expect to achieve in Fort Wayne. The photographs show the importance of form panel layouts and the necessity for well constructed clean straight forms. Please notice the fine straight unchamfered inside and outside corners of columns and walls, the fine smooth consistent surface of the concrete and the accuracy of the form panels."[134]

"Forms and concrete finish" is a recurring and fundamental subject in the communications between Kahn's office, the Hagerman Construction Corporation and Shoaff during the months of August and September, shortly before the start of the work. Besides the detailed definition of the areas of the base to be built with or without the precautions to be applied for exposed concrete,[135] and preparation of the samples for Kahn's inspection in order to choose the color and texture of the concrete,[136] the main topic of discussion is the type of formwork.

The Hagerman Construction Corporation proposes, as an alternative to the formwork of the Evans Products Company, the use of the Allen Form

[132] Anon., handwritten notes, n.d., LIK, 030.II.18.32.
[133] Anthony Pellecchia, letter to Richard T. Shoaff, 10 August 1970, LIK, 030.II.19.20.
[134] Anthony Pellecchia, letter to James Schafenacker, 4 November 1970, LIK, 030.II.19.1.
[135] Richard T. Shoaff, letter to James Schafenacker, 21 August 1970, LIK, 030.II.20.
[136] James Schafenacker, letter to Richard T. Shoaff, 25 August 1970, LIK, 030.II.19.1.

System, and the sealing of the holes with the Plastic Poly Plugs produced by the Joaillier Construction Products, Inc.[137] Pellecchia replies that no one in Kahn's office knows the qualities of the Allen Form System, and thus requests a "list of buildings where it has been used."[138] As for the hypothesis of replacing the lead seals, the contractor is informed of Kahn's opinion: "not the plastic plugs."[139] In the meantime, Pellecchia specifies the type of tape to use to seal the formwork panels, the Hi-Lo Foam Tape produced by the Porter Engineering Company, based in West Barnstable, Massachusetts.[140]

In the second half of September construction of the basement walls begins. In the Performing Arts Theater, as in other buildings by Kahn, starting with the Salk Institute, this part of the work becomes the phase in which the contractor's workers begin to familiarize themselves with the rules and methods for the formwork and concrete, in order to achieve the level of quality desired by Kahn for the most significant parts of the exposed concrete. Also for this reason, the inspections of the first works generate detailed indications on defects, causes and remedies.

Plywood panels of the Allen Form System are used for the basement walls, made without tie cones and without joint taping. During the first inspection of the "concrete finish," that of the North Wall, done on 25 September, the defects are listed: some panels have gone "out of plane – particularly at horizontal joints"; the concrete surfaces have "voids or holes" larger than 1/2 inch in diameter; "rough honeycomb" is found in the corners of the pillars and "at some panel joints"; "discoloration of surface" and "variations in color" are visible; finally, the "general color" is deemed "probably too blue."[141]

There are various causes for the defects identified during the inspection: "improper aggregate gradation - possibly grade gap - 1/2" or No. 40 sand"; "inadequate care in lining up panels & inadequate studs and/or walers"; the reuse of panels ("old plywood surface - too rough & color variations"); "form joints not tight - too much leakage at joints"; "corners of west pier improperly formed – obvious leakage"; "improper vibration – at least in some areas."

The advised remedies are: to use new panels "with treated surface to hide grain, etc."; to apply "double vertical braces" at the joints between panels; to insert tapes in the joints; to use groups of ganged panels, or "gangforms," for high walls; to make use of "coil type ties with cones"; finally, for the compound, it is recommended to try other mix designs, besides the standard "Erie-Haven blend mix," with or without airtraining. Also during the inspection of 25 September, it is decided not to seal the tie holes on the walls facing the interior of the theater, and instead to seal them on the other faces below ground level; on the first floor, "coil type ties with cones" are to be used, and the holes should be plugged with lead "as soon as possible."[142]

The use of the Allen Form System is disputed by Kahn's team, not only due to the results achieved in the basement walls, but also because it does not permit the use of "gang forms," which according to Pellecchia's view is

[137] For the Plastic Poly Plugs see Joseph C. Joaillier, of Joaillier Construction Products, Inc., letter to James Schafenacker, 18 August 1970, LIK, 030.II.19.20, and T. Richard Shoaff, letter to Louis I. Kahn, 21 August 1970, LIK, 030.II.19.20.

[138] Anthony Pellecchia, letter to Richard T. Shoaff, 10 August 1970, cit.

[139] T. Richard Shoaff, letter to James Schafenacker, 21 August 1970, LIK, 030.II.19.20.

[140] Anthony Pellecchia, letter to T. Richard Shoaff, 2 September 1970, LIK, 030.II.19.1.

[141] T. Richard Shoaff, "Construction inspection & discussion of problems, etc. [...], September 25, 1970," 1 October 1970, LIK, 030.II.18.32.

[142] *Ibid.*

[143] Anthony Pellecchia, letter to James Schafenacker, 4 November 1970,

a good way to achieve regular concrete surfaces. It is worth noting that the gang formwork had already been proposed by Palmbaum on the worksite of the National Assembly Building, and that Gustav Langford recommends them in 1969 at Exeter. “It is almost impossible,” Pellecchia writes to Schafenacker, “to construct straight well lined panels, both horizontally and vertically, without gang forming. The Allen Form seems to not allow one to gang form, therefore, we must insist that this form type not be used in areas specified to have a No. 1 quality finish.”[143]

It is precisely to eliminate all the defects that have emerged in the first phase of the work that in March 1971 the text of the specifications is changed to call for definitive use of the system of the Evans Products Company, and in particular of Evans Concrete Form 107 (this type specification is discussed precisely in March).[144]

Alongside Pellecchia, for the supervision of the Fort Wayne worksite Palmbaum also works on a continuous basis, not limiting his activity to the calculation of the structural members. His interventions in the preliminary phases of the worksite are aimed at improving the quality of the poured in place concrete surfaces. The experience he has gained on the Dhaka worksite is fundamental for the technical methods necessary to achieve the correct quality of Kahn’s architectural concrete.

Following several tests performed by Michigan Testing in June 1971, Palmbaum criticizes the quality of the compound. “It is his opinion that these inconsistencies will result in variations in color and texture of concrete surfaces.”[145] “[…] variations in slump, air content and unit weight” are, in his view, “the reasons for variable concrete appearance.”[146] Other defects come from the vibration of the concrete, which produces a “bubbly appearance on the surface,” while the excessive “slumps mean wet mixes with subsequent leakage at form joints.”[147] After having observed, during the worksite inspection of 22 June, a series of flaws in the formwork system that generate other defects in the placing of the concrete, Palmbaum makes some general recommendations and proposes some specific corrections: “Contractor was requested to restudy his inadequate method of holding bulkheads. […]. Tape would still be used at the seam between the bulkhead and the main run of the wall or column. […]. All concrete construction joints are to be keyed; for those cases where the key is inadvertently omitted, the joint is to be completely roughened to improve bonding.”[148]

Besides the exposed concrete, the concrete of the blocks and the mortar for the joints of the masonry in concrete blocks and bricks are, as usual, carefully controlled by Kahn and his staff by means of samples made on the site. Kahn approves the blocks produced by Masolite Concrete Products, Inc., based in Fort Wayne.[149] The blocks have the usual 8 × 8 inches size (with the exception of special corner pieces measuring 8 × 12 inches). For the mortar of the masonry, to tune with the color of the bricks, Kahn wants

LIK, 030.II.19.1.
144 Anthony Pellecchia, letter to James Schafenacker, 22 March 1971, LIK, 030.II.19.1.
145 Richard T. Shoaff, “Memo on Concrete Tests,” 10 June 1971, LIK, 030.II.18.32.
146 Harry M. Palmbaum, letter to David P. Wisdom, 11 June 1971, LIK, 030.II.19.29.
147 *Ibid.*
148 Harry M. Palmbaum, letter to Anthony Pellecchia, 24 June 1971, LIK, 030.II.19.25.
149 David P. Wisdom, letter to Milford M. Miller, Jr., 12 January 1970, LIK, 030.II.19.25. On the blocks see also Chalk Netherland, letter to Anthony Pellecchia, 14 July 1970, LIK, 030.II.19.1, and Ralph W. Walb, president of the Masolite Division, letter to Chalk Netherland, 10 August 1970, LIK, 030.II.19.1.

a dark tone, for which he indicates an example: that of the Nazareth Type I produced by the Nazareth Cement Company, in Pennsylvania, and already used by him for the Philadelphia Psychiatric Hospital and the Richards Laboratories.[150] But the "dark mortar by using a dark masonry cement" is not a feasible solution, given the fact that Nazareth Type I is no longer available on the market. It is then suggested to Kahn to obtain the same tone by using local products with the addition of "ferric oxide."[151] "[...] This method does not provide adequate control of color," is the observation of Kahn's staff.[152]

Precast concrete planks for hollow floor slabs

Also in the Performing Arts Theater, as in many other buildings by Kahn, the floor slab becomes a kind of complex organ equipped with cavities for the passage of physical plant systems. At the start of 1970 the decisive modification of the floor slab design is introduced, precisely to create an integrated system of structure and physical plant. Kahn decides to use "precast concrete planks" instead of "concrete slabs."[153] The configuration of these planks makes it possible for them to contain a series of parallel cavities, which in the technical drawings are shown as being both rectangular and circular.[154] The main types of hollow-core prestressed concrete planks available on the market, as shown in the drawings, are Span-Deck produced by American Precast Concrete, Inc., with rectangular passages; Flexicore, a patented product by the Price Brothers Company; and the Tube Slab of the Tube Slab Company based in Hartford, Connecticut.[155] Certain cavities inside the planks are used for the circulation of conditioned air, so they must be manufactured so as to have smooth surfaces, as indicated in the section "Precast plank" that is now part of the specifications. Given the fact that the precast planks are to be installed without any painting or cladding, Kahn wants the surface of their concrete, forming the ceiling of the various spaces, to be "uniform" in "color and texture." In order to transform the planks, which are usually covered, into fundamental parts of the architectural concrete, the specifications indicate that for their installation the joints should be sealed with mortar, and that "any grout that seeps down to ceiling below shall be promptly cleaned off."[156]

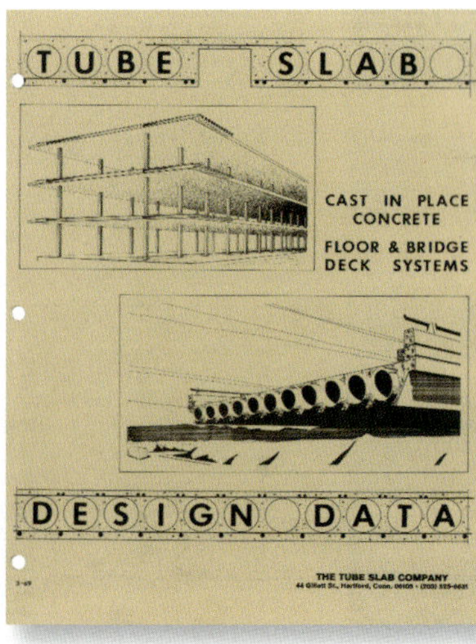

434-435. The Tube Slab Company, *Tube Slab flat plate concrete floors and decks using tubes as voids*, n.d., (LIK, 030.II.17.38.001; 002).

436-437. Span-Decks booklet (LIK, 030.II.A.17.23.001; 002).

Shoaff asks Kahn's office to provide information on the form of the cavities in the planks, reporting that he has not been able to locate, near Fort Wayne, any manufacturer capable of making "rectangular cores."[157] "The rectangular voided precast plank is the plank to be used," is Pellecchia's reply, and he also indicates the producers specializing in this type of plank.[158] But starting in the summer, the hypothesis of using Flexicore with circular passages is considered, though the standard type would have to be modified to adapt to the air conditioning and other physical plant systems.[159]

Several Flexicore Flat Planks are produced at the start of 1971 by the Price Brothers Company, but without the openings for the plant systems, which therefore have to be cut inside the already cast pieces.[160] Given the difficulties involved in the production of custom Flexicore Flat Planks, the research shifts towards the idea of using the planks produced by American Precast Concrete, Inc.[161] The Span-Deck precast concrete planks are easily perforated by hand for the vertical passage of tubes, even though this type

[150] Chalk Netherland, letter to Anthony Pellecchia, 9 November 1970, LIK, 030.II.18.32.
[151] J. B. Lendrum, vice-president of the Peninsular Portland Cement Division of General Portland Cement Company, letter to Louis I. Kahn, 3 May 1971, LIK, 030.II.19.1.
[152] *Ibid.*
[153] David P. Wisdom, letter to Milford M. Miller, Jr., 12 January 1970, LIK, 030.II.19.25.
[154] Richard T. Shoaff, "Memo on specifications & drawings - general contract," 9 April 1970, LIK, 030.II.19.20.
[155] See the booklet, Tube Slab Company, *Tube Slab flat plate concrete floors and decks using tubes as voids*, n.d., (LIK, 030.II.17.38).
[156] "Section 3A. Concrete," excerpt from the specifications, n.d., [end of March, early April 1970], LIK, 030.II.18.29.
[157] Richard T. Shoaff, "Memo on specifications & drawings – general contract," 9 April 1970, LIK, 030.II.19.20.
[158] Anthony Pellecchia, "Answers to your memo on specifications & drawings – general contract," 14 April 1970, LIK, 030.II.19.20. The manufacturers are Breeko Industries, based in Nashville, Tennessee, Cleveland Builder Supplies, located in Cleveland, Ohio, and Midwest Prestress, of Springfield, Illinois.
[159] Anthony Pellecchia, letter to James Schafenacker, 25 August 1970, LIK, 030.II.19.1. For the Flexicore the Price Brothers Company, located in Dayton, Ohio, is also consulted (it is the company that owns the rights to Flexicore since 1938).
[160] Anthony Pellecchia, letter to James Schafenacker, 15 March 1971, LIK, 030.II.19.1.
[161] James Schafenacker, letter to Anthony Pellecchia, 22 March 1971, LIK, 030.II.17.23.

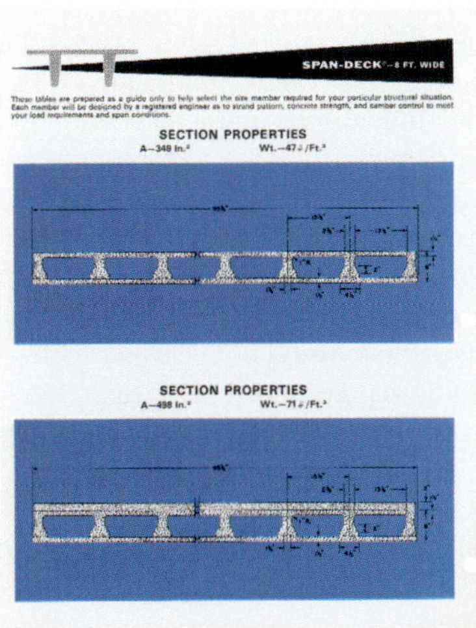

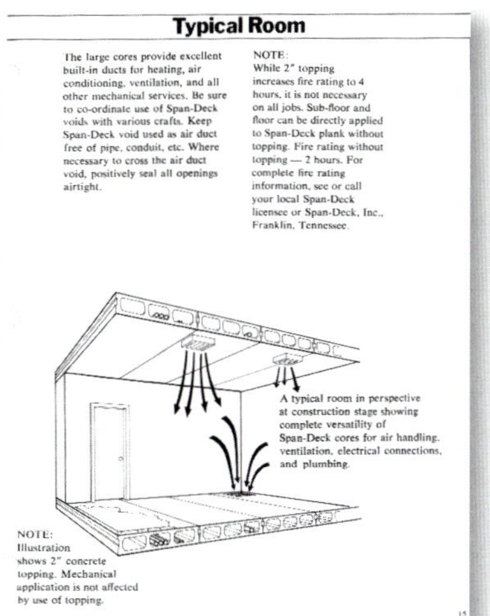

438. Fine Arts Center, School, and Performing Arts Theatre (now Performing Art Center), Fort Wayne, Indiana, 1959-73. Site construction photograph (LIK, 030.IV.A.605.3.6).

of hole can be made in the prefabrication stage. The finish of the surface that forms the ceiling is available either with acoustic cladding, or in "gray concrete."[162] Pellecchia informs the Hagerman Construction Corporation of Kahn's willingness to use Span-Deck precast concrete planks, but he insists that the holes for the ducts and vertical conduits must be "cast into the plank when it is manufactured," and that the planks should have a "hard gray concrete finish on the underside."[163]

The floor slabs of the Performing Arts Theater are made with the precast concrete planks of the Span-Deck type, with rectangular cavities, produced by Span-Deck of Indiana, based in Indianapolis.[164] The Span-Deck planks, in any case, are not free of defects, in both their casting and their installation. During a worksite inspection on 10 August 1971, Pellecchia and Palmbaum find a series of flaws, ranging from the lack of solidity of certain planks, at times not only corresponding to the cavities, to the chipping of corners and marks on the concrete produced by the molds. Besides requesting repairs of all the "holes and cracks" visible on the lower surface of the installed planks, Pellecchia and Palmbaum recommend certain procedures Span-Deck should follow in the production phase: "The manufacturer is to eliminate all oil stains from the material. The plank should be manufactured in such a way as to not have oil stains rather than being finished out in the field. The manufacturer is to clean and demonstrate to the Architect that certain stains can be removed without leaving noticeable differentiations on the surface."[165]

Kahn's decision to use certain cavities in the planks for the passage of air conditioning leads to precautions in the Span-Deck production to improve the surface finish. Any roughness left by the molds inside the cavities for the circulation of air has to be removed by Span-Deck by blasting compressed air.[166]

During construction certain concrete blocks prove to have whitish marks probably caused by rainwater passing through the precast planks.[167] When the planks are installed, many lower surfaces have concrete with "very irregular" color.[168]

Formwork for the walls of the Auditorium: the limits of carpentry

After the meeting with Harris, Kahn always explains the form of the Auditorium of the Performing Arts Theater with the image of the violin. Only in one version – later corrected – of a text on the theater does he mention a figure that has become familiar among those of engineering constructions in reinforced concrete, also documented by Angerer, namely the shell: "The auditorium built in concrete was shaped to construct itself like a shell, independent of the roof and the outside walls and shaped to send and reinforce the sounds emanating from the stage."[169] Angerer's book contains the schemes of "folded structures," "conical folds" and "folded frames,"[170] a family to which Kahn's system belongs, as well as the similar systems of Pier Luigi Nervi with Marcel Breuer and Bernard Zehrfuss for the Conference Room of the UNESCO Building in Paris, completed in 1958, and of Doshi and Raj for the Tagore Memorial Theater in Ahmedabad, completed in 1966. One possible source for the conception of the structure of the Auditorium might also be found in the folded structures created by Anthony P. Coppa, an engineer of the General Electric Company, at King of Prussia, Pennsylvania, given the fact that Kahn has been in contact with Coppa since 1968.[171]

For the walls of the Auditorium, which are the most important and characteristic sectors of the architectural concrete of the Performing Arts Theater, Kahn's office prepares a series of drawings from September 1970 to April 1971 with indications of the height of the panels, the pattern of the spreaders of the formwork and of the "construction joints."[172] Given the hypothesis of constructing a sample concrete panel for one sector of the segmented and folded wall of the Auditorium, on 23 September 1970 a special table of drawings is made.[173] This is the most complex detail of the Performing Arts Theater, given the fact that those wall segments are designed to form folds, like those of a curtain, through which the audience can enter and exit from the steps placed up against the Auditorium.[174] The system is made complex by the angles of the two wall sectors, of which one is triangular with the summit coinciding with the beams of the roof.

[162] Ralph W. Walb, of Masolite Concrete Products, Inc., letter to James Schafenacker, 22 March 1971, LIK, 030.II.17.23. For a drawing of the Span-Decks see the sketches on Price Brothers Company stationery, received by Kahn's office on 3 November 1973, LIK, 030.II.17.23.
[163] Anthony Pellecchia, letter to James Schafenacker, 2 April 1971, LIK, 030.II.19.1. The drawings of American Precast Concrete, Inc., are sent to Arthur Liphard of Span-Deck of Indiana to include the indications supplied by the physical plant experts and regarding the various holes required in the planks for the conduits (Ralph W. Walb, letter to James Schafenacker, 30 April 1971, LIK, 030.II.17.23).
[164] See the worksite photograph, LIK, 030.IV.A.605.3.6.
[165] Anthony Pellecchia, letter to James Schafenacker, 13 August 1971, LIK, 030.II.19.1.
[166] William R. Kniptash, vice-president of Span-Deck of Indiana, letter to Ralph W. Walb, 4 October 1971, LIK, 030.II.17.23. But the operation of cleaning of the cavities of the planks is not done (William E. Swale, of Dubin-Mindell-Bloome Associates, Consulting Engineers, "Field Report," 2 February 1972, LIK, 030.II.19.7).
[167] Anthony Pellecchia, letter to James Schafenacker, 19 June 1972, LIK, 030.II.19.16.
[168] Anon., handwritten notes, 31 October 1972, LIK, 030.II.18.32.
[169] Louis I. Kahn, "Fort Wayne Community Center for the Performing Arts, Fort Wayne, Indiana," 20 September 1973, LIK, 030.II.19.22.
[170] Angerer, *Surface Structures in Buildings* cit., pp. 123 and seb.
[171] See Anthony P. Coppa, letter to Louis I. Kahn, 9 April 1968, LIK, 030.II.1.8.
[172] Anthony Pellecchia, letter to James Schafenacker, 6 January 1971, LIK, 030.II.19.1.
[173] The drawing is marked FW3 (LIK, 030.I.C.605.016). It is discussed at the time of the worksite inspection of 25 September (Richard T. Shoaff, "Construction inspection & discussion of problems, etc. […], September 25, 1970," 1 October 1970, LIK, 030.II.18.32).
[174] See also the drawing "Elevation Auditorium," marked FW1, 23 September 1970, LIK, 030.I.C.605.016. On 10 January 1971 several drawings are made for the concrete walls ("Elevation south wall of stagehouse," marked FW4, LIK, 030.I.C.605.016, and "Elevation north wall of stage house," marked FW5, LIK, 030.I.C.605.016). On 6 March 1971 the detail of the formwork for the walls of the theater hall is drawn, marked FW6, LIK, 030.I.C.605.016. The drawing "Elevation lobby stair," marked FW6, and that of the "Isometric auditorium wall," marked FW7, date back to 12 April 1971 (LIK, 030.I.C.605.016).

438-441. Fine Arts Center, School, and Performing Arts Theatre (now Performing Art Center), Fort Wayne, Indiana, 1959-73.

439. "Elevation Lobby Stair," 12 April 1971 (LIK, 030.I.C.605.13).

440. "Elevation Auditorium," 23 September 1970 (LIK, 030.I.C.605.13).

441. "Details," 23 September 1970 (LIK, 030.I.C.605.13).

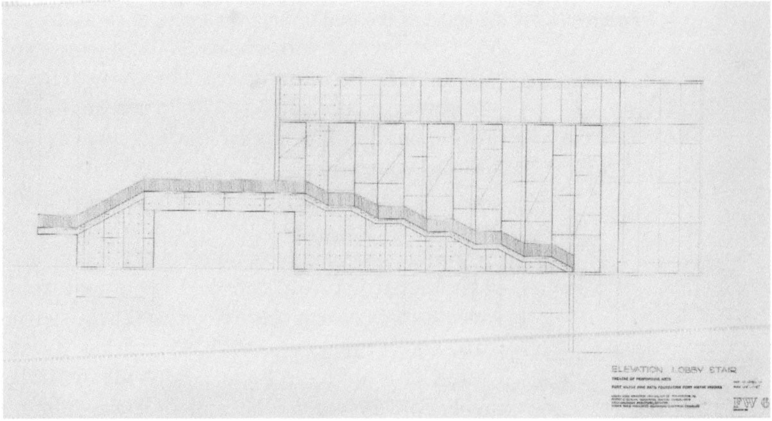

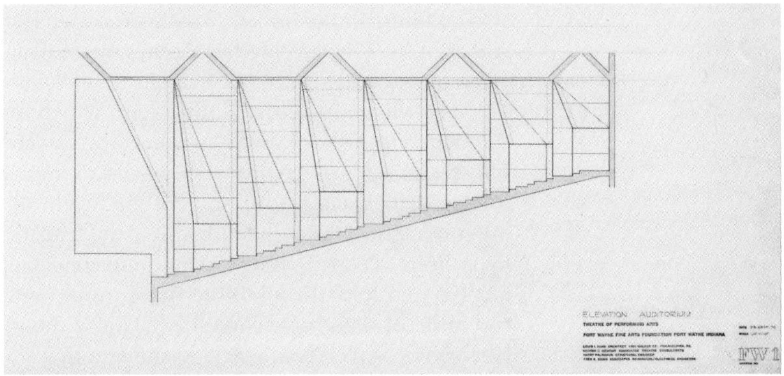

In February 1971 the Hagerman Construction Corporation studies the possible structure of the formwork for the wall segments of the Auditorium.[175] But those walls are not built until 1972. Precisely when their construction is approaching, Pellecchia complains that the "quality of concrete has become noticeably worse, particularly in the most important areas," and he prepares a list of the defects observed during various visits to the worksite together with Palmbaum, from leakage in the junction between panels and corners, to leakage around the cone ties.[176] He then appeals to the representative of the Hagerman Construction Corporation to raise awareness among the workmen of the importance of making high-quality concrete, such as that which is required for the walls of the Auditorium:

"Overall, the greatest cause for the bad concrete, etcetera, seems to be the attitude of the workmen and contractor. This demonstrates a lack of understanding of the problems that exist. We need a close and concerned eye to oversee the work. It must be remembered that the good workmanship, as well as the bad, will exist for people to see for many years."[177]

The question at this time is the definition of the technical details of the formwork system for the walls of the Auditorium. After the preliminary drawings, now Kahn and his staff delegate the definition of the details to the Hagerman Construction Corporation. If we consider the fact that on the worksite for the Exeter Library Kahn's office is also relying to a significant extent on the expertise of the office of Protze Materials Technologist, we can hypothesize that after the experiences of the Salk Institute, the Kimbell

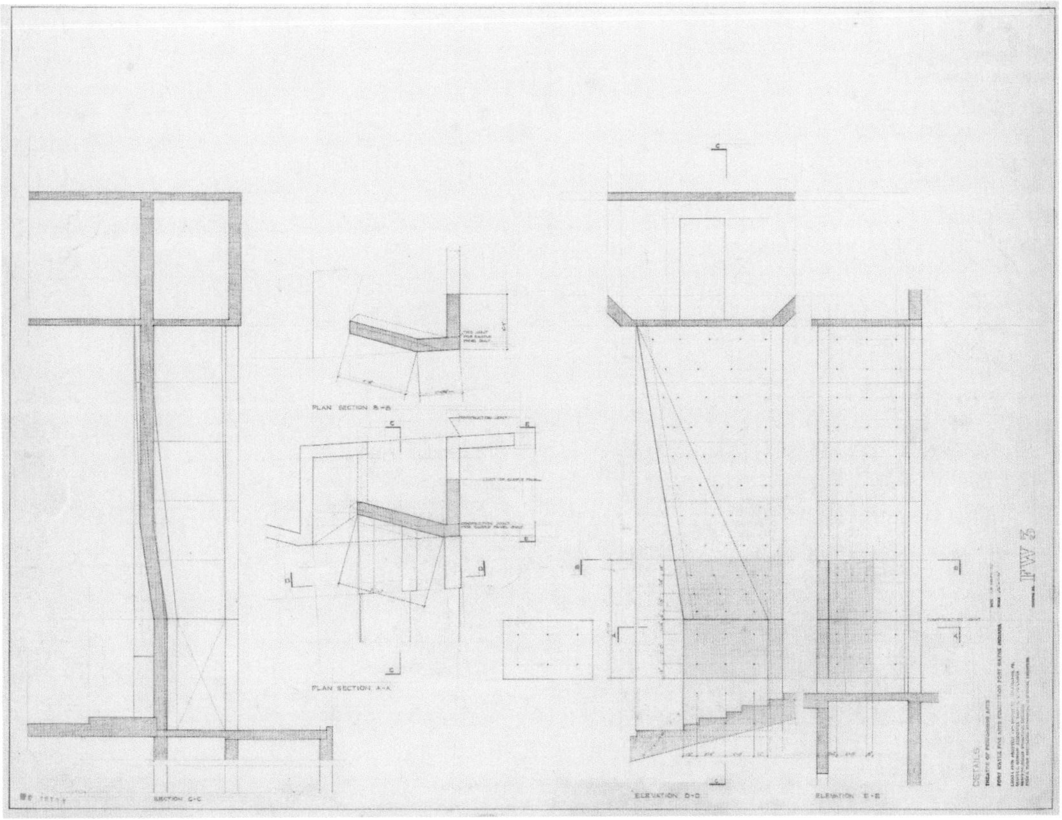

Art Museum or the National Assembly Building, and also following the unsuccessful attempts to create an independent professional office run by Frederick and Gustav Langford, specializing in formwork design, Kahn is assigning this aspect of the design to the contractors involved in the construction, simply offering indications on the quality of the concrete to be achieved, the usual patterns of the spreaders and the position of the pour joints. In practice, for the development of the details of the formwork of the Auditorium of the Performing Arts Theater the intervention of the Hagerman Construction Corporation becomes decisive.

For the formwork of the "auditorium walls" the plan is to use, for the corners with angles larger than 90 degrees, specially shaped solid wood wedges to insert between the plywood panels, according to the indications of Pellecchia to Schafenacker.[178] The introduction of special wedges in the system formed by plywood panels generates, as Pellecchia himself admits,

[175] On 12 February 1971 Hagerman prepares the drawing of the formwork system for the wall of the Auditorium (the drawing is attached to, James Schafenacker, letter to Anthony Pellecchia, 31 January 1972, LIK, 030.II.18.32)

[176] Anthony Pellecchia, letter to James Schafenacker, 18 January 1972, LIK, 030.II.19.1.

[177] *Ibid.*

[178] "A solid milled continuous wood insert must be used to form corners where the vertical/sloped planes meet. A milled continuous piece of wood must also be used to form corners between the two sloped planes. The milled wood must have a smooth finish and be coated with a form coat such as 'Nox-Crete Pre-Form Co-Polymer Resin.' Without this method, forming good concrete will be impossible" (Anthony Pellecchia, letter to James Schafenacker, 18 January 1972, LIK, 030.II.18.1).

442-443. Fine Arts Center, School, and Performing Arts Theatre (now Performing Art Center), Fort Wayne, Indiana, 1959-73.

442. "Isometric Auditorium Wall," 12 April 1971 (LIK, 030.I.C.605.13).

443. "Detail of Form Construction," 6 March 1971 (LIK, 030.I.C.605.13).

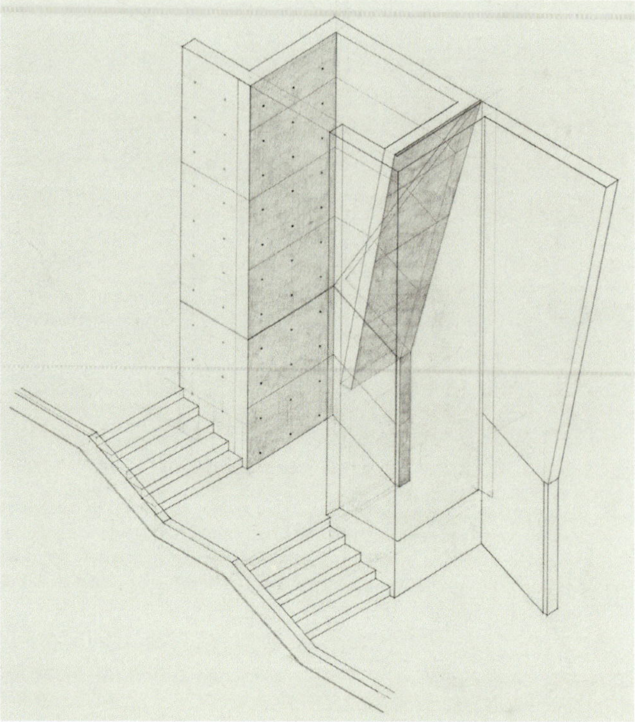

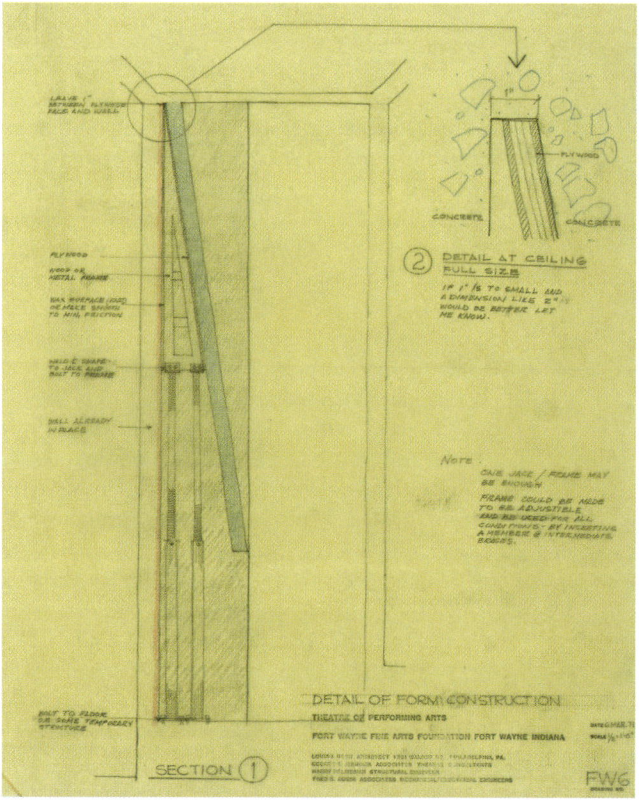

different textures on the surface of the formwork, which if not suitably treated may leave diversified and visible imprints on the concrete. To cope with this type of problem, and due to the high cost of making the special pieces of wood, Schafenacker does not simply develop a formwork system based on Pellecchia's indications, but also prepares an alternative solution that does not require the pieces of wood.

Kahn's office receives the detail drawings of the two proposals for the formwork of the walls of the Auditorium from the Hagerman Construction Corporation, accompanied by Schafenacker's explanation of all the details and their logic of construction. What Schafenacker writes constitutes a meaningful summary of the technical methods required to achieve the goal of perfection sought by Kahn in the design of the exposed concrete surface after stripping. In any case, in multiple areas, Schafenacker himself sees this as arduous, if not impossible, if use is made of the usual available materials for the construction of formwork with a wooden panels. It even seems possible to state, in the light of these documents, that works like the Performing Arts Theater offer a glimpse, and not only to Kahn, of the impossibility of getting beyond elementary geometries for the structures of exposed reinforced concrete without abandoning the use of carpentry. On this issue, it is worth recalling that the spectacular curved surfaces in reinforced concrete of the TWA Terminal of the John F. Kennedy International Airport in New York, completed in 1962, were clad with cement to avoid leaving the irregular imprints of the formwork boards visible. Kahn is well aware of the technical implications that lie between the formal conception of a reinforced concrete structure and the constructive logic of its formwork. Such implications orient him in the choice of the final solution for the structure of the Auditorium of the Performing Arts Theater, since he does not want to apply the expedient of cladding, as done by Saarinen, nor does he want to accept, for aesthetic reasons, a freer, less rigorous assembly of the boards leaving visible imprints.

Schafenacker's observations on the complexity of the assembly of the panels of the formwork for the walls of the Auditorium point to the technical limits reached by Kahn in the implementation of his exposed concrete structures without finishing after stripping. We can hear the voice of worksite experience when Schafenacker lists, in a pithy but blunt way, certain aspects that make the perfection sought by Kahn impossible:

"As you know, all of the form materials we have to work with have certain variances in them. The 2 × 4 lumber is twisted, warped and varies in dimension. The plywood is warped and varies in thickness. The ties are not perpendicular with the face of the cone. None of the members is perfect. [...]. As you can see, with most of the surfaces not being plumb, the ties not being horizontal and the plywood being cut on an angle and a bevel, building the form to perfect alignment and tightness, which is what is wanted, is highly impractical if it is even possible. [...]. The materials which we work with are not accurate enough for a perfect job, and, as you remember, everyone questioned the possibility of building a satisfactory form for the difficult sloping and angular walls of the Auditorium at the start of the project."[179]

After having developed the two solutions for the formwork of the walls of the Auditorium, with and without the special pieces in solid wood,

[179] James Schafenacker, letter to Anthony Pellecchia, 31 January 1972, LIK, 030.II.18.32.

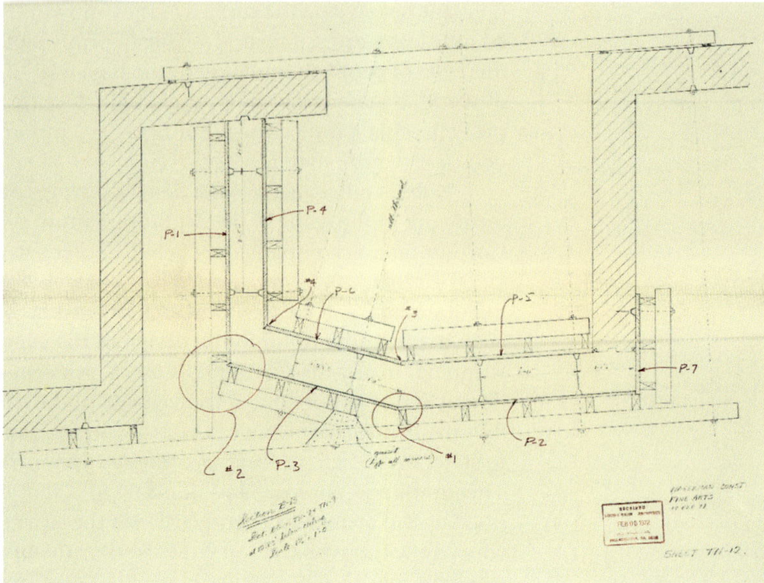

444-446. Fine Arts Center, School, and Performing Arts Theatre (now Performing Art Center), Fort Wayne, Indiana, 1959-73.

444. Hagerman Construction Corporation, formwork detail, 17 February 1971 (LIK, 030.II.A.18.32).

Schafenacker explains the reasoning behind his choices: "We chose a laminated member because it is too difficult to mill a large solid member; the laminated member will remain more stable and the laminated plywood will hold the joint tighter since it will not split as a solid lumber member could do. [...] the cut edge of the plywood will not match the edge of the milled member since we cannot cut the plywood perfectly straight and the gasket tape cannot be compressed enough to take up any variations even with the use of cabinet clamps [...]. The contact area of the solid wood must be coated to resemble the plywood surface, but, as you know, these surfaces will be different and will cause a different texture in the concrete."[180]

Nevertheless, Schafenacker himself also mentions the possibility of a formwork system that breaks with tradition and would be more suitable to cope with the geometry of the wall segments of the Auditorium, though he considers it impossible to put into practice, at least with the current technical expertise of the various local manufacturers and contractors: "We might say that after we build the form for one wall the rest would be easier and fit better, but each one is different and requires its own individual fitting. One might think that building the entire form in one piece would be the answer but no one in this area can see that this could possibly work due all the variables."[181]

The walls of the shell of the Auditorium are completed by the beams, which form the roof, with the large cavities for the insertion of spotlights pointed at the stage. The beams are poured in place, using plywood formwork. During the inspection of the work to assess its quality, "staining" is observed on the concrete of the first beam. But before intervening to correct this defect, the opinion of Frederick Langford[182] is requested, who after a visit to the site attributes the cause of the marks to the plywood panels, "usually the worst when used for the first time," and recommends various cleaning

[180] *Ibid.*
[181] *Ibid.*
[182] Anon., handwritten notes, 31 October 1972, LIK, 030.II.18.32.

445-446. Hagerman Construction Corporation, formwork details, 26 January 1972 (LIK, 030.II.A.18.32).

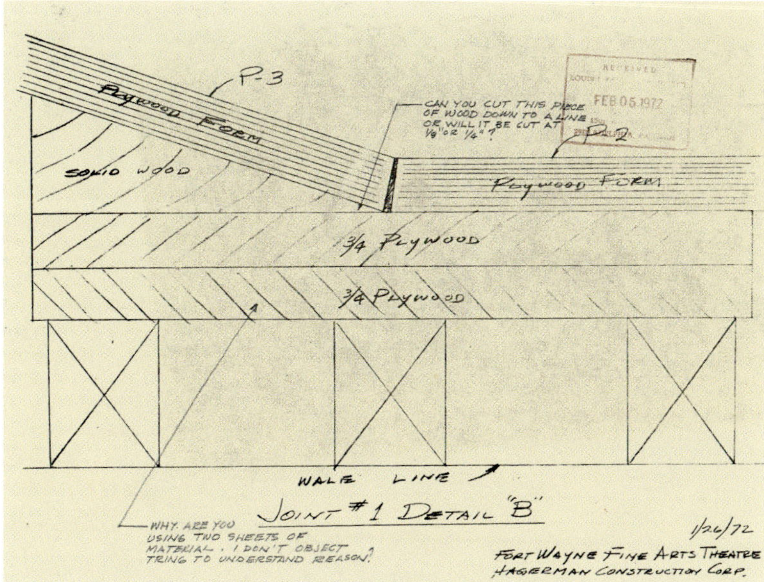

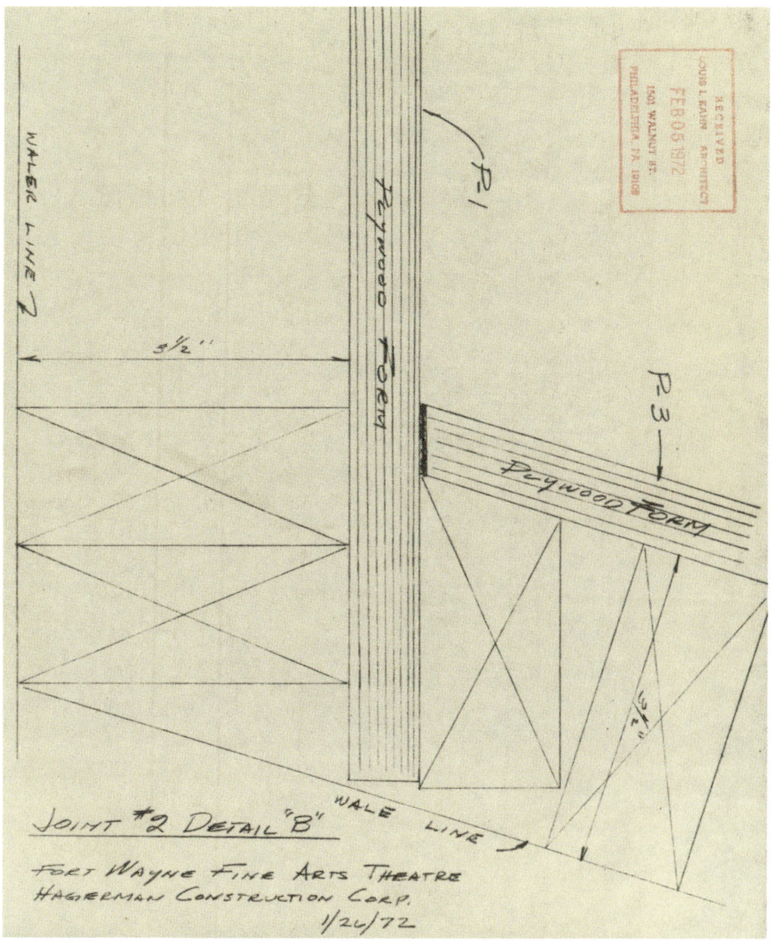

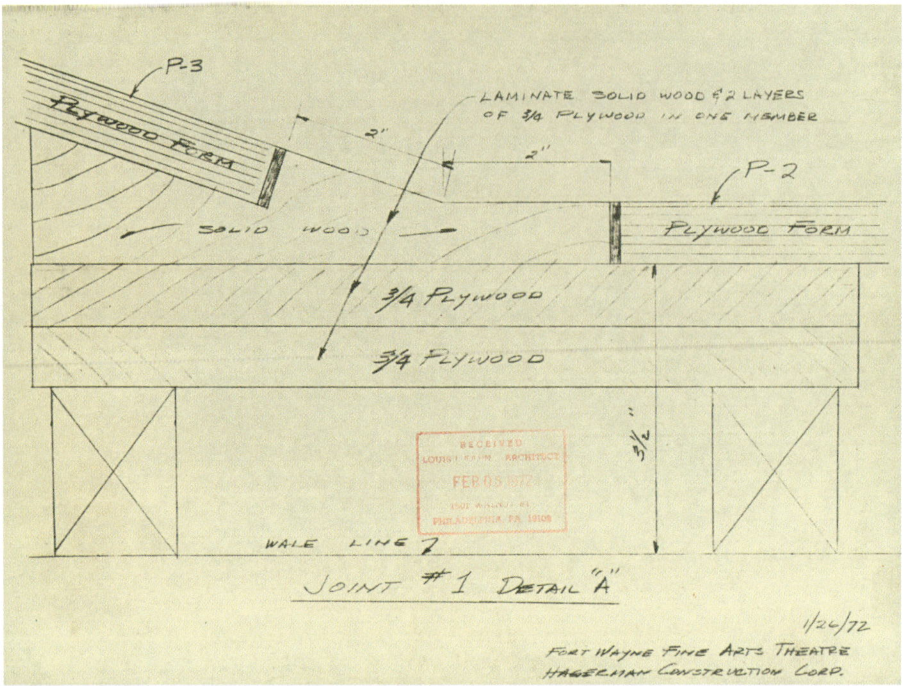
P-3
PLYWOOD FORM
LAMINATE SOLID WOOD & 2 LAYERS OF 3/4 PLYWOOD IN ONE MEMBER
2"
2"
P-2
PLYWOOD FORM
SOLID WOOD
3/4 PLYWOOD
3/4 PLYWOOD
3 1/2"
RECEIVED LOUIS I. KAHN ARCHITECT FEB 05 1972 PHILADELPHIA PA 19103
WALE LINE
JOINT #1 DETAIL "A"
1/26/72
FORT WAYNE FINE ARTS THEATRE
HAGERMAN CONSTRUCTION CORP.

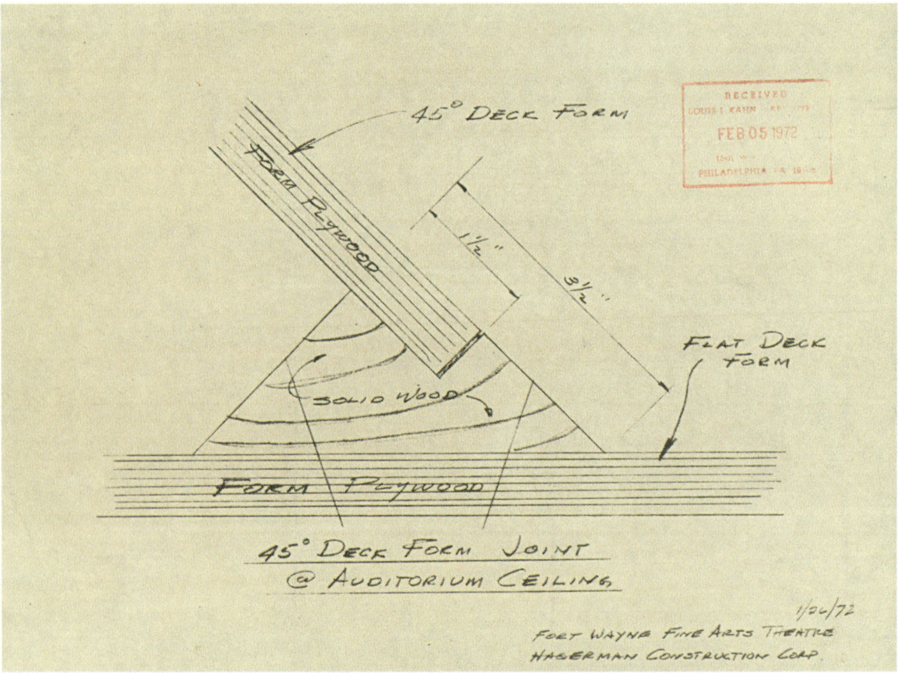
45° DECK FORM
RECEIVED LOUIS I. KAHN ARCHITECT FEB 05 1972 PHILADELPHIA PA
FORM PLYWOOD
1 1/2"
3 1/2"
FLAT DECK FORM
SOLID WOOD
FORM PLYWOOD
45° DECK FORM JOINT
@ AUDITORIUM CEILING
1/26/72
FORT WAYNE FINE ARTS THEATRE
HAGERMAN CONSTRUCTION CORP.

447-450. Fine Arts Center, School, and Performing Arts Theatre (now Performing Art Center), Fort Wayne, Indiana, 1959-73.

447-449. Hagerman Construction Corporation, formwork details, 26 January 1972 (LIK, 030.II.A.18.32).

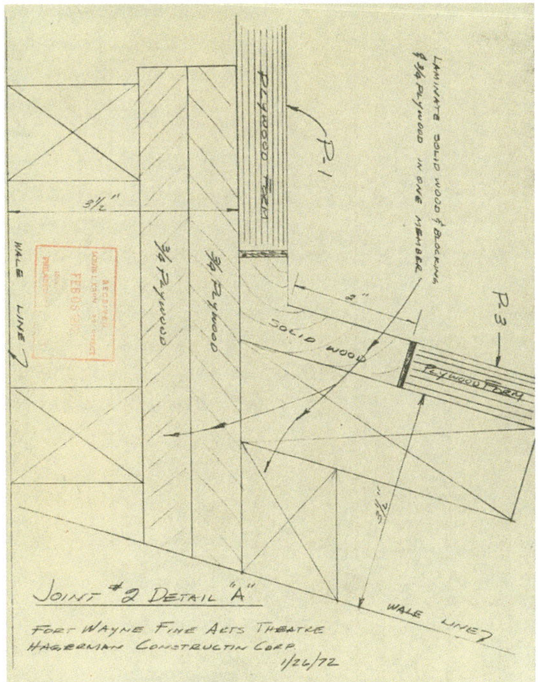

450. Hagerman Construction Corporation, formwork detail, 26 May 1971 (LIK, 030.II.A.19.1).

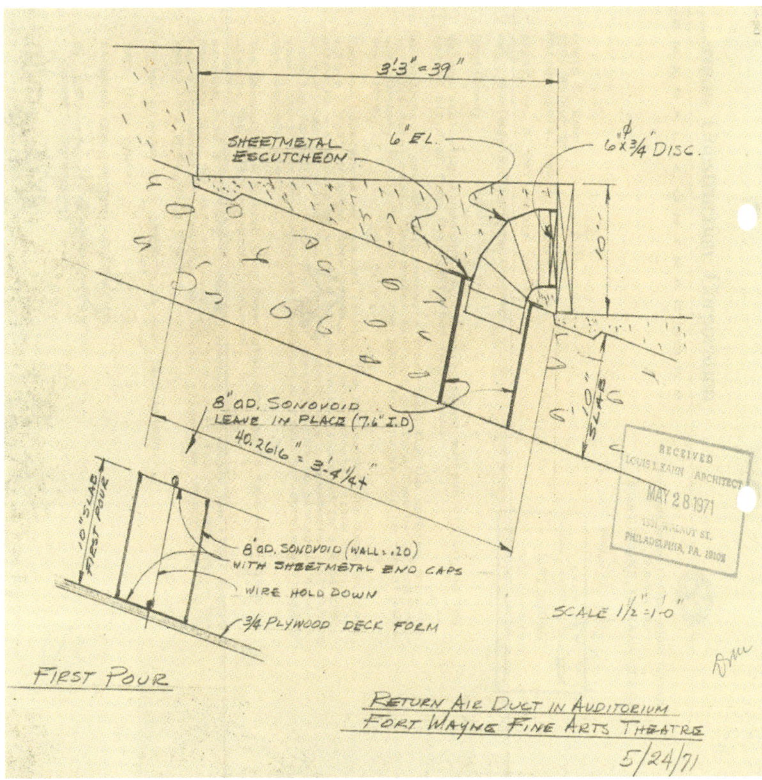

451-455. Fine Arts Center, School, and Performing Arts Theatre (now Performing Art Center), Fort Wayne, Indiana, 1959-73.

451-452. Construction site photographs (LIK, 030.IV.A.605.3.11; 21).

methods, from sand-blasting to abrasive paper, carborundum, and detergents.[183] The cleaning is eventually done with carborundum.

Grand reinforced concrete T-beams form the roof around the Auditorium, visible from the foyer and the banquet hall. These beams are prefabricated by Masolite Concrete Products, Inc., based in Fort Wayne, and installed on the worksite with a crane.[184] The spaces around the Auditorium are powerfully characterized by the textures and colors of the various materials of the structure, always left visible, without cladding or coating: concrete block walls, clay brick walls, beams in precast concrete. The effect is still that of a primordial construction. Yet again, in the Performing Arts Theater, Kahn's aspiration to find the true nature of concrete clashes, as happened previously in the Tower X and the service towers of the Richards Laboratories, with the choice of disguising the reinforced concrete walls of the colossal fly tower behind a layer of clay bricks.[185]

[183] "Fred Langford. Concrete (remove red color & finish)," handwritten notes, 1 November 1972, LIK, 030.II.18.32.
[184] Bruce R. Wolfe, of Masolite Concrete Products, Inc., letter to James Schafenacker, 16 February 1972, LIK, 030.II.A.19.16. For the booklet on the T-beams see, Prestressed Concrete Division, of the Construction Products Corporation, Lafayette, Indiana, *Precast-Prestressed concrete tees. Long Span, Roof and Floor Construction*, n.d., (LIK, 030.II.18.32).
[185] See the worksite photograph, LIK, 030.IV.A.605.3.21.

453-454. Views (photographs be the author, 2010).

455. View during the construction site
(LIK, 030.IV.A.605.3.39).

chapter six
different forms of exposed concrete

Towards the "glass-concrete" for the Memorial to the Six Million Jewish Martyrs

The color and texture of concrete and the reaction of its surfaces to light have been explored by Kahn in a range of extraordinarily varied solutions. When Kahn says that the colors of the architect are the materials, deployed with primal, essential workmanship, he always refers to their capacity to shape light. Often the concrete surface becomes essential to reflect artificial or natural light, from the tetrahedral coffers of the Yale Art Gallery to the domes of the Washington Library, to the cruciform roof of the church in Rochester, and the cycloid shells of the Kimbell Art Museum. When he is designing the cycloid vaults, in a lecture on 14 November 1967 Kahn describes the particular color of the light reflected on the concrete as "luminosity of silver."[1] The succinct considerations regarding the Kimbell Art Museum in that lecture focus on the "glow of silver" generated by the concrete and the counterpoint of light created with the three courts cut out of the vaults, the Green, Yellow and Blue Courts, "named for the kind of light" that "their foliation, or their sky reflections on surfaces, or on water, will give."[2] That same year, Kahn is making the first development drawings for the most exceptional monument he has ever designed, one with a lyrical intensity comparable to that of the Danteum of Terragni: the Memorial to the Six Million Jewish Martyrs, to build in Battery Park, at the tip of Manhattan, in the form of prisms to erect with glass blocks (the project is abandoned in 1972). If smooth concrete, in Kahn's hands, becomes a surface reflecting

[1] Louis I. Kahn, *Space and Inspirations*, 14 November 1967, now in, Latour, *Louis I. Kahn: writings, lectures, interviews* cit., (pp. 224-230), p. 228.
[2] *Ibid.*

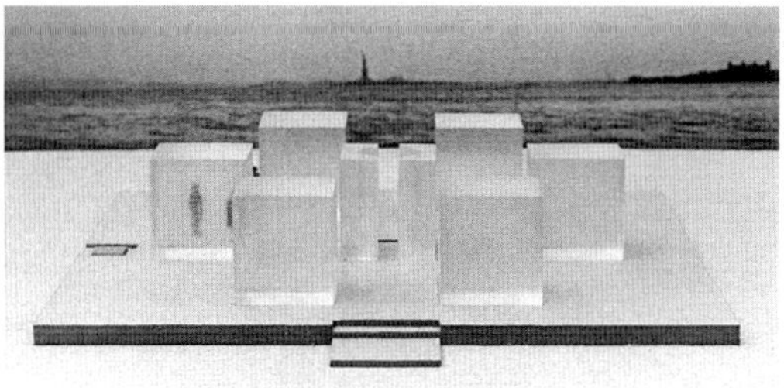

456. Project for Memorial to the Six Million Jewish Martyrs, New York, New York, 1966-72.
Model (LIK, 030.IV.A.690.1.33).

457. Slab found in the catacombs of the Necropolis of Beth She'arim in Israel (R. H. Brill, J. F. Wosinski, *A Huge Slab of Glass in the Ancient Necropolis of Beth She'arim*, in *International Congress on Glass, Comptes Rendus*, Charleroi, Brussels, Institut National du Verre, Fédération de l'Industrie du Verre, 1965, LIK, 030.II.A.36.7).

natural light, now – thanks to the glass blocks of the Memorial – the construction itself becomes a source of light, transparency and evocative impact, achieving a new form of block and hollow column. In the lecture in 1967, immediately after the colors of the light of the Kimbell Art Museum, Kahn investigates the "shadow of white light." The Memorial is the quintessence of the conception of light pursued for the Kimbell Art Museum; it is a true device for contemplation of the phenomena of natural light. "Natural light has the varied mood of the time of the day and the seasons of the year," he says in the lecture.[3]

The glass blocks of the Memorial, for Kahn, also represent the goal towards which concrete must strive in order not to be limited to being a surface that reflects light; they are the sublimation of the concrete blocks. Kahn considers producing a special kind of translucent material, to develop in collaboration with Corning Glass Works, and equipped with a secret archaeological meaning. He is very interested in the analysis conducted by Dan Barag, of the Hebrew University, on a slab found in the catacombs of the Necropolis of Beth She'arim, in Israel, research spread and discussed by the chemist and archaeologist Robert H. Brill, an employee of the Corning Museum of Glass in Corning, and by John F. Wosinski, of the staff of the Refractories Services of Corning Glass Works. The analysis shows that the slab was made with a glass compound. Brill and Wosinski also produce specimens, starting with the discovery of the composition of the slab.[4]

In the report on the analysis of the material from Beth She'arim, dated 27 April 1964 and also sent to Kahn, Brill and Wosinski expressly write about a "sample of 'glass/concrete'" and offer a disquisition on mortars, cements and concrete, from those of the ancient Egyptians to the *opus puteolanus* of the Romans, to demonstrate the differences between the Israeli slab from those products and from contemporary concrete, in spite of the apparent similarities of the compounds. But it is precisely that ambiguous "little resemblance" between the "Beth Shearim slab" and the "modern cements and concretes"[5] that prompts Kahn to manufacture the glass blocks that take the place of the concrete blocks used in works like the Bath House.

Crude appearance of the concrete for the Olivetti Underwood Corporation

The invention of a form of structure in reinforced concrete, always different to be the most suitable for every program and every landscape, becomes the goal of the new project of Kahn. Because it possesses no form, concrete constantly returns as the ideal material to find the identity of a program and a building. In this perspective, the formwork increasingly takes on the value of a tool for the "design" of the idea of the architecture itself.

For the headquarters in Harrisburg of one of the most prestigious manufacturers of typewriters and computers, the Olivetti Underwood Corpora-

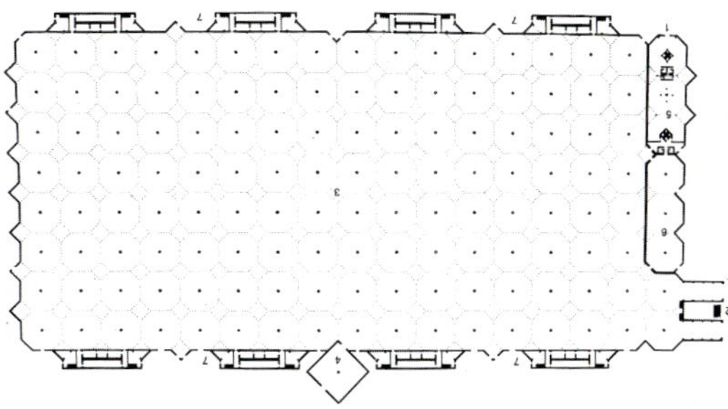

458. Project for Hurva Synagogue, Jerusalem, 1967-74.
Section sketch, showing an open vaulted version of the inner chamber (LIK, 030.I.A.755.33).

459-460. Olivetti-Underwood Factory, Harrisburg, Pennsylvania, 1966-70.

459. Reflected ceiling plan (H. Ronner, S. Jhaveri, eds., *Louis I. Kahn Complete Work 1935-1974*, Basel, Birkhäuser, 1987, p. 326, No. 17).

460. Study sketch (LIK, 030.IV.A.735.4.1).

tion, a commission received in 1966, Kahn invents a structural unit in reinforced concrete that will "be flexible enough to fit the company's sense of alertness to change."[6] This unit takes the form of an umbrella, similar to that of the Parasol House studied with Oscar Stonorov. Thanks to the collaboration of Komendant and Gianopulos, still operating on behalf of Keast & Hood, the definition of the structural unit, during the course of 1967, takes on definitive features, though at the start of the year the material – steel or reinforced concrete – has not yet been decided.[7] For the solution of the structure formed by columns positioned in a grid of 20 × 20 inches, and each topped by a "concrete reversed umbrella,"[8] an estimate is requested from Barclay White & Company of Mechanicsburg, near Harrisburg. This is the contractor hired for the construction.[9] It is no coincidence that in the project for the Hurva Synagogue, designed in 1967 and 1968, Kahn makes use of an umbrella structure in exposed concrete, conceived as a monumental variation on the one for the headquarters of the Olivetti Underwood Corporation.

[3] *Ibid.* The phrase is repeated in the text for the Memorial (see Louis I. Kahn, "The Memorial to Commemorate the Six Million Jewish Martyrs," 3 May 1968, LIK, 030.II.36.6).
[4] Robert H. Brill, John F. Wosinski, *A Huge Slab of Glass in the Ancient Necropolis of Beth She'arim*, in, *International Congress on Glass, Comptes Rendus*, Charleroi, Brussels, Institut National du Verre, Fédération de l'Industrie du Verre, 1965, pp. 1-11. Kahn also has two photographic prints of the slab of glass in the catacomb, received from the The Corning Museum of Glass of Corning, New York, where Brill and Wosinski work (LIK, 030.II.36.6).
[5] "All of the evidence above leads one to the conclusion that at least this part of the slab was originally manufactured as a glassy material" (Robert H. Brill, John F. Wosinski, "Laboratory examination of sample of 'glass/concrete' slab from Beth Shearim, Israel," 27 April 1964, LIK, 030.II.36.7).
[6] *Cf.* Ronner, Jhaveri, Vasella, *op. cit.*, p. 335.
[7] Anon., "Factory construction," n.d., [January 1967], LIK, 030.II.7.4. At the meeting of 4 January, "Mr. Kahn presented both a concrete and a steel proposal and reviewed, in general, the characteristics of each" (John E. MacAllister, "Meeting of January 4, 1967," LIK, 030.II.7.14).
[8] Luigi Gandi, "Meeting with Messers. Kahn, McAllister and Scott in Philadelphia on February 20, 1967," LIK, 030.II.7.32.
[9] "Outline Specifications," n.d., [February 1967], LIK, 030.II.7.93.

461-462. Olivotti Undorwood Factory, Harrisburg, Pennsylvania, 1966-70.

461. "Wall Sections," 5 June 1967 (LIK, 030.I.C.735.4).

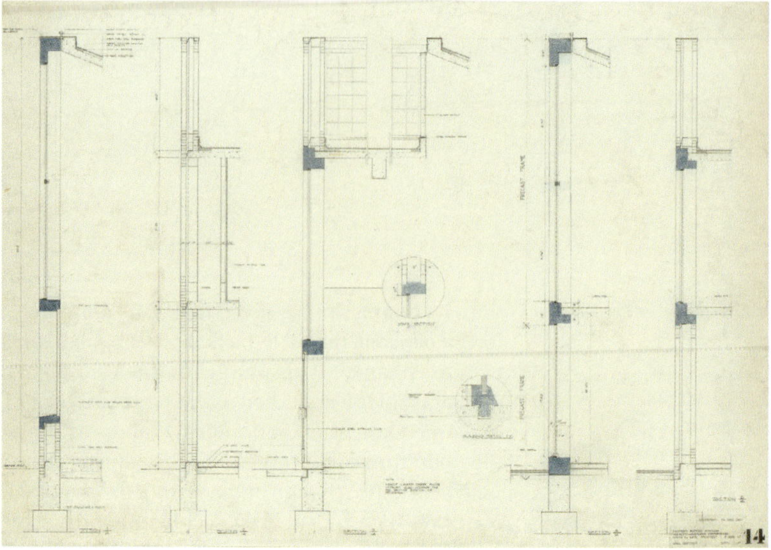

The first significant indications on what will become the poured in place and precast concrete of the Olivetti Underwood building in Harrisburg date back to the start of June 1967 and are contained in the "Outline Specification," which describes a building wrapped by red brick, with the frames for the openings in reinforced concrete to be prefabricated on the worksite, the partitions of the rooms to be made in Waylite concrete blocks, and a grand structure in exposed reinforced concrete for the vast area of the factory, with precast elements for the roof. The alternative of using "standard size Waylite concrete brick" instead of clay brick for the enclosure already points to the final chromatic effect.[10]

The writers of the "Outline Specification" cover the fabrication of the formwork for the concrete to be poured in place, describing certain principles that would be unusual in other works by Kahn for their accent on the practical aspects of the work, even with the indication to avoid any drawings for the assembly of the panels: "The most economical and structural layout possible without regard to symmetry should be submitted first."[11] It is clear that at least in the preliminary phase of the design of the concrete, Kahn wants to explore the expressive potential of a primordial formwork system, to make without any logic of pattern, on the basis of drawings prepared by the contractor, also to express the character of the factory directly in the work, with the imprints on the concrete.

[10] "Olivetti Underwood Corporation. Documents prepared for design development phase. Outline specification," 16 June 1967, LIK, 030.II.1.7.

[11] "Plywood with polyurethane coating," we read. "No chamfer in forms. Plywood panels to be planned on a drawing by contractor and submitted for architect's comments and approval as to location of joints between panels. The most economical and structural layout possible without regard to symmetry should be submitted first. The construction of forms is to be precise and careful for the purpose of speeding erection, stripping and relocating. After stripping, the concrete is not to be touched by concrete finishers. No honeycombs are to be filled, no fins are to be ground nor erased. Fins may be broken accidentally in places while stripping. The natural appearance produced by leaving concrete as stripped is what is wanted for appearance as well as monetary savings" ("Olivetti Underwood Corporation. Documents prepared for design development phase. Outline specification," 16 June 1967, LIK, 030.II.1.7). In the version of the specifications dated 5 June 1967 and revised on 14 June other indications have to do with the placing of the precast concrete, and more information is added on the formwork.

[12] See "Static Computations, Factory Building for the Olivetti Corp. Prismatic Shells," May and June 1967, LIK, 030.II.8.31.

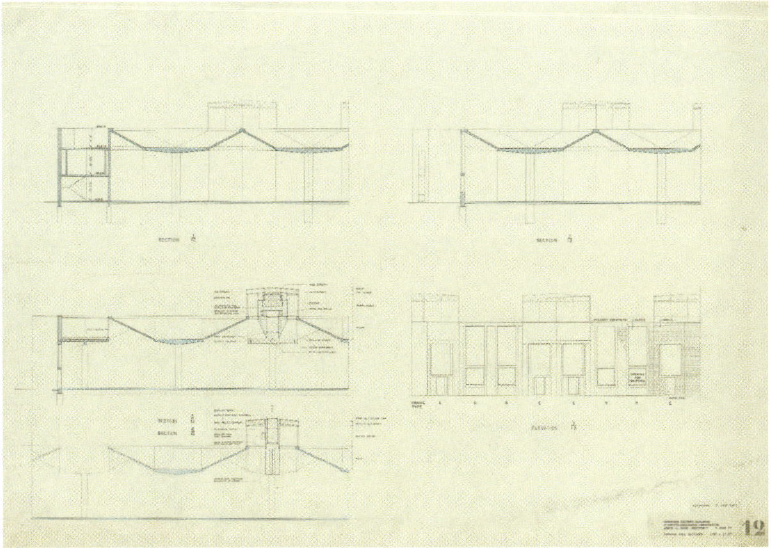

462. "Service Wall Sections," 5 June 1967 (LIK, 030.I.C.735.4).

The fact that the word "first" is written regarding the economical layout would seem to indicate that Kahn intends to intervene, secondly, with his own design, but one that is based on the technical parameters supplied by the contractor.

It is significant that what is defined as a "natural appearance" in the draft of the "Outline Specification" is instead indicated as "somewhat crude appearance," using this expression to underline the fundamental difference between the concrete Kahn wants for the Olivetti Underwood factory and his architectural concrete. Perhaps by pointing to the goal of a "somewhat crude appearance," Kahn wants to explore every absence of expressivity in the assembly of the formwork, ranging well beyond that degree zero already pursued in other works.

For the structure of the envelope, Kahn studies a system of pillars and beams in precast concrete, to panel with glass or with exposed brick. The precast parts are not the individual pieces of the structure (beams and pillars), but large frames that can be positioned on the worksite to create a pattern of compartments for the infills. Kahn has already invented similar precast composite structural elements, for the design of the Washington University Library. Although in the constructed work made in Harrisburg the frames are replaced by single posts, the proposal for the factory of the Olivetti Underwood Corporation confirms that after the experience of the Richards Laboratories, Kahn considers precasting a construction process based on the use of cranes on the worksite, making it possible to design parts of the skeleton with a different articulation than what is made possible by poured in place concrete.

Research on formwork

Kahn, Komendant and Gianopulos define the structural unit of the Olivetti Underwood factory in detail, indicating the type of metal reinforcements and the type of formwork.[12] The question of the mode of construction of the concrete, poured in place or precast, continues to be examined in mid-1967. At that time the solutions for the "roof structure" to be made in 64 modules

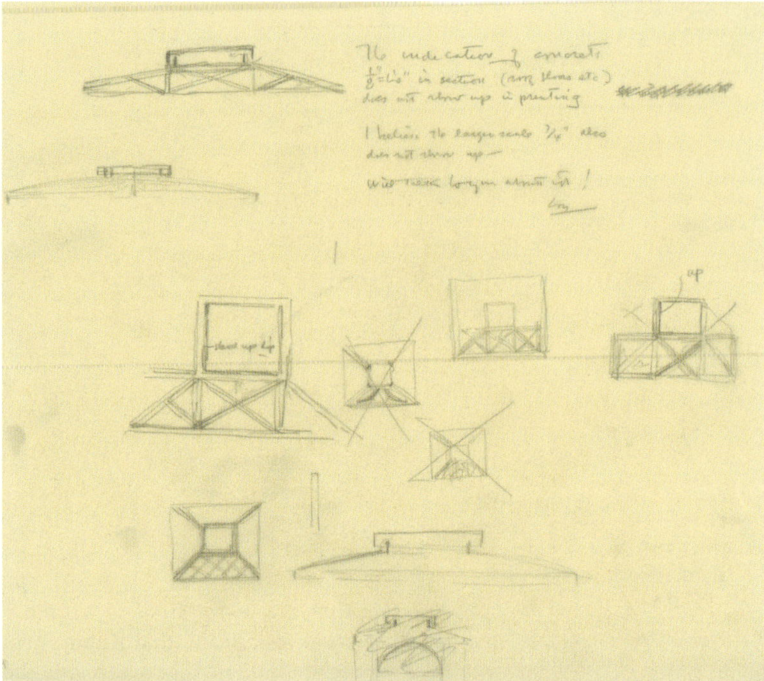

163. Olivetti Underwood Factory, Harrisburg, Pennsylvania, 1966-70. Study sketches for the umbrella structure (LIK, 030.I.A.735.4).

call for the alternative between a roofing unit called the "Shell Scheme," like an overturned umbrella, and one called the "Pyramidal Scheme," derived from the design for the Jewish Community Center[13]. The Shell Scheme and the Pyramidal Scheme correspond to two different concrete construction systems, the first to be poured directly in place, the second to fabricate in the workshop and assemble at the site – it includes "precast prestressed elements."[14] Barclay White & Company is charged with investigating the feasibility of both units, by contacting various contractors specialized in concrete construction. The cost estimate for the Shell Scheme covers an operation of finishing of the "underside of concrete" after stripping, which nevertheless could be avoided "if the Architect is willing to accept the surface available with the use of plywood forms uncoated and unlined as the stripping operation leaves it."[15]

For the cost estimate of the Pyramidal Scheme, Schuylkill Products, Inc., based in Cressona, Pennsylvania, specializing in precast concrete, is contacted, as well as the Formigli Corporation, which has already produced the precast parts of the Tribune Review Building and the Biology Building.[16]

According to Komendant, in the case of the Shell Scheme the "columns and capital panels can be precast," while all the elements of the Pyramidal Scheme are part of a "precast-prestressed system": "The pyramids are poured on the floor level, using only a few molds, and erected by crane."[17]

Kahn, Komendant and Gianopulos choose the Shell Scheme and develop all the technical details of the "Basic Factory Roof Unit." The structural unit is composed of a square column, a "distribution slab," also square, and a "roof shell" with inclined planes inserted in an octagonal perimeter.[18] The "distribution slabs" are also called "caps" or "column caps." The column, capital and shell contain the conduits for the post-tensioning bars. The configuration of the capital, which has the upper part recessed along the border, is specially developed in order to do the pour of the inclined planes after the construction of the column and capital system (the irons of connection with the inclined planes and the ends of the post-tensioning bars emerge from that recessed strip). Thanks to this setback, the joint of the pouring of the inclined planes can be controlled, while after their construction the load-bearing part of the fundamental structural unit of the whole building remains visible.

The construction of the Roof Unit happens in two phases: first the erection of the columns with the capital, their post-tensioning and stripping; then the construction of the shell, with its post-tensioning and stripping (the post-tensioning phase in both cases is done when the concrete has reached a strength of 4000 psi).

A specification on the formwork of the structure offers a glimpse of the intention of Kahn and his staff to make use of a particular system of precast parts: "A proposed sequence of planned construction based upon the use of a limited number of prefabricated forms is to be made available for the contractor on 20 June 1967. It is anticipated that this information will assist the contractor in devising the most economical sequence of work."[19]

We cannot rule out the possibility that in the phases of development of the construction system for the structure of the factory, Kahn has considered the hypothesis of using the fiberglass molds produced by Kreier, since among the documents on the project for the Olivetti Underwood Corporation we find a booklet, *George Kreier, Jr., Incorporated Concrete Molds*, which shows some of the most important architectural structures built in the United States using fiberglass molds. Such works done by Kreier's firm include the ones designed by Edward Durell Stone for the State University of New York at Albany, with a mushroom structure similar to the one designed by Kahn – the Service Building Complex, Student-Faculty Building and Dormitory Unit One.

A drawing titled "Time estimate and forming schedule for roof construction," dated 23 June 1967, shows the elements of the scaffolding and formwork, and illustrates the construction phases of the "column and distribution slab" and the shell, called "Roof Slabs." The mobile compartments of the formwork of the distribution slab are four in number; those of the inclined planes of the shell are four squares and four triangles.

In October and November the specifications[20] are further developed, while Komendant works on the design of the reinforcement of the "Prismatic shell," confirming the two construction phases (column with capital and shell).[21]

[13] See *The Louis I. Kahn archive: personal drawings* cit., vol. V, p. 290, No. 735.4.

[14] Barclay White, Jr., "Memorandum re: Olivetti Underwood Corporation. Cost Estimates for Concrete Roof Structural Schemes Based on August E. Komendant Drawing dated 5/21/67," 12 June 1967, LIK, 030.II.7.35.

[15] *Ibid.*

[16] *Ibid.* For the various solutions of the structural system see Komendant, *Contemporary Concrete Structures* cit., p. 493.

[17] Komendant, *18 Years with Architect Louis I. Kahn* cit., p. 96.

[18] This is the terminology used by Gianopulos (Nicholas L. Gianopulos, letter to Barclay White & Company, 29 April 1968, LIK, 030.II.7.24).

[19] "Olivetti Underwood Corporation. Documents prepared for design development phase. Outline specification," 16 June 1967, LIK, 030.II.1.7. See also "Outline of Construction requirements for Basic Factory Roof Unit," n.d., [June 1967], handwritten manuscript, LIK, 030.II.7.97.

[20] See "Construction Specifications. Factory Building for Olivetti-Underwood Corporation," 4 October 1967, LIK, 030.II.8.46.

[21] See the drawing "Prismatic shell," 25 October 1967, LIK, 030.I.C.735.012. The indications supplied by Gianopulos confirm the sequence of the pour of the load-bearing elements of the structural unit: "Columns and Distribution Slabs for Roof Shells are to be poured in a single operation. A 'tremie-like' device will be necessary for placing concrete in the columns" (Nicholas L. Gianopulos, letter to Barclay White & Company, 29 April 1968, LIK, 030.II.7.24).

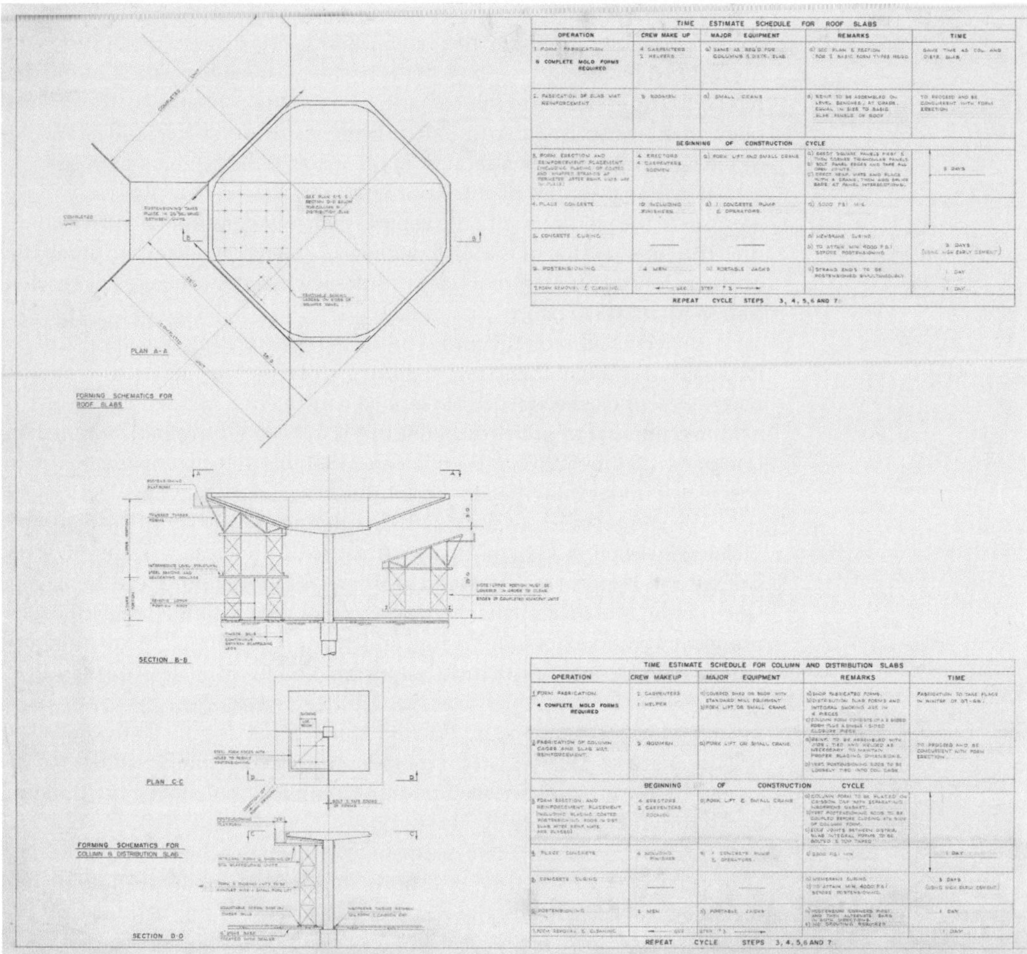

For the formwork, it is decided to use plywood for the "primary roof structure," while for the "office slabs" and the columns the choice is to treat the plywood with "Toch Brothers RIW plastic form coating (polyurethane)."[22]

The construction of the factory is assigned to Barclay White & Company. The poured in place concrete will be done by Lambert & Intreri, Inc., based in Harrisburg, while for the precast concrete the Camden Lime Company is hired, based in Camden, New Jersey.

Unlike all other constructions by Kahn, in this case the mix design of the concrete is not done to obtain a particular type of color (the compound includes the use of Pozzolith). It is the industrial function of the building that prompts this unusual silence on Kahn's part regarding an aspect that has always been considered crucial in his other works. The trials for the strength of the concrete are done in December, by Astrotech, Inc., located in Harrisburg.[23]

[22] "Addendum No. 1. Factory Building for Olivetti-Underwood Corporation," 8 November 1967, LIK, 030.II.7.70.

[23] Edward A. Meister, of Astrotech, Inc., letter to Lambert & Intreri Contractors, 2 December 1967, LIK, 030.II.7.55.

464-466. Olivetti-Underwood Factory, Harrisburg, Pennsylvania, 1966-70.

464. "Time Estimate and Forming Schedule for Roof Construction," 23 June 1967 (LIK, 030.I.C.735.3).

465. "Prismatic Shell," 25 October 1967 (LIK, 030.I.C.735.12).

466. "Reflected Plan of Concrete Shell Showing Plywood Sizes," 20 May 1968 (LIK, 030.I.C.735.14).

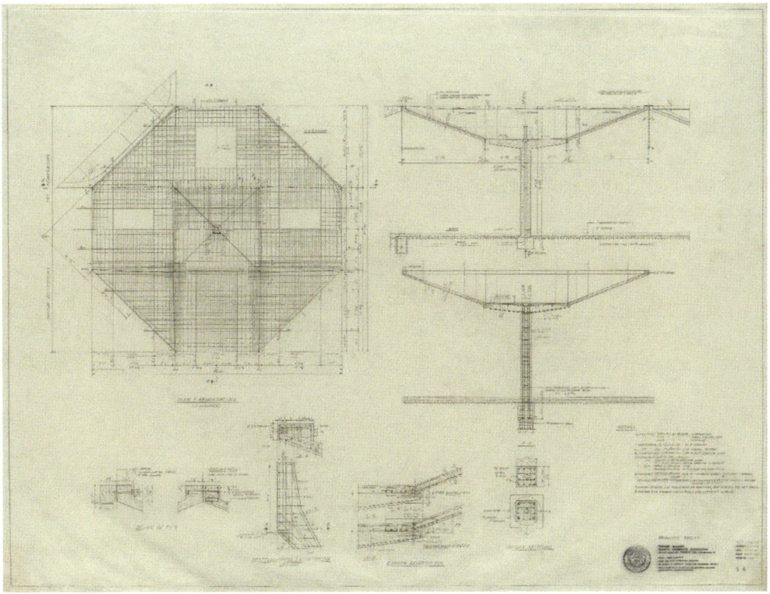

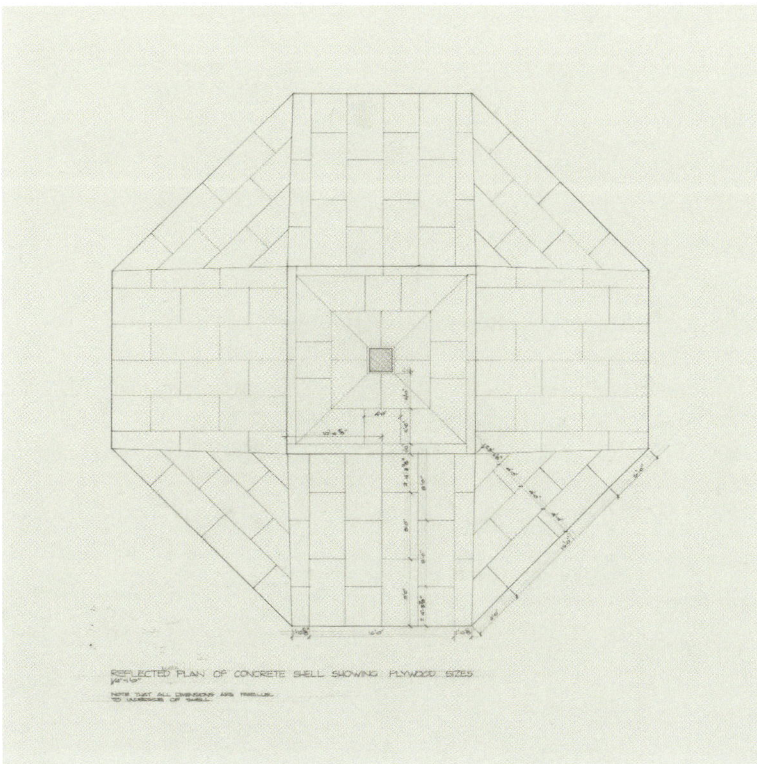

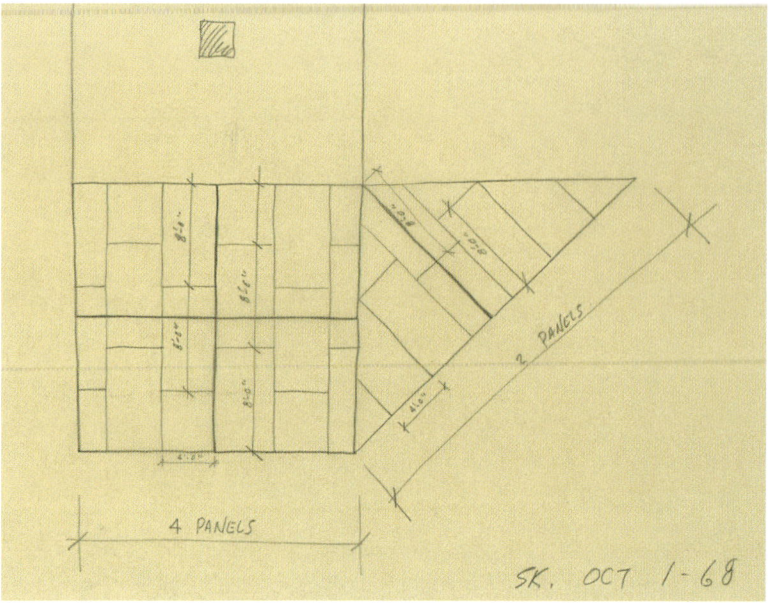

467-471. Olivetti-Underwood Factory, Harrisburg, Pennsylvania, 1966-70.

467. "Reflected Ceiling Plan of Concrete Shell," 1 October 1968 (LIK, 030.I.C.735.9).

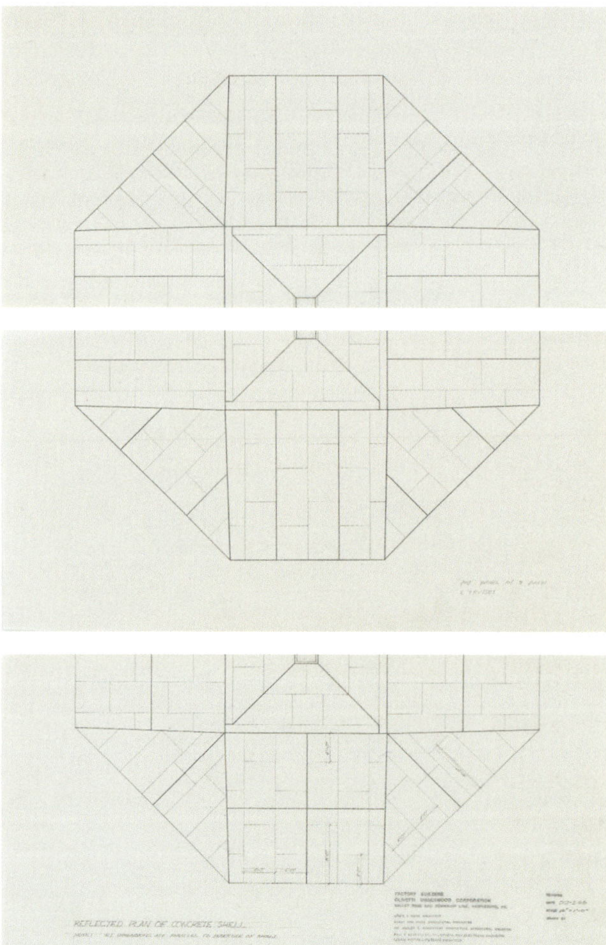

468-470. "Reflected Plan of Concrete Shell," 2 October 1968 (LIK, 030.I.C.735.4).

471. Photograph showing the beams with threaded bolts and nuts for the housing and attachment of the bases of the prefabricated pillars (LIK, 030.IV.A.735.3.134).

Kahn agrees to allow Barclay White & Company to use the special formwork produced by the Simons Corporation and composed of plywood panels mounted on a steel structure. These formwork is used to make the "grade beams," columns and "column caps," according to the positions of joints between the panels indicated by Kahn.[24]

As opposed to the initial indications, at this point Kahn is interested in specifying a pattern for the assembly of the panels. From May to October 1968 his team examines the possible patterns of the assembly of the formwork panels, starting with 8 × 4 feet panels. The variants have to do with the grouping of the plywood panels that form the large compartments of the inclined planes (in the drawings these compartments are also called "panels"), generating joints that are different from those of the assembly of the plywood panels. The two fundamental cases depend on the position of the joints between the compartments: running along the junctions made by the plywood panels, or following the logic of subdivision of the surface of the inclined planes.

Post-tensioned columns and capitals: defects in the concrete

The works in concrete for the factory of the Olivetti Underwood Corporation begin with the construction, in the spring of 1968, of the edge beams that form the visible summit of the walls. The beams have hollows, with threaded bolts and nuts, for the housing and attachment of the bases of the prefabricated pillars.[25] The detail is similar to the one utilized for the Richards Laboratories. The pouring of the beams requires particular care to obtain surfaces as similar as possible to those of the precast concrete. On 9 May, Kahn makes a special trip to the worksite to inspect the formwork of the first segments of the edge beams.[26]

During June the last decisions are made on the construction of the concrete columns and capitals. Barclay White & Company informs Lambert & Intreri, Inc., on the operations to be done, also forbidding any "external vibration." Each pour is to have a height of 3 feet, and has to be vibrated from the inside.[27]

The pouring of the columns begins on 1 July. The first capital is made on 26 July and corresponds to column G-1.[28] The stripping of this capital and this column becomes the example of the operations to follow in the other cases. Gianopulos explains the phase sequence to Barclay White & Company, in keeping with methods already envisioned since June 1967: "The capital form for Column G-1 may be presently removed in quarters, provided that each section is firmly reshored with two adjustable posts at each side."[29] The temporary supports positioned near the corners are composed of steel tubes (not planned in 1967).

The first tests of the strength of the concrete pose a series of questions that keep the engineers and technicians busy for months to come. Following the tests done at the start of August on the G1 capital and column,

[24] Maury E. Bassett, Jr., letter to Louis I. Kahn, 5 April 1968, LIK, 030.II.7.55. A document regarding the project of the extension of the factory describes "3/4" plastic surfaced plywood" of the Georgia-Pacific Plywood, Co., "G.P.X." ("Specifications for Additions to the Present Building Olivetti Corporation of America," n.d., LIK, 030.II.83.33).

[25] See the worksite photographs, LIK, 030.IV.A.735.3.

[26] Maury E. Bassett, Jr., "Minutes of Meeting No. 6 held Thursday, May 9, 1968," LIK, 030.II.7.42.

[27] Maury E. Bassett, Jr., letter to Lambert & Intreri Contractors, 28 June 1968, LIK, 030.II.8.28.

[28] Lambert & Intreri Contractors, "Olivetti Underwood Corp. Factory Building, Table A. Pouring dates for 6500 psi columns and column capitals," 6 August 1968, LIK, 030.II.8.30.

[29] Nicholas L. Gianopulos, letter to Barclay White & Co., 6 August 1968, LIK, 030.II.7.56.

Astrotech Laboratory reports that the concrete needs more time to reach adequate strength for the post-tensioning of the bars. One of the causes is identified as the increase of the quantity of water, over what was stipulated in the mix design.[30] It is decided to extend curing time in already poured columns and capitals, and to replace the Pozzolith with Plastiment in the mix design of the concrete as yet to be poured.[31] The trials with this new retarder are done on 7 August, and pouring of the columns resumes on the 12th; the first capital with the new mix design, in column G-7, is made on 13 August.[32]

Also to check the strength, still considered inadequate, of the concrete of capitals G-1 and E-3, Gianopulos has other testing done on those capitals, and has core bores taken in the upper part of the columns.[33] The tests continue through the month of August, demonstrating that the concrete does not reach the expected strength, even when curing is prolonged (only 16 of the 65 tests report satisfactory results).[34]

As a result of the inadequate strength of the concrete, the hypothesis is considered of reinforcing the column shafts by means of the addition of a "concrete casing around col[umn]s,"[35] with a thickness of 5 inches on each side, and the resulting increase in the size of the "defective columns" from 30 × 30 inches to 40 × 40 inches.[36]

Kahn is worried about the consequences of such a decision for the overall design of the factory: "It is important that all columns be of the same dimension. The new dimension adopted will make conditions around them consistent."[37] The Olivetti Underwood Company authorizes the corrections, though it emphasizes the loss of "usable floor space" and the "problems with walls and conveyors," while agreeing with what Kahn asserts, and therefore the company wants "all columns in the building," and not just the "defective" ones, to have an "equal size by adding the 5" mantle."[38] The drawing for the mantle to build around the columns, identified with the code SK 492-10, dates back to 29 August, and is done by Gianopulos. "[…] but formwork joints should be reviewed with the Architect before proceeding with that work," Gianopulos indicates.[39]

The problem of the lack of strength of the concrete of columns and capitals to be post-tensioned continues to remain unsolved.

A company specialized in concrete, Hempt Bros., Inc., based in Camp Hill, Pennsylvania, is called in to analyze the defective compound used and to suggest improvements. Hempt Bros., Inc., advises not simply replacing the Pozzolith with Plastiment, though this has "brought new concrete up to required strength,"[40] but also altering the percentages of the various mix design components (starting with the cement), in order to make the concrete stronger.[41]

472-474. Olivetti-Underwood Factory, Harrisburg, Pennsylvania, 1966-70.

472. Construction site photograph (LIK, 030.IV.A.735.3.88).

473. Construction site photograph (A. E. Komendant Archive, AAUP, 027.III.33a).

474. Construction site photograph (LIK, 030.IV.A.735.3.1).

On 19 September, at a meeting attended by Komendant, Gianopulos and representatives of Barclay White & Company, Lambert & Intreri, Inc., and the Olivetti Underwood Company, it is decided that among the columns made prior to 12 August, and therefore with the compound still containing Pozzolith, only those whose concrete has a strength of less than 5500 psi should be reinforced with the 5 inches mantle, to be made with concrete with a strength of 6000 psi. Keast & Hood is assigned the task of specifying the technical methods for the construction of this mantle.[42] The decision does not take the requests of Kahn and of the vice-president of the Olivetti Underwood Company to have all columns of equal size into account.

The consultation with laboratories specialized in analysis of concrete continues in October, when new results reported by Certified Testing Laboratories, Inc., of Bordentown, New Jersey, confirm that the 21 columns with concrete below 5500 psi "should be mantled with composite encasements of 6000 psi reinforced concrete," while the three "capitals with less than 4500 psi are not acceptable and should be removed."[43]

The question of the mix design is again addressed by Hempt Bros., Inc., proposing a 6500 psi concrete for the mantle and 5000 psi for the shell, to make with Pozzolith.[44] Hempt Bros., Inc., itself has proven that the retarder used to replace the Pozzolith, namely Plastiment, is still not suitable to proceed with post-tensioning operations.[45]

[30] Nicholas L. Gianopulos, letter to Barclay White & Co., 12 August 1968, LIK, 030.II.8.30.
[31] Nicholas L. Gianopulos, letter to Barclay White & Co., 6 August 1968, cit.
[32] Maury E. Bassett, Jr., "Minutes of Meeting No. 19 held Wednesday, August 15, 1968," LIK, 030.II.7.42.
[33] Nicholas L. Gianopulos, letter to Barclay White & Co., 15 August 1968, LIK, 030.II.7.56.
[34] Nicholas L. Gianopulos, letter to Barclay White & Co., 27 August 1968, LIK, 030.II.7.56.
[35] Anon., handwritten notes, n.d., [end of August 1968], LIK, 030.II.7.43.
[36] Vice-President of the Olivetti Underwood Company, letter to Barclay White & Co., 16 September 1968, LIK, 030.II.7.58.
[37] Louis I. Kahn, letter to Maury E. Bassett, Jr., 30 August 1968, LIK, 030.II.7.43. The letter bears the handwritten note "hold. Not sent."
[38] Vice-President of the Olivetti Underwood Company, letter to Barclay White & Co., 16 September 1968, LIK, 030.II.7.58.
[39] Nicholas L. Gianopulos, letter to Barclay White & Co., 6 November 1968, LIK, 030.II.8.30.
[40] Anon., handwritten notes, n.d., [end of August 1968], LIK, 030.II.7.43. 3362
[41] Maury E. Bassett, Jr., "Minutes of Meeting No. 20 held Thursday, September 5, 1968," LIK, 030.II.7.42.
[42] Nicholas L. Gianopulos, letter to Barclay White & Co., 19 September 1968, LIK, 030.II.7.56. Komendant approves the planned construction sequences and procedures (Maury E. Bassett, Jr., "Minutes of Meeting No. 22 held Thursday, September 19, 1968," LIK, 030.II.7.42).
[43] Nicholas L. Gianopulos, letter to Barclay White & Co., 6 November 1968, LIK, 030.II.8.30.
[44] Maury E. Bassett, Jr., "Minutes of Meeting No. 28 held Thursday, October 31, 1968," LIK, 030.II.7.42.
[45] Harold A. Maxwell, of Hempt Bros., Inc., letter to Anthony T. Intreri, 28 October 1968, LIK, 030.II.28.

The construction of the shells

At the start of September 1968 not one of the shells of the Olivetti Underwood factory has been made, due to the delays caused by the strength testing of the concrete of the columns and capitals. The question of the mix design also extends to the shells. Barclay White & Company asks Astrotech, Inc., and Hempt Bros., Inc., to "produce a revised design mix for the 5000 psi concrete," suitable for pumping, to apply in the shells. But even with the new mix the pumping operations meets with difficulties. After having considered the hypothesis of going back to the compound used towards the middle of August,[46] to solve the problem of the pumping Lambert & Intreri, Inc., purchases a machine capable of using the mix design proposed by Hempt Bros., Inc.[47]

The start of the construction of the shells also depends on the completion of the post-tensioning of the capitals of the columns that have already been poured and stripped. This operation begins with column G-7, on 17 September. On that occasion it is decided that the construction of capitals and shells should proceed on all the columns, including the defective ones, with the precaution of not performing post-tensioning on the columns that require the added mantle. The post-tensioning has to be done in the capitals prior to the construction of the shells.[48]

For the construction of the eight inchlined planes that form what is called a shell in the worksite documents, special scaffolding of steel pipes is prepared, anchored to the column by means of ties that start at a collar attached at the top of the shaft.[49] Triangular reticular beams are mounted on this scaffolding with wooden posts attached by means of metal plates – the plywood panels are attached on the beams.[50] The complex framework has to be assembled and knocked down for each shell. For the curing, at the top of the scaffolding, on each side of the inclined planes, large tarpaulins are hung.

475-478. Olivetti-Underwood Factory, Harrisburg, Pennsylvania, 1966-70.

475. Construction site photograph (LIK, 030.IV.A.735.3.51).

On 10 October, when 69 columns and 25 capitals have been made, the first five shells "are in the process of being formed."[51] But the completion of the first shell, that above column G-13, demonstrates the many problems regarding the carrying out and supervision of the work: "Manpower, method of construction, and material handling problems must be anticipated and worked out so that shell construction can attain an acceptable rate of production."[52] To cut down on production times, Lambert & Intreri, Inc., is urged to subcontract the work, to concentrate manpower on the construction of the shells.

After curing, post-tensioning operations are performed on the bars of shell G-13, on 23 October, but they cannot be completed due to the presence of three "honeycomb areas." The repairs of the defective pour are done by removing the three areas, and then patching them with mortar or cement, depending on the size of the holes, followed by curing with burlap and Visqueen.[53] To prevent the formation of "honeycomb areas" in the shells

[46] Maury E. Bassett, Jr., "Minutes of Meeting No. 21 held Thursday, September 12, 1968," LIK, 030.II.7.42.
[47] Maury E. Bassett, Jr., "Minutes of Meeting No. 22 held Thursday, September 19, 1968," LIK, 030.II.7.42.
[48] Nicholas L. Gianopulos, letter to Barclay White & Co., 19 September 1968, LIK, 030.II.7.56.
[49] See the worksite photograph, LIK, 030.IV.A.735.2.
[50] For the design of the structural system see LIK, 030.I.C.735.009. See also the worksite photograph, LIK, 030.IV.A.735.3.
[51] Maury E. Bassett, Jr., "Minutes of Meeting No. 25 held Thursday, October 10, 1968," LIK, 030.II.7.42. At the end of October 39 capitals have been built, while three shells have already been poured and one is slated for pouring on 1 November (Maury E. Bassett, Jr., "Minutes of Meeting No. 28 held Thursday, October 31, 1968," LIK, 030.II.7.42).
[52] Maury E. Bassett, Jr., "Minutes of Meeting No. 25 held Thursday, October 10, 1968," LIK, 030.II.7.42.

476-478. Construction site photographs (A. E. Komendant Archive, AAUP, 027.III.33b; 33c; 33d).

479. Olivetti Underwood Factory, Harrisburg, Pennsylvania, 1966-70. Construction site photograph (LIK, 030.I.IV.A.735.3.7).

not yet built, Lambert & Intreri, Inc., is instructed to do the pouring of the concrete from the bottom up.

When the scaffolding and the formwork have been removed, only the supports positioned near the edges of the capital remain in place, composed of two pairs of wooden posts firmly bonded together. "As work progresses and shell units are completed with adjacent curbs pinned together," Gianopulos explains, "temporary shoring is to be installed at the four contact edges between shells until the concrete floors are in place."[54]

But the formation of gravel clusters in the shells continues to happen, and on 23 October the post-tensioning of shell G-7 also has to be interrupted to intervene, with the same operations applied to shell G-13. Again in this case the defect is blamed on the pouring method of Lambert & Intreri, Inc., which has not taken the recommended precautions. The pour has again been done from above downward.[55]

To be able to continue with the making of the shells, on the basis of analysis it is determined which of the columns will require the added mantle. On this occasion, it is also decided to demolish and rebuild three capitals.[56]

At the end of 1968 two tests are conducted on the concrete of the 21 columns and 3 capitals being challenged, one by the engineers Jacob Feld and Jacob J. Creskoff by order of Lambert & Intreri, Inc.,[57] the other by Ambric Testing & Engineering Associates, Inc., of Philadelphia, by order of Barclay White & Company.[58]

In January 1969 two columns, two capitals and four shells are completed. For the mantle of the defective columns, Gianopulos accepts the pro-

[53] Maury E. Bassett, Jr., "Minutes of Meeting No. 27 held Thursday, October 24, 1968," LIK, 030.II.7.42.

[54] Nicholas L. Gianopulos, letter to Barclay White & Co., 19 September 1968, LIK, 030.II.7.56.

[55] Nicholas L. Gianopulos, letter to Barclay White & Co., 12 November 1968, LIK, 030.II.8.30. See also Maury E. Bassett, Jr., "Minutes of Meeting No. 29 held Thursday, November 7, 1968," LIK, 030.II.7.42.

[56] Following the analysis, the columns requiring the mantle are indicated as those with the names: C1, C4, C5, C6, C7, from D1 to D7, D9, E8, E9, F4, F5, G4, G7, H2 and H4; instead, the distribution slabs E1, E3 and G5 have to be demolished and built again (Maury E. Bassett, Jr., "Minutes of Meeting No. 27 held Thursday, October 24, 1968," LIK, 030.II.7.42).

[57] Anthony T. Intreri, letter to Barclay White & Co., 9 December 1968, LIK, 030.II.7.56.

[58] D. D. Meisel, "Olivetti Underwood Factory Building. Metropolitan Harrisburg, Pa.," n.d., [received by Kahn's office on 11 December 1968], LIK, 030.II.8.28. See also Nicholas L. Gianopulos, letter to Louis I. Kahn, 19 December 1968, LIK, 030.II.7.56.

[59] Maury E. Bassett, Jr., "Minutes of Meeting No. 36 held Thursday, January 9, 1969," LIK, 030.II.7.27.

posal of Lambert & Intreri, Inc., to use a "hammer to chip the surface," instead of the envisioned "sand blasting," though he warns that only an indispensable minimum of material should be removed from the edges of the columns, and that dust should be eliminated with the use of compressed air.[59] He then provides a series of instructions to follow for the pouring of the mantle:

"Columns must be wetted thoroughly so that no suction will take place. Top tie must be more than 1 1/2" from top of column. Curing - Columns must be kept at a temperature of 55° F for at least 6 days and then coated with a curing compound. Where sleeves have been installed, the mantle can be poured through the sleeve up to the capital. The mantle will then be inspected to see if the mantle is tight to the bottom of the capital. If it is not tight the space between the mantle and the bottom of the capital will have to be dry-packed. The dry-packing will have to be done with enough men to do it in one operation. In the case where no sleeves are available, the top of the mantle will have to be dry-packed."[60]

In May 1969 the construction of the reinforced concrete structure is complete; the staircase has yet to be built. During inspections to check on the quality of the work, Gianopulos indicates all the gravel pockets that can have an effect on the shell structure, suggesting methods of intervention.[61] In spite of the fact that the specifications ban any retouching of the concrete after stripping, the contractor makes a list of patching operations to be performed, also for the precast concrete.[62]

Envelope in precast concrete and concrete blocks

While remedies are being studied for the defective columns and pouring continues of the columns and capitals, the construction of the perimeter walls of the Olivetti Underwood factory begins. The tall, slender pillars that simultaneously form the structures of the concrete block walls and the casements for the openings are precast with a base to insert in special housings made in the edge beams. The bases are then bolted to the beams. The erection of the prefabricated struts is planned for the end of August 1968, starting with those of the Mechanical Room,[63] but the first elements produced by the Camden Lime Company do not reach the worksite until 12 September, and are installed that same day.[64] During the inspection of the work done on 19 September, it is observed that "struts and tees erected in the Mechanical Room are not correct."[65]

For the construction of the walls, after having approved the concrete blocks produced by Samson Industries, Inc., based in Media, Pennsylvania (though their excessive cost is then challenged), Kahn approves the new samples made with blocks produced by the York Building Products Company.[66] The "sample block panel" is approved by Kahn and Gustav Langford on 26 September 1968, but with certain reservations regarding the design of the mortar in the joints, to be obtained with a suitable tool:

[60] E. Maury Bassett, Jr., "Minutes of Meeting No. 37 held Thursday, January 16, 1969," LIK, 030.II.7.42.

[61] Nicholas L. Gianopulos, letter to Barklay White & Co., 1 May 1969, LIK, 030.II.7.56.

[62] E. Maury Bassett, Jr., "Minutes of Meeting No. 64 held Thursday, August 14, 1969," LIK, 030.II.7.42. For the treatment of the surfaces with special products see M. A. Bruder & Sons, Inc., "Acra-Bond Exterior Coating," n.d., [1969], LIK, 030.II.7.29, and Robert Marsch, letter to Louis I. Kahn, 7 October 1969, LIK, 030.II.7.29. For the operations to perform on the concrete see Gustav Langford, letter to E. Maury Bassett, Jr., 4 August 1969, LIK, 030.II.10.14.

[63] E. Maury Bassett, Jr., "Minutes of Meeting No. 19 held Wednesday, August 15, 1968," LIK, 030.II.7.42.

[64] E. Maury Bassett, Jr., "Minutes of Meeting No. 21 held Thursday, September 12, 1968," LIK, 030.II.7.42.

[65] E. Maury Bassett, Jr., "Minutes of Meeting No. 22 held Thursday, September 19, 1968," LIK, 030.II.7.42.

[66] E. Maury Bassett, Jr., "Minutes of Meeting No. 19 held Wednesday, August 15, 1968," cit.

480-485. Olivetti-Underwood Factory, Harrisburg, Pennsylvania, 1966-70.

480-482. Construction site photographs (LIK, 030.I.IV.A.735.3.77; 19; 31).

483. View of the canteen
(LIK, 030.IV.A.735.1.15).

484. View of the structure
(photograph by the author, 2010).

485. View of the roof
(LIK, 030.IV.A.735.2.11).

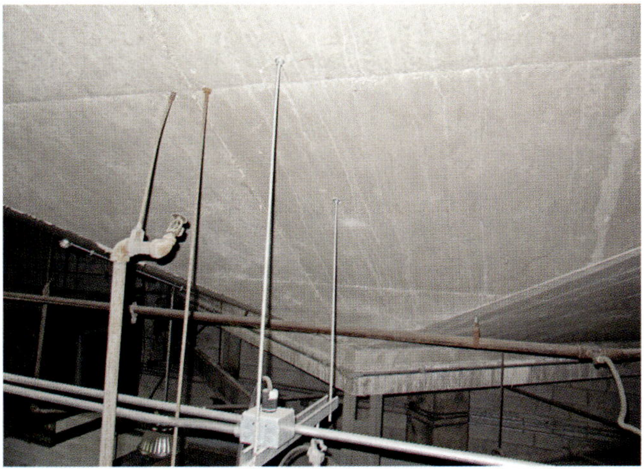

"The horizontal impression to be continuous through the vertical impressions. Also, the end vertical joints between the corner blocks and the precast members are to be routed to a depth of 1/2 inch."[67]

After the erection of the posts, the concrete blocks are laid, and the slender upper beams are mounted over the pillars. The envelope of the Olivetti Underwood factory becomes a chromatic and textural variation of the gray of the concrete.

Frames of life: skeleton and infill for the Inner Harbor complex

Kahn assigns the characterization of the different towers with a square, rectangular and octagonal plan, arranged to create the picturesque effects of European streets and squares, built facing the sea in Baltimore, Maryland, to contain apartments, offices and a hotel, with parking areas and shops, to a skeleton of simple grids with diagonal posts to bear great loads at the basement levels. The complex is called Inner Harbor and it is designed from 1970 to 1973 for a group of investment companies headed by Owen & Company (in this project Kahn includes Keast & Hood for the engineering[68]). The fact that Kahn does not apply the typical American curtain wall solution for the towers is dictated by his constant concern with the visibility of the structure.

The particular pattern of the skeleton is based on the nature of the terrain and the program, which impose – on the lower levels which contain the parking areas and shops – a limited number of columns: "the columns tend to have a certain kind of inhibition to the use of space." What Meyers indicates as the "poetic" principle pursued by Kahn in the Inner Harbor design lies in giving form, on the lower levels, to a "brutal structure" or "long span structure" on which to place a "delicate structure" shaped to correspond to the "nature of rooms," thus obtaining a "composite structure."[69] This is the character that sets the skeleton of the Inner Harbor complex apart from other frameworks left similarly exposed.

Again for this project, Kahn imposes exposed concrete, and starting with the first meetings he discusses the construction details, calling for reuse of the formwork "five to seven times,"[70] describing the buildings in "concrete frame work left in its natural state."[71] Kahn also considers the hypothesis of using the slip-forming system for the concrete of the Office Building.

The discussions involving Meyers, Wisdom and the contractor called in by the developer Howard R. Owen, Morton Bender of the Blake Construction Company of Washington, reveal certain founding principles of Kahn's idea of concrete. Meyers' statements on this subject are significant: "We don't mean bushhammered concrete, we don't mean exposed

[67] Gustav Langford, letter to E. Maury Bassett, Jr., 2 October 1968, LIK, 030.II.8.28.

[68] Marshall D. Meyers, "Baltimore Inner Harbor Project I. Meeting with Howard Owen 22 October 1971," LIK, 030.II.12.15. Kahn's office contracts the Concrete Consulting Corporation of Berkeley, specialized in concrete design, from mix design to samples, to organization and management of the worksite.

[69] "Inner Harbor Project, Baltimore, Maryland," 25 February 1971, LIK, 030.II.12.12.

[70] "Gateway Developers, Inc., Inner Harbor Project I, Baltimore, Maryland," 25 January and 1 February 1971, LIK, 030.II.12.19.

[71] Dubin-Mindell-Bloome Associates Consulting Engineers, "Edited minutes of meeting," 18 August 1971, LIK, 030.II.12.5.

[72] "Inner Harbor Project, Baltimore, Maryland," 25 February 1971, LIK, 030.II.12.12.

[73] *Ibid.*

[74] *Ibid.*

[75] Among the documents in the Inner Harbor dossier there is a series of photographs of buildings in brick and concrete, including the Fourteen Rack Warehouse L of the Calvert Plant, Jos. E. Seagram and Sons, Inc., designed by the engineers Smith, Hinchman and Grylls, Inc.,

aggregate concrete; we don't even mean doctored concrete in the sense it is rubbed and grouted."[72] "Poured and done," he explains, meaning that the "forms are very carefully attended to." After having declared that "we are not afraid of the marks of construction," Robert Kier, representing Owen & Company, exclaims: "On the outside?" When Bender objects that the post-tensioned beams required for the large spans envisioned by Kahn would cause problems with the idea of the "exposed exterior facade" due to the holes for the ends of the cables ("you have all those holes"), Meyers responds: "we love them." "We express it in some way but we are not afraid of it," he states, regarding the holes.[73]

The choice of continuous paneling, for Kahn, depends not on economic reasons of the deployment of materials, but on the effect in the viewing of the works in the landscape, and the relationship between the materials and the exposed concrete skeleton. For the walls of the buildings of Inner Harbor he first suggests brick – he is against metal panels and he "hates aluminum."[74] In the brick version the towers would have taken on the character of variations not only of the Richards Laboratories, but also of the Promontory Apartments built by Mies van der Rohe in Chicago from 1946 to 1949.[75] Following the various objections raised regarding the costs and difficulties of masonry work, Kahn opts for slabs of slate. As an alternative, he suggests using bricks of "an equal patina to that of the slate,"[76] or sheets of granite.[77] Envisioning the layout of the type of closure of the spans, he takes into account not only the differing qualities of the spaces, but also the final pictorial effect of the individual towers in the landscape. To this end, he suggests that the surface of the infill be divided equally between opaque slabs and transparent panes of glass.[78] "The project will have its own distinction with the general effect being white and gray," he says.[79]

The perspective sketches for Inner Harbor show that the various spans of the reinforced concrete skeleton would have created an equal number of diversified frames or paintings of life, changing throughout the hours of the day. The essential lines and grids of the exposed concrete would take on their own force and meaning thanks to the "endless variety" of the individual frames of life.

"Essentially the exteriors of the buildings are concrete walls with window penetrations which follow around the entire periphery relentlessly avoiding any blank walls. This will display the constant feeling of life which is in the buildings expressed in the endless variety of windows in light, windows dark. At no time will the light in the windows present a predictable pattern – it will always vary – like a swarm of fireflies on a summer night."[80]

and built by Baltimore Contractors, Inc., (LIK, 030.II.12.14).
[76] Dubin-Mindell-Bloome Associates Consulting Engineers, "Edited minutes of meeting," 18 August 1971, LIK, 030.II.12.5.
[77] "Gateway Developers. Inner Harbor Project 1," meeting on 18 August 1971, LIK, 030.II.A.11.9.
[78] Dubin-Mindell-Bloome Associates Consulting Engineers, "Edited minutes of meeting," 18 August 1971, LIK, 030.II.12.5.
[79] "Gateway Developers. Inner Harbor Project 1," meeting on 18 August 1971, cit.
[80] "Baltimore Inner Harbor Project I, telephoned to Owen," 1 December 1972, LIK, 030.II.12.21.

486. Inner Harbor project, Baltimore, Maryland, 1969-73 (LIK, 030.I.A.790.153).

The ark in wood and concrete for Temple Beth El

Planned as the headquarters of a Jewish community to contain spaces for meetings, worship and children, the Temple Beth El of Northern Westchester, in Chappaqua, New York, is designed and built by Kahn from 1966 to 1972 on a lot hidden in the woods. Kahn decides to use wood as the main material of the envelope, in unpainted redwood, for economic reasons and also to create a work rooted on its site, immersed in that landscape, filled with narratives aimed at the congregation, as in the case of similar temples from the past. A reinforced concrete skeleton surfaces from that wooden enclosure, then materializing in the box that forms the temple entrance, the Entrance Building, a true contemporary portico made with that degree of artistic abstraction Kahn is capable of granting to historical memory. The box and skeleton are simply fragments of that reinforced concrete structure that constitutes the secret platform of the wooden temple. The mystical geometry of the main volume – with an octagonal plan – concludes at the summit with an unexpected object, elementary to the point of being primal, a true representation of the original house, all in wood, in the form of a prism with a square plan perforated by large quadrangular windows and covered with a basic pitched roof. This volume at the top is called the "clerestory space."[81] With that elementary icon, Kahn wants to stage a secret metaphor of the floating house that offers survival through the universal flood: Noah's Ark.

The fact that the reinforced concrete structure is not just the support for the envelope is clearly stated in the documents, when modifications are made regarding that essential part of the temple: "The nature of this structure is such that, that which is structure is also its architecture. When re-design took place with the structural components, its architecture of Mr. Kahn's original conception continued to serve and be served."[82]

In 1970 the project enters the phase of development of the details, prior to the start of construction in November, done by the contractor Cuzzi Bros. & Singer, Inc., headed by Alexander Singer and based in Mount Vernon, New York. The discussions on the cost of the reinforced concrete skeleton lead to limitation of the use of this construction system to the platform with the Entrance Building. "Very expensive concrete frames. Extremely expensive forms ('millwork')" are the considerations expressed by Singer.[83]

The description of the final state of the concrete structure of Temple Beth El, as outlined in the documents prepared by Kahn's staff for the start of the construction, formulates a true poetics of the controlled defect; it is also an important reflection of the experiences of Kahn and his team on previous projects, in the face of irremediable flaws observed after stripping.

"The concrete as it comes out of the forms will constitute the final finish," a document of Kahn's office explains. "Exposed concrete surfaces are to show no wood grain markings, no chamfers, no honeycombings, no markings of broken or chipped form surfaces. The exposed surfaces are expected to show certain planned and architectural markings of fins at joints, some random broken parts of fins caused by stripping, some positive markings left by joints between form panels and some small voids caused by very

[81] David P. Wisdom, Henry N. Wilcots, letter to Herbert L. Rosedale, 14 April 1972, LIK, 030.II.97.2.

[82] Kahn's office, "Analysis of Changes in Scope of Contract as of 7 December 1971," LIK, 030.II.40.11.

[83] "Temple Beth El, Chappaqua, N.Y., Meeting," 22 January 1970, LIK, 030.II.42.4.

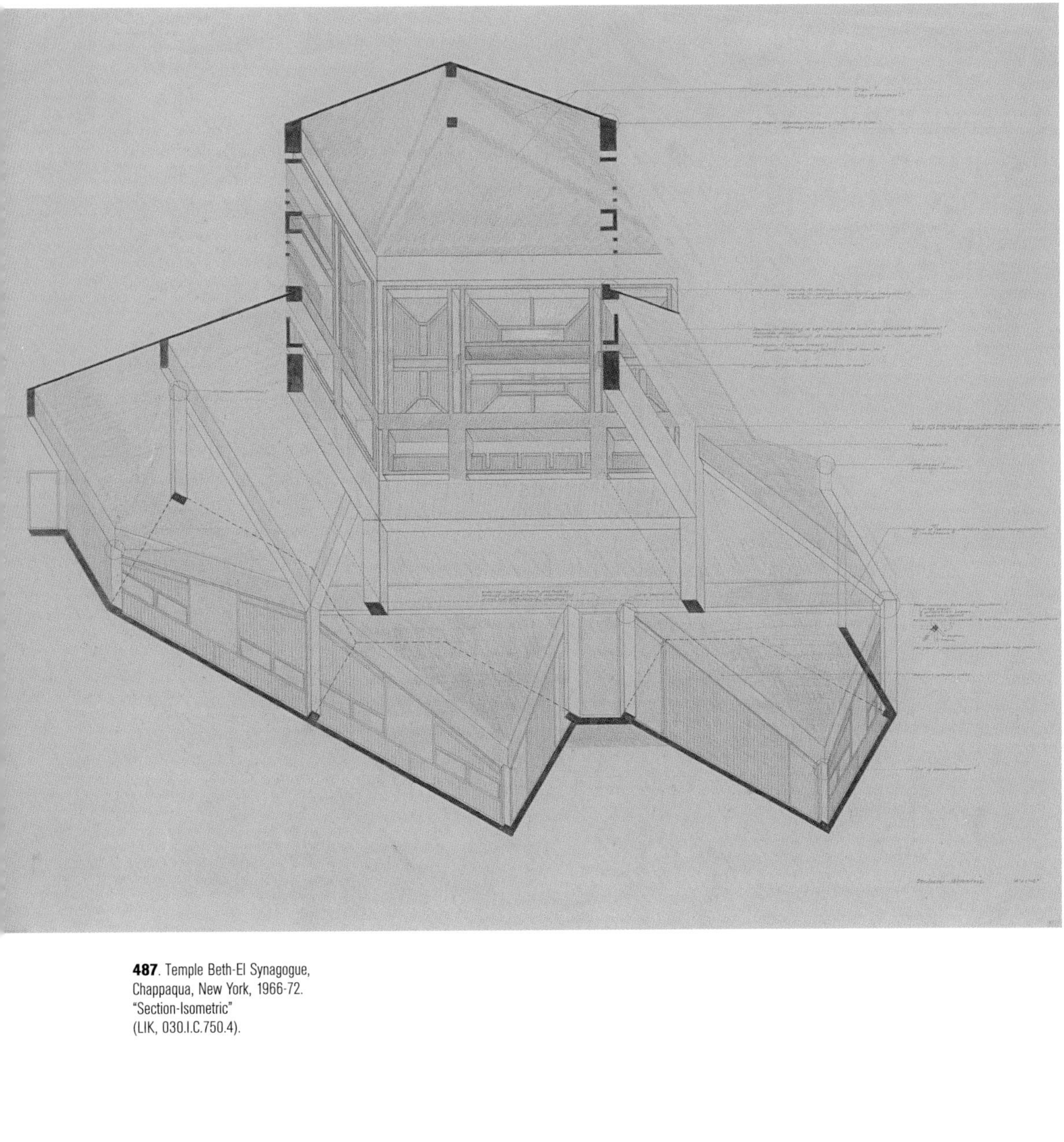

487. Temple Beth-El Synagogue,
Chappaqua, New York, 1966-72.
"Section-Isometric"
(LIK, 030.I.C.750.4).

488-492. Temple Beth-El Synagogue, Chappaqua, New York, 1966-72.

488. "Upper Level Form Layout," 9 March 1971 (LIK, 030.I.C.750.11).

489. "Column Form Panel Position," 22 January 1971 (LIK, 030.II.A.40.15).

490. Detail (photograph by the author, 2010).

491. "Form Panel Layout. South Wall Entrance Building," 17 December 1970 (LIK, 030.II.A.40.15).

492. "Form Panel Layout. North Stair Wall Entrance Building," 17 December 1970 (LIK, 030.II.A.40.15).

small air bubbles trapped against the formwork. Exposed concrete is to show uniform color. Cone holes are to remain visible on the surface just as stripped. The metal part of remaining tie is to be covered by lead caulking only where exposed to view on exterior work."[84]

In the preparation of the "Outline specifications" of January 1970, Kahn addresses the contractor to explain his vision of the sacred nature of the traces visible on the concrete after stripping, from the gravel clusters to the chipping along the joints, which will bear everlasting witness to the skill (or lack thereof) of the workers and masons involved in the fabrication of the concrete, in the performance of the most suitable operations during the various phases. For this reason, every gesture of the workmen must be controlled and accurate, given the fact that it will have visible consequences on the final quality of the concrete.

"The appearance of formed concrete surfaces will be important in finished portions of the work," we read in "Division 3 Concrete" of the specifications. "Accordingly good workmanship will be required, but not any unusual degree of finish. The intention is simply to require the exercise of reasonable skill and care in constructing forms and in mixing, placing and curing concrete. If such care is properly exercised, the concrete as poured will be acceptable as a finished surface for all formed concrete areas without rubbing, grinding, patching or any other finishing or remedial procedures."[85]

Kahn's staff develops the drawings of the formwork of the main parts of the Temple Beth El.[86] On 6 November 1970, Cuzzi Bros. & Singer, Inc., begins assembly of the formwork, which is completed on 17 December. During 1971 the work is interrupted for about three months when the met-

[84] "Technical sections for general construction," draft for the specifications, n.d., LIK, 030.II.40.12.
[85] Louis I. Kahn, "Outline specifications. Temple Beth El, Chappaqua, New York," 22 January 1970, LIK, 030.II.40.12.
[86] For the pattern of the panels of the staircase see the drawing "Form panel layout" prepared on 6 September 1970 and revised on 8 February 1971, LIK, 030.II.42.17.

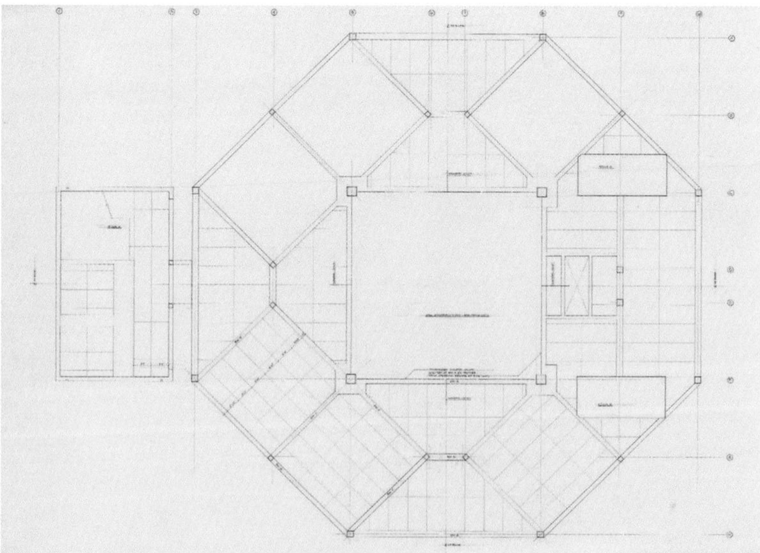

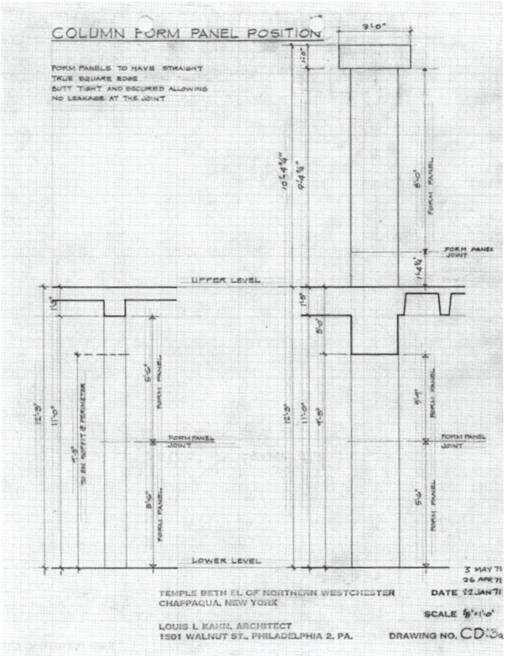
COLUMN FORM PANEL POSITION
FORM PANELS TO HAVE STRAIGHT TRUE SQUARE EDGE BUTT TIGHT AND SECURED ALLOWING NO LEAKAGE AT THE JOINT
FORM PANEL
FORM PANEL JOINT
UPPER LEVEL
LOWER LEVEL
3 MAY 71
26 APR 71
TEMPLE BETH EL OF NORTHERN WESTCHESTER
CHAPPAQUA, NEW YORK
DATE 22 JAN 71
SCALE
LOUIS I. KAHN, ARCHITECT
1501 WALNUT ST., PHILADELPHIA 2, PA.
DRAWING NO.

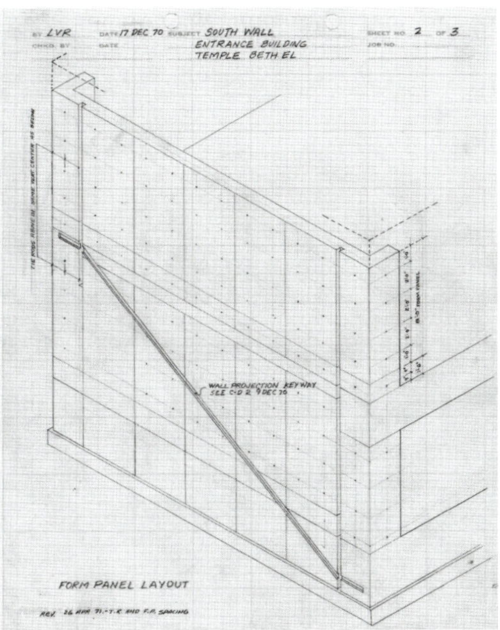
LVR
17 DEC 70
SOUTH WALL
ENTRANCE BUILDING
TEMPLE BETH EL
2 OF 3
WALL PROJECTION KEYWAY
FORM PANEL LAYOUT

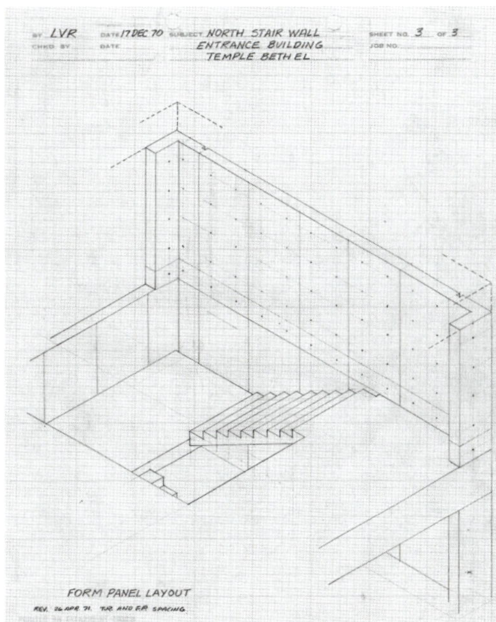
LVR
17 DEC 70
NORTH STAIR WALL
ENTRANCE BUILDING
TEMPLE BETH EL
3 OF 3
FORM PANEL LAYOUT

al formwork of the floor slabs is already in place, due to decisions that have to be made regarding the physical plant systems to install prior to the pouring. The floor slabs are completed on 8 July; that same month, the wooden structures are assembled.[87] The floor slab with square coffers in exposed reinforced concrete becomes a fundamental feature in the definition of the essential structure of Temple Beth El.

During the pause in the work the decision is made by the contractor to eliminate the detail of the groove indicated by Kahn between the slabs and beams to prevent staining of the surfaces of already placed structures during the pour. This same detail had been carefully considered by Kahn and his staff, including Palmbaum, for the projects in Dhaka, Exeter and Fort Wayne. "This recess," Wilcots explains, "had been designed to help control the flow of laitance, which spoils the face of beams and walls during the pouring of the slab, and served as a sealing element of the soffit forms."[88]

The Entrance Building

The metamorphosis of the structure of the Entrance Building in the various phases of the design of Temple Beth El, up to the last phase during the excavation work, confirms the meaning Kahn attributes to this iconic introduction to the temple.

"After on site observations the architect redesigned these walls for both levels to be of wood with all areaway and retaining walls to hold back the earth from the building. This grants the building a more restful position. The decision was then made to 'hold' the design until after the general contractor had excavated enough earth and rock, which would help towards formulating the proper design at the least cost. The final design decision made was not to use wood, but to make these walls completely of concrete, eliminating the retaining walls altogether. The decision voided all previously developed designs by the architect's consulting engineer, for the requirement now was for wall bearing foundations in lieu of grade beams and footings."[89]

Kahn designs the final version of the Entrance Building in December 1970, as a kind of theorem of exposed concrete, where every line is coherent with the structural system.[90] It is significant that in the project of the "form panel layout" on the two north and south walls, the extension of the landing of the staircase is visible, differing in the two cases. Only the corrections in red, made on 30 April 1971 by Singer, radically modify Kahn's conception, introducing two continuous parallel recesses on both walls, inside and outside, which wind up giving those signs the value of decorative lines, as happened with the grooves at the base of the columns in the portico of the Kimbell Art Museum, added by the contractor without Kahn's control.[91] Paradoxically, the purveyors of reinforced concrete are responsible for transforming the logical recesses of Kahn into decorations and lines of graphic composition.

[87] Alexander Singer, letter to Richard Bleier, 3 December 1971, LIK, 030.II.40.11.
[88] Henry N. Wilcots, letter to Alexander Singer, 24 March 1971, LIK, 030.II.40.6.
[89] Kahn's office, "Analysis of Changes in Scope of Contract as of 7 December 1971," handwritten version of the text, LIK, 030.II.A.40.11. See also Louis I. Kahn, "General Contractor's Request for Extension of Time and Added Construction Cost," 11 February 1972, LIK, 030.II.97.1.
[90] See the drawings of the formwork for the Entrance Building, prepared on 17 December 1970 and revised on 26 April 1971, LIK, 030.II.40.15.
[91] See the drawing "Form panel layout," LIK, 030.II.A.40.7. Kahn receives the drawing with corrections on 14 May 1971.

493-494. Temple Beth-El Synagogue, Chappaqua, New York, 1966-72. View of the Entrance Building demolished in 2010 (photographs by the author, 2010).

chapter seven
delirious formwork: slip-form method for skyscrapers

First designs for the Office Building in Kansas City: precast or poured in place

By the end of the 1960s, Kahn has developed, through experience on construction sites, the technical criteria and formal objectives for the design of formwork that is always different, to obtain different qualities of the concrete surfaces in his works. The National Assembly Building being built in Dhaka is giving him the status of the true protagonist in the international debate on a second era of *béton brut*. It is precisely in this period that Kahn seems to be confronted with the possibility of finally experimenting with exposed reinforced concrete for the construction of a skyscraper, after the failures of the projects of the City Hall Building and the City Tower variant developed for the Universal Atlas Cement Company.

The question of the processes of fabrication of reinforced concrete for the construction of skyscrapers becomes crucial in the United States at the start of the 1960s, when the technical challenge to the widespread use of metal structures takes form thanks to three works of an experimental character: the office tower of the Columbia Broadcasting System in New York, designed by Eero Saarinen in 1960 and constructed by the George A. Fuller Company under the supervision of Kevin Roche and John Dinkeloo, and with Harold Roth after Saarinen's death in 1961 (the skyscraper is completed in 1965); the two twin towers for the apartments of Marina City, in Chicago, designed by Bertrand Goldberg in 1959 and built in 1964-67; and the tower of the headquarters of the Catholic organization the Knights of Columbus, in New Haven, built from 1965 to 1969 and designed by Roche and Dinkeloo. Two of these works – Marina City and the Knights of Columbus tower – are erected thanks to slip-forming, a sliding formwork system that is increasingly widespread through the course of the 1950s. Perhaps it is no coincidence that both works have a cylindrical form, which is starting to gain success in works made with the slip-form, as in the case of silos. While Saarinen and Roche-Dinkeloo conceal the reinforced concrete structure with cladding in Canadian black granite (Columbia Broadcasting System), or in terracotta bricks (Knights of Columbus), Goldberg studies the use of concrete with parts poured in place through slip-forming and precast parts to be assembled and left visible, to obtain a smooth concrete surface.[1] Kahn's designs for an exposed reinforced concrete skyscraper should be seen in the context of these investigations of the potential of new systems of construction and finishing of concrete.

The opportunity to develop an innovative reinforced concrete structure, capable of expressing the almost natural process of growth of that material applied to the construction of a skyscraper,[2] arises for Kahn during the course of 1966, when he is contacted – in the month of June – by members of the Board of Trustees of the Broadway United Church of Christ to design a skyscraper for apartments and offices, with a connected church, to build in New York between Broadway, 7th Avenue, 56th Street and 57th Street. In the month of July he is also contacted by Arnold Garfinkel and Richard Altman of Altgar Enterprises of Kansas City, who commission him for the design of a tower for offices, stores and entertainment facilities, to be built in Kansas City (Polk participates on this project). For both assignments, Kahn works with Komendant to develop a system, similar in the two cases, free of a traditional skeleton, with the goal of creating a reinforced concrete structure equipped with a single powerful central core, corner piers and a platform at the summit from which to hang the floor slabs. This structure represents the extreme evolution of the breakdown of the skeleton already announced by Saarinen's invention for the Columbia Broadcasting System, with its central load-bearing core and perimeter pillars.

In the intertwined evolution of the designs for the two skyscrapers, besides proposing often similar variations, Kahn reaches the point of developing, during the course of 1967, a solution for the tower in Kansas City that will be revised only after the building in New York has taken on a definitive configuration, precisely when the project is abandoned at the end of that same year.

In the design for Kansas City, Kahn wants to demonstrate that exposed concrete can become the emblematic material of an image building, with fashionable venues, in a city with burgeoning economic growth. The clients want the skyscraper to have "an image of compelling beauty and quality" in order to welcome the "first class tenants" and to contain the "finest shops in the city," attracting the "best tenants for luxury office space."[3] Unlike the projects for the City Hall Building and the City Tower, left in the state of an icon of a possible skyscraper to come, the project for Kansas City, in its various phases from 1966 to 1973, is developed in all its constructive details, up to the working drawings, from the steel reinforcement of the con-

[1] See Igor Marjanovic, *Marina City, Bertrand Goldberg's Urban Vision*, New York, Princeton Architectural Press, 2010.
[2] Komendant, *18 Years with Architect Louis I. Kahn* cit., p. 155.
[3] "Meeting notes – Held in Architect's Office. Kansas City – Office Building," 5 January 1967, LIK, 030.II.A.39.22.
[4] *The Louis I. Kahn archive: personal drawings* cit., vol. V, pp. 289-294, No. 725.3-11.
[5] "Meeting notes – Held in Architect's Office. Kansas City – Office Building," 5 January 1967, LIK, 030.II.A.39.22.
[6] *Ibid.*
[7] "The order of structure can make conscious the crane. The crane that can lift twenty-five tons should appear in a specification

495. Project for Palazzo dei Congressi, Venice, 1968-74.
Study drawing (LIK, 030.I.A.780.36).

crete to the formwork system, against the background of a concrete industry that has identified new possibilities, at this point, for the construction of tall buildings, besides that of prefabrication. The project is also the culmination of Kahn's constant thinking about exposed concrete, which first began to take systematic form in the Yale Art Gallery.

In the first design schemes for the skyscraper in Kansas City the crucial question is the invention of a structural system that would be so exceptional as to itself become the expression of that "something else" requested by the clients as a quality of the new building. Kahn and Komendant propose a structure with support points concentrated inside the perimeter of the building, in order to have floor slabs that are cantilevered and in part suspended from the roof slab which is supported by the central core containing the service spaces, the elevators and the emergency staircases.[4] The genealogy of this structural system, familiar to Komendant, can be traced back in the case of skyscrapers to the projects of Wright in the 1920s.

The two schemes presented by Kahn to the clients on 5 January 1967 are variations of the overhanging structural system in which the sequence of the cantilevered floors can generate a "kind of horizontal series of layers," or can be concealed by a envelope of bricks, or "possibly other facades." The question of the visibility of the structural system is discussed in terms of the psychological effects caused by the overhanging floor slabs – or the "problem of possible restlessness of feeling."[5]

In both schemes the concrete, where visible, would bear decorations reflecting the logic of the construction process, introducing horizontal strips in marble or slate to mark the pour joints – a "kind of concrete webbing with marble and slate bands at pour joints."[6] This decoration is of the type developed by Kahn for the National Assembly Building, which he would also propose in 1968 for the Congress Hall in Venice. Evidently at this date the construction system is still not that of slip-forming, which appears in the documents only towards April. Kahn also describes a gilded joint to accentuate the symbolic element of the new construction process made possible by the crane.[7]

After the meeting on 5 January 1967, the document no longer contains traces of the brick cladding, and all the reasoning on the skyscraper gravitates around the form of the structure and the ways of making the exposed concrete.

The significant variation, with respect to the initial schemes, has to do with the configuration of the structure, which Kahn now designs with the slip-form method in mind.[8] Komendant knows about the application of the

of present-day architecture which does not appear now. [...] the crane is a designer; that you can make something that's twenty-five tons coming to something that's twenty-five tons, and you can make a joint that's so magnificent, because that joint is no little thing. In fact, if you'd put gold into it, you wouldn't be spending too much money, because it's so big. So the realization that joint making, which is the beginning of ornament [...] comes into being again, you see" (Louis I. Kahn, *Silence and Light*, 1969, now in, Latour, *Louis I. Kahn: writings, lectures, interviews* cit., (pp. 234-246), p. 245).

[8] Komendant, *18 Years with Architect Louis I. Kahn* cit., p. 157.

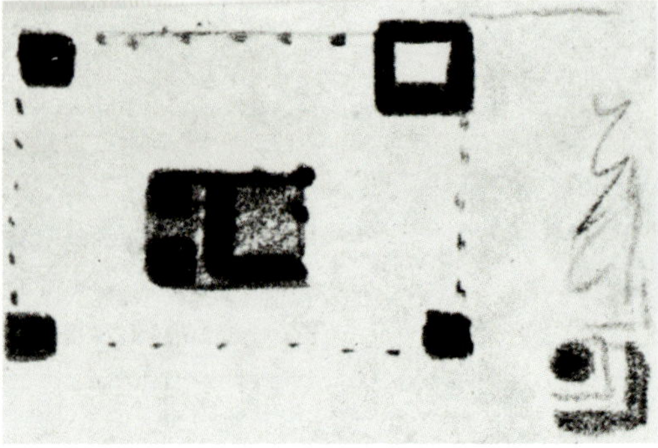

496-501. Kansas City Office Building, Kansas City, Missouri, 1966-73.

496. Study drawing (LIK, 030.I.A.725.4).

497. Plan sketch showing the concept of a service core and four corner columns (H. Ronner, S. Jhaveri, eds., *Louis I. Kahn Complete Work 1935-1974*, Basel, Birkhäuser, 1987, p. 318, No. 3).

498. Elevation sketch showing the expression of a building with suspended floor slabs, April 1967 (LIK, 030.I.A.725.223).

499-501. Notes and sketches taken during the meeting of 7 April 1967 (LIK, 030.II.A.39.22).

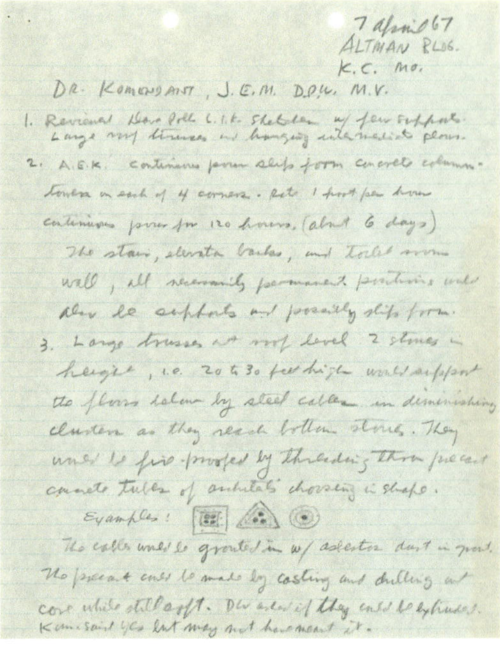

7 April 67
ALTMAN BLDG.
K.C. MO.

DR. KOMENDANT, J.E.M. D.P.W. M.V.

1. Reviewed Dave Polk's L.I.K. Sketches w/ few supports. Large roof trusses and hanging intermediate floors.
2. A.E.K. continuous pour slip form concrete column-towers in each of 4 corners. Rate 1 foot per hour continuous pour for 120 hours. (about 6 days)
The stair, elevator banks, and toilet room wall, all necessarily permanent partitions would also be supports and possibly slip form.
3. Large trusses at roof level 2 stories in height, i.e. 20 to 30 feet high would support the floors below by steel cables in diminishing clusters as they reach bottom stories. They would be fire-proofed by threading thru precast concrete tubes of architect's choosing in shape.

Examples:

The cables would be grouted in w/ asbestos dust in grout. The precast could be made by casting and drilling out core while still soft. DPW asked if they could be cylinders. Kom. said yes but may not have meant it.

slip-form in the construction of buildings of different kinds, from silos to skyscrapers. For the choice of building the skyscraper in Kansas City with the slip-form, the decisive factor could be not only Komendant's suggestion, but also the construction of the Knights of Columbus building. The building by Roche and Dinkeloo is one of the masterpieces of the application of the slip-form in the simultaneous erection of four colossal cylindrical corner towers and the central prismatic core that form the load-bearing structure of steel beams and joists of the floor slabs. But Roche and Dinkeloo are using slip-forming to build concrete columns that will be faced, and they do not include the construction of the slabs in the same system.

At the corners of the rectangular plan of Kahn's skyscraper for Kansas City four powerful pillars are positioned to support the roof beams from which the floor slabs are still hung by means of cables (the anchoring to the corner pillars serves to stiffen the floor slabs). The core of the service spaces becomes the central column of the structure. This overall structural scheme will be confirmed in the successive versions, including the definitive design.

Kahn discusses the indications of Komendant on the system of the "continuous pour slip-form columns,"[9] the floor slab in "precast U units" with cavities for conduits, and the system of steel ties to which to attach the precast elements of the floor slabs.[10] Precisely the structure of the floor slabs becomes the essential point in the developments of the project and of the slip-form system. The objective is still the one indicated by Kahn's organicist vision of the integration of all the parts in the construction process. Thus

[9] See the notes taken during the meeting of 7 April 1967, LIK, 030.II.A.39.22.

[10] The structure of the floor slabs proposed by Komendant is composed of prefabricated concrete parts with the form of a U ("precast U units"), arranged with the open part upward, which function as formwork left in place for the continuous upper floor slab (notes taken during the meeting of 7 April 1967, LIK, 030.II.A.39.22).

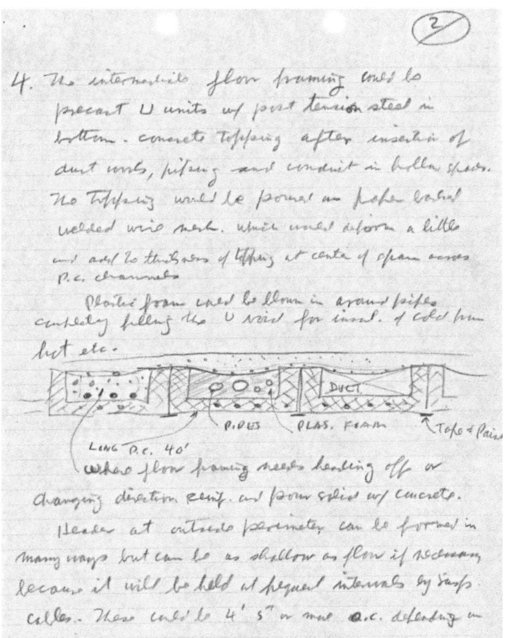

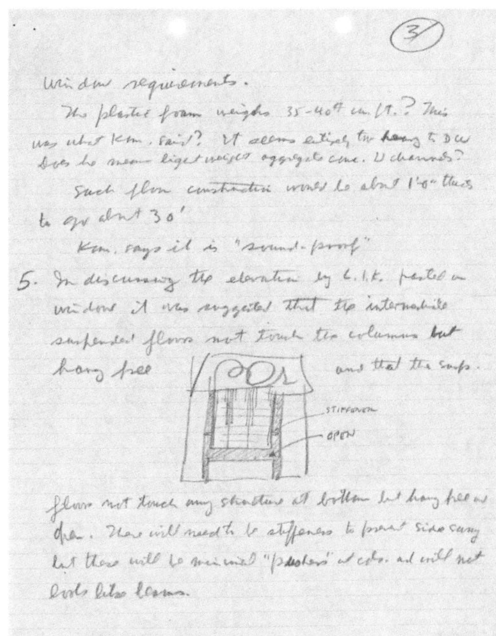

Komendant makes provisions so that the cables, from which to hang the edge beams of the floor slabs, can also have a role in the worksite phase and be part of the formwork system for the pouring of the floor slabs from those above, proceeding downward.

Skyscraper variant in precast concrete with marble joints

The system of pouring concrete with slip-forms, though leading to the idea of the powerful corner structures of the skyscraper for Kansas City, raises certain doubts in Kahn concerning the possibility of controlling the quality of the surfaces, so much so that he proposes, as an alternative construction system, the assembly of precast parts in reinforced concrete. After having discussed the formwork system with Komendant, Kahn indicates his preferences for the construction of the corner columns, namely "not slip-form" but elements in "p.[pre]c.[cast]," to obtain a perfect finish. The assembly of those elements would easily permit the insertion of marble strips at the joints. Even from the documents, it is clear that for Kahn the placement of the concrete for the skyscraper in Kansas City, apart from the system used to make the material, in place or in the workshop, has to justify the presence of what he calls the "pour joints" or "marble joints,"[11] as if those decorations had become, after Dhaka, the emblem of his exposed concrete.

From this point on, with Polk, Kahn concentrates until the summer of 1967 on the development of skyscraper variants with the structural system invented with Komendant, no longer to be made with the slip-form but with precast parts enhanced, as the detail drawings show, by marble strips to insert in special recesses created straddling the joints. The piers are doubled to obtain pairs at the corners, and they can have a triangular footprint, to contain a cavity (the version from June[12]), or a square, full layout (the version from August[13]). At the summit, colossal beams are placed from which to hang the ties supporting the edge beams of the floor slabs. The ground floor is left open to clearly display the system of the suspended floors.

[11] See the notes taken during the meeting of 7 April 1967, LIK, 030.II.A.39.22.
[12] See, among others, the drawing "Street Level Plan," 15 June 1967, LIK, 030.I.C.725.001.
[13] See the drawing "Var. Floor Plan. Office Building for Kansas City," 25 August – 15 September 1967, 030.I.C.725.001.

502. Bertrand Goldberg Associates, Marina City, Chicago, Illinois, 1959-67. Photomontage published in "Chicago Daily News," 2 October 1962 (I. Marjanovic, K. Rüedi Ray, *Marina City, Bertrand Goldberg's Urban Vision*, New York, Princeton Architectural Press, 2010, p. 147).

503. Silos construction site (A. E. Komendant Archive, AAUP, 027.III.15a).

504. K. Roche and J. Dinkeloo, Knights of Columbus, New Haven, Connecticut, 1965-69. Construction site photograph ("Architectural record," August 1970, No. 2, p. 114).

505-507. Kansas City Office Building, Kansas City, Missouri, 1966-73.

505. "Bar Floor Plan," 25 August 1967 (LIK, 030.I.C.725.1).

506. "Section B-B Looking North," 25 August 1967 (LIK, 030.I.C.725.1).

507. "Walnut Street Elevation West," 25 August 1967 (LIK, 030.I.C.725.1).

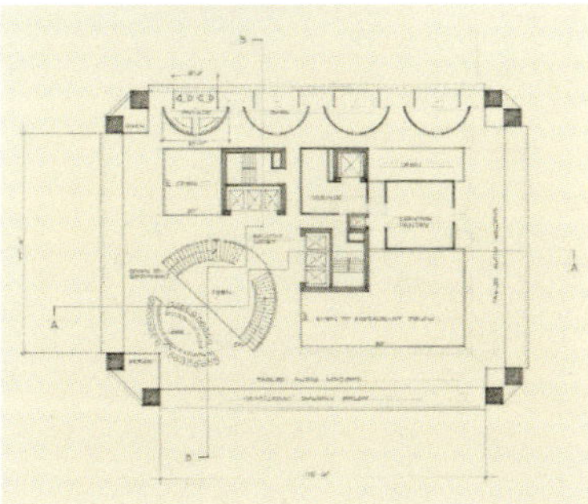

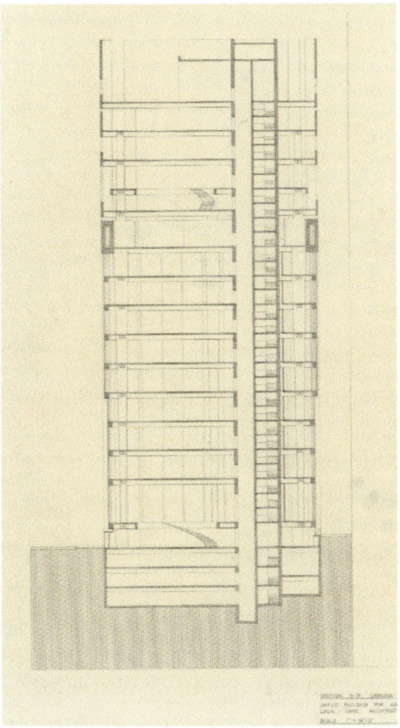

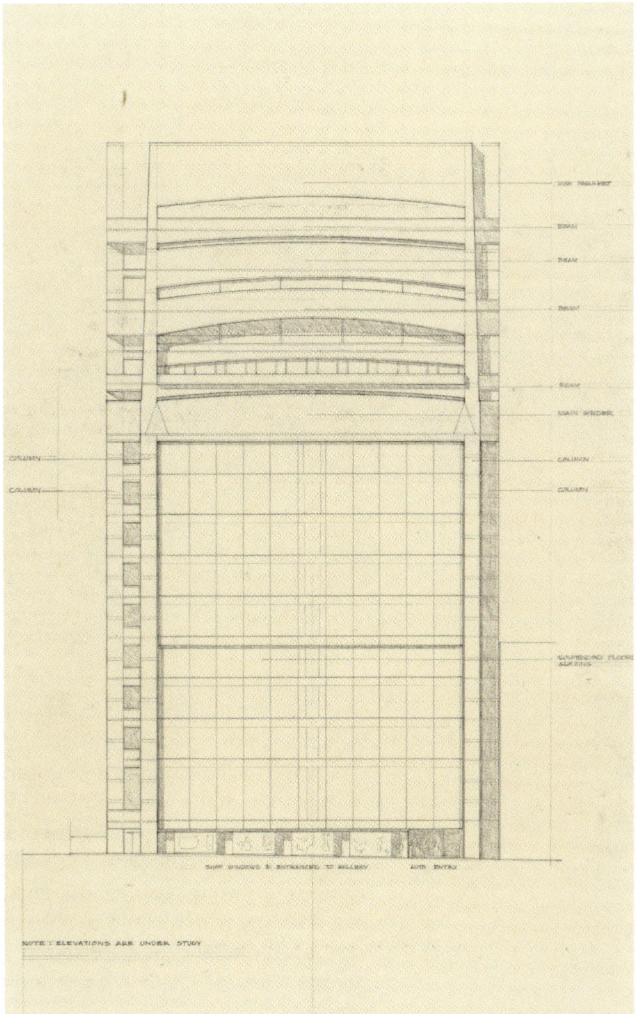

508-511 Kansas City Offioo Building, Kansas City, Missouri, 1966-73.

508. "Sketch Wall Section. Showing Floors 10-12," 25 August 1967 (LIK, 030.I.C.725.1).

509. Model (A. E. Komendant Archive, AAUP, 027.III.35a).

510. "Typical Office Plan" (LIK, 030.I.C.725.19).

511. Elevation study (LIK, 030.I.C.725.19).

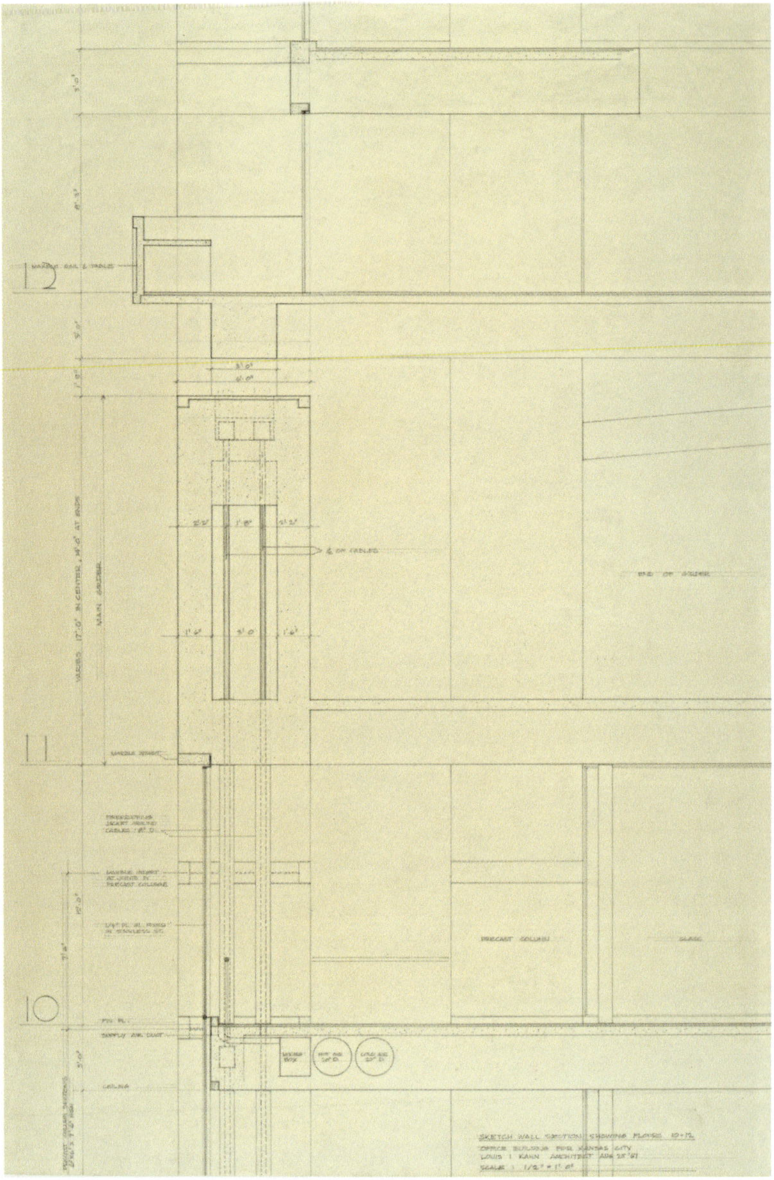

The reasons behind this radical change of the construction method from slip-forming to prefabrication[14] have to do with Kahn's desire to obtain perfect visible concrete decorated with his marble strips, as is shown in what Polk writes to Komendant. It is surprising to observe how much the presence of those strips influences Kahn's choices.

"Lou feels," Polk writes, "that if we were to precast the columns in short sections, we could control the finish of the concrete better and have an urban building which could remain unclad. [...]. The idea of slip-forming is a very good one but the control of the finish is bothersome and the joints in precast work give rise to opportunities for the development of marble strips as inserts."[15]

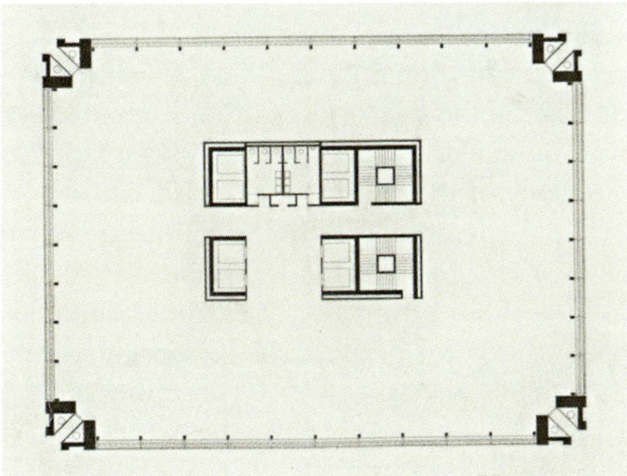

The question of the visibility or concealment of the reinforced concrete structure of the floor slab, and therefore of its geometry, is the focus of studies and discussions towards the middle of 1967. The ceiling and the space frame emerge as conflicting solutions. Komendant studies an elementary structure of beams concealed by a ceiling.[16] The exposed concrete space frame, instead, is the solution urged by Kahn, because it corresponds to the goal of a meaningful structural form, while also providing an ideal "flexibility" and a homogeneous layout of lighting and electricity systems, with "easily

[14] See the drawings of 25 August 1967: "Walnut St. Elevation. West," LIK, 030.I.C.725.001; "Sketch Wall Section Showing Floors 10-12," LIK, 030.I.C.725.001; "Eleventh St. Elevation North," LIK, 030.I.C.725.001. See also David C. S. Polk, letter to August E. Komendant, 25 August 1967, LIK, 030.II.A.39.16.

[15] David C. S. Polk, letter to August E. Komendant, 25 August 1967, LIK, 030.II.A.39.16.

[16] See the drawings: "Section B.B Looking North," n.d., [June 1967], LIK, 030.I.C.725.001; "Section A.A Looking West," 25 August 1967, LIK, 030.I.C.725.001; "Section B.B Looking North," 25 August 1967, LIK, 030.I.C.725.001. See also, David C. S. Polk, letter to August E. Komendant, 13 June 1967, LIK, 030.II.A.39.16.

[17] David C. S. Polk, letter to August E. Komendant, 25 August 1967, LIK, 030.II.A.39.16.

512-513, 515 Project for Broadway United Church of Christ and Office Building, New York, 1966-68.

512. Entrance lobby level plan (A. E. Komendant Archive, AAUP, 027.unprocessed-j).

513. Elevation sketch (LIK, 030.I.A.720.44).

514. Drawing with the work-platform for slip-forms (A. E. Komendant Archive, AAUP, 027.II.392-3).

515. Elevation study (LIK, 030.I.A.720.49).

accessible" conduits.[17] "It occurs to us," Polk writes to Komendant on 13 June, "that the floors, because of their typically irregular configuration, do not lend themselves too well to a system of beams and framing but that a hollow concrete space frame sort of structure – something on the order of the ceiling of the Yale Art Gallery which would provide spaces for air ducts to run where needed, and providing spaces for lighting fixtures – would be more natural to the situation."[18]

The first phase of study of the project for Kansas City can be said to conclude in summer-fall 1967, with the preparation of a wooden model[19] conceived as the expression of the "idea," and as that "tool for helping to explain the project" required by the client to seek financial backing.[20] The model documents one variation of the structure in which the central services column rises beyond the height of the corner piers and carries a platform for a heliport. On the back of a photographic print, the structural system and construction process are tersely explained. But what is described does not correspond to the model, given the fact that it refers to the slip-form method for the "Columns & Core," which instead has been put aside by Kahn.[21] The design documented by the model becomes the starting point for Kahn in the work done starting in 1968, perhaps when the note is added to the photograph, for the version of the Kansas City tower to be built with slip-forming.

Discovery of the slip-formed cylindrical column: the project for New York

In spite of the perplexities and fears regarding the possibility of achieving a satisfactory finish for the concrete using the slip-form method, Kahn decides to use that system for the development of the design of the skyscraper in New York for the Board of Trustees of the Broadway United Church of Christ. The organizer of the financial operation calls in the architecture firm of Emery Roth and Sons of New York, represented by Richard Roth, Jr. The program is comparable to that of the Knights of Columbus headquarters in New Haven. In both cases, the forms of the architecture pay tribute to religious institutions, with skyscrapers erected in the city, like monumental contemporary steeples.

In the various phases of the project for New York, after the studies with profiles similar to the geometry of the City Hall Building and the City Tower, Kahn reaches a configuration of the structure in the form of six colossal hollow cylindrical columns arranged along the edge of the rectangular plan – four at the corners and two in the middle zone – stretching the full height of the skyscraper. The towers for the elevators and services are in the middle part of the plan. A pair of similar hollow towers is also designed for the church.

The choice of the cylindrical form for the columns makes the project for New York so similar to the tower in New Haven that we can suppose either that Kahn is still not aware of the work of Roche and Dinkeloo (presently under construction), or – much more probably – that he has been

[18] David C. S. Polk, letter to August E. Komendant, 13 June 1967, LIK, 030.II.A.39.16.

[19] For the dating of the model see David C. S. Polk, letter to August E. Komendant, 25 August 1967, LIK, 030.II.A.39.16.

[20] "Meeting notes – Held in Architect's Office. Kansas City – Office Building," 5 January 1967, LIK, 030.II.A.39.22.

[21] Photograph of the model, August E. Komendant Archive, AAUP, Philadelphia, 027.III.27.2.

[22] David C. S. Polk, "Broadway United Church of Christ. Meeting in Emery Roth's Office," 19 September 1967, LIK, 030.II.A.33.1.

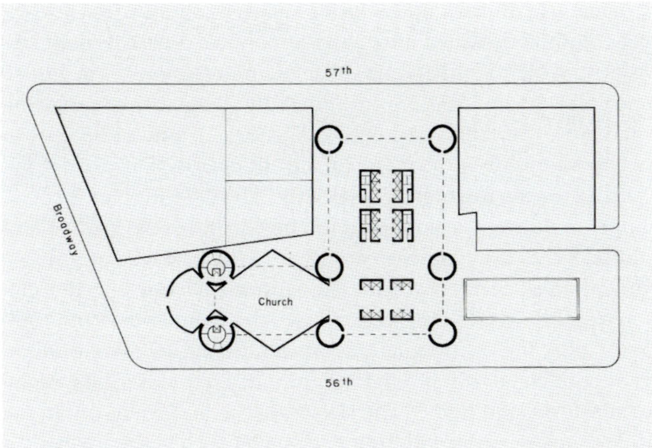

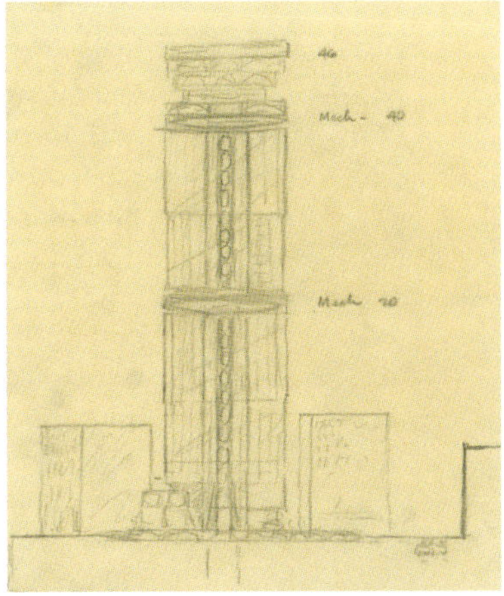

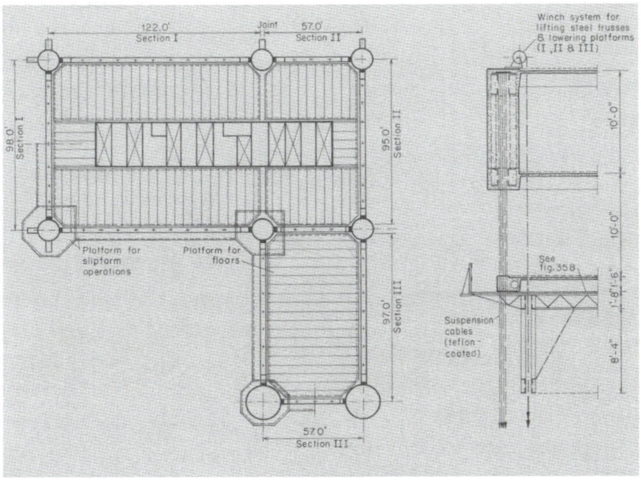

so impressed by it as to propose his own variation, with significant differences where the material is concerned.

The reasons for the form of the column can be sought in the construction system proposed by Kahn and Komendant. Starting in September 1967, the question of the use of the slip-form is the main topic on the agenda. Roth asserts that "slip-form construction [is] not possible in New York," because the unions would demand significant extra pay due to the continuous work cycle required by this system. Following the discussions, Roth develops a skyscraper design with a "steel core structure,"[22] while Kahn and Komendant work on the slip-form system, still suggested by Komendant for its potential for speedy construction. For the development of the system, Kahn and Komendant make contact with Heede International, based in Greenwich, a company in the avant-garde of construction machinery, producer of slip-forms since the 1950s and inventor of the system of sliding formwork by means of "hydraulic jacks," for which the firm holds the patent.

The report prepared by Komendant starts by setting out to debunk the belief that only a steel structure can be economical for the construction of skyscrapers. The reasoning focuses on the potential of new forms of structure and placement of reinforced concrete, including prestressing and slip-forming. Logical use of these potentialities leads to a new idea of structure: of colossal columns designed as organs to contain physical plant systems and functions, which Komendant calls "column-stations," "acting simultaneously as columns and mechanical ducts" and capable of replacing the "conventional large number of columns." This organ, similar by now to the columns of the Metabolists or of Roche and Dinkeloo, has not yet appeared in the studies for the tower in Kansas City. The definition of the "column-station" appears during the course of the project for New York. Kahn and Komendant take the static function and that of the conduits as guiding principles. Komendant does not associate the cylindrical form with the sliding formwork system; he states that in any case, the definitive form is still under study. Kahn sees the cylinder as the "most acceptable" form of abutment, and this is confirmed when "circular columns" are described, to be made with the "slip-form" method.[23] The skyscraper would stand out in the New York skyline thanks to its grand hollow cylindrical columns, rising like the towers of a medieval fortress, yet logical in terms of their construction process. Like the stone columns that were cylindrical, according to the interpretation of Berardo Galiani or Viollet-le-Duc, to make it possible to roll their drums from the quarry to the worksite, so those of Kahn have the form most appropriate for the new construction system, taking on monumental proportions and sculptural forms.

In the design for the tower in New York, the floor slabs are hung from the structure, as in the prototype for Kansas City. The floor slab is a "coffer type" with exposed intrados. The air conduits run through the columns, the floor slabs and the core.

Given the fact that the concrete poured in the slip-forms is not to be worked or covered, as is instead done by Roche and Dinkeloo, Kahn and Komendant study modifications to the "existing forming and casting method."[24] Their goal is also to apply the same slip-form system for the construction of the floor slabs, since those slabs too are in reinforced concrete, not in steel as in the design by Roche and Dinkeloo.

Kahn and Komendant predict that the quality of the surface of the concrete poured with the slip-form method can be satisfactory once the formwork is made to slip upward. "During downward movement," Komendant explains, "the exposed concrete will be thoroughly cleaned, and will have two coats of highly concentrated water-repellent silicone application, in this way treated concrete surface will be uniform and durable."[25] About twenty workers are assigned to control the "surface finish," as well as the reinforcement of the columns – "climbing rods."[26]

The introduction of this correction of the surfaces seems to reveal the concern regarding the quality of the concrete that had convinced Kahn, in the project for Kansas City, to avoid the slip-form method and opt for precasting. But it is no coincidence that this procedure is then eliminated from the next slip-form design, when the system is extended, with improvements, to the skyscraper of Kansas City as well. Because for Kahn intervention

[23] August E. Komendant, "Static Computation for 50-Story Highrise Located between 56-57-Broadway and Seventh Ave, New York City," November 1967, LIK, 030.II.A.39.16.
[24] *Ibid.*
[25] *Ibid.*
[26] August E. Komendant, "Work Arrangement, Cost Estimate and Time Schedule for the Proposed 50-Story Highrise," December 1967, LIK, 030.II.A.33.1.
[27] *Ibid.*
[28] August E. Komendant, letter to B. H. Fritz, 1 December 1967, LIK, 030.II.A.33.1.

after the movement of the formwork means negation of the principle of the concrete surface defined only by the formwork, a principle he intends to conserve even in the case of the slip-form method.

While the "huge column-stations" are in "exposed concrete," the edge beams of the floor slabs can be clad, but in keeping with a method that incorporates the slabs of marble or stone into the process of construction of the structure. The slabs with the metal hardware attached to the reinforcement rods are to be applied in the formwork prior to pouring, "cast as integral part of concrete," like formwork to remain in place,[27] as in the initial designs for the Salk Institute, confirming Kahn's will to absorb every cladding into the phases of the making of the concrete.

Komendant's work-platform

To be able to apply the particular slip-form system also for the construction of the floor slabs of the skyscraper in New York, Komendant and Kahn envision installing a "work-platform" at the site, to be made with steel joists and wooden boards,[28] on which to attach boxed steel elements, or "coffers" that shape the floor slab, then pouring the concrete after having positioned the steel reinforcements. The work-platform with the coffers is attached to the slip-form only when the system is at the top of the columns and the core, while the lower levels are anchored in the desired positions thanks to special housings left in the columns and thanks to steel cables that attach it to the already poured floor slab immediately above it. The fact that the cables of the ends of the floor slabs are utilized for the temporary anchoring of the work-platform is yet another demonstration of Kahn's desire to incorporate the temporary elements required in certain construction phases within the permanent structural system. Even the cavities of the core, where the elevators are inserted, become part of the construction process of the concrete: it is here that the freight elevators are installed to lift the concrete to be poured at the various levels. Only the temporary steel structures, assembled diagonally along the corners and the various levels to consolidate the system prior to the assembly of the floor slabs, are not incorporated in the definitive form of the structure when the pouring of the columns and the core is completed.

In the development of the system of mobile formwork, it is decided to extend the work-platform beyond the perimeter of the floor slab, so as to have a surface available for the workmen during the pouring operations.[29] To cut fabrication time, the work-platform can be equipped with a heating unit to warm the coffers by means of tubes, in order to "accelerate the hydration and hardening of concrete."[30] Komendant studies a work system with a three-way subdivision of the work-platform, to be able to proceed continuously with the installation of the metal reinforcements and the concrete. The "method of construction" with the slip-form is described in detail by Komendant in a fundamental document of the evolution of the formwork of Kahn's concrete. Many passages of the document are reprised for the specifications of the skyscraper for Kansas City.[31]

[29] August E. Komendant, "Work Arrangement, Cost Estimate and Time Schedule for the Proposed 50-Story Highrise," December 1967, LIK, 030.II.A.33.1.

[30] August E. Komendant, "Static Computation for 50-Story Highrise Located between 56-57-Broadway and Seventh Ave, New York City," November 1967, LIK, 030.II.A.39.16.

[31] By request of Komendant, the engineer of the George A. Fuller Company, Peter F. Dejongh, offers his opinion on the system developed for the construction of the skyscraper in reinforced concrete, stating that "the concrete finish provides the architectural finish" – he is surprised to find no indication "for rubbing of exterior walls" (Peter F. Dejongh, letter to August E. Komendant, 29 December 1967, LIK, 030.II.A.33.2).

Slip-forms for the skyscraper in Kansas City

The project for New York, with its hollow cylindrical towers to erect with the slip-form system, marks a turning point in the development of the construction system for the skyscraper in Kansas City. The project in 1968 for the Congress Hall, in Venice, also seems like a corollary of the idea of the suspended structure envisioned for the two skyscrapers, with the horizontal bands in marble.[32] The fundamental passage in the conception of the fabrication of the concrete of the tower in Kansas City happens when the studies are resumed during the course of 1968, when Kahn – after the project for New York – decides to eliminate the precast elements. The fact that Kahn decides, again in this case, to use the slip-form confirms that the design process for New York has left him convinced of the possibility of generating new logical structural forms and also of controlling the quality of the concrete.

The design of the Kansas City tower documented by the wooden model of 1967 is examined with the aim of converting it to the slip-form system. The pairs of corner piers, the core and the floor slabs are to be fabricated with a system that has evolved with respect to those presently in use, and is equipped with the work-platform, as in the case of the skyscraper in New York.[33] The discussions with Komendant allow Kahn to identify the formal and spatial characteristics capable of expressing all the potential of the slip-form system for the pouring of the concrete.

One of the modifications to the design caused by the use of the new system has to do with the form of the corner piers, which shift from being rectangular, with different shapes for the cavities of the conduits, to cylindrical, as in the project for New York. Nevertheless, they remain in pairs. Kahn sets a protruding element against each column, also a cylinder, in order to have a large load-bearing column and another smaller one, probably for the conduits.

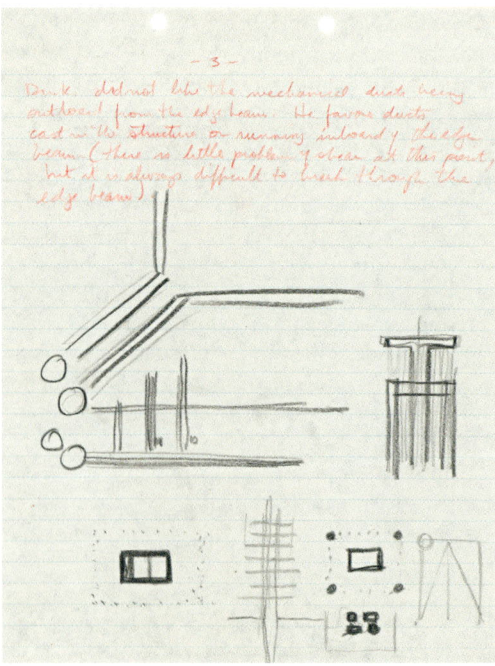

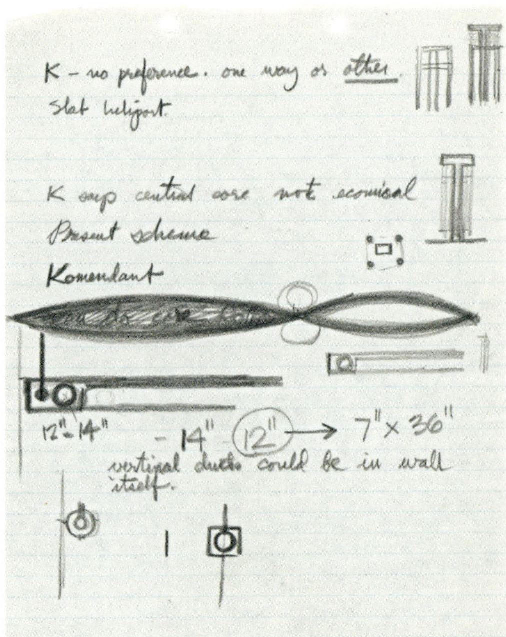

516-518. Kansas City Office Building, Kansas City, Missouri, 1966-73.

516-517. Sketches and notes (LIK, 030.II.A.39.22).

518. Elevation study (LIK, 030.I.C.725.18).

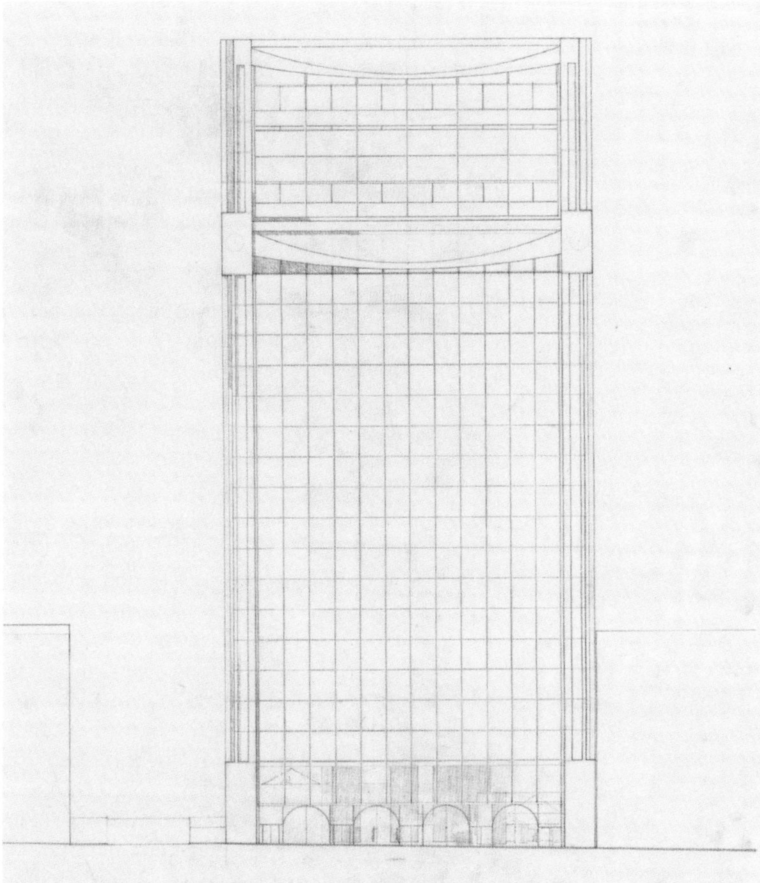

The choice of the slip-form system also leads to revision of the standard floor, given the fact that to make the construction process more economical, Komendant advises positioning of the nucleus at the center of the plan. Again as a result of the slip-form method, Kahn decides to modify the configuration of the structure itself in the upper part of the tower. One of the fundamental questions discussed with Komendant has to do with the precise way of building both the central core and the platform at the top, when the pairs of corner columns stop short at a lower level. To bridge the gap between the top of the columns and the platform level, Komendant suggests using "temporary steel columns"[34] to place on top of the already poured corner columns, in order to be able to use the slip-form system all the way to the summit of the tower. This expedient demonstrates to Kahn that the form of the structure in precast parts documented in the model cannot be transposed

[32] Kahn does not supply detailed descriptions on the making of the reinforced concrete of the Congress Hall; he simply indicates the presence of "marble details" that are probably of the type used in the National Assembly Building, in Dhaka, therefore marking the overlaying of the formwork and the pour phases ("Louis I. Kahn talks about his project," n.d., [received by Kahn's office on 20 February 1969], LIK, 030.II.A.36.1).

[33] "Outline specification for Office Building Kansas City Missouri," 14 October 1968, LIK, 030.II.A.39.22.

[34] See Anon., handwritten notes, n.d., [1968], LIK, 030.II.A.39.22.

into the slip-form system without being modified in its upper part, still in pursuit of the objective of the visibility of all the elements and phases of construction. Thus Kahn suggests making those columns permanent, extending the paired columns of reinforced concrete, and possibly giving them a smaller diameter, to make it possible to recognize the different function of the two column segments. It is in this phase of the project that the need also emerges for a connection, using beams, between the pairs of corner piers and the central core – this is an issue that will have an influence on the study of the structural configuration of the floor slabs of the various levels.

The comment on Komendant's observation regarding the need for temporary columns contains an expression that helps us to grasp the radical constructive nature of Kahn's creative process. "Since we need *temporary* columns let's make them permanent. Lou doesn't like temporaryness," we find written after a meeting between Kahn and Komendant.[35] In the case of the structure of the model, the shoring required to make the construction possible, for subsequent removal, is the proof that the system of pouring of the concrete has not been integrated, because no imprint would remain from those columns in the finished work. The temporary nature to which Kahn refers is not that of the formwork, be it mobile or removable, because each type of formwork leaves the imprint of its presence during the worksite on the concrete.[36]

So the first design for the Kansas City tower documented by the model is gradually modified based on the systems of construction of the concrete and its formwork, reaching a version that is developed and refined until the spring of 1971: the pairs of cylindrical columns in the corners are extended to the platform at the top of the building, seamlessly; as a result, the prestressed inverted beams are anchored at the top of the pairs of columns, to prevent the upper segment proposed by Komendant from having only a temporary role during the construction phase; the core of the service spaces is shifted to the center of the rectangular plan.

The purpose of obtaining the best functioning in every part of the structure leads Kahn and his team to investigate the geometric configuration of the floor slab, to see if it should be identical in all the parts, favoring the concept of the floor slab as a homogeneous distribution system of fluids in space, or split into sectors, in this case favoring diversified static behavior of the floor slab near the pairs of columns.

The hypothesis of the space frame in the form of the Tetrahedron Floor System, which appears in 1967, is no longer taken into consideration, but it seems to represent Kahn's goal for a visible concrete truss structure with incorporated conduits. The position of the core of services at the center of the plan permits study of a type of floor slab with a homogeneous structure for the whole surface, like that of the Tetrahedron Floor System. In the pursuit of a structure capable of rigidly connecting the pairs of columns to the core, Kahn's staff select the waffle type slab, in exposed reinforced concrete. For the passage of the conduits, special openings are to be left between the coffers of the waffle.[37]

[35] Anon., handwritten notes on a meeting with Komendant, n.d., [1968], LIK, 030.II.A.39.22.

[36] "The stopping of the columns at the mech level also stops the slip-form/platform process," we read in a handwritten note of Kahn's staff. "To continue it, temporary steel columns would be used at the corners. This would allow the platform to continue to rise from the core and to provide a work place to put in place the prefabed [sic] cages for the upper girders. The cages would be made to be structural and the floors would be hung from them before the concrete was placed. For reasons of speed and cost the upper floors might be made of precast elements with PIP topping" (Anon., handwritten notes, n.d., [1968], LIK, 030.II.A.39.22).

[37] See "Outline Specification for Office Building Kansas City, Missouri," 14 October 1968, LIK, 030.II.A.39.22.

519-520. Kansas City Office Building, Kansas City, Missouri, 1966-73.

519. "Reflected Ceiling Plan," 18 September 1968 (LIK, 030.I.C.725.2).

520. North Elevation, 18 October 1968 (LIK, 030.I.C.725.2).

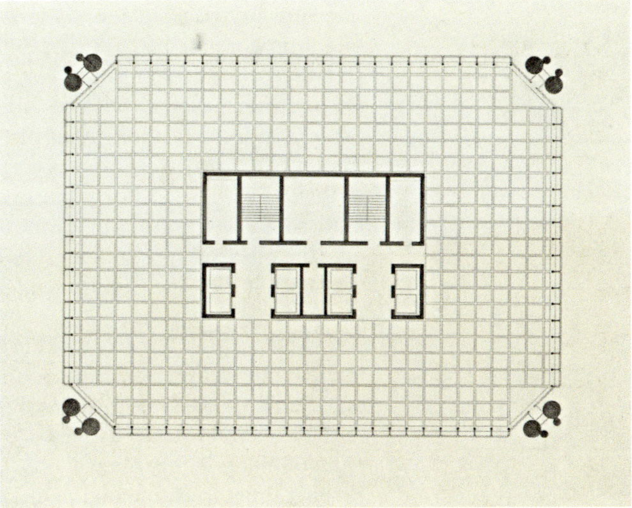

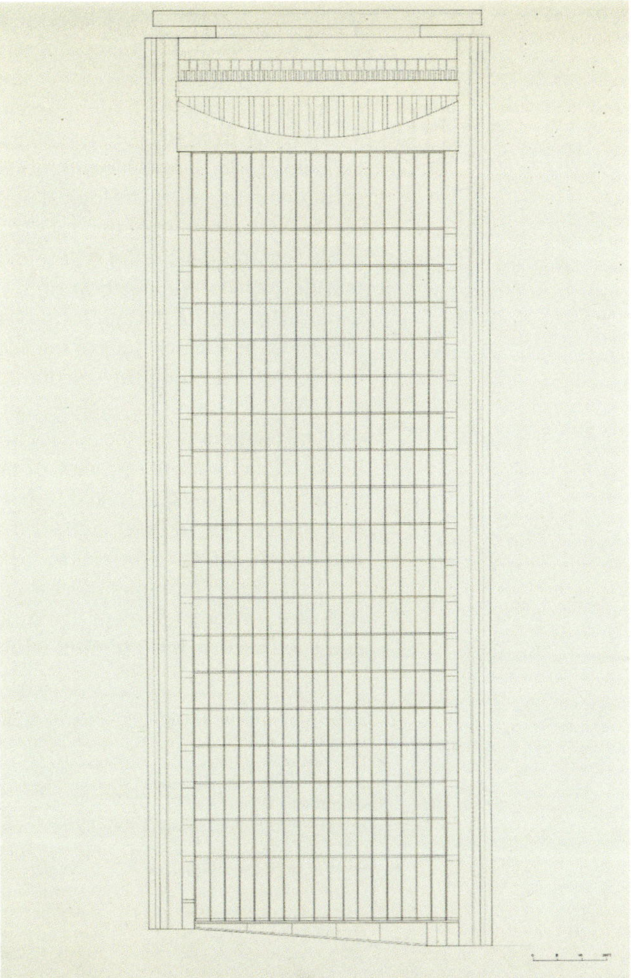

521-522. Kansas City Office Building, Kansas City, Missouri, 1966-73.

521. Study for typical floor plan (LIK, 030.I.C.725.19).

522. "Typical Office Floor Plan," 17 May 1971 (LIK, 030.I.C.725.4).

The "slip-form method" and phases of pouring of the concrete of the structure are specified and planned in October 1968. The parts of the mobile formwork for the columns and the core, specified with stainless steel cladding and 4 feet in height, are studied to obtain a surface finish during the pouring phases, also taking advantage of the typical slipping movement of the slip-form method. Again in this case, Kahn wants to integrate the phases of the construction process to determine the quality of the concrete, so that the work will leave visible traces on its surfaces. The possible polishing of the concrete by means of the movement of the steel of the slip-form would take the place of the initial marble strips. The functioning of the work-platform, also called the "slip-form/platform," is indicated in the specifications for the New York project.[38]

The fact that Kahn is looking for the most logical form of concrete for the two fabrication systems – poured in place or precasting – and the various formwork systems is proven by the fact that the pairs of corner columns of the tower in Kansas City have a different surface from that of the same pairs of pillars in the first version. The slip-form system permits Kahn, for the first time, to eliminate the question that until then had been crucial in the making of the concrete: the joint. The technical nature of slip-form systems implies a process of organic growth, pour after pour, made with the same form that is gradually slipped upward, as soon as the layer of concrete has set. This process cancels out the joint in the forms investigated by Kahn in the past, as the result of the assembly of precast elements or of formwork. The passage from the enhancement of the joint with marble inserts in the first versions, to the smooth continuous surface in the version with the pairs of columns, is therefore the logical outcome of a different construction system. It is as if the question of the "tubular forms" welded without visible joints discussed in *Monumentality* had returned to a timely status. To introduce the marble strips at the position of the pour joints would not express the particular character of the slip-form system. In such details, Kahn demonstrates that he is the greatest heir of the *béton brut* of Le Corbusier, bringing about its evolution through the use of different formwork systems.

All the documents confirm that the structure of the skyscraper in Kansas City, its configuration, the phases of its construction and its very finish would be correlated with the process of pouring the concrete using the "slip-form method," in a way that would be hard to find in other works by Kahn and his contemporaries. In the design for that skyscraper, Kahn shows that he wants to reduce the creative process to a dialectic of shape, formwork and slippage.

The specifications of October 1968 set the sequence of the construction of the various parts of the building: the making of the corner columns and central core, both poured with the slip-form method; the construction

[38] See Anon., handwritten notes, n.d., [1968], LIK, 030.II.A.39.22

[39] "The corner columns and central core area constructed from the ground up, using the slip-form method," we read in the specifications. "The slip-forms are lined with stainless steel, thereby finishing the concrete by being slipped up into position for the next pour. As these structures are built up they will be braced with steel trusses for lateral stability. These trusses will be removed in reverse sequence as the floors are poured from the top down. During upward movement of the slip-forms, openings are provided for slab and ribs, beams between elevator shafts, miscellaneous openings and dowels placed. Average casting speed is 15 feet every 24 hours with about 10 feet in 8 hours for discontinuous pouring. Concrete girders will be poured at the top level of the building. To these girders will be secured the parallel strand cables consisting of individual strands, prearranged, measured, marked and provided with end anchors so that each floor is hung independently from the

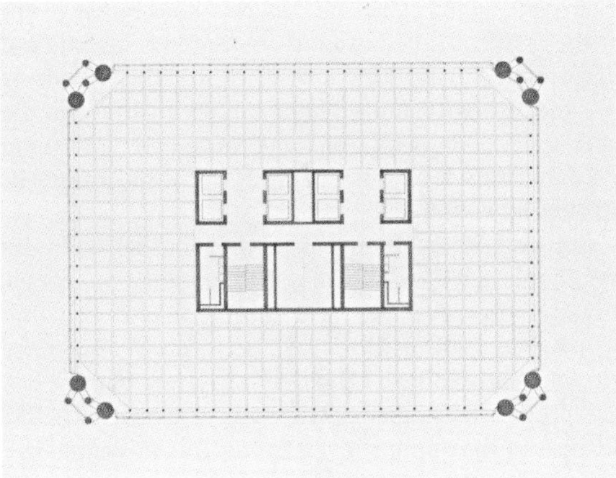

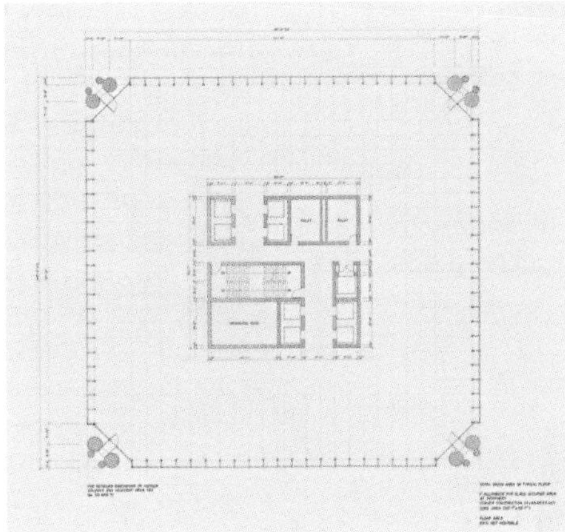

of the beams at the top; the positioning of the cables from which to hang the floor slabs, anchored to the upper beams; the pouring of the floor slabs. During the construction of the columns, housings are made to contain the floor slabs.[39]

The surface quality of the concrete is controlled by means of the material of the slip-form method, stainless steel. For other concrete parts, the formwork is in plywood panels coated with polyurethane, for which Kahn would supply the design of the "patterns of joints"[40] and indications on the alignment of the spreaders.[41] As usual, all the traces of the process have to be visible on the surface of the concrete, as indicated in the handwritten draft of the specifications. Confirmation of Kahn's desire to make all the traces of the concrete construction process visible can be seen in the decision not to seal the tie holes with "lead tamped in place," as originally envisioned.[42] "Do not fill any honeycombs or form tie holes," one entry reads. Kahn reiterates the choices already made for other projects, such as that of not using, apart from exceptional cases, any cosmetic finishing of the concrete ("concrete will not be painted"[43]). The only treatment allowed is a double coat of a protective silicon film on the surface of the concrete.[44] The concrete is entirely exposed in the functional zones of the skyscraper, from the parking areas to the heliport platform, or in the stairwells. The walls of the central core, certain spaces in the Health Club and the floors of the main rooms, on the other hand, are to be clad with white marble slabs to deliver the "luxury" requested by the client ("white Georgia Marble").

After having obtained an estimate on the cost of the skyscraper from the Macomber B. H. Company,[45] at the start of 1969 Kahn's office consults

girders. The work platform is hung from the strands designated to support the floor at the work platform level. The floor is then poured and when design strength is reached, the work platform is disconnected from the strand, the floor slab is connected to the strands, and the work platform, hung only by lowering devices, is lowered to the floor level where it is secured to the strands designed to support the floor at that level. In this manner, all floors are poured from the top down, removing the stabilizing trusses on the way. Floor slab reinforcement is prefabricated for rapid placement in the form. The elevator shafts will be used to lift concrete during slip-form and floor lowering operation" ("Outline Specification for Office Building Kansas City, Missouri," 14 October 1968, LIK, 030.II.A.39.22).

[40] Specifications for contract, n.d., [1968], LIK, 030.II.A.39.22.

[41] See "Outline Specification for Office Building Kansas City, Missouri," 14 October 1968, LIK, 030.II.A.39.22.

[42] Specifications for contract, handwritten manuscript, n.d., [1968], LIK, 030.II.A.39.22.

[43] See "Outline Specification for Office Building Kansas City, Missouri," 14 October 1968, cit.

[44] Specifications for contract, n.d., [1968], cit.

[45] George Macomber Company, letter to Carlos Enrique Vallhonrat, 18 December 1968, LIK, 030.II.A.39.22.

523-526. Kansas City Office Building, Kansas City, Missouri, 1966-73.

523. "Typical Office Floor Plans, Details and Sections," 29 August 1971 (LIK, 030.I.C.725.8).

other contractors, also specialized in reinforced concrete, to create a group for the construction of the tower. The contractors considered include the Massman Construction Company, based in Kansas City, and the George A. Fuller Company, located in Chicago.[46] Both companies are involved in the final project phases, while the Macomber Company is no longer part of the project. In 1969 other solutions for the corner supports are also examined, still based on groups of circular columns.[47]

During 1971, Kahn and Komendant, in collaboration with the Massman Construction Company,[48] develop a new version of the project for Kansas City, destined to undergo only a few variations in the years to come (the lot also changes, now situated in Block 92, between Main Street, Baltimore Avenue, 11th and 12th Streets). The first important change has to do with the plan, which becomes square, with the core at the center, also to facilitate application of the slip-form system on the worksite. The supports return to being paired pillars with a square section, probably to distinguish the tower from the work which was one of Kahn's main sources of inspiration, also for the construction system: the Knights of Columbus headquarters.[49] The structure with prestressed parts of the floor slab is organized into four corner grids, or "coffer type" sectors, and intermediate sectors with parallel beams and compartments for the physical plant conduits.[50]

The question of the fabrication of the concrete structure on the worksite continues to be the focus of technical studies.[51] The core and the pairs of corner piers are to be poured using the slip-form method, always with the work-platform, and it is calculated that a round-the-clock worksite can erect the piers in 25 days. Apart from the heliport platform and the floor slab immediately below it, built with conventional methods making use of the large upper beams, the other floor slabs are made by proceeding from the top down. With respect to the previous versions, the detail of the positioning of the work-platform during the movement of descent for the construction of the floor slabs is changed.

For the machinery and development of the various systems for the construction of the reinforced concrete of the Kansas City tower, Kahn and

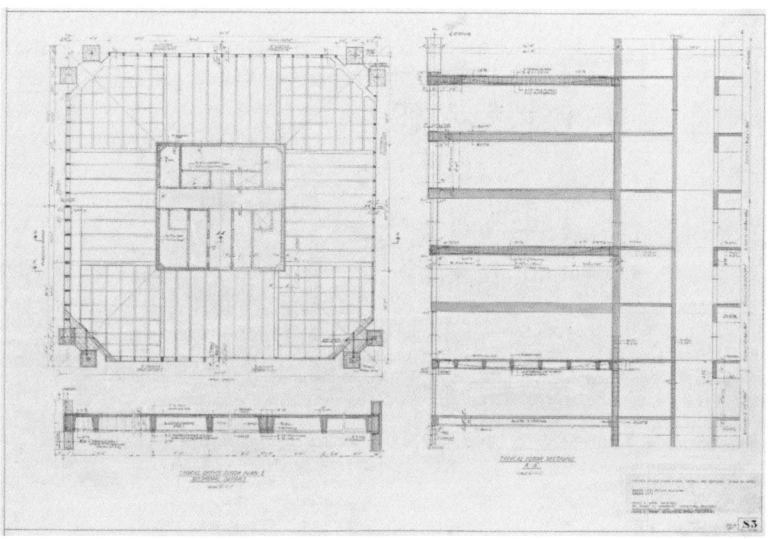

524. "Plan and Sections Above Floor Level 26," 31 August 1971 (LIK, 030.I.C.725.8).

525. "Structural Hanging Section and Details," September 1971 (LIK, 030.I.C.725.8).

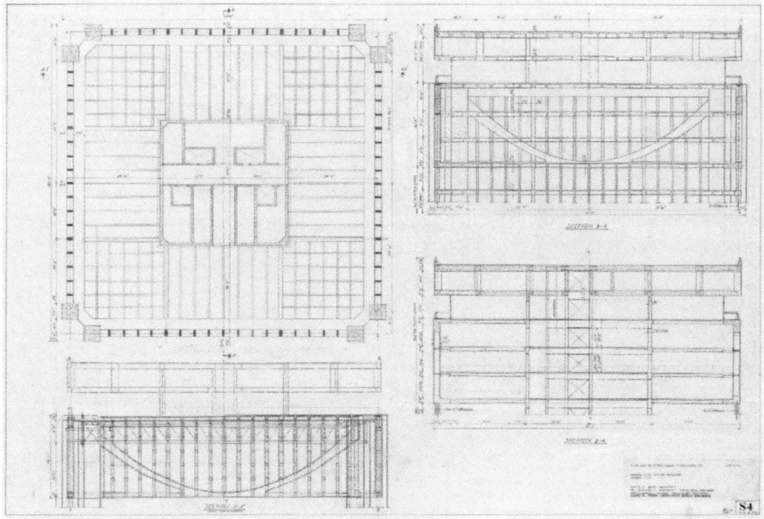

[46] See Carlos Enrique Vallhonrat, letter to Arnold Garfinkel and Richard Altman, 1 April 1969, LIK, 030.II.A.39.22.
[47] See the drawing, n.d., [1969], LIK, 030.I.C.725.019.
[48] See Gary W. Moye, letter to Arnold Garfinkel and Richard Altman, 20 May 1971, LIK, 030.II.A.39.23. The "prestressing units" are supplied by the Stressteel Corporation (August E. Komendant, letter to Louis I. Kahn, 16 September 1971, LIK, 030.II.A.39.16).
[49] See among others, the drawing "Typical office building," 29 August 1971, LIK, 030.I.C.725.008, and August E. Komendant, "Preliminary Computations and Cost Estimate," September 1971, LIK, 030.II.A.39.16.
[50] In the explanation of the construction process of the floor slab contained in a document from September 1971, Komendant describes another structure, composed of precast box elements containing the conduits and the insulation, and two continuous floor slabs (August E. Komendant, "Preliminary Computations and Cost Estimate," September 1971, LIK, 030.II.A.39.16). For the disagreements between Kahn and Komendant on certain details of the structure see Ray Arnold, letter to John Lucas, 12 July 1972, LIK, 030.II.A.39.23, and Ray Arnold, letter to Louis I. Kahn, 18 July 1972, LIK, 030.II.A.39.23.
[51] See August E. Komendant, "Preliminary Computations and Cost Estimate," September 1971, LIK, 030.II.A.39.16.

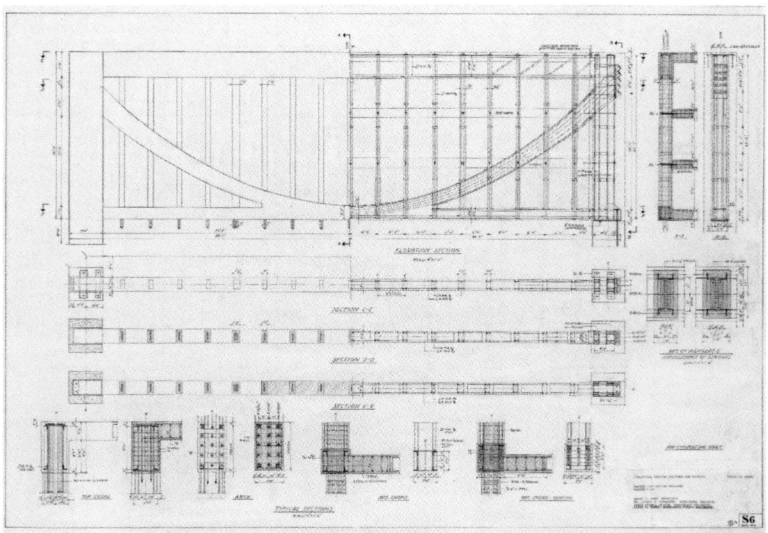

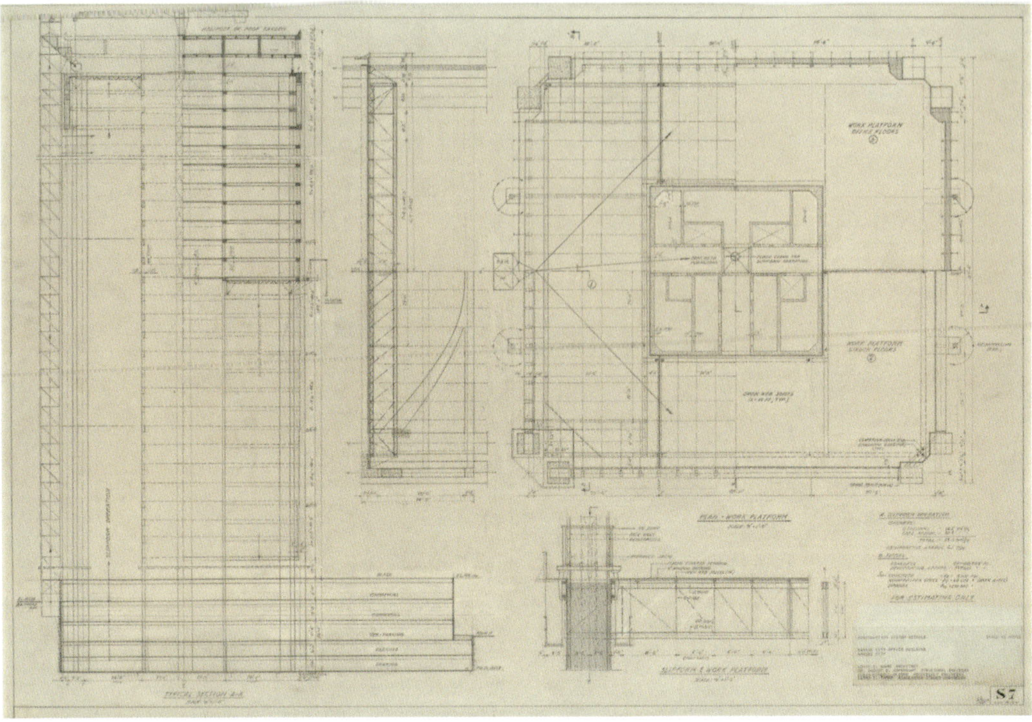

Komendant call in Heede International. Besides the slip-form system produced by Heede International, other machines will be supplied by the same company for the worksite: the mobile tower crane patented by the Swede Knut Elis Lindén in 1962, and normally used on slip-form worksites, such as that of the Marina City Towers;[52] the metal tower with hopper for the raising of materials, known as the "Yo-Yo," invented by Heede International "to accomplish the vertical hoisting of concrete without the necessity of erecting a hoisting tower from ground level";[53] finally, the elevator of passengers, known as the Scando, with a central trestle and double cabin, produced by the Swedish company Lindén-Alimak and distributed in the United States by Heede International.

According to the forecasts of Kahn and Komendant for the worksite organization, the materials reach the work level by means of two mechanical hoisting systems, the Lindén crane and a metal tower with platforms; both are erected to follow the progress of the worksite. The crane is installed at the center of the core and ensures movement of the concrete above all in the phase of erection of the vertical structures. The elevator, or "Tower Hoist," is attached to one side of the skyscraper and guarantees transport of the concrete to the level of the platform for the construction of the floor slabs

[52] It is the "Heede-Linden Climbing Crane Model L-30/75HS" (Heede International, letter to August E. Komendant, 17 September 1971, LIK, 030.II.A.39.15).

[53] Heede International, letter to August E. Komendant, 17 September 1971, LIK, 030.II.A.39.15.

[54] See the drawing "Construction system details," S7, 18 August 1971-March 1973, LIK, 030.I.C.725.008.

[55] August E. Komendant, "Preliminary Computations and Cost Estimate," September 1971, LIK, 030.II.A.39.16.

526-527. Kansas City Office Building, Kansas City, Missouri, 1966-73.

526. "Construction System Details," 18 August 1971 (LIK, 030.I.C.725.8).

527. Perspective drawing (D. B. Brownlee, D. G. De Long, *Louis I. Kahn: In the Realm of Architecture*, New York, Rizzoli, 1991, p. 42, orig., MoMA).

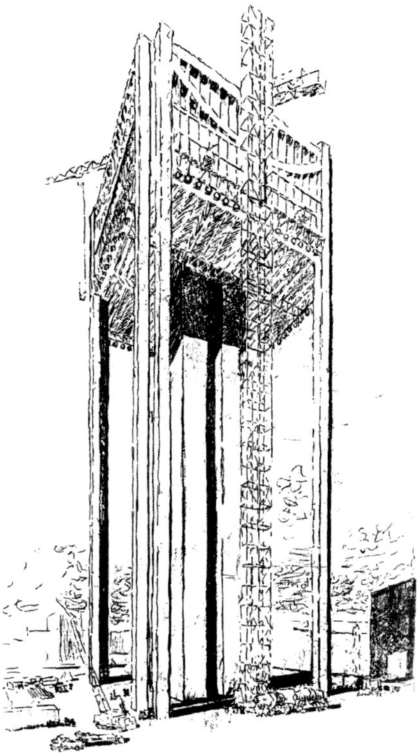

(once at platform level, the concrete is distributed by means of buggies). Two small cranes installed at the overhanging extremity of the work-platform hoist the reinforcements to assemble in the floor slabs.

The top of the tower detached from the corner supports is studied to find the most suitable form to express the worksite phases and the different pouring systems for the concrete. It is planned, in fact, to make the upper floor slabs, including the heliport platform, with traditional methods and not with the work-platform, thus resting the scaffolding on the first floor slab poured with the work-platform.

The work-platform represents the most innovative and complex part of the slip-form method, and Kahn and Komendant introduce other modifications to its design, with respect to its configuration since 1967. The work-platform is composed of a structure of reticular beams in steel, "wooden decking" and a thin final cladding layer, in "plastic covered plywood," in contact with the poured concrete.[54] The cables for anchoring of the edge beams of the floor slabs are used to keep the work-platform in position during the phases of construction of a slab. The work-platform is detached from them once the concrete has set, and then moved thanks to the system of "hydraulic jacks" attached to the pairs of corner piers[55].

Composition and surface treatments of the concrete

In the development of the project of 1971 for Kansas City, the documents clearly explain Kahn's position regarding the role of concrete in the definition of the form of the architecture of the skyscraper. The work is to be an abstract expression of the process, not only by means of the suspended floor slabs and the particular construction using the slip-form method. The finishes

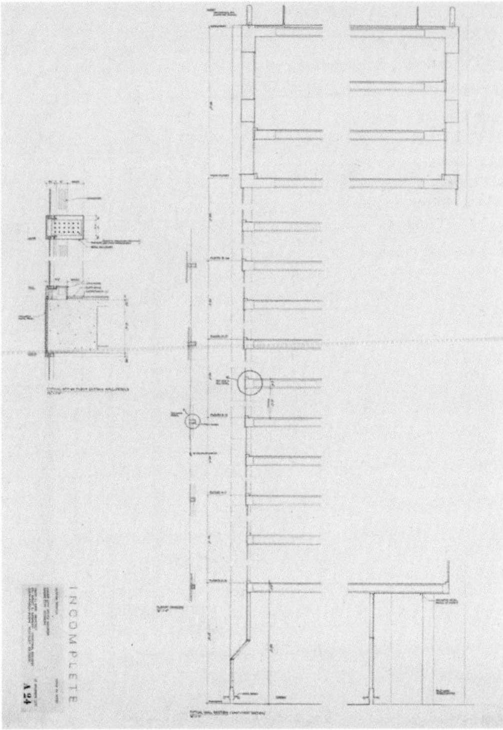

of the surfaces also have to make a contribution, and a decisive one in Kahn's expectations, to the creation of the image of the first grand skyscraper in an exposed concrete marked by that expressive force pursued by Action Painting in the potential of technique. Just as in Action Painting it is the energy of movement that is made visible in the physical nature of the dripping of the paint on the canvas, so Kahn, for the skyscraper in Kansas City, aims at creating a concrete surface shaped by the ascending process of the slip-forms.

The guiding principle in the research on the quality of the concrete is explained in the report written in September 1971 by Komendant: "since this is a special type of concrete work it will largely determine the quality of the architecture. Accordingly, concrete must have not only the required strength but also the desired appearance."[56]

Kahn, together with Komendant, has trials conducted to assess the "color," "texture" and "finish" of the concrete, and to study its artistic potential, always starting with its essential natural ingredients, without the use of additives to alter its hue. He chooses a "light colored" cement, and "natural quartz sand" and "crushed stone or natural" as aggregates.[57] His conviction that to discover the nature of concrete it is necessary to avoid adding powders or chemical products to obtain surprising colors or shades that imitate the look of stones reflects his place in the debate among American artists, striving for a painting of color without figures, based on the use of pure color without shadings.

As on the worksite of the Kimbell Art Museum, for the skyscraper in Kansas City Kahn intends to use the composition of "pozzolan blend cement" applied for the Salk Institute, to be ordered from the American Cement Corporation.[58]

528-529. Kansas City Office Building, Kansas City, Missouri, 1966-73.

528. Perspective, 11 December 1971 (LIK, 030.I.C.725.16).

529. "Glazing details," 28 December 1971 (LIK, 030.I.C.725.4).

During 1972 and until the end of 1973, when the project is definitively abandoned, Kahn and Komendant proceed with the development of the structure and the study of the organization of the worksite, slated to begin in March 1973, in order to complete the pouring with the slip-form method before the winter begins.[59] The differing ideas of Kahn and Komendant on the form of the floor slab raised doubts as to whether Komendant "represent[s] the architect or the mechanical engineers."[60] The documents show that Kahn wants "coffers" as the exposed structure of the floor slabs.[61]

As in the case of the City Tower and the first version of the skyscraper for Kansas City in 1967, a model is built, accompanied by a small publication to promote the operation and find financial backing,[62] while to evaluate the effects of "concrete construction on continuous slip-forming" a visit is urged to worksites like that of the chimney of the Sibley Power Station of the Missouri Public Service Company.[63]

When the project seems to be entering its final phase, Kahn focuses on a type of cement produced by the General Portland Cement Company, the so-called "Warmtone R," capable of supplying "the first true tan colored cement without a tinge of gray,"[64] to be left exposed and waterproofed by means of hydrorepellant products like silicon.

The technicians of the Fruin-Colon Corporation of the Massachusetts Mutual Insurance Company, consulted to estimate the construction costs of Kahn's design, underline the meaning of the choice of constructing a skyscraper in exposed reinforced concrete: "The exposed concrete, protected with proper sealers, will provide a pleasing exterior facade which will require little or no maintenance except for reapplication of sealers at predetermined intervals (usually five to eight years) [...]. The boldness of the exposed concrete exterior to the tower coupled with the utilization of the exposed concrete suspension trusses at the top of the building will produce a structure of outstanding beauty and eye appeal."[65] The structure of the floor slabs, with the exposed concrete beams, appears as a "sculptured ceiling pattern which, with proper form detailing, and placement and stripping procedures, will produce a very pleasing diversion from the usual flat ceiling with lights of typical offices."[66] For the protective treatment of the surfaces, the technicians call for the use of a single coat of "concentrated silicone of approximate 8 percent concentration."[67] These are the last technical specifications of Kahn's dream, pursued since the 1950s and destined to remain on paper: the dream of building a great monument to exposed concrete.

[56] *Ibid.*
[57] *Ibid.*
[58] Given the fact that one of the ingredients of the concrete of the Salk Institute, the "Oro Grande Type III," is no longer available, Kahn is advised, to obtain "the proper color," to use the following mixture: "Oro Grande Type II, 40%; Crestmore Type III, 48%; Airox Pozzolan, 12%" (Jim K. Hicks, letter to Art Lynch, 22 November 1971, LIK, 030.II.A.39.30). See also Bruce Morgan, letter to Carlos Johnson, 11 January 1972, LIK, 030.II.A.39.30.
[59] Massman Construction, letter to Altgar Entreprises, 19 February 1973, LIK, 030.II.A.39.23.
[60] Anon., handwritten notes on a telephone call with Arnold Garfinkel, 28 January 1972, LIK, 030.II.A.39.23.
[61] "Review of Geo. A. Fuller proposal," handwritten notes, n.d., [February 1972], LIK, 030.II.A.39.32. In March 1973 the solution for the floor slabs is defined, using the waffle type "at the corner regions" to guarantee structural stiffness, and the "beam-slab" type in the other sectors; in the spaces between the beams there are the "mechanical & electrical installations, lighting fixtures, etc.," to be closed off by means of a grille, as in the service spaces of the Yale Art Gallery (August E. Komendant, "Kansas City Office Building, Principles of Design, Method of Construction, Specifications, Preliminary Computations and Cost Estimate," March 1973, LIK, 030.II.A.39.15).
[62] Ray Arnold, letter to Louis I. Kahn, 5 June 1972, LIK, 030.II.A.39.23.
[63] Laurence Poisner, letter to Altgar Entreprises, 5 March 1973, LIK, 030.II.A.39.23.
[64] "Trinity Warmtone R," the company explains, "derives its natural color from a unique manufacturing process [...] and contains no pigment, so its color is uniform and lasting" (Don Westcott, letter to Robert Massman, 18 April 1973, LIK, 030.II.A.39.21).
[65] Massachusetts Mutual Insurance Company, Fruin-Colon, "Supplement No. 1 to April 2, 1973," 18 April 1973, LIK, 030.II.A.39.21.
[66] *Ibid.*
[67] Fruin-Colon, letter to Arnold Garfinkel, 21 August 1973, LIK, 030.II.A.39.31.

chapter eight
towards the monolith: the yale center for british art

Harmonies between concrete and other materials

The design of the skyscraper, be it that with the Tetrahedron Floor System for Philadelphia or that with the slip-form system for Kansas City, appears in Kahn's work as an exercise to experiment with the full potential of reinforced concrete, either precast or poured in place, nurturing visions of extraordinary structures imagined by Kahn for other projects. There are undeniable affinities between the Yale Art Gallery and the City Tower for the Universal Atlas Cement Company, or between the skyscraper for Kansas City and the Congress Hall for Venice, always under the aegis of a visionary impulse towards the creation of a spectacular structure that can become a true ode to exposed concrete. So even when Kahn is commissioned to design buildings of a smaller size, their structures reflect the drive for grandeur contained in the projects for the towers. One such work is the last masterpiece by Kahn, the facility of the Paul Mellon Center for British Art and British Studies in New Haven, on Chapel Street, built on a lot facing the Yale Art Gallery. For the design of the building, to contain Paul Mellon's collection of British art donated to Yale University in 1966, the professor Jules David Prown, appointed as director of the Mellon Center, turns to Kahn in February 1969.

In the "Preliminary Thoughts on Architecture" prepared a few weeks before the commission, Prown identifies the characteristics of the materials of the future Mellon Center, almost as if to ward off not only any type of curtain wall, but also the Brutalism of the Yale Art Gallery and even its "austere street wall"[1]: "The character of the collections of British art suggests the inappropriateness of brutal, rough-textured sheathings; of fanciful gossamer screens; or of cold, sheer, impersonal glass and steel surfaces. In terms

530-533 Yale Center for British Art, New Haven, Connecticut, 1969-74.

530. "Chapel Street Elevation," 10 June 1970 (LIK, 030.I.C.805.1).

531. Plan (LIK, 030.I.C.805.2).

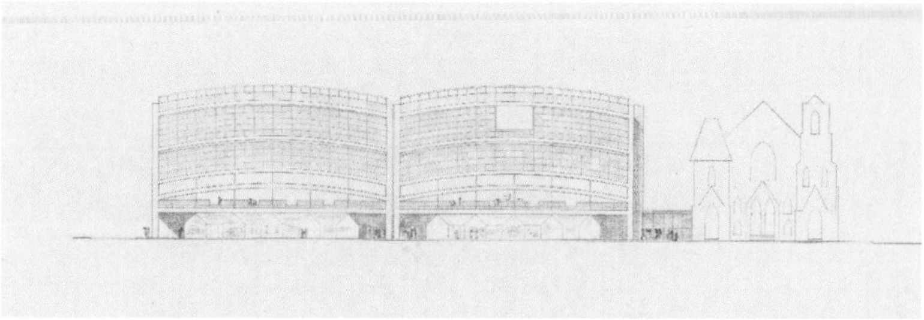

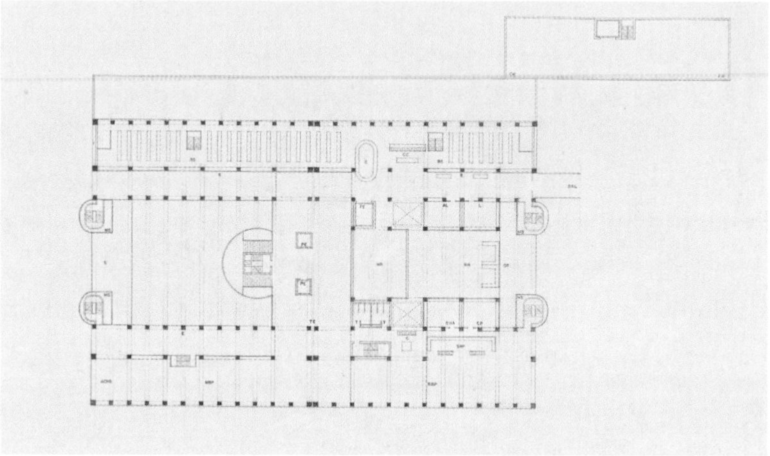

of English Romantic art theory, perhaps it might be said that the building should be 'Beautiful' rather than 'Sublime' or 'Picturesque.'"[2]

In April, Prown and Mellon take a journey of initiation, accompanied by Kahn, to La Jolla to see the Salk Institute[3], and to realize that exposed concrete can achieve the degrees of beauty invoked by Prown.

In the designs of 1970 the Mellon Center appears under the guise of a grand structural machine, like the one designed for the skyscraper in Kansas City or the Congress Hall in Venice, with colossal arches or spectacular Vierendeel trusses, organized in two juxtaposed volumes. The commercial activities planned for the ground floor, the auditorium, the halls for university seminars, the offices, workshops, library and exhibition rooms are arranged around two courtyards, of which one has a monumental outdoor flight of steps, in the initial studies – "Palazzo Mellon," as Kahn himself indicates the Mellon Center, to underscore the domestic character desired for the rooms.

Kahn's constant reflection on the value of the joints in the panels of the formwork, to be opened to insert a special shaped strip, also lies behind the solution of widening the expansion joint between the two structures of the building. "Another characteristic," he says, "is the expansion joint between

[1] Anon., "For release to morning papers of Monday, October 27, 1969," LIK, 030.II.A.111.1.

[2] Jules David Prown, "Memorandum to Mr. Kingman Brewster, Jr., The Paul Mellon Center of British Art and British Studies. Preliminary Thoughts on Architecture," 22 January 1969, LIK, 030.II.A.111.29.

[3] Jules David Prown, letter to Louis I. Kahn, 30 April 1969, LIK, 109, cit. in, Patricia Cummings Loud, *The Art Museums of Louis I. Kahn*, Duke University Press, 1989, p. 186.

[4] Brownlee, De Long, *op. cit.*, p. 411. See also Ronner, Jhaveri, Vasella, *op. cit.*, p. 381.

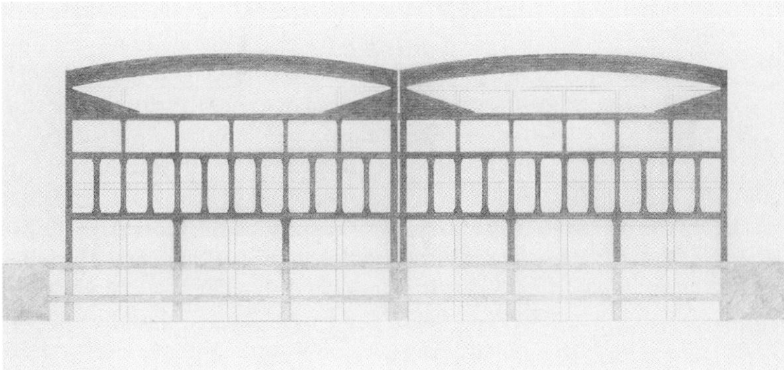

532. Study of the structure, 8 September 1970 (LIK, 030.I.C.805.2).

533. "Perspective View of the Prints and Rare Books Room (1)," 21 December 1970 (LIK, 030.I.C.805.4).

the two buildings. I emphasized and dramatized the expansion joint, causing it to have two entrances at this point instead of the usual one."[4]

Kahn begins to indicate the characteristics of the materials of the Mellon Center with a document in October 1970 containing the first generic version of the specification guidelines, prepared in collaboration with the engineers chosen by Kahn, Pfeisterer, Tor & Associates Consulting Engineers, based in New Haven and New York, and the physical plant specialists of the office of Van Zelm, Heywood & Shadford. The image of the Mellon Center that emerges from the list of materials is that of a building with the colors of steel, of granite slabs, polished and probably gray, of the exposed concrete of the structure and the reflections produced by the glazings of the openings. It is already clear that Kahn's building will form a contrast with the ribbed concrete, shaped by emphatic chiaroscuro, of the nearby Art and Architecture Building by Rudolph, built from 1959 to 1963. Already in the first drawings, Kahn develops the chromatic range of the building starting with the most conventional color of concrete (gray) selecting materials with related tones to produce an effect of shading. Beyond the

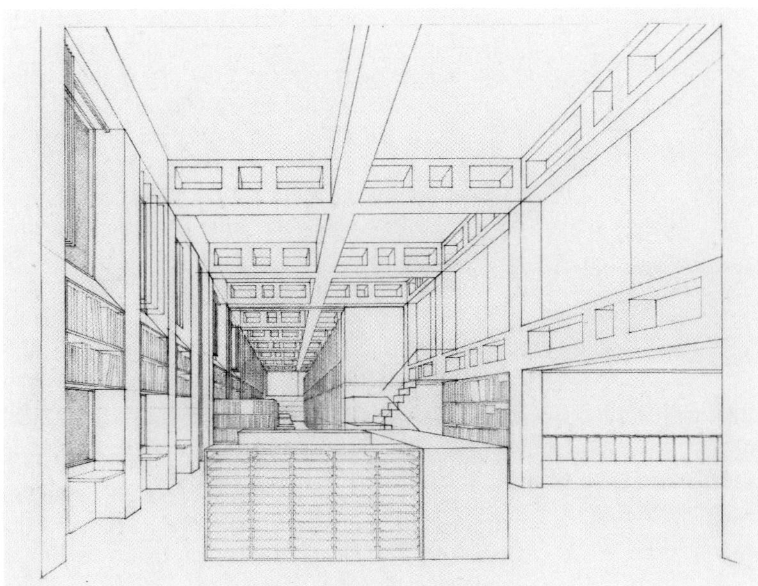

entrance, visitors and scholars would be immersed in rooms featuring surfaces in wood paneling, with plaster claddings, or the texture of concrete blocks. In spite of the initial recommendations of Prown, Kahn continues to favor the use of materials that express their own processes of fabrication: "Exposed concrete and masonry will not be finished, the natural material as it comes out of the form or as it is laid constituting the final finish."[5]

Kahn has already addressed what at this point is the crucial part of the specifications of his works, the "concrete formwork." In the list of the general characteristics, plywood panels with polyurethane coating are mentioned; it is indicated that the "formal arrangement" will be supplied by Kahn, "using full length sheets"; that the screw ties "will be lined up"; and, finally, any kind of "inserts" not included in the design of the formwork are to be ruled out.

The first draft of the text contains the expression already used by Kahn in the specifications for the Exeter Library, prepared in 1967: "The building is designed for maximum reuse of forms." In the revised version the phrase is modified: "Forms will be designed for maximum reuse."[6]

Starting with the example of the Salk Institute and the samples being made for the Kimbell Art Museum, Kahn also indicates "recessed snap type or screw type" ties in the preliminary designs for the Mellon Center, whose holes are to be sealed with lead disks only on the surfaces exposed to weather.[7] In the documents for the Mellon Center the distinction between walls and skeleton parts based on the presence or absence of tie holes takes on the value of a principle of Kahn's architectural concrete, also in the wake of the errors committed by the contractor on the Kimbell Art Museum worksite: "There will be no ties permitted in columns or in Vierendeel trusses."[8]

The concept of a surface obtained directly by means of control of the design and assembly of the formwork has by now been extended to a ban on any type of retouching of the concrete after stripping, insisting that the material should be "untouched after the forms are stripped": "Do not fill any honeycombs or formtie holes. Do not rub the concrete or grind any fins."[9]

The first major change to the initial range of materials envisioned for the Mellon Center dates back to November 1970, when Kahn eliminates the "polished granite."[10] Starting with this choice, the meaning of the metal panels changes with respect to the first design, where they were limited to the envelopes of the volumes containing functional and technological organs.[11] From now on, those panels take the place of the granite slabs. So after the brick of the Richards Laboratories, the wood of the Salk Institute or the Travertine of the Kimbell Art Museum, Kahn experiments in the Mellon Center with yet another variation of the dialectic between the exposed concrete of the skeleton and the materials of the infill, demonstrating that he is the main heir to the lessons of Perret.

Concrete for the skeleton and steel panels for the infills

In the studies from December 1970 to March 1971, Kahn eliminates the grand structures of the previous versions, assigning the definition of the Mellon Center to the presence of an elementary structural grid of pillars and beams. A vaulted roof also appears, based on that of the Kimbell Art Museum.

The beams for the floor system are of the Vierendeel type, derived from those of the Richards Laboratories, and equipped with cavities for the passage of the physical plant conduits. The influence of the Richards Laboratories is not limited to the type of beam, but also has to do with the concrete fabrication

534-535. Yale Center for British Art, New Haven, Connecticut, 1969-74.

534. "Transverse Section Through Court," 3 March 1971 (LIK, 030.I.C.805.004).

535. Elevation, 14 January 1971 (LIK, 030.I.A.805.212.1).

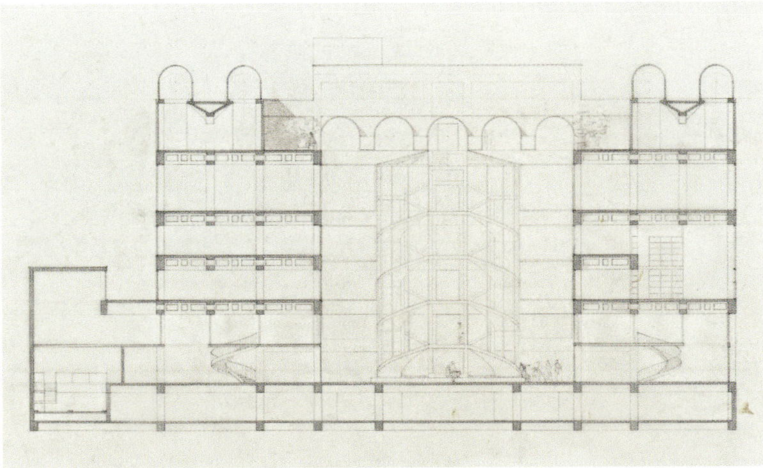

technique itself. Kahn's staff, in fact, discusses the possibility of using precast concrete for the lower sectors of the Vierendeel trusses with Abba A. Tor. "Pouring condition" is the expression most often used in their telephone calls.[12]

The uncertainty regarding the use of precasting is suddenly cleared up in the "Outline specifications" of 12 March 1971: "All concrete is to be poured in place."[13] For Kahn, the guiding principle continues to be based on the "imprint of the forms in the concrete" as an "integral element of the architectural design." In spite of the choice of poured in place concrete, which in any case is later reconsidered, Kahn is looking for concrete with a finish similar to that of the precast parts of the Richards Laboratories. The

[5] Kahn, Pfeisterer, Tor & Associates, Van Zelm, Heywood & Shadford, "The Paul Mellon Center for British Art and British Studies," 5 October 1970, LIK, 030.II.A.112.52.

[6] Kahn, Pfeisterer, Tor & Associates, Van Zelm, Heywood & Shadford, "The Paul Mellon Center for British Art and British Studies," 5 October 1970, with modifications of 29 October 1970, LIK, 030.II.A.112.22.

[7] *Ibid.*

[8] *Ibid.*

[9] Kahn, Pfeisterer, Tor & Associates, Van Zelm, Heywood & Shadford, "The Paul Mellon Center for British Art and British Studies," 5 October 1970, LIK, 030.II.A.112.52.

[10] In the further revision of 10 November to the text of 5 October, the "polished granite" for the facades is eliminated (Kahn, Pfeisterer, Tor & Associates, Van Zelm, Heywood & Shadford, "The Paul Mellon Center for British Art and British Studies," 5 October 1970, with modifications of 10 November 1970, LIK, 030.II.A.112.29).

[11] Cummings Loud, *op. cit.*, p. 219. *Cf.* Latour, ed., *Louis I. Kahn, l'uomo, il maestro* cit., p. 141.

[12] "Tel. Tor, Mellon Yale," pages of notes, 28 January 1971, LIK, 030.II.A.112.17.

[13] Kahn, Pfeisterer, Tor & Associates, Van Zelm, Heywood & Shadford, "Outline specifications. The Paul Mellon Center for British Art and British Studies," 12 March 1971, LIK, 030.II.A.112.21.

536-537. Yale Center for British Art, New Haven, Connecticut, 1969-74.

536. Graphic study of natural lighting (LIK, 030.I.C.805.005).

537. "Typical Wall Section," 28 July 1971 (LIK,030.I.C.805.6).

fact that the Mellon Center is mostly a skeleton, and that the wall surfaces are concentrated in the basement level, leads Kahn to envision use of the largest possible formwork, to eliminate joints and thus "the imprint of the forms."[14] For the prospect of the elimination of the concrete walls, the first design hypotheses for the volumes wrapped in metal and not in concrete are significant. In the project in March, the panels with sheets of 22-gauge stainless steel enclose the four corner towers with air ducts at the top, and form the surfaces of many of the structural spans (taking the place of the initial granite sheets), while closing the walls on the upper level, where the light comes from the vaults, making a decisive contribution to the chromatic definition of the envelope.

To be able to control all the elements of the building, it is decided to construct a special prototype on the worksite, a "demonstration structure" composed of the combination of multiple elements ("wall, column, slab and stair"[15]), therefore of the type made at the time of the construction of the Carpenter Center.

In the summer of 1971 the definitive design of the Mellon Center takes form, based on an even more elementary structure than the one in the project from March, no longer divided into two sectors abutted along an expansion joint, but continuing along a uniform grid with a square module (the size will be 20 × 20 feet). It is no coincidence that skylights appear at the top of the building, again consistent with the square grid of the structure. Each skylight is closed by a concrete dome clad with sheets of lead. The elimination of the expansion joint between the two structures coincides with the shift of the entrance to the corner of the building.

After the experience of the relationship between pozzolanic concrete and Travertine of the Salk Institute, and the variation of the Kimbell Art Museum,

[14] *Ibid.*

[15] *Ibid.*

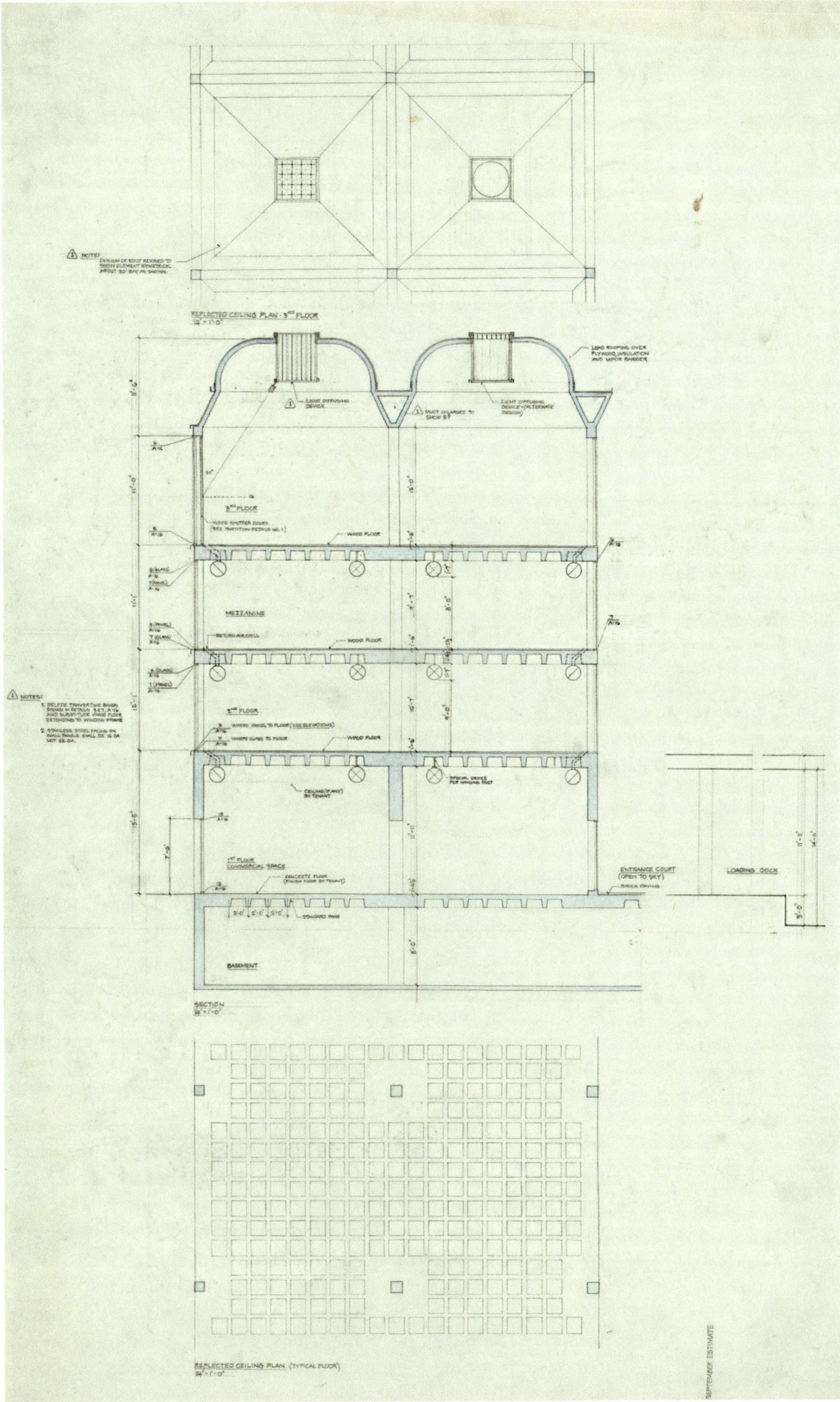
REFLECTED CEILING PLAN · 3RD FLOOR
LIGHT DIFFUSING DEVICE
3RD FLOOR
WOOD FLOOR
MEZZANINE
RETURN AIR GRILL
2ND FLOOR
1ST FLOOR COMMERCIAL SPACE
CONCRETE FLOOR
ENTRANCE COURT (OPEN TO SKY)
LOADING DOCK
BASEMENT
SECTION
REFLECTED CEILING PLAN (TYPICAL FLOOR)
SEPTEMBER ESTIMATE

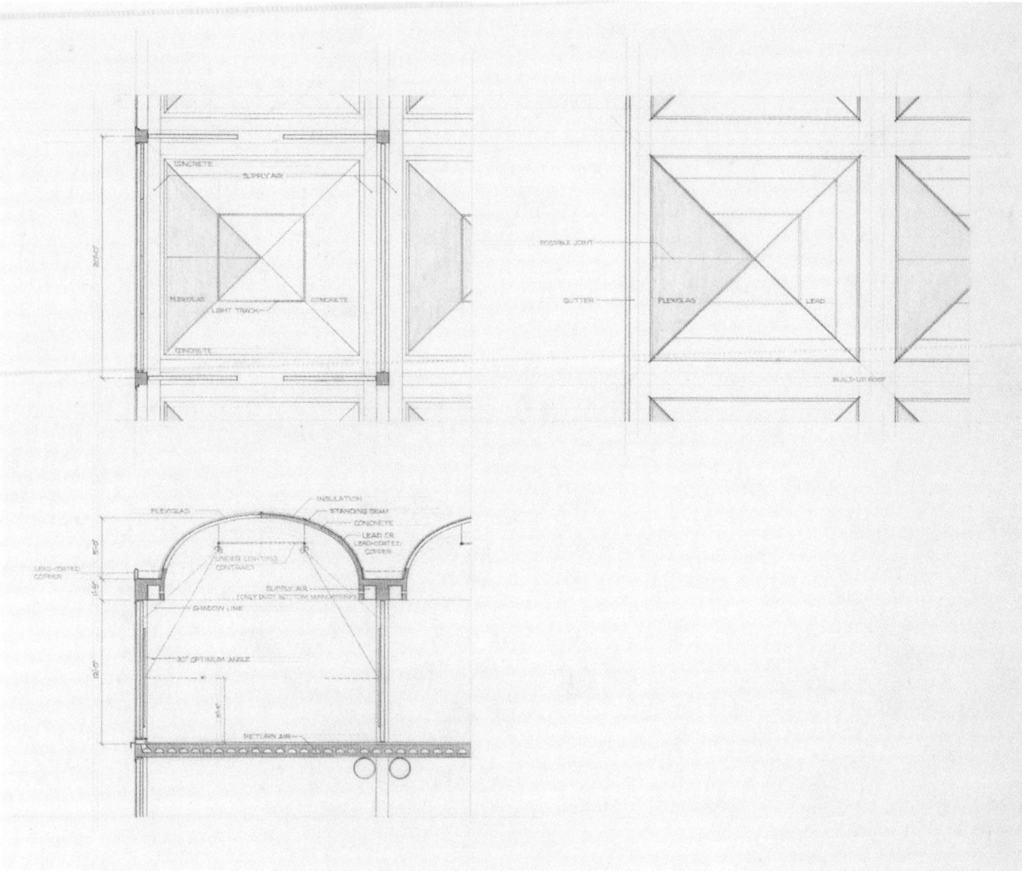

also with the addition of lead, Kahn moves towards a new chromatic range in which metal, in the types of steel and aluminium, takes on a leading role. Reflecting the desire to identify the metal surfaces most suitable to be combined with the matte surfaces of the concrete, for the Mellon Center Kahn proposes metal worked in such a way as to obtain an effect similar to "lead" or "pewter."[16] "On a gray day it will look like a moth; on a sunny day like a butterfly," is how he explains the effect of the metal in the Mellon Center.[17]

Kahn envisions a particular type of finish for the steel of the panels, not shiny and polished like the typical American curtain wall. His steel, to be worked with the same kind of care that goes into the concrete, is supposed to create "a 'sgraffito' effect of 'writing.'"[18] Kahn's intention to investigate the graphic effects of the panels and the glass between the posts of the structure can be seen in several development sketches for the envelope, including the representation of what he considers a "written" effect. The fact that the Mellon Center contains an art collection has its consequences on the particular treatment of the steel selected by Kahn, who probably intends to transform the claddings into a pictorial panel of abstract graphic signs, in the manner of De Kooning or Gorky, bordered by the smooth concrete frames of the skeleton. As in the case of every detail of the formwork of his concrete, from the V-joints to the pour joints, for the panels the degree of artistic expression is also directly based on the fabrication process, without

538. Yale Center for British Art, New Haven, Connecticut, 1969-74. Section and plan details, 20 September 1971 (LIK, 030.I.C.805.6).

539. Airfloor Company of California, Inc., booklet (*Build Comfort [...] by Airfloor*, n.d., LIK, 030.II.A.110.3).

any intention of obtaining a given ornamental design. The posts of the structure themselves, for Kahn, become lines with which to design fields of different sizes, corresponding to the different functions of the building, and taking on a degree of process-based artistic abstraction.

Kahn's original conception of the metal as an envelope of the monumental sheaths of the physical plant can still be seen, within the Mellon Center, in the two volumes that contain the vertical conduits of the ventilation system, to be made with the same steel panels that form the walls inside the skeleton.

The choice of a uniform pattern of columns as the structural system for the Mellon Center is explained by Prown in terms of Kahn's desire to create a museum that is like a "comfortable house."[19] It is significant that while in the Olivetti Underwood factory Kahn reduces the columns to a minimum to interpret the character of the space of a production facility, in the Mellon Center, on the other hand, he uses the columns to define the spatial coordinates of the underlying unit of the whole works: the room. Travertine becomes the fundamental material for the insertion, in the floors, of the bands corresponding to the structure, and to design the whole grid inside the space, as in the Exeter Library.

The elimination of the Vierendeel trusses, and therefore of the visible conduits, to obtain that domestic character of the rooms also desired by Prown, is possible thanks to the modification of the slab structure, shifting from the waffle type with exposed coffers envisioned in July-August to the definitive type, identified at the end of August: the Airfloor produced by the Airfloor Company of California, Inc. An advertising booklet sent to Kahn describes the Airfloor as a "hollow floor construction used to distribute heated and cooled ventilating air to perimeter registers."[20]

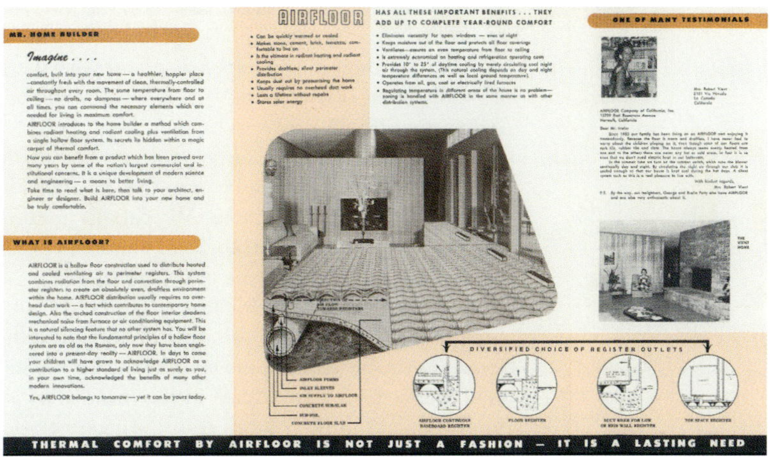

[16] Jules David Prown, *The Architectural of the Yale Center for British Art*, New Haven, Yale University, 2009, p. 46.
[17] *Ibid.*, p. 45.
[18] *Ibid.*, p. 48.
[19] JDP, [Jules David Prown], "Release preliminary design. The Paul Mellon Center for British Art and British Studies," LIK, 030.II.A.111.1. The text is received by Kahn's office on 29 November 1971.
[20] Airfloor Company of California, Inc., *Build Comfort [...] by Airfloor*, n.d., (LIK, 030.II.A.110.3).

Estrangement of materials: color research on metallic concrete

The Mellon Center confirms the fact that for Kahn the question of the color of concrete always depends on the identification of chromatic harmonies to be matched with the landscape, the local construction materials or the meanings of a work. The Mellon Center, like the Kimbell Art Museum, is another storehouse form transfigured thanks to the effect of estrangement generated by the materials and their sophisticated chromatic harmonies. But in the selection of the materials for the Mellon Center, Kahn goes further than in the past, to the point of using not stone slabs, but the steel panels usually applied in shopping malls or department stores. Thus the estrangement becomes double, given the fact that it also involves the skin of the storehouse, which becomes surprising because it is made with a banal industrial material to give it an unusual surface.

Henceforth the experimentation on the color tone of the concrete is based on the choice of using steel panels as the infills framework. The steel introduces different parameters in the choice of the fabrication of the concrete, with respect to the Travertine of the Kimbell Art Museum, whose porous surface established a dialogue with the small honeycombs of the material poured in place.

The concept of the monolith as a result of combination of different materials, united by similar surface textures and chromatic tones, is explained by Kahn regarding the Kimbell Art Museum, in an interview conducted by Meyers when the museum had just been opened, on 11 August 1972. But there is no doubt that the concept also lies behind the construction of the Mellon Center and the search for chromatic harmonies and textural blends of the joints and grooves of its materials.

According to what is stated by Kahn in the interview, the pursuit of chromatic harmonies between materials, from the concrete compound to the type of Travertine, is translated in the idea that "natural materials" have "a way of blending together" and that they should not be altered by painting, though their natural color is not always the same and therefore creates a feeling of "apprehension." "[...] the color will be discontinued so you can't renew it or it will prove to be that the gray of one manufacturer is not the same gray, even with the same name, as that of another manufacturer," Kahn says, regarding one of the colors that has been an earmark of his architecture for decades, at this point. We should point out that by "natural materials" Kahn also means the artificial materials produced by a "manufacturer," like the concrete having the gray tone of which he speaks. Sensitivity to the grain and the characteristic hue of materials can make the building into an organic whole, which Kahn now defines with the term "monolithic." If we consider the fact that at the start of the 1950s he had described the impossibility of creating buildings in the form of monoliths, due to the various joints present in their fabrication, it becomes clear that the concept of the monolith, as it is formulated in 1972, in any case includes a sequence controlled and free of contrasts, of materials, joints, textures and colors. What

[21] "Marshall D. Meyers interview with Louis I. Kahn, 11 August 1972, 'The Wonder of the Natural Thing,'" LIK, 030.II.A.37.19.
[22] "Concrete. Concrete Work," 24 February 1972, LIK, 030.II.A.110.4.
[23] Pat Morss, letter to Abba A. Tor, 30 March 1972, LIK, 030.II.A.106.7. Tor points out that "the Salk specifications *do not* call for coated plywood" and asks to add a section on "cleaning of exposed surfaces" (Abba A. Tor, letter to David P. Wisdom, 7 April 1972, LIK, 030.II.A.106.7). He also addresses the question of the joints

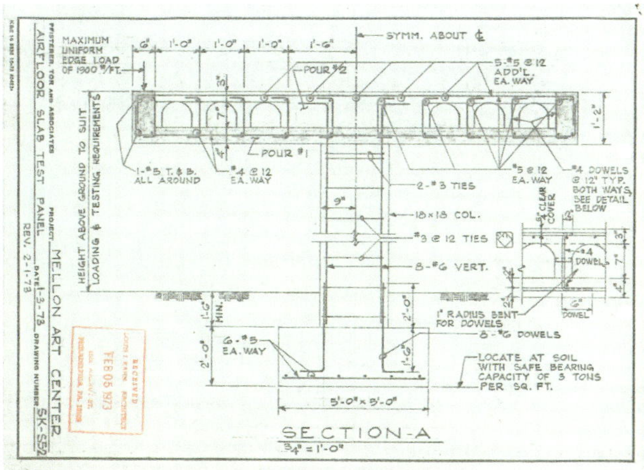

540. Yale Center for British Art, New Haven, Connecticut, 1969-74. "Airfloor Slab Test Panel," 3 January 1973 (LIK, 030.II.A.110.4).

Kahn says regarding the affinities between Travertine and concrete is significant, after the observation that in nature there are no "polished materials."

"[...] Travertine in the walking areas of the museum is a very excellent material because it is naturally not a slippery material. Marble is certainly very nice in a museum, but it is slippery and besides anything glaringly white, let's say, or very polished against concrete [added by hand: has too much contrast – while that] which has a similar tone to concrete – let us say for instance Travertine, which is very much like concrete – has almost the character at times so that concrete and Travertine look like the same material. That makes the whole building again monolithic and it doesn't separate things. If you had a contrasting material the concrete would look dirty, muddy – whereas – white would make concrete look very bad because the concrete would look really dirty and the white would be very white and the contrast between the materials would make both materials rather not belong to each other. [...] Travertine and concrete belong beautifully together, because concrete must be taken for whatever irregularities or accidents in the pouring that reveal themselves. One must [...] avoid any kind of irregularities in a purposeful, 'artistic' kind of way. Artistic being put in quotation marks as being rather a fault than a virtue."[21]

The choice of the material to place beside the concrete to achieve the monolithic effect is the question that concerns Kahn in the case of the Mellon Center. The specifications continue to be revelatory documents, in this sense.

At the start of 1972 the latest version of the specifications of the previous year for the construction of the Mellon Center is taken and copied identically, with particular focus on the section on "Concrete Work," and submitted for the comments of the office of Pfisterer, Tor & Associates.[22] On this occasion Kahn proposes also looking at the specifications of the Salk Institute, "since the concrete work was so successful."[23] For this reason, his office also

indicated in the specifications of the Mellon Center: "Concerning pour sequence and construction joints, we suggest that instead of us telling the contractor where to locate joints, we specify that he has to submit proposed locations of joints for our approval, and then we change locations by mutual agreement. Reference to 'shrinkage joints' is not clear. If the intent is to cover the slab on grade, we should ask for a checkerboard pattern of pour, the individual pours not to exceed 400 s.f. and without re-entrant corners."

sends Tor several documents on the Salk Institute.[24] A new version of the specifications, taking the observations of Pfisterer, Tor & Associates into account, is prepared on 10 May and contains, among other things, the indication of a "pouring plan, showing pour sequence and construction joints" and a paragraph on cleaning, to be done with "stiff bristle brushes and clean water."[25] This is the version that first mentions the Airfloor as the construction system of the floor slabs.[26] For the panels a minimum 5/8 inch thickness is indicated.

When the specifications for the concrete have been altered, a representative of the Concrete Consulting Corporation, based in Oakland, California, who is in contact with Wisdom regarding participation in Kahn's project for the Inner Harbor complex in Baltimore, proposes the involvement of the company also in the project for the Mellon Center, to cover all the aspects of the concrete. The Concrete Consulting Corporation offers its services for research and experimentation on components, organization of the responsibilities of the concrete suppliers, the making of samples, assistance in the preparation of drawings and the creation of mock-ups.[27] This is the same kind of professional consulting Frederick Langford had sought to propose for the worksites of Dhaka and Fort Worth, though his focus was more concentrated on formwork.

The contractor hired for the construction of the Mellon Center, George B. H. Macomber Company Builders, also selected because it had done the Yale Art Gallery, starts to make the first trials in May 1972 to test the color effects in the concrete caused by the inclusion of "pozzolan additives."[28] The indications on this type of additive probably come directly from Kahn, who from the earliest contacts with the Mellon Center clients had indicated the concrete of the Salk Institute as the model to follow. But the Macomber Company also recommends, regarding the color of the "concrete work," the use of the cement of the Penn-Dixie Cement Corporation, obtained with stones from the Hudson River Valley (and in particular at Howe Caverns), with a "controlled color," a "light buff."[29] At the same time, different types of stainless steel panels are examined for the making of the walls inside the skeleton.[30]

Guidelines and requirements for architectural concrete

The work meeting held on 12 October 1972 is a fundamental moment, not just for the development of the Mellon Center but also, more generally, for the definition of the formal qualities of Kahn's concrete. On this occasion, the "guidelines and requirements of Mr. Kahn for the architectural concrete" of the Mellon Center are specified.[31] It is the first time Kahn supplies such clear indications of principles to follow in the assembly of the formwork, a sort of testament of his vocabulary of imprints for the concrete. The minutes of the meeting show that Kahn again makes use of the concept of the monolith, though with a rather different interpretation than what was explained a

[24] Pat Morss, letter to Abba A. Tor, 30 March 1972, LIK, 030.II.A.106.7.
[25] "Division 3. Concrete. Concrete Work," 10 May 1972, LIK, 030.II.A.113.30.
[26] *Ibid.*
[27] Larry C. Washburn, letter to David P. Wisdom, 12 May 1972, LIK, 030.II.A.95.24.
[28] Frederick P. Wales, of George B. H. Macomber Company, "Notes of meeting held 4/26/72 at the Office of Louis Kahn," LIK, 030.II.A.110.20.
[29] George B. H. Macomber Company, "Estimate Digest. Mellon Center - Yale University," 22 May 1972, LIK, 030.II.A.112.1. In 1967 the Penn-Dixie Cement Corporation had acquired Callanan Industries, Inc., of Albany, New York, a company specialized in quarrying and trade of stones from the Hudson River Valley. The Macomber Company, besides the "superties" for the formwork in the exposed concrete parts (suitable to avoid the

few months earlier regarding the Kimbell Art Museum: "Form material. No grain to show, use absolutely smooth surfaces such as plastic-covered plywood. No board marks on concrete anywhere in building. Large cyclopic monolithic surfaces desired."[32] The fact that the monolithic concept is associated with the lack of any imprints of joints and texture of the formwork means that the term now refers not so much to the combination of related materials, as in the interview on the Kimbell Art Museum, as to the individual components of the structure, which are to be poured in formwork with panels of the proper size to convey the impression of unified elements, like those that can be obtained by precasting. Only in the light of the idea of "large cyclopic monolithic surfaces" can we fully grasp the meaning of details like the pillars in the two courtyards, which Kahn sets back from the edge beam of the various levels, so that there is no longer even a groove to indicate the pour phases, since the aim is to make each pillar look like a monolith.[33]

Other principles indicated by Kahn confirm the pursuit of the "large cyclopic monolithic surfaces." On this topic, it is significant that for the walls Kahn recommends, as is his habit by now, the use of "panel size as large as possible." The absence of joints is associated with the idea of a "handsome wall": "To make a handsome wall, joints must be kept to a minimum."[34]

Though the expression of the joint between formwork panels is limited to just a few wall sectors, Kahn indicates the form, measurements and meaning of his typical V-joints. "Joints to be raised V's with a base 1/4" wide and projection no more than 1/8," he explains.[35] Concerned about the alternations to his formwork system made by the contractors in Dhaka and by the consultants and contractor at Fort Worth, Kahn explains the technical nature of those V-joints, which cannot be reduced to grooves engraved on the wood in order to imitate his system: "Raised joints must be formed by beveling slightly the edges of the plywood and not by artificial scoring or other mechanical imitations."[36] This is a fundamental remark in order to grasp Kahn's critical position against the alterations made to the logic of his formwork system on the worksites in Dhaka and Fort Worth and, in definitive terms, against the concept of "imitation" of a detail that has to be the result of a process, producing the only ornament he believes is admissible. It is at this point in his tenacious thinking on the concept of ornament and the process of fabrication of concrete that Kahn reaches the point of even specifying the fundamental role of the imperfections of the V-joint imprints. The flaws that can happen during the pour along the V-joints are the guarantee of the accurate implementation on the part of the contractor of that joint as a line of contact between two panels, not a groove engraved on the plywood: "Small breaks or chips in the raised joints are necessary to show the real life nature of the material."[37] On the other hand, irregular corners and all interventions after stripping are to be considered errors:

formation of rust marks), calls for "5/8" plastic faced plywood" for the exposed concrete and "5/8" wood faced plywood" for the "non-exposed concrete" (George B. H. Macomber Company, "Estimate Digest. Mellon Center - Yale University," 22 May 1972, LIK, 030.II.A.112.1).

[30] Frederick P. Wales, "Notes of meeting held 5/26/72 at the Office of Louis I. Kahn, architect," LIK, 030.II.A.110.28.

[31] Frederick P. Wales, "Notes of meeting held 10/12/72," LIK, 030.II.A.111.60.

[32] *Ibid.*

[33] Graphic diagram on the margins of the page (Frederick P. Wales, "Notes of meeting held 10/12/72," LIK, 030.II.A.111.60).

[34] Frederick P. Wales, "Notes of meeting held 10/12/72," LIK, 030.II.A.111.60.

[35] *Ibid.*

[36] *Ibid.* "Try for V-joint protrusion – take eye away from blotching out joint variances – 'embossing,'" is written in the handwritten version (minutes of the meeting, handwritten manuscript, 12 October 1972, LIK, 030.II.A.106.7a).

[37] Frederick P. Wales, "Notes of meeting held 10/12/72," LIK, 030.II.A.111.60.

541. Yale Center for British Art, New Haven, Connecticut, 1969-74. "Proposed Formwork Layout Elevation (Basement Level)," 20 November 1972 (LIK, 030.I.C.805.18).

"Bleeding and sanding are definitely not desired as these are mistakes." Honeycombs are also "absolutely intolerable."

Other indications have to do with the tie holes. At this point Kahn has identified a precise range of holes to be left exposed or plugged: "Holes to be left infilled must be small, precise and clean. 1/2" diameter would be maximum dimension for this case. If larger holes have to be used then they must be plugged with lead, hammered smooth and recessed slightly. Supertie she-bolt system should produce the desired unplugged effect."

The specifications for the design of the ceilings are no less precise. Again in this case, all the imprints have to be coordinated in a design that takes the interlocks with the columns into account. In Kahn's reasoning on the imprints, the assembly of the panels of the formwork for the ceilings actually justifies the square form of the columns. The rhythm of the imprints has to be spare and solemn, never imitating that of laid masonry. In the case of the insertion of special pieces, they have to be placed at the edges of the surface field to be designed by the formwork.

"Keep joints to a minimum," is one of Kahn's principles. "Joints must be patterned. Use columns as reference for edge joints. Joints must intersect column corners not within a face. Long joints to be parallel to the long building axis. Do not use a staggered brick type pattern anywhere. Cross-section of raised joints to be the same as wall joints. If panel sizes within the field must vary, then vary only perimeter panels. Ceiling outlets, junction boxes, detectors, etc. should be recessed if possible."[38]

In the section on "column and slab intersection," Kahn clarifies the design of the joint between the two structures. It is only at this point, following the logic of the process of pouring and construction, that Kahn allows for the possibility of a groove. This is also perhaps a warning, to avoid repetition of the mistakes made by the contractor in the isolated columns of the portico of the Kimbell Art Museum, which are decorated at the base by two grooves made in optical continuity with the thickness of the floor slab, without any constructive logic whatsoever. The imprint indicated for the columns of the Mellon Center has to be made with a strip to leave in place when the column has been poured, to prevent marks from the concrete during the phase of construction of the floor slab. "The only joints in columns are at the slab intersections. Keep the joints at columns as small as possible to prevent filling. Joints to be recessed, not raised."

Where the color of the concrete is concerned, Kahn states that it should be a "warm gray tone" in order to "complement stainless steel." The indication of the tone confirms the goal of a monolithic appearance pursued through chromatic harmonies of the materials and the new power granted by Kahn to the materials associated with the concrete in the definition of that tone. But at the same time, revealing the hesitations regarding the chromatic nature of the concrete of the Mellon Center, Kahn continues to point to the concrete of the Salk Institute as a model. It is not by chance that the Macomber Company is instructed to "investigate mix and use of pozzolan to produce color change." In the handwritten version, we see Kahn's orientation towards a particular color: "tend toward lavender (not buff)."[39] This remark clearly shows the desire to obtain a concrete with the same tone as that of the Salk Institute. The substance to add to obtain that lavender tone is also indicated: "pozzilon [sic: pozzolan] – volcanic ash."[40]

Kahn chooses sample number two from the three steel panels as having the "best finish"; but it is also decided to continue the research on the

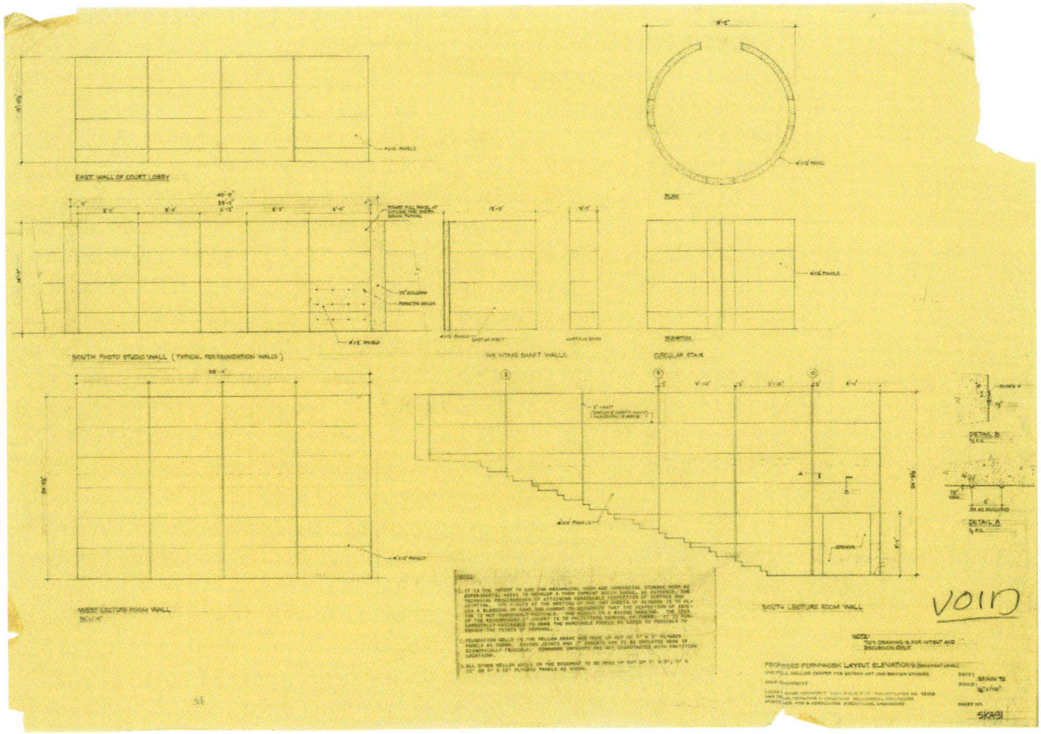

panel finish in Kahn's presence. Kahn also inspects the mock-up and indicates changes to be made.

While the worksite gets under way in November 1972 with the start of excavation for the foundation and the basement level, on 9 November other important decisions are made regarding the making of the concrete. To familiarize the workmen of the Macomber Company with Kahn's formwork system, certain walls and the columns of the Mechanical Room are earmarked as samples of the architectural concrete that will remain exposed in other parts of the Mellon Center.[41] On this occasion the Macomber Company agrees to prepare the drawing of the formwork for the floor slab.

The set of drawings entitled "Proposed formwork layout elevations" of the "basement level" and dated 20 November is prepared by Kahn's office after the meeting on 9 November, to serve not as a working document but as an illustration of the various cases to be discussed with the contractor – "for intent and discussion only," is indicated.[42] Kahn depicts the south and west walls of the Lecture Room (panels of 4 × 10 feet), the "south Photo Studio wall" (panels 4 × 8 feet), the "east wall of Court Lobby" (panels 4 × 10 feet), the wall of the "circular stair" (panels 4 × 12 feet), and the "air intake shaft walls" (panels 4 × 12 feet), with all the various details concerning joints and patterns.

[38] The handwritten version shows that Kahn gives indications to coordinate the soffit formwork with the position of the column and also with the bands of the "Travertine floor" (minutes of the meeting, handwritten manuscript, 12 October 1972, LIK, 030.II.A.106.7a).

[39] Minutes of the meeting, handwritten manuscript, 12 October 1972, LIK, 030.II.A.106.7a.

[40] *Ibid.*

[41] Frederick P. Wales, "Notes of meeting held 11/9/72," LIK, 030.II.A.111.40, and handwritten copy, LIK, 030.II.A.106.7a.

[42] See the drawing "Proposed Formwork Layout Elevations (Basement Level)," SKA-31, 20 November 1972, LIK, 030.I.C.805.018.

The drawings also show the V-joints in detail – the base of the triangle is 1/4 inch with maximum rise of 1/8 inch, as indicated by Kahn on 12 October – and the double vertical V joints or "2" inserts" are shown, warning that they are only to be applied in the horizontal joints, "if useful." Other indications include the arrangement of the panels with respect to the floor slabs: "start full panel at ceiling and work down." Given the fact that the formwork system will determine the character of the staircase in the courtyard of the library, which is the only element where the exposed concrete surfaces will become powerful presences visible from the ground floor to the roof, the "Proposed formwork layout elevations" also show the cylindrical wall of that staircase, to demonstrate the construction with 4 × 12 feet panels alternating with other, narrower panels.

A note in the set of drawings ratifies the decision made on 9 November to make samples at the basement level. The note contains another important consideration on the V-joint, further confirming the theoretical value assigned to that detail by Kahn. "It is the intent to use the mechanical room and commercial storage room as experimental areas to develop a form imprint which shows, as evidence, the technical requirements of attaining reasonable perfection of surface and jointing. The V-cuts at the meeting of the two sheets of plywood is to allow a bleeding of sand and cement to recognize that the perfection of sealing is not thoroughly possible. The result is a raised marking. The idea of the recommended 2" insert is to facilitate removal of forms. It is fundamentally desirable to make the removable panels as large as possible to reduce the prints of removal."[43]

After the first indications of Kahn to the Macomber Company on the color of the concrete, the firm assigns the research to Stanley G. Kledaras of the New Haven Testing Laboratory. The trials are to take Kahn's desire to obtain the color of the concrete with pozzolan additives into account. The New Haven Testing Laboratory has to ascertain if the pozzolan is available locally, to immediately determine the cost of the making of the concrete. At the time of the assignment of the job the chromatic objective pursued by Kahn is again confirmed, summed up in the words written by Froeberg to Kledaras: "to produce a warm gray color concrete which will be compatible to the stainless steel window wall."[44] The composition of the concrete, developed by the New Haven Testing Laboratory for the Mellon Center, has to cover aspects of workability, to avoid the formation of honeycombs or marks. And the mix design of the same color also has to be used for the prefabricated parts that include, in this phase of the design, not only the V-beams but also the columns of the upper level. During the course of November 1972 the New Haven Testing Laboratory is already preparing "sample concrete mixes for color" to be examined by Kahn.[45]

The choice of the light gray tone of the concrete is based on the color of the steel panels. The importance assigned by Kahn to the materials of the Mellon Center in the research on the concrete compound comes from the fact that what is at stake is not obtaining different colors, but matching the tones of gray of the concrete and the steel. It is significant that when the

[43] For the discussions on the concrete and the type of joint of the "slab formwork" see "Minutes of Meeting held November 21, 1972," LIK, 030.II.A.111.40.
[44] Kenneth E. Froeberg, letter to Stanley G. Kledaras, 30 October 1972, LIK, 030.II.A.109.8.
[45] "Minutes of Meeting held November 21, 1972," LIK, 030.II.A.111.40.
[46] Kenneth E. Froeberg, letter to Stanley G. Kledaras, 30 October 1972, LIK, 030.II.A.109.8.
[47] Edward L. Cairns, of Trio Industries, Inc., letter to Frederick P. Wales, 5 December 1972, LIK, 030.II.A.111.40.
[48] "He also feels that another panel with a different finish

Macomber Company assigns the task of developing the color of the compound to the New Haven Testing Laboratory, it also handles the shipment of steel samples so that Kledaras can adjust the gray tone of the concrete in the preparation of the samples to be shown to Kahn.[46]

At the start of December, Kahn is engaged in choosing the color tones and ways of treating the steel and concrete, and the type of panels for the formwork.

Trio Industries, Inc., based in Bridgeport, Connecticut, specializing in metal casements and panels, is asked to provide samples of steel sheets for the Mellon Center. The production of the panels is done by Republic Steel, in Canton, Ohio. On 1 December 1972 the manager of the Erection Department of Trio Industries, Inc., Edward L. Cairns, goes to Philadelphia to discuss the "finish on stainless steel" of panels and casements with Kahn. On this occasion, Cairns brings sixteen steel samples, of which twelve are made with normal manufacturing processes.[47] The brief report on this meeting shows that Kahn is interested in incorporating the indications of Cairns on the characteristics of the making of the panels into the project, confirming that he is looking for that essential working method that reveals the "natural thing," its degree zero, not only for concrete but also for all other materials. To better assess the color tones, Kahn advises placing the panel and casement samples inside the mock-up.[48]

"This did not seem to present a problem to Mr. Kahn," it is written, referring to the limitations explained by Cairns regarding the manufacturing processes of steel. "He also accepted the rolling lines, indicating that they should run horizontally. He also did not consider the 48" width limitation a problem that could not be readily solved by varying the design drawings as necessary within these width limitations."[49]

Kahn chooses a sample that does not have a "'striated' finish." "Louis was directed to see if this finish can be produced or something close to it. L.I.K. wants frames and panel of same finish, not standard 2D for frames. Does not want '2 tone' effect."[50] In the choice of the type of finish for the steel, Kahn is seeking a matte surface effect, to tune the hue to that of the concrete. "[...] Kahn wanted the steel surface of the Center to be absolutely non-reflective," Prown recalls. "He often referred to the material metaphorically as 'lead' or 'pewter.' In considering samples of steel, he would reject any that reflected his hand."[51]

After the trials conducted by the New Haven Testing Laboratory, Kahn is shown samples of concrete made with the cement of the Universal Atlas Cement Company. To obtain a color in tune with the metal, one of the samples has been prepared with the addition of fly ash, a product that can nevertheless be brought to New Haven only at a high cost. The color of this sample is too dark, while that of the sample made without fly ash is too green.[52] "Concrete mix design samples were submitted. L.I.K. felt they were still too 'greenish' in cast and rejected these. Samples are to be resubmitted. Froeberg would like to use Pozzolith in concrete: gives a workable mix with less slump. Also use larger aggregate so as to reduce cement quantity and

should be placed on another side of the mock-up so that there will be some choice" (Edward L. Cairns, letter to Frederick P. Wales, 5 December 1972, LIK, 030.II.A.111.40).
[49] Edward L. Cairns, letter to Frederick P. Wales, 5 December 1972, LIK, 030.II.A.111.40.
[50] "Mellon Job Meeting," 7 December 1972, LIK, 030.II.A.111.40.
[51] Prown, *The Architecture of the Yale Center for British Art* cit., p. 48.
[52] Frederick P. Wales, "Notes of meeting held 12/7/72," LIK, 030.II.A.111.40.

lessen shrinkage."[53] For the fabrication of the new trials, the samples prepared by the New Haven Testing Laboratory remain Kahn's model: "Sample will be cast to attain the same color as the concrete used at the Design Lab and submitted next week."[54] As for the formwork, the Macomber Company provides Kahn and the engineers with brochures on the "form lock tie system," while representatives of Macomber Company, Pfisterer, Tor & Associates and Yale University discuss the thicknesses of the panels.[55]

Specifications of Pfisterer, Tor & Associates and Kahn

At the end of November 1972 the studio of Pfisterer, Tor & Associates develop the sections of the specifications on poured in place and precast concrete of the Mellon Center, without using the standard expressions of the specifications of Kahn's office on architectural concrete. The list of the ingredients of the concrete does not include additives for color. There is not even any mention of the pozzolan, which has always been indicated by Kahn as a fundamental part of the mix. For the construction of the floor slabs, the Airfloor type is confirmed. Certain indications on the formwork have to do with the control of the panel structure during pouring and the size of the holes left by the spreaders, which in the exposed concrete are to be no larger than 1/2 inch in diameter. To obtain those "large cyclopic monolithic surfaces" desired by Kahn, the specifications recommend the necessary techniques during the pour phases to avoid leaving unwanted marks.[56] Only on the Dhaka worksite had Gustav Langford attempted to also stipulate this operation, when he attempted to calculate the position of the vertical strips of marble.

The specifications of Pfisterer, Tor & Associates contain a section, "Finishing of concrete surfaces," prepared according to the habitual criteria of working after stripping, calling for intervention on the surfaces, including patching, as well as instructions to "cut out honeycomb, poor joints, and voids," and to fill tie holes. Kahn cancels the entire paragraph. This decision too corresponds to the ideal goal of the "wonder of the natural thing." By now, the concept of roughness has been purified of any remaining hint of imperfect or artistic finishing still surviving in the culture of late European Brutalism, and is taken towards a primordial essence that seeks its results in the fabrication process, so much so that Kahn, with the Mellon Center, achieves that expression of concrete he began to seek with tenacity and courage in the last decades of his activity, starting with the Salk Institute: a process concrete that reduces chance to a minimum.

In about the same period as the preparation of the specifications, in November and December the connections between the V-beams of the skylights and the two types of concrete on the upper level of the Mellon Center are under study.[57]

Towards the start of 1973, in preparation for the imminent launch of the concrete work, certain modifications are made to the text of the specifications of Pfisterer, Tor & Associates. A whole new section is added by Kahn's office to provide indications on the architectural concrete. This is the

[53] "Mellon Job Meeting," 7 December 1972, LIK, 030.II.A.111.40.

[54] Frederick P. Wales, "Notes of meeting held 12/7/72," LIK, 030.II.A.111.40.

[55] "Samples of 1/2" plywood (from Finland [?]) for formwork were submitted. [Kenneth] Borst protested that 1/2" was too thin. Froeberg insisted that calculations proved that this was not so. Tor stated that calculations should be relied upon rather than a subjective position screw" ("Mellon Job Meeting," 7 December 1972, LIK, 030.II.A.111.40).

[56] Pfisterer, Tor & Associates, "Section 3A. Cast-in-place concrete," n.d., [received by Kahn's office on 1 December 1972], LIK, 030.II.A.111.23.

title of the section that discusses themes ranging from color to formwork, to techniques to prevent the formation of rust on the concrete caused by reinforcements exposed to weather. Regarding the formwork, it is established that the drawings with indications of all the elements that leave imprints on the concrete, from the measurements of the panels to the positioning of the ties, should be made in collaboration between Kahn and the Macomber Company. Kahn's staff explain the reasons behind the V-joints, pointing to the need to identify, also in collaboration with the contractor, their position in keeping with the technical logic of the pour. "Since there is," they write, among other things, "a natural tendency for water and sand to bleed at form joints, the edges of form panels are to be slightly beveled to provide a place where this bleeding may occur and permit fins to form."[58] Beyond the quality of their making, it is in any case specified that the "fins shall be left untouched," as they are the result of controlled bleeding. As for the tie holes, they are not to be filled unless otherwise indicated by Kahn. If filling is required, lead is to be used as the material. The indications on the assembly of the formwork call for every column of a height up to 12 feet to have no signs of joining of the panels along the trunk, while in the beams the vertical joint of the panels should be positioned at the center. The plywood is the Filmply produced by the North American Plywood Corporation - "Exterior Type Plyform, C.I, per U.S. product Standard PS-1, edge-sealed, but not oiled, and finished with polyurethane varnish […]."

Prior to the start of construction of the architectural concrete, it is decided to make "samples of form plywood" also with the application of "form coating," including "form ties" and all other elements in contact with the pour. Kahn insists on the making of that "demonstration structure," including all the most characteristic features of the architectural concrete, from the columns to the walls to the stairs, already called for in the "Outline specifications" of March 1971. But the trials are limited to the making of the walls and columns at the basement level, in the Mechanical Room, along High Street, as already suggested in September, and in the Commercial Storage area along Chapel Street. Both trials are covered by the same modifications to the specifications, with the indication of "further experimentation if required," to be done above and beyond the "demonstration structure."

In the section of the specifications on the precast concrete, Kahn, adds the requirement of three samples of 12 × 12 inches with different materials for the molds, in order to evaluate the "finish appearance" of the concrete. Once the material has been chosen, a sample is to be made of a beam section 4 feet in length.[59] The color of the two types of concrete – precast and poured in place – has to be the same. Several later modifications to the specifications on the precast concrete insist on the composition of the concrete to achieve precast parts that "match the exposed cast-in-place concrete."[60] But it is significant, above all, that even in the case of the precast parts, Kahn and his team make modifications to the specifications to prevent any intervention after the pieces have been removed from the molds.

[57] "All connections of precast to cast in place concrete do not appear to be problems. The cast in place concrete will be cast around the erected precast sections" (Frederick P. Wales, "Memo of meeting held December 11, 1972," LIK, 030.II.A.110.20).

[58] "PART 4. Architectural Concrete," in, Louis I. Kahn, "Specifications. The Paul Mellon Center for British Art and British Studies, Yale University," n.d., attached to, "Addendum No. 2. The Paul Mellon Center for British Art and British Studies, Yale University. Louis I. Kahn Architect," 28 March 1973, LIK, 030.II.A.112.16.

[59] *Ibid.*

[60] William S. Kaminski, letter to Louis I. Kahn, 9 March 1973, LIK, 030.II.A.112.16.

The choice of the color of the concrete has not been definitively made when construction is about to begin. Other trials with other cements, after those of the Universal Atlas Cement Company, are ordered to prepare samples for Kahn's inspection. In January 1973 two samples are made, one with the cement of the Atlantic Cement Company – "Type 1 light cement" – and one with the cement the Macomber Company had already suggested in May 1972 – that of the Penn-Dixie Cement Corporation, "buff" color, made with stones from the Howe Caverns or, as indicated in the minutes of the meeting, "Type 11, Buff Howe's Cavern."[61] Kahn approves the cement of the Atlantic Cement Company, for the making of a mock-up sample.[62] This is the type of cement used on the Mellon Center worksite for the mixing of the concrete, produced by the Foxon Concrete Corporation of New Haven.[63]

The first formwork for the architectural concrete

In January 1973 assembly of the formwork begins for the sample walls and pillars in the Commercial Storage area and the Mechanical Room of the Mellon Center.

Control of the design of the patterns of the formwork, with indications of all the joints with "structural significance,"[64] becomes complex due to the more technical interventions, those of Kahn's office, of Pfisterer, Tor & Associates and of the Macomber Company. The exchange of views regarding the design of the formwork for the walls of the Lecture Room is emblematic. After the complex procedure of correction of the drawings for these formwork done by Macomber Company, revised by Pfisterer, Tor & Associates and then corrected by Kahn, who then has to submit them again for the examination of the engineers, Tor proposes a method that would give priority to Kahn's decisions. "It seems more logical," he writes to Kahn on 10 April, "that we be requested by your office to review construction joint locations after you find the balance of the drawing to be acceptable, or that

542-544. Yale Center for British Art, New Haven, Connecticut, 1969-74.

542-543. Construction site photographs, 1 March 1973 (LIK, 030.II.A.110.18.1; 2).

544. Construction site photograph, 1 August 1973 (LIK, 030.II.A.110.18.3).

[61] "Memorandum. Agenda for meeting 11 January 1973," 10 January 1973, LIK, 030.II.A.110.20.

[62] Kenneth E. Froeberg, "Job Meeting January 11, 1973," LIK, 030.II.A.111.40. For the sand to use in the concrete compound, the New Haven Testing Laboratory, after tests, advises against the red sand of the Elam City Construction Company of North Haven, because it has "considerable variation," as well as "problems in maintaining consistency in gradation" (William S. Kaminski, letter to David P. Wisdom, 10 April 1973, LIK, 030.II.A.110.4).

[63] The company had also provided the mix design for the Art and Architecture Building of Rudolph, again constructed by the Macomber Company.

[64] Abba A. Tor, letter to Louis I. Kahn, 10 April 1973, LIK, 030.II.A.110.4.

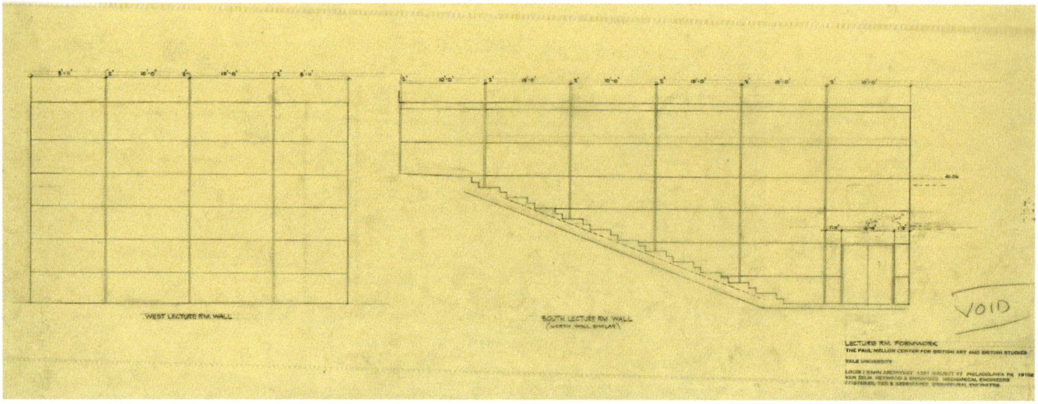

545-546. Yale Center for British Art, New Haven, Connecticut, 1969-74.

545. "Lecture RM. Formwork," 4 November 1973 (LIK, 030.I.C.805.19).

we receive only one print at the time drawings are submitted to your office. We could review joint locations and joint details and send the print on to your office for final approval."[65]

The day after Tor's letter, Kahn designs the "Lecture Room Formwork" for the west and south walls (the north wall is identical to the south wall), with rectangular panels 10 feet long, no longer measuring 9 feet and 10 inches as indicated in 1972, still assembled horizontally with aligned joints; the vertical joints are the double V-joints, as envisioned in 1972.[66]

In the context of continuous steps taken to reduce construction costs, including Tor's recommendations to save time in the checking of the drawings, the possibility is also considered of modifying the construction system of the cylindrical staircase in the library courtyard, replacing the poured in place concrete with concrete blocks. The last indication provided by Kahn on this point is that "the stair towers would remain concrete," though the Macomber Company continues to insist that "substantial savings can be effected by changing the stair towers from concrete to block even the cost of engineering, and some added concrete columns are taken into account."[67] To illustrate the system of panels of the formwork of the staircase, Kahn's staff provides detailed drawings of the continuous development of the wall, with indications of formwork, joints and tie positions.[68]

While work continues on the basement, when he can already have an idea of the making of the construction details of the holes or the V-joints, Prown writes a confidential letter, sent to Kahn for his information, in which he describes the poor quality of the concrete in the works in progress:

"The quality of the concrete poured thus far for the Mellon Center is quite erratic. Parts are quite good, but there are some very ragged details. The concrete work is not up to the standard of the Salk Institute, and far below the superb work at the Kimbell Art Museum. I suspect there has not been adequate Yale representation on the site to maintain a high standard of quality control. Are there steps we can take to insure better on-site supervision on behalf of the owner, to be certain that we get first-rate detailing as the building rises above ground level?"[69]

[65] *Ibid.*

[66] The drawing, in any case, is invalidated ("Lecture RM. Formwork," SKA-56, 4 November 1973, LIK, 030.I.C.805.019). See also the detail for the insertion of the electrical equipment next to the door, "Lecture Room Elec. outlet locations," 1 May 1973, with the indication: "position plates away from raised V joints" (LIK, 030.I.C.805.019).

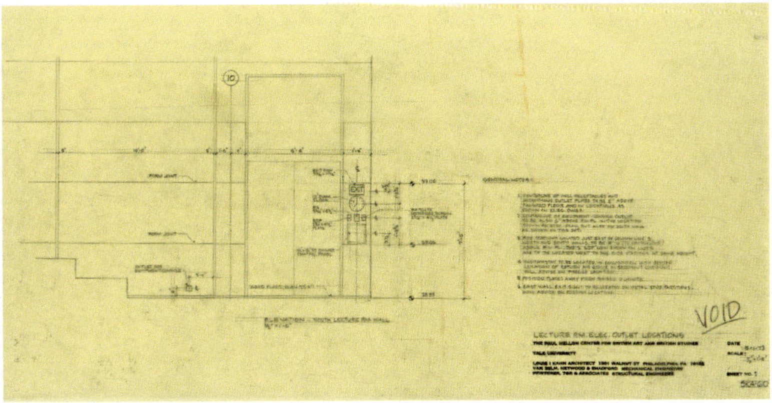

546. "Lecture RM. Electronic Outlet Location," 1 May 1973 (LIK, 030.I.C.805.19).

The correspondence and the minutes of the meetings throughout 1973 show that on the worksite of the Mellon Center, as on that of the Kimbell Art Museum, delays are caused by lack of coordination of the contributions of the various specialists, along with the usual delays of Kahn's office in providing the shop drawings, including those for the formwork, so much so that Macomber Company raises the possibility of suspending work to avoid keeping the work force idle on the site.

Also in response to the concerns expressed by Prown, the Macomber Company, besides threatening Kahn by declaring its intention to proceed with the erection of the west wall of the Lecture Room without the "panel joint layout" corrected by him, unless the shop drawing is not sent by the end of July, underlines the need for the presence on the worksite of a member of Kahn's staff to assist the contractor in the preparation of the drawings of all the "panel joints in poured concrete surfaces [...]."[70] On the margin of the request from the Macomber Company, Kahn writes the name of his expert of choice, Frederick Langford, although over the short term the making of the drawings requested by the Macomber Company of the "panel joints for slabs, beams, spandrels, balance of poured concrete walls" is assigned to another member of his team, Arthur Owen. Only starting in September is Kahn able to ensure the weekly presence on the worksite of one of his collaborators, Meyers, who has gained experience with concrete on the worksites of the Salk Institute and the Kimbell Art Museum, and in the meantime has opened his own professional practice in Philadelphia, with another one of Kahn's collaborators, Pellecchia.

The formwork of the Lecture Room are assembled a few days before the complaints voiced by the Macomber Company, and the walls are poured in July 1973. On 1 August most of the walls of the basement level have been made, including the floor slab of the sector along Chapel Street. On 1 September the formwork is being assembled for the inclined plane of the Lecture Room, along with the formwork for another floor slab sector; and the rows of pillars along Chapel Street, with a sector of the wall of the entrance courtyard, have already been erected. On 1 October the structures for the floor slabs around three sides of the entrance courtyard have already been installed.

[67] Theodore R. Burghart, letter to Louis I. Kahn, 22 May 1973, LIK, 030.II.A.110.22.
[68] See the drawing "circular stair form work," 31 July 1973, LIK, 030.I.C.805.019.
[69] JDP [Jules David Prown], "Confidential," 13 June 1973, LIK, 030.II.A.105.9.
[70] Theodore R. Burghart, "Agenda. Meeting July 5, 1973," LIK, 030.II.A.112.23.

547-548. Yale Center for British Art, New Haven, Connecticut, 1969-74.

547. Construction site photograph (LIK, 030.II.A.110.18.4).

548. Construction site photograph, 1 September 1973 (LIK, 030.II.A.110.18.5).

The Airfloor groove in the entrance courtyard

Pour grooves continue to determine the character of Kahn's concrete, also in the Mellon Center. The drawing called "interior unrolled elevation of stair" dated 22 October 1973 and made based on the drawing from July documents the level already built of the cylindrical wall of the staircase, that of the formwork in place at that date and the sector of the cylindrical wall still to be built.[71] The heights of the pours indicated on the drawing go from floor slab to floor slab and are shown, in the construction and not in the drawing, by a continuous groove.

During the course of an inspection on 23 October, Kahn makes the decision to imprint a horizontal pour groove in the entrance courtyard as well, along three sides only, corresponding to the faces of the first floor slab, with the exception of the long beam on the west side, due to the lack of the pillar at the center of that side. This recess expresses the construction process of the Airfloor, which happens first with the construction of a sub-slab 4 inches thick, and then with the positioning, over all its surface, of the shaped vaulted boxes that make it possible to obtain the cavities for the passage of conduits, on which the last pour of the top slab is spread. The groove desired by Kahn corresponds to the height of the sub-slab. But Kahn's team, Meyers, the technicians of Macomber Company and the engineers wonder about the reasons behind Kahn's decision to limit the presence of that groove to the first level of the entrance courtyard.

"Louis Kahn verbally approved the use of a horizontal pour joint (see field sketch FSK-1 dated 29 Sept 73) at the 14" deep beams on the south, east and north sides of the 2nd floor of court S-19," Meyers reports. "The pour joint will reflect the first pour of the lower 4" air floor slab and will be 4" above the bottom of the 14" beam. There will be no such pour joint at the 40" deep beam on the west side of the court. Mr. Kahn requests that the joint be done carefully and precisely. Mr. Kahn was also informed, at the same time, that on the exterior facade where beams span 40'-0," there would not be a horizontal pour joint."[72]

Also on 23 October, Meyers informs Walter Shapiro of Pfisterer, Tor & Associates that "we could avoid this horizontal pour joint at the exterior 14" deep beams." "Because this exterior condition is similar to that at court S-19 and court 217, I will obtain a decision from Mr. Kahn as to whether he wishes to continue using this horizontal pour joint at the remaining floors in court S-19 and in court 217."[73] Kahn's reply arrives on 3 December: "Mr. Kahn has decided *not* to use the horizontal pour joint discussed in the Oct. 23rd memo at any other locations where the 14" high beam occurs. This means that this pour joint will only occur at the North, East and South sides of court S-19 at the second floor slab, which has already been poured. We also reconfirm that no such joints will appear on any other floors at court S-19, or on any floor in court 217 or at any place on the exterior face of the building including the deep beams."[74]

The recess of the Airfloor is one of the signs of the sophisticated expressive game staged with the structural elements that are not always visi-

[71] "Interior Unrolled Elevation of Stair," 22 October 1973, LIK, 030.I.C.805.019.
[72] Marshall D. Meyers, "Speed Message," addressed to Frederick P. Wales, 23 October 1973, LIK, 030.II.A.110.9, and LIK, Macomber/I to 5 Dec 73.
[73] *Ibid.*
[74] Marshall D. Meyers, "Speed Message," addressed to Frederick P. Wales, 3 December 1973, LIK, 030.II.A.110.9, and LIK, Macomber/I to 5 Dec 73.

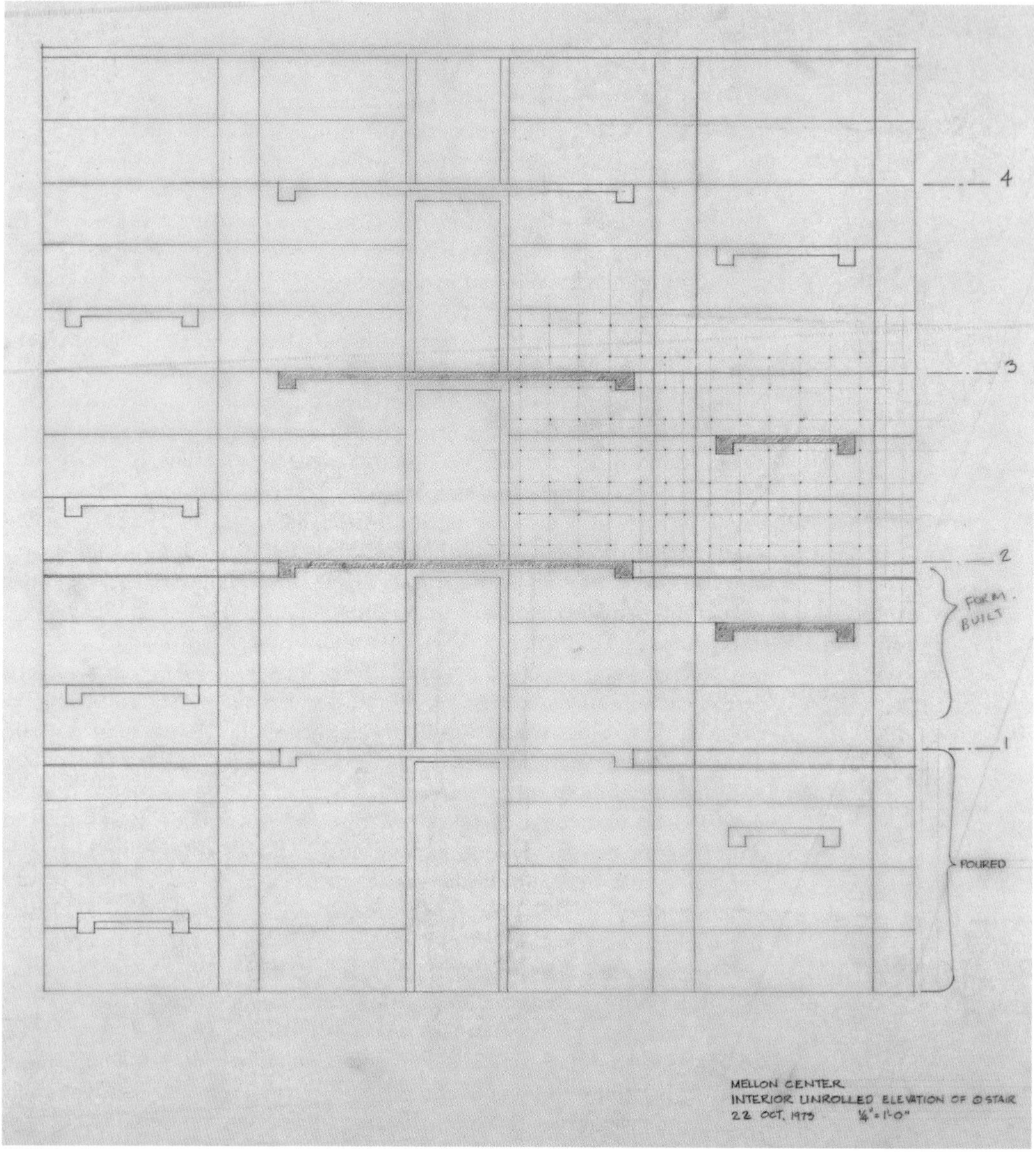
4
3
2
1
FORM. BUILT
POURED
MELLON CENTER
INTERIOR UNROLLED ELEVATION OF ⊙ STAIR
22 OCT. 1975 ¼" = 1'-0"

549-551. Yale Center for British Art, New Haven, Connecticut, 1969-74.

549. "Interior Unrolled Elevation of Stair," 22 October 1973 (LIK, 030.I.C.805.19).

550. Detail of the concrete surface (photograph by the author, 2010).

551. Stair wall and ceiling (photograph by the author, 2010).

ble, as happens with the pillars in the corners of the courtyard. It is difficult to identify the reasons that induce Kahn to limit the expression of the construction phases of the Airfloor to the first floor slab of the entrance court, where that groove, for those who understand the pouring system required by the Airfloor, takes on an enigmatic value, almost like an arbitrary ornament, like the grooves at the base of the columns in the portico of the Kimbell Art Museum (perhaps it is the impossibility of understanding the technical reasons behind the recess that prompts Kahn to give up on that detail on the other levels).

Other decisions on the finishes of the materials are made by Kahn at the start of 1974, when he approves the type of joints for the concrete blocks in the sample wall, although Meyers advises waiting to start the work, given the fact that Kahn has examined the samples only by means of a photograph, and not directly.[75] On 8 January, Kahn, during a visit to the worksite, approves the "stainless steel sample provided by Trio Industries from the run of steel previously inspected at Republic's plant in Ohio." "Grain or steel shall run with long dimension of panel work." On this occasion he also approves the sample of "mansard roof paneling" and that of the concrete blocks: "Mr Kahn confirmed approval of the use of lightweight concrete masonry units made with Solite lightweight aggregate by Plasticrete Corp. He also confirmed approval of the jointing that was used in that portion of the sample panel where the Solite blocks were used."[76]

The construction and installation of the precast V-beams of the roof, which characterize the lighting system on the upper level of the Mellon Center, are assigned to Allied Building Systems, Inc., based in Manchester, Connecticut. Kahn examines and approves, on 11 December 1973, the color of the precast sample sent to him on the previous day.[77] The start of

[75] Marshall D. Meyers, "Speed Message," addressed to Frederick P. Wales, 7 January 1974, LIK, 030.II.A.110.8.

[76] Theodore R. Burghart, letter to David P. Wisdom, 16 January 1974, LIK, 030.II.A.110.8.

[77] Marshall D. Meyers, "Speed Message," addressed to Frederick P. Wales, 11 December 1973, LIK, Macomber II/from 5 Dec 73 to 31 Jan 74.

552-554. Yale Center for British Art, New Haven, Connecticut, 1969-74.

552-553. Construction site photographs (LIK, 030.IV.A.805.3.6; 4).

554. Construction site photograph (M. D. Meyers Archive, AAUP, 136.144.28).

prefabrication is planned after the making of a prototype.[78] The exchanges between Meyers and the Macomber Company cover the type of V-beam and joint to be precast, to submit to Kahn prior to batch production.[79] The trial beam is examined by Kahn on 30 January 1974.[80] Following his approval,[81] Allied Building Systems organizes its workshop in Newington, also in Connecticut and not far from New Haven, for the production of the V-beams of the Mellon Center. After various uncertainties, it is decided to use precast concrete, rather than poured in place, also for the beam near the elevator, "in order to provide better match for color and texture of entire roof concrete."[82]

Production of the beams is already in progress when Kahn dies of a heart attack on 17 March 1974. The beams reach the worksite on long trucks, from which they are directly hoisted by the crane installed on the small plaza at the back corner of the building. The installation of the beams has reached almost the halfway point on 1 May, when the work is proceeding under the supervision of Pellecchia and Meyers Architects. On 4 June all the beams have been installed and the deployment of the steel panels can begin, starting with those of the facade along High Street, which is completed at the start of July. The Mellon Center, renamed as the Yale Center, is opened on 17 April 1977. All the imperfections of the pour, including the jagged edges of the V joints, are left intact. "Random

[78] "The first precast sample containing typical details and inserts will be begun on 12/12/73" (Robert Ducibella of Macomber Company, "Job n. 7139. Minutes of mechanical coordination meeting No. 14 - December 5, 1973," LIK, 030.II.A.110.19).
[79] "You assured me that the sample piece of precast will be a 20' long section that could, if approved, be used as a production piece in the building. You also understand that this is a 'calculated risk' on your part and that if rejected, this piece will be abandoned. You also stated that two short end pieces would be fabricated to show the joint at the juncture of perpendicular members when Mr. Kahn is to view the sample, so he can include these joints in his approval" (Marshall D. Meyers, letter to Theodore R. Burghart, 27 December 1973, LIK, Macomber II/ from 5 Dec 73 to 31 Jan 74).
[80] Theodore R. Burghart, letter to Frank Lore, vice-president of , 25 January 1974, LIK, 030.II.A.110.8.
[81] Theodore R. Burghart, letter to David P. Wisdom, 25 January 174, LIK, Macomber IV. See also, Jan Vidmar, letter to Tim Kellerher, 25 January 1974, LIK, 030.II.A.110.4.
[82] George Macomber "Tabulation of bids for subcontract work," 4 March 1974, LIK, 030.II.A.110.21.

555-557. Yale Center for British Art, New Haven, Connecticut, 1969-74.

555. View (photograph by the author, 2010).

discolorations," Prown explains in his text for the opening, "flaws or breaks in the concrete and elsewhere were not, in Kahn's opinion, to be hidden. They are, he felt, natural expressions of the way in which the building was made and should be allowed to remain rather than be cosmeticized."[83]

The Mellon Center concludes the cycle of Kahn's passionate research on the nature of concrete, which began at the time of the construction of the building on the other side of Chapel Street, with the particular Brutalism of the Yale Art Gallery. Starting with that work, Kahn sets out to experiment with the expression of concrete by means of subtraction of every decorative imprint made possible by the formwork. The passage from wooden boards to plywood panels is significant, in this perspective. In the process of erasure of the expressive potential of the formwork, pursued by Kahn with rare tenacity, the imprints of the wood grain vanish, thanks to special treatments of the surface of the formwork, and the panels themselves become increasingly larger in order to reduce the imprints of the joints to a minimum, or even to eliminate them altogether. The Salk Institute is the lyrical point of arrival of Kahn's research on concrete. The monolithic surfaces of the Mellon Center bear witness to the aspiration to discover its hidden essence in the rigorous expression of the sole process of fabrication of the compound: concrete, and "nothing else." This is one of the goals of Kahn's pursuit of the mythical degree zero.

"[...] if you are dealing with concrete," he states in a lecture in 1974, "what you have to do is to keep in mind that it is not brick or stone or anything else. You are dealing with a material which is marvelous in itself. It is a cross between steel and molten stone, and you can make it do things which no other material can do. Therefore concrete forms should not bear any resemblance to structures made of any other material. The way you have used it and the way it looks must let people see that this is the material you have been using and nothing else."[84]

[83] Prown, *The Architecture of the Yale Center for British Art* cit., p. 63.
[84] Louis I. Kahn, *Harmony between Man and Architecture*, in, "Design Incorporating Indian Builder," vol. XVIII, March 1974, No. 3, pp. 23-28, now in, Latour, *Louis I. Kahn: writings, lectures, interviews* cit., (pp. 333-343), p. 340.

556-557. Interior views (photographs by the author, 2010).

558-564. Yale Center for British Art, New Haven, Connecticut, 1969-74. Details (photographs by the author, 2010).

Acknowledgements

The books *Exposed Concrete and Hollow Stones, 1949-1959*, by Roberto Gargiani, and *Towards the Zero Degree of Concrete, 1960-1974*, by Anna Rosellini, are the result of direct research into all of Kahn's works in concrete from the 1940s onwards and all of the main archival materials. Without the support, the availability, and the welcome of the Architectural Archives of the University of Pennsylvania in Philadelphia it wouldn't have been possible to write these books. We thank all the members of the archives in Philadelphia and in particular the collection manager William Whitaker who always supported each initiative that has contributed to the publication of these books, and the archivist Nancy Thorne who helped us consult the archival documents. Our thanks also goes to Nathaniel Kahn, Sue Ann Kahn, Samantha Sizemore of the Kimbell Museum, Kathleen Mylen-Coulombe of the Yale Art Gallery for having put at our disposition the material conserved in their respective archives, and to Patricia Cummings Loud, the Kimbell's curator of architecture, who accompanied Anna during the visit to the museum and facilitated the consultations of the documents conserved in the museum archives.
Our understanding of some important concrete construction processes in Kahn's work come from the fundamental discussions Anna had with Nicholas L. Gianopulos and Frederick Langford – to whom we are indebted for their contributions and willingness to help.
There are many people who have facilitated our visit to Kahn's architectures. We would like to thank in particular the personnel of the Swiss Embassy in Dhaka and the architect Nurur Rahman Khan, and Garry Van Gerpen and Bob Lizarraga who accompanied us during our visit to the Salk Institute and who have made some archival materials available.
The fabrication of the books, the translation of texts and the page layout were opportunities to collaborate with the Presses polytechniques et universitaires romandes (PPUR) of the Ecole polytechnique fédérale de Lausanne (EPFL). We thank in particular Frederick Fenter and Nicholas Tappin and our most sincere acknowledgement goes to the director Olivier Babel.
An affectionate thanks goes also to our friend Paola Ranzini Pallavicini for taking care of all graphic design of the book.
In particular our thanks goes to deans of the Environnement Naturel, Architectural et Construit faculty (ENAC) of EPFL, for the support and the impetus that on several occasions we have wished to honour.
Finally, we are grateful to the European Research Council for a grant that has supported our research on the topic of exposed concrete.

Name index

Figure Credits

Louis I. Kahn Collection, University of Pennsylvania and the Pennsylvania Historical and Museum Commission. Housed in the Harvey and Irwin Kroiz Gallery, the resources of the Kahn Collection are used by the permission of the Architectural Archives, University of Pennsylvania (hereafter cited as Kahn Collection).
Fig.#: 1, 2, 4-19, 21, 23, 25-31, 33-46, 61-89, 92-94, 98, 100-102, 104-126, 154-156, 160-164, 168, 169, 171, 176-178, 180-216, 240-290, 309-331, 334, 341, 342, 353, 358-361, 373, 393-402, 405-411, 416-422, 425, 431-452, 455, 456-458, 460-472, 474, 475, 479-483, 485-489, 491, 492, 495, 496, 498-501, 505-508, 510, 511, 513, 515-526, 528-549, 552, 553

The Architectural Archives, University of Pennsylvania, Henry Wilcots Collection
Fig.#: 217-239

The Architectural Archives, University of Pennsylvania, Marshall D. Meyers Collection
Fig.#: 377-382, 385, 554

The Architectural Archives, University of Pennsylvania, August Komendant Collection
Fig.#: 473, 476-478, 503, 509, 512, 514